CALCIUM AND CELL FUNCTION

Volume II

MOLECULAR BIOLOGY

An International Series of Monographs and Textbooks

Editors: BERNARD HORECKER, NATHAN O. KAPLAN, JULIUS MARMUR, AND HAROLD A. SCHERAGA

A complete list of titles in this series appears at the end of this volume.

CALCIUM AND CELL FUNCTION

Volume II

Edited by

WAI YIU CHEUNG

Department of Biochemistry
St. Jude Children's Research Hospital
Memphis, Tennessee

1982

ACADEMIC PRESS
A Subsidiary of Harcourt Brace Jovanovich, Publishers
New York   London
Paris   San Diego   San Francisco   São Paulo   Sydney   Tokyo   Toronto

ACADEMIC PRESS, INC.
111 Fifth Avenue, New York, New York 10003

United Kingdom Edition published by
ACADEMIC PRESS, INC. (LONDON) LTD.
24/28 Oval Road, London NW1 7DX

Library of Congress Cataloging in Publication Data
Main entry under title:

Calcium and cell function.

  (Molecular biology, an international series of mono-
graphs and textbooks)
   Vol. 1-   edited by W. Y. Cheung.
   Includes bibliographies and index.
   CONTENTS: v. 1. Calmodulin.
   1. Calcium--Physiological effect. 2. Calcium
metabolism. 3. Cell physiology. I. Cheung, Wai Yiu.
II. Series. [DNLM: 1. Calcium. 2. Calcium--Binding
proteins. QU55 C144]
QP535.C2C26   612'.3924   80-985
ISBN 0-12-171402-0 (v. 2)   AACR1

PRINTED IN THE UNITED STATES OF AMERICA

82 83 84 85   9 8 7 6 5 4 3 2 1

Contents

Contributors   xi

Preface   xv

Contents of Previous Volume   xvii

Chapter 1   Calcium Binding to Proteins and Other Large Biological Anion Centers
B. A. Levine and R. J. P. Williams

I.	Introduction	2
II.	Extracellular Calcium-Binding Proteins	8
III.	Calcium Ion Binding to Large Particles, Membranes, and Surfaces	17
IV.	Calcium Transport	19
V.	Vesicles: Calcium Stores	23
VI.	Intracellular Proteins	23
VII.	Calcium-Binding Intestinal Proteins	32
VIII.	Calcium Proteins in Membranes	34
IX.	Summary: Intracellular Calcium	35
	References	35

Chapter 2   Mitochondrial Regulation of Intracellular Calcium
Gary Fiskum and Albert L. Lehninger

I.	Introduction and Scope	39
II.	Mechanisms, Kinetics, and Regulation of Mitochondrial Ca^{2+} Transport	41
III.	Evidence for Mitochondrial Regulation of Cellular Ca^{2+}	62
IV.	Steady-State Buffering of Free Ca^{2+} by Mitochondria	67
	References	71

Chapter 3   Calcium Movement and Regulation in Presynaptic
Nerve Terminals
*Catherine F. McGraw, Daniel A. Nachshen, and
Mordecai P. Blaustein*

I.	Introduction	81
II.	Calcium Content and Intraterminal Ca^{2+} Distribution	41
III.	Ca^{2+} Entry Mechanisms	92
IV.	Ca^{2+} Efflux from Nerve Terminals	100
V.	Ca^{2+} Movements and Transmitter Release	101
VI.	Summary	102
	References	103

Chapter 4   Calmodulin and Calcium-Binding Proteins:
Evolutionary Diversification of Structure
and Function
Jacques G. Demaille

I.	Introduction	111
II.	Calcium Ions as Second Messengers	112
III.	The Evolution of the Ca^{2+}-Binding Protein Family	113
IV.	Parvalbumin as the Prototype of Suppressor Molecules	120
V.	Calmodulin as the Prototype of Sensor Molecules	124
VI.	Sequential Activation–Deactivation of Ca^{2+}-Dependent Enzymes	135
VII.	Conclusion	138
	References	139

Chapter 5   Troponin
James D. Potter and J. David Johnson

I.	Introduction	145
II.	Ca^{2+}-Binding to Troponin and the Regulation of Muscle Contraction	147
III.	Thin-Filament Protein Interactions in the Regulation of Muscle Contraction	151
IV.	Structure and Ca^{2+}-Induced Structural Changes in Troponin	155
V.	Propagation of the Ca^{2+}-Induced Structural Changes in Troponin C to Thin-Filament Proteins	163
VI.	Rates of Ca^{2+}-Exchange and Structural Changes in Troponin C	164
VII.	Rates of Ca^{2+}-Exchange in Troponin	167
VIII.	Conclusion	168
	References	169

Chapter 6   Vitamin D-Induced Calcium-Binding Proteins
R. H. Wasserman and C. S. Fullmer

I.	Introduction	175
II.	Species and Tissue Distribution	176
III.	Properties of Calcium-Binding Proteins	181
IV.	Cellular Localization of Calcium-Binding Proteins	187
V.	Physiological Factors Affecting CaBP	188
VI.	*In Vitro* Synthesis of CaBP	198
VII.	Embryonic Development	201
VIII.	CaBP and Calcium Reabsorption in the Kidney	202
IX.	Temporal Responses of CaBP to Acute Doses of $1,25(OH)_2D_3$	202
X.	Discussion	205
	References	207

Chapter 7 γ-Carboxyglutamic Acid-Containing Ca^{2+}-Binding Proteins
Barbara C. Furie, Marianne Borowski, Bruce Keyt, and Bruce Furie

I.	Introduction	217
II.	γ-Carboxyglutamic Acid-Containing Proteins of Blood Plasma	221
III.	γ-Carboxyglutamic Acid-Containing Protein of Bone	236
IV.	Other γ-Carboxyglutamic Acid-Containing Proteins	237
V.	Summary	238
	References	238

Chapter 8   Parvalbumins and Other Soluble High-Affinity Calcium-Binding Proteins from Muscle
Wlodzimierz Wnuk, Jos A. Cox, and Eric A. Stein

I.	Introduction	243
II.	Historical Review	244
III.	Distribution of Sarcoplasmic Calcium-Binding Proteins in the Animal Kingdom	245
IV.	Parvalbumins	250
V.	Sarcoplasmic Calcium-Binding Proteins from Invertebrates	261
VI.	Physiological Implications	270
	References	273

Chapter 9   **Myosin Light Chain Kinase in Skinned Fibers**
             W. Glenn L. Kerrick

   I. Introduction   279
  II. Usefulness of Skinned Fibers as a Model for Contraction   283
 III. Evidence for a Light Chain Kinase–Phosphatase System
       in Skinned Fibers   285
  IV. Summary   292
       References   293

Chapter 10   **Possible Roles of Calmodulin in a Ciliated
             Protozoan *Tetrahymena***
             Yoshio Watanabe and Yoshinori Nozawa

   I. Introduction   297
  II. Properties of *Tetrahymena* Calmodulin   299
 III. Activation of Membrane-Bound Guanylate Cyclase of *Tetrahymena*   303
  IV. Search for New Functions of Calmodulin in *Tetrahymena*   308
   V. Concluding Remarks   319
       References   319

Chapter 11   **Calcium Control of Actin Network Structure
             by Gelsolin**
             Helen L. Yin and Thomas P. Stossel

   I. Introduction   325
  II. Structure of the Cortical Cytoplasm   326
 III. Regulation of Actin Gel–Sol Transformation   327
  IV. Calcium Regulation of Actin Filament by Gelsolin   328
   V. Mechanism of Action of Gelsolin   330
  VI. Gelsolin Is an Important Physiological Regulator   331
 VII. Effect of Other Calcium-Dependent Proteins on Actin   333
VIII. Discussion   333
       References   335

Chapter 12   **Calcium and the Metabolic Activation
             of Spermatozoa**
             Robert W. Schackmann and Bennett M. Shapiro

   I. Introduction   339
  II. Nature of the Activation Process   341
 III. Triggers for Sperm Activation   346
  IV. Ca^{2+} Functions in Sperm–Egg Association   348
   V. Conclusions   348
       References   349

Chapter 13   **The Physiology and Chemistry of Calcium
during the Fertilization of Eggs**
David Epel

I.	Introduction	356
II.	Evidence That Free Calcium Content Changes at Fertilization	357
III.	Calcium Permeability at Fertilization	358
IV.	The Rise in Intracellular Calcium as the Cause of Activation of the Egg	360
V.	Egg Activation as a Result of the Release of Calcium from Intracellular Stores	361
VI.	What Is the Role of the Calcium Influx after Fertilization?	363
VII.	Extracellular Calcium as a Requirement for Activation in Eggs	365
VIII.	Nature of the Cytoplasmic Calcium Stores	367
IX.	Calcium-Binding Proteins and Calcium Buffers of the Egg	368
X.	Control of Metabolism by Calcium	369
XI.	Summary and Overview	378
	References	379

Chapter 14   **Calcium and Phospholipid Turnover as
Transmembrane Signaling for
Protein Phosphorylation**
*Yoshimi Takai, Akira Kishimoto, and
Yasutomi Nishizuka*

I.	Introduction	386
II.	Enzymology of Calcium-Activated, Phospholipid-Dependent Protein Kinase	387
III.	Mode of Enzyme Activation	389
IV.	Phospholipid Metabolism and Receptor Function	396
V.	Physiological Implication in Transmembrane Control	401
VI.	Coda and Prospectives	405
	References	406
	Index	413

Contributors

Numbers in parentheses indicate the pages on which the authors' contributions begin.

Mordecai P. Blaustein (81), Department of Physiology, University of Maryland School of Medicine, Baltimore, Maryland 21201

Marianne Borowski (217), Division of Hematology–Onocology, Tufts–New England Medical Center, and Departments of Medicine, Biochemistry, and Pharmacology, Tufts University School of Medicine, Boston, Massachusetts 02111

Jos A. Cox (243), Department of Biochemistry, University of Geneva, 1211 Geneva, Switzerland

Jacques G. Demaille (111), Centre de Recherches de Biochimie Macromoléculaire du CNRS et U-249 INSERM, 34033 Montpellier, France

David Epel (355), Hopkins Marine Station, Department of Biological Sciences, Stanford University, Pacific Grove, California 93950

Gary Fiskum* (39), Department of Physiological Chemistry, The Johns Hopkins University School of Medicine, Baltimore, Maryland 21205

C. S. Fullmer (175), Department of Physiology, New York State College of Veterinary Medicine, Cornell University, Ithaca, New York 14853

Barbara C. Furie (217), Division of Hematology–Onocology, Tufts–New England Medical Center, and Departments of Medicine, Biochemistry, and Pharmacology, Tufts University School of Medicine, Boston, Massachusetts 02111

Bruce Furie (217), Division of Hematology–Onocology, Tufts–New England Medical Center, and Departments of Medicine, Biochemistry, and Pharmacology, Tufts University School of Medicine, Boston, Massachusetts 02111

---

* Present address: Department of Biochemistry, George Washington University School of Medicine, Washington, D.C. 20037.

J. David Johnson (145), Departments of Pharmacology and Cell Biophysics, University of Cincinnati College of Medicine, Cincinnati, Ohio 45267

W. Glenn L. Kerrick* (279), Department of Physiology and Biophysics, University of Washington, Seattle, Washington 98195

Bruce Keyt (217), Division of Hematology–Onocology, Tufts–New England Medical Center, and Departments of Medicine, Biochemistry, and Pharmacology, Tufts University School of Medicine, Boston, Massachusetts 02111

Akira Kishimoto (385), Department of Biochemistry, Kobe University School of Medicine, Kobe 650, Japan, and Department of Cell Biology, National Institute for Basic Biology, Okazaki 444, Japan

Albert L. Lehninger (39), Department of Physiological Chemistry, The Johns Hopkins University School of Medicine, Baltimore, Maryland 21205

B. A. Levine (1), Department of Inorganic Chemistry, Oxford University, Oxford OX1 3QR, England

Catherine F. McGraw (81), Department of Ophthalmology, Washington University School of Medicine, St. Louis, Missouri 63110

Daniel A. Nachshen (81), Department of Physiology, University of Maryland School of Medicine, Baltimore, Maryland 21201

Yasutomi Nishizuka (385), Department of Biochemistry, Kobe University School of Medicine, Kobe 650, Japan

Yoshinori Nozawa (297), Department of Biochemistry, Gifu University School of Medicine, Tsukasamachi-40, Gifu 500, Japan

James D. Potter (145), Departments of Pharmacology and Cell Biophysics, University of Cincinnati College of Medicine, Cincinnati, Ohio 45267

Robert W. Schackmann (339), Department of Biochemistry, University of Washington, Seattle, Washington 98195

Bennett M. Shapiro (339), Department of Biochemistry, University of Washington, Seattle, Washington 98195

Eric A. Stein (243), Department of Biochemistry, University of Geneva, 1211 Geneva, Switzerland

Thomas P. Stossel (325), Hematology–Oncology Unit, Massachusetts General Hospital, Boston, Massachusetts 02114, and Department of Medicine, Harvard Medical School, Boston, Massachusetts 02115

Yoshimi Takai (385), Department of Biochemistry, Kobe University School of Medicine, Kobe 650, Japan, and Department of Cell Biology, National Institute for Basic Biology, Okazaki 444, Japan

---

* Present address: Department of Physiology and Biophysics, University of Miami School of Medicine, Miami, Florida 33101.

R. H. Wasserman (175), Department of Physiology, New York State College of Veterinary Medicine, Cornell University, Ithaca, New York 14853

Yoshio Watanabe (297), Institute of Biological Sciences, The University of Tsukuba, Niihari-gun, Sakura-mura, Ibaraki 305, Japan

R. J. P. Williams (1), Department of Inorganic Chemistry, University of Oxford, Oxford OX1 3QR, England

Wlodzimierz Wnuk (243), Department of Biochemistry, University of Geneva, 1211 Geneva, Switzerland

Helen L. Yin (325), Hematology–Oncology Unit, Massachusetts General Hospital, Boston, Massachusetts 02114, and Department of Medicine, Harvard Medical School, Boston, Massachusetts 02115

Preface

The theme of this volume continues that of this open-ended treatise: a timely assessment of the current status of the multifunctional role of Ca^{2+} in cell function. The first volume focuses on calmodulin; this one extends the coverage to the metabolism of Ca^{2+}, other Ca^{2+}-binding proteins, and various Ca^{2+} functions; future volumes will address appropriate topics under active investigation.

The organization of Volume II is divided into three sections. The first three chapters deal with the chemistry and metabolism of Ca^{2+}; the next five describe various Ca^{2+}-binding proteins in addition to calmodulin. The functions of Ca^{2+}, some mediated by calmodulin and some by other proteins, are discussed in the last six chapters. As in the first volume, each chapter reflects the style and interest of the contributors. The length of each chapter varies somewhat, depending on the need and the extent of coverage that was felt necessary.

The field of Ca^{2+} research continues to accelerate noticeably, with a good number of articles focusing on calmodulin. According to a computer search published by a recent Current Contents, the number of articles bearing calmodulin in their titles in 1979 was 213, the term calmodulin having been introduced the year before; by 1980, it quintupled to 1013. The increase appears unlikely to abate in the near future.

One of the aims of this treatise is to keep students and investigators in all disciplines of biological research abreast of the developments in this rapidly expanding field; another is to stimulate new research for a better understanding of the intricate regulatory mechanisms underlying cellular function. I thank all the contributors for their splendid efforts in this endeavor.

This volume is dedicated to my brother, who spared no effort to see a young lad receive a proper education.

Wai Yiu Cheung

Contents of Previous Volume

Volume I

1. **Calmodulin—An Introduction**
   Wai Yiu Cheung

2. **Assay, Preparation, and Properties of Calmodulin**
   Robert W. Wallace, E. Ann Tallant, and Wai Yiu Cheung

3. **Structure, Function, and Evolution of Calmodulin**
   Thomas C. Vanaman

4. **Calmodulin: Structure–Function Relationships**
   Claude B. Klee

5. **Ca^{2+}-Dependent Cyclic Nucleotide Phosphodiesterase**
   Ying Ming Lin and Wai Yiu Cheung

6. **Calmodulin-Dependent Adenylate Cyclase**
   Lawrence S. Bradham and Wai Yiu Cheung

7. **Calmodulin and Plasma Membrane Calcium Transport**
   Frank F. Vincenzi and Thomas R. Hinds

8. **Smooth Muscle Myosin Light Chain Kinase**
   Robert S. Adelstein and Claude B. Klee

9. **The Role of Calmodulin and Troponin in the Regulation of Phosphorylase Kinase from Mammalian Skeletal Muscle**
   Philip Cohen

10. **Plant and Fungal Calmodulin and the Regulation of Plant NAD Kinase**
Milton J. Cormier, James M. Anderson, Harry Charbonneau, Harold P. Jones, and Richard O. McCann

11. **Calcium-Dependent Protein Phosphorylation in Mammalian Brain and Other Tissues**
Howard Schulman, Wieland B. Huttner, and Paul Greengard

12. **Role of Calmodulin in Dopaminergic Transmission**
I. Hanbauer and E. Costa

13. **Immunocytochemical Localization of Calmodulin in Rat Tissues**
Jeffrey F. Harper, Wai Yiu Cheung, Robert W. Wallace, Steven N. Levine, and Alton L. Steiner

14. **Immunocytochemical Studies of the Localization of Calmodulin and CaM-BP$_{80}$ in Brain**
John G. Wood, Robert W. Wallace, and Wai Yiu Cheung

15. **Calmodulin-Binding Proteins**
Jerry H. Wang, Rajendra K. Sharma, and Stanley W. Tam

16. **Mechanisms and Pharmacological Implications of Altering Calmodulin Activity**
Benjamin Weiss and Thomas L. Wallace

Index

Chapter 1

Calcium Binding to Proteins and Other Large Biological Anion Centers

B. A. LEVINE
R. J. P. WILLIAMS

I. Introduction 2
  A. Use of Nuclear Magnetic Resonance Spectroscopy 3
  B. The Distinction between Proteins and Small Molecules
    as Ligands 5
  C. A Note on the Relative Function of Magnesium
    and Calcium 6
  D. Calcium-Binding Proteins 8
II. Extracellular Calcium-Binding Proteins 8
  A. External Enzymes 8
  B. Allosteric Control 12
  C. Calcium-Dependent Enzymes and Metalloenzymes 12
  D. Other Extracellular Calcium Activities 14
  E. Calcium Bone and Tooth Proteins 16
III. Calcium Ion Binding to Large Particles, Membranes,
  and Surfaces 17
  A. Polysaccharide–Calcium Interactions 18
  B. Lipid Binding 19
IV. Extracellular Calcium Transport 19
  A. Proteins 19
  B. Calcium Concentration in Extracellular Fluids and
    Its Control 20
  C. General Summary of Extracellular Calcium Proteins 22
V. Vesicles: Calcium Stores 23
VI. Intracellular Proteins 23
  A. Troponin C and Calmodulins (Triggers) 23
  B. Calcium Proteins *in Situ* 30
  C. Rates of Calcium Binding 31

1

CALCIUM AND CELL FUNCTION, VOL. II
Copyright © 1982 by Academic Press, Inc.
ISBN 0-12-171402-0

VII. Calcium-Binding Intestinal Proteins   32
      Possible Function of Intestinal Calcium-Binding
      Protein   32
VIII. Calcium Proteins in Membranes   34
  IX. Summary: Intracellular Calcium   35
      References   35

I. INTRODUCTION

In a recent article (Levine and Williams, 1981) we described the inorganic chemistry of the calcium ion based upon work with small complex ions. We used the general Eq. (1) to refer to biological activity:

$$\text{Activity} \propto [Ca^{2+}]K_{aq}p(\text{structure factors}) \tag{1}$$

where $[Ca^{2+}]$ is the concentration of the free calcium ion in the compartment under consideration (e.g., in general $>10^{-3}\ M$ outside cells and $<10^{-7}\ M$ in cells at rest), K_{aq} is the binding constant of the calcium ion to any free aqueous ligand L to give the complex CaL, p is a partition coefficient which modifies K_{aq} to give the binding in the phase (membrane) or structure where the complex CaL acts. This partition coefficient will also describe the effect of fields, mechanical or electrical, on the stability of CaL in the structure. The product $[Ca^{2+}]K_{aq}p$ therefore describes the binding of calcium to L but does not describe the activity, since activity is related to binding through certain rate constants. The rate constants are a function of the structure and energy of the ground and excited states of CaL. The relevant structures were given in the previous article. In that article data on small calcium complexes were used to describe all four terms in Eq. (1). We also described data for Na^+, K^+, and Mg^{2+}, since the activity of calcium is modulated by the (competitive) activities of these ions (and the proton).

This chapter reviews the conclusions of our previous article before describing the properties of calcium bound to proteins, especially as revealed by our nuclear magnetic resonance (NMR) studies:

1. $[Ca^{2+}]$ can be at any level from about 10^{-3} to $10^{-8}\ M^{-1}$ liter in different biological compartments.

2. The binding of calcium to complexes occurs through carboxylate and neutral oxygen donor centers. The binding strength can be varied readily from 10^3 to 10^{12} by varying the number of donor centers and their stereochemical arrangement. Competition from Mg^{2+}, Na^+, K^+, and H^+

can be set at any chosen level by suitable choice of ligand no matter how large K_{aq} is.

3. The partition coefficient p is difficult to describe, but a simple part of it is the effect of an applied potential ψ when p is proportional to $e^{\psi/RT}$; see below.

4. The structure of calcium complexes varies from 6- to 12-coordinate, grouping at about 8. The stereochemistry differs strikingly from that of magnesium in that the *geometry is irregular* both in bond length and bond angle. The calcium ion does not have a fixed geometry and readily forms cross-links.

5. The rates of exchange of ligands, i.e., the ability of the calcium ion to change structure, both in on/off reactions and fluctuational rearrangements of the ligands on the surface of the calcium ion, are fast—much faster than the corresponding rates for the magnesium ion. The energy of ''excited'' structures is often low.

In many ways, especially related to points (4) and (5), sodium and potassium ions are much more like the calcium ion than the magnesium ion is.

We shall assume that this information from model studies is immediately relevant to the description of calcium activity in biology. This chapter will then be divided into three major sections: a description of calcium-binding proteins, a short description of calcium binding to lipids and saccharides, and a survey of the relationship of these data to calcium activity in biology.

A. Use of Nuclear Magnetic Resonance Spectroscopy

Elsewhere we have described the use of NMR spectroscopy in the study of proteins (Campbell *et al.*, 1975; Levine *et al.*, 1979). Here we give an outline of the method, since many of the observations described below depend directly on an understanding of the procedure. The proton NMR spectrum of troponin C is shown in Fig. 1. Different regions of the spectrum have been assigned to particular types of amino acids, and for some resonances the assignments are to particular amino acids in the sequence (Levine *et al.*, 1977a). The assignment of peaks in such detail allows us to follow the effect on the protein of changes in solution conditions such as those involving pH, [Ca^{2+}], salt concentration, and temperature. Now we can interpret the changes in position of the resonances in terms of changes in structure (Levine *et al.*, 1977b). This is possible because the energy of a transition, an NMR absorption peak, depends upon the chemical groups that are nearest-neighbors to the atom which has absorbed the energy. It is especially helpful to an understanding of solution structure at this stage

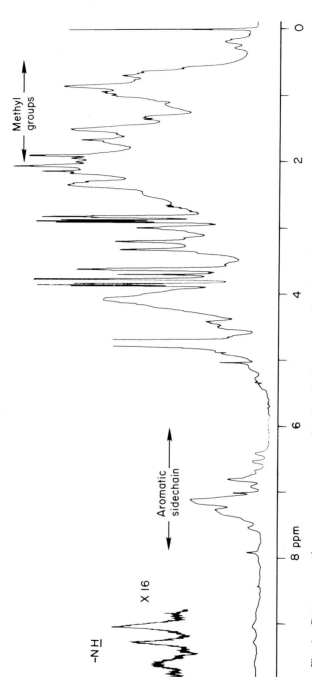

Fig. 1. Proton magnetic resonance spectrum of native (Ca₄P) troponin C in D₂O solution at neutral pH. Resonances in the spectral range 6.5–8 ppm derive from groups of aromatic side chains; in the range 0–2 ppm are methyl group resonances. For assignments see Levine *et al.*, 1977a.

if a crystal structure is available, even though the two structures may not be too similar. For example, the NMR resonance energies of groups in phospholipase A_2 are entirely consistent with the fold found in the crystal structure (Aguiar *et al.*, 1979). Various techniques are available for augmenting the structural evidence from direct absorption NMR spectroscopy, for example, by studies in the presence of (lanthanum) shift probes (Levine *et al.*, 1979). Lanthanide ions usually replace calcium ions fairly exactly.

Apart from evidence from line positions we can use the line width or relaxation properties. Especially valuable are nuclear Overhauser effects (NOEs) (Noggle and Schirmer, 1971) which are seen as changes in line intensity on irradiation of another line belonging to a nearest-neighbor amino acid. These NOE data give distances in molecules directly. Line widths can also be affected by relaxation probes, e.g., Gd^{3+} or Mn^{2+}, cations which readily replace calcium and give structural information (Campbell and Dobson, 1979; Levine *et al.*, 1979).

Considerable information about molecular tumbling and internal segmental or side-chain motion is also available from the NMR spectra. Again without going into detail, differences in relaxation times of different lines often seen in line widths can be used to assess (1) surface residue motion (such as that of lysines), (2) restricted motions (e.g., flipping of aromatic rings, valines, and leucines), and (3) motion of the main chain based on studies of α-CH or NH protons (Williams, 1978; Levine *et al.*, 1979). A major finding is that many calcium proteins have mobile interiors.

In this chapter we shall rarely refer to the primary NMR data, since we prefer to illustrate the major conclusions of our work, but a detailed appreciation does require reference to the original NMR studies.

B. The Distinction between Proteins and Small Molecules as Ligands

It is important to observe that proteins, as ligands, have specific features. Because of their size, their fold energy may equal or exceed that of the binding energy of the metal to the protein. It follows that the way in which the metal binds, its energy and stereochemistry, and the way in which the protein folds are mutually dependent (Williams, 1977; Levine and Williams, 1981). Furthermore the mobility of the protein is constrained by the metal. One way of seeing this is to consider a ligand such as EGTA, with four carboxylates on a highly mobile chain, in comparison to four glutamates in a protein. When EGTA binds to a metal, the binding has a stereochemistry and energy dictated by the metal ion and the entropy loss of the ligand. In the protein the four glutamates can only bind if they are (1) not on a random protein and (2) somewhat close together in a preferred form of the apoprotein fold. The entropy loss on binding is smaller than for EGTA, but the stereochemistry is decided in part by the interaction energies of different protein folds or conformations. This last energy is not largely dependent on the first coordination sphere of the metal as it is in EGTA, i.e., the metal–carboxylate links, but depends additionally on many interactions running deeply into the protein. If we could cut out the $M(CH_2CO_2^-)_4$ unit from the hypothetical protein, then the transfer energy of this preferred unit from water to the protein would involve many energy terms involving long-range interactions. The hypothetical complex "partitions" into the protein. Hence, if K'_{aq} is the stability for $M + [CH_2CO_2^-]_4$ (leaving these anions as they were in the protein), then K_{aq} observed for M binding to a protein could be either much larger or smaller than K'_{aq} by this partition energy p'. $K'_{aq}p'$ could be very metal ion-specific in surprising ways.

Because we cannot determine K' and p' experimentally, we shall refer to K_{aq} both for proteins and model ligands, but we must keep the above distinction firmly in mind.

C. A Note on the Relative Function of Magnesium and Calcium

Our previous article (Levine and Williams, 1981) stressed that Mg^{2+} and Ca^{2+} were very different cations. The differences stem from the difference in ion size: Mg^{2+}, 0.60 Å; Ca^{2+}, 0.95 Å. This difference does not seem to affect the structure of the simplest complex, the aquo cations, greatly since in strong solutions both cations can be $[M(H_2O)_6]^{2+}$ octahedra (Cummings et al., 1980). Combination with ligands reveals large differences in structural, thermodynamic, and kinetic properties. First, large anions of low surface change density, i.e., strong acid anions such as SO_4^{2-}, $(RO)_2PO_2^-$, ClO_4^-, usually do not displace water from around Mg^{2+}. In fact with such anions Mg^{2+} tends to form outer sphere complexes (compare Ni^{2+}). In contrast, calcium allows the anion to enter the coordination sphere (compare Ln^{3+}). As far as the stability of the complexes or the insolubility of salts is concerned $Ca^{2+} > Mg^{2+}$ and calcium behaves as the *smaller* cation. Magnesium will not block a calcium channel composed of strong acid donors.

With weak acid donors, e.g., RCO_2^- and $(RO)PO_3^{2-}$, both Mg^{2+} and Ca^{2+} form inner sphere complexes. Steric factors control stability constants and insolubility, since calcium is usually more than 6-coordinate.

The on–off rates of calcium are fast—perhaps because of an S_N2 reaction, while those of Mg^{2+} are slow, certainly involving an S_N1 attack. The difference is parallel to the difference between silicon and carbon chemistry.

Biological systems have turned these chemical differences to functional use and have added to them by separating the ions in outer (calcium) rather than in both outer and inner (magnesium) cell compartments. However, the ions are so different even when present in the same compartment that direct competition is unusual.

TABLE I
Classes of Calcium Proteins

Class	Example	Conformation change	$\log K_{aq}{}^a$	$\log k_{ex}$
Extracellular enzyme	Phospholipase A_2	Very small	3–4	~3
Extracellular trigger	Prothrombin	Considerable	3–4	>5
Extracellular bridge swivel	Fibrin	Fluctuational	~6	—
Extracellular structure	Bone proteins	Considerable	3–4	>5
Intracellular trigger	Troponin calmodulin	Considerable	~6	~3
Intracellular bridge swivel	Calmodulin troponin	Considerable	~8	Slow
Intracellular transport (aqueous)	Intestinal calcium-binding parvalbumin (?)	Considerable	~6	~3
Membrane ATPase	ATPase pump outer membranes	?	~7	Fast?
Membrane channels	Nerve cells at synapses and Na^+-Ca^{2+} Exchange	None	<3	>6
Intracellular signal	Aequorin	Considerable	>6	~3
Extracellular protein stabilization	Bacterial proteases	Small	3–4	?
Extracellular transport	Vitellin	—	3–4	>5
Intracellular store (vesicle)	Calsequestrin	—	5	Fast
Proenzyme destabilization	Trypsinogen	Considerable at the tail	3–4	—

^{a} These data are from Wasserman et al., eds. (1977).

D. Calcium-Binding Proteins

Table I lists some calcium-binding proteins and their functions. It will be observed that calcium serves many functions, i.e., it is a current carrier. It is undoubtedly a major signal inside the cell; it is a major receptor of messages, i.e., exported proteins, outside the cell; alternatively both inside and outside the cell it can act as a fixed part of protein structures; finally in the membrane it has special paths and pumps. Not mentioned in Table I are the proteins locked in hard structures and the formation and re-solution of these structures, as they are associated with the bonding of proteins to solid-state calcium ions rather than to free calcium ions. Using the divisions of Table I we see that the reaction of calcium with a protein covers a wide range of log K_{aq} values (a wide variety of structures are known to result), a wide range of exchange rates are observed, and the conformational effect of the binding of calcium to a protein can be either very small or very large. It is against this multiple use of the calcium ion that we must see its functional significance. We shall start with the extracellular proteins.

II. EXTRACELLULAR CALCIUM-BINDING PROTEINS

A. External Enzymes

1. Extracellular Enzymes Requiring Calcium

The calcium-containing extracellular enzymes of Table II are of known crystal structure. The enzymes require calcium absolutely in that even strontium gives altered catalytic activity and selectivity. The binding of the calcium ion is to one or two carboxylate ions plus several neutral oxygen donor ligands. The binding constants are usually about 10^3–$10^4 M^{-1}$ liter. We have used NMR spectroscopy to show that, in the case of one enzyme, phospholipase A_2, the change in conformation on binding calcium is extremely small (Aguiar et al., 1979). We suspect that the cavity which binds the calcium ion is well-defined before calcium enters. In such a case we suppose that the on rate of the calcium ion would not be diffusion-controlled and would be relatively slow. In fact, the off rate is 10^{-3} sec^{-1} (Forsen et al., 1980), which means that the on rate is much slower (10^{-6} sec^{-1}) than diffusion (10^{-10} sec^{-1}), since the binding constant is about 10^3. In a later part this chapter we shall compare the trigger proteins, but we note here that digestive enzymes do not have to be activated in milliseconds. In general a high speed of combination implies lower selectivity. The sites of the external enzymes do not accept other ions readily, and

TABLE II

Some Extracellular Enzymes Which Require Calcium

Enzyme	Site of calcium binding
Phospholipase A_2	Very close to active site
Staphyloccal nuclease	Very close to active site
Amylase	Far from active site (structure unknown)
Trypsin	Far from active site
Carboxypeptidase (*Streptococcus griseus*)	Cross-linking far from active site
Subtilisin Carlsberg	Stabilizing action far from active site

even Mg^{2+} does not bind well to the calcium site of phospholipase A_2. We suspect that many enzymes triggered by calcium are similar in selectivity and kinetics (for calcium binding) to phospholipase A_2.

Since digestive enzymes must act only outside the cell, extreme selectivity is required for calcium and against magnesium for the following reason. All proteins are produced inside cells, where the concentration of free magnesium ions is $>10^{-3} M$ but calcium ions are at $10^{-8} M$. If triggering of activity were just dependent on divalent metal ions, then magnesium ions could activate digestion inside cells. This must not happen, even to a level of 1%. Even if magnesium binds with very low affinity, it should also not activate. Not only is a binding energy difference between calcium and magnesium required, but the effects of binding the two ions must be such that calcium (at $10^{-3} M$) activates but magnesium must not. The extreme selectivity for calcium is seen to carry over to strontium which works only as a partial activator of calcium-dependent enzymes and often alters substrate selectivity. As an additional form of protection the enzymes are often in the form of proenzymes in cells, which are activated only by Ca^{2+}.

The actual site of calcium binding to an enzyme may be close to the catalytic site or remote from it (Table II). In the first case it may well act to bind substrate, although it influences the catalytic step poorly, e.g., in phospholipase A_2. Alternatively it may bind far from the catalytic site when it serves to tighten the fold of the protein generally and to protect it from proteolysis, e.g., in bacterial proteases (see below).

Finally many enzymes are released as precursor proenzymes, and calcium often assists conversion (by digestion) to the enzymes. A good example is trypsinogen conversion to trypsin. Here calcium unfolds the protective tail of the proenzyme so that it can be removed. Both trypsin and trypsinogen have another calcium site which acts to stabilize this protease (Table II).

2. Mobility within Enzymes

Mobility in enzymes can be examined by many methods (Blake *et al.*, 1978):

1. The diffuse character and temperature factors associated with x-ray diffraction electron density maps.
2. Rate processes observable from relaxation times, line widths, and the temperature dependence of line shifts in NMR spectra.
3. Fluorescence lifetimes and effects of oxygen on fluorescence.
4. Hydrogen–deuterium (H-D) exchange rates from peptide bonds or tryptophan NH.

Table III lists some observations on small movements in certain enzymes on binding substrates. The active site of an enzyme must have some slight mobility in order that the reactants can be converted to the products via a series of intermediates, which directly involve the enzyme active site side chains. In most isolated enzymes some, though little, motion is seen elsewhere within the body of the protein.

An exceptional finding is the mobility in the C-terminal region of phospholipase A_2. Here NMR studies in particular indicate that there is greater flexibility in a nonactive site region of the protein than elsewhere in the protein (Aguiar *et al.*, 1979; Aguiar, 1979). In another calcium-dependent enzyme, staphylococcal nuclease, it appears that there is mobility in the sequence from residues 1 to 5, from 44 to 53, and from 143 to 149 (Cotton *et*

TABLE III

Some Observed Movements of Active Site Residues of Enzymes

Enzyme	Movement observed on substrate binding	Ref.
Lysozyme	Trp-108 moves relative to Glu-35; Trp-63 stops oscillating; Val-110 flips rapidly at all times	Blake *et al.*, 1978
Carboxypeptidase	Tyrosine moves >10 Å	Johansen and Vallee, 1975
Cytochrome *c*	Very small changes in Trp-59; alterations in Met-80 (and Ile-57)	Moore and Williams, 1980a,b
Nuclease (staphylococcal)	Considerable readjustment of active site	Cotton *et al.*, 1971
Carbonic anhydrase	Adjustment of zinc coordination sphere	Kannan *et al.*, 1975

al., 1971). We note in passing (see later) that these mobile sequences have a relatively high content of charged as opposed to hydrophobic amino acids.

Elsewhere in this chapter we shall make frequent reference to the mobility of a protein. Almost invariably our comments will be based upon NMR data, especially the temperature dependence of ring current shifts, line widths, and relaxation data, and qualitative data on H-D exchange. In the future the methods, both NMR and H-D exchange, will be made quantitative, but this depends on resolving individual rate processes. Even with the limited data at our disposal the general conclusions regarding the mobility of different proteins, both side chain and main chain, are quite definite (Williams, 1979a). We do not anticipate that studies on crystals will necessarily give results parallel to those obtained in solution, and in the case of phospholipase A_2 major differences have already been observed (Aguiar, 1981).

Mobility must be distinguished from two-state equilibria—allosteric switches (Williams, 1978, 1979a). A mobile polymer has many states open to it. It is particularly important to note the extreme cases of an $A \rightleftarrows B$ system, e.g., a snap shutter, and an $\Sigma_0^i A_i$ system, where there are a large number of different states, e.g., a "revolving door" or a "hinged door." Although all intermediate situations are possible, the distinction in energetics is vitally important for the functional potential of the two systems. Calcium proteins of both kinds are known.

3. Calcium Protection from External Proteolysis

The discharge of proteins for digestive purposes means that, in certain regions of an organism or external to a single-cell organism, many proteases are present. In this media other proteins must function, for example, phospholipases, saccharases, and nucleases, as well as the proteases themselves, *without* being destroyed by proteases. Protection can be achieved by limiting the exposure of the protein peptide bonds. Two means are known: (1) the formation of disulfide bridges and (2) the binding of calcium ion. Both are cross-linking reactions, but both give relatively mobile cross-links. Mobility inside an enzyme must be maintained. The mobility of a protein is easily followed by NMR, and the effect of calcium on protein stability can be studied by following changes in NMR secondary shifts with temperature (Levine *et al.*, 1977a,b; Williams, 1978). It has been known for a long time that calcium raises the "melting point" of proteins, but the effect of calcium at low temperatures is to reduce the population of states which can be acted upon by proteases; i.e., partial unfolding is of much reduced probability or of higher free energy. The effects are seen throughout wide regions of the protein, and they are reflected in >NH/>ND peptide exchange rates.

B. Allosteric Control

The control over the catalytic site by a regulating agent, e.g., a coenzyme, which binds at a distant site is termed allosteric control. It often operates through a change in conformation which is generated by binding at the control site and which gives a cooperative change at the active site. An alternative mode of allosteric control which might not be seen in any appreciable change in the ground state conformation would result if, on binding the allosteric effector at the control site, there were a change in the dynamics of the interior of the protein through the active site, i.e., a general slight tightening of the protein. The reaction *energetics* at the active site would then be altered, even though the binding of substrates was hardly affected. Our NMR studies on several enzymes, lysozyme, cytochrome c, and phospholipase A_2 (Williams, 1978), have revealed that very small changes in dynamics as well as in structure occur deep inside these proteins on binding inhibitors or by chemical modification of side chains. Through such an effect on motion within a protein frame the allosteric effect could act on the catalytic process itself, rate constant k_{cat}, and not on the binding constant K_M where the overall rate is $K_M k_{cat}$. Any such so-called allosteric control might have a barely detectable ground state *steric* consequence. Calcium ions could act as regulators in this way, which is quite distinct from allosteric control as seen in hemoglobin. However, even here (R $\rightleftarrows$ T changes, where R is relaxed and T is tense) the transmission of change must be through motions in the protein and cannot be simply a two-state cooperative rearrangement, for this would have a very high activation energy.

C. Calcium-Dependent Enzymes and Metalloenzymes

A direct comparison between calcium enzymes, which are often metal-activated enzymes, and metalloenzymes, such as those of zinc, reveals the following differences:

1. Metals in metalloenzymes are usually directly involved in the catalytic act and not in control, e.g., carbonic anhydrase.

2. Metals do not exchange from metalloenzymes but calcium exchange from its enzymes is relatively fast. *Calcium is usually involved in control and substrate binding, not catalysis.* Compare the extracellular proteases of zinc and of calcium.

3. Bound metals such as zinc considerably alter the acidity of coordinated water or other ligands. Calcium has little effect; e.g., compare phospholipase A_2 with carbonic anhydrase.

4. Where metals in metalloenzymes serve only as cross-linking agents, they maintain a relatively rigid geometry. Calcium allows gross fluctuations within its coordination sphere; i.e., calcium restricts movement to a much smaller degree, since it is a mobile cross-linking agent. Compare transcarbamylase (zinc) and amylase (calcium).

5. It has been proposed that *calcium as a cross-link predates in biological evolution S—S bridges.* Calcium cross-links are particular prevalent in extracellular bacterial enzymes.

Another interesting difference between calcium and zinc (iron or cobalt) sites is that the calcium site is often formed from a continuous part of the sequence (Table IV), except when it is at the active site. This is not the

TABLE IV

Calcium Sites in Some Proteins

Protein	Sequence	Continuous
Phospholipase A_2	Tyr-28, Glu-30, Gly-32, Asp-49, $2H_2O$	No
Trypsin	Glu-70, Asn-72, Glu-80, $2H_2O$	Yes
Staphylococcal nuclease	Asp-19, Asp-21, Asp-40, Thr-41, Glu-43, $1H_2O$	No
Con A	Asp-10, Tyr-12, Asn-14, Asp-19, $2H_2O$	Yes
Thermolysin	Asp-138, Glu-177, Asp-185, Glu-187, Glu-190, $1H_2O$	No
	Glu-177, Asn-183, Asp-185, Glu-190, $2H_2O$	
	Asp-57, Asp-59, Glu-61, $3H_2O$	Yes
	Tyr-193, Thr-194, Ile-197, Asp-200, $2H_2O$	
Parvalbumin		
CD	Asp-51, Asp-53, Ser-55, Phe-57, Glu-59, Glu-62	Yes
EF	Asp-90, Asp-92, Asp-94, Lys-96, Glu-101, $1H_2O$	
Troponin C, rabbit skeletal		
I	Asp-27, Asp-29, Asp-33, Ser-35, Glu-38, $1H_2O$	
II	Asp-63, Asp-65, Ser-67, Thr-69, Asp-71, Glu-74	
III	Asp-103, Asn-105, Asp-107, Tyr-109, Asp-111, Glu-114	Yes
IV	Asp-139, Asn-141, Asp-143, Arg-145, Asp-147, Glu-150	
Calmodulin		
I	Asp-20, Asp-22, AsX-24, Thr-26, Thr-28, Glu-31	
II	Asp-56, Asp-58, AsX-60, Thr-62, Asp-64, Glu-67	Yes
III	Asp-93, Asp-95, AsX-97, Tyr-99, Ser-101, Glu-104	
IV	Asn-129, Asp-131, Asp-133, Glu-135, Asn-137, Glu-140	
S-100	Asp-61, Asp-63, Asp-65, Glu-67, Asp-69, Glu-72	Yes
Wasserman[a]		
I?	Ala-15, Glu-17, Asp-19, Glu-21, Leu-23, Glu-26	Yes
II	Asp-54, Asn-56, Asp-58, Glu-60, Ser-62, Glu-65	

[a] From NMR. The crystal structure published after this paper was submitted gives for hand I (Moffatt, personal communication) Ala-15, Glu-17, Asp-19, Asn-22, Ser-24, Glu-27. Hand II is as shown.

case for zinc or other truly catalytic metal ions. This means that the protein folds around zinc, but uptake of calcium only causes a protein loop to alter its shape in the first instance. The different binding devices have obviously different restrictions on metal on–off reactions; zinc is never lost, whereas calcium easily dissociates. While the binding to an unfolded loop of a *hydrophobic* enzyme may tighten the whole protein somewhat, as well as fold the loop, the same binding to a less folded *hydrophilic* protein may cause dramatic protein changes (Section VI). The effect of calcium binding to a protein loop depends upon the rigidity of the whole protein, which depends upon the whole protein sequence (Williams, 1977; Levine and Williams, 1981).

D. Other Extracellular Calcium Activities

1. Triggers

Many extracellular nonenzymatic activities are triggered by calcium ions. Here the calcium ion only brings about a conformation change in a protein which then may or may not act on an enzyme. An example is the action of calcium on prothrombin. This protein contains several binding groups for calcium, which use the special amino acid γ-carboxyglutamate (Gla) (Esnouf, 1977). Gla itself does not bind calcium very strongly—in fact no more strongly than does malonate does. However, the Gla units are clustered in the first Kringle (the name of a special cross-linked structure found only in these proteins) of prothrombin, and at least two are held relatively close together by a disulfide bridge (Fig. 2). The binding constants for the six calcium ions then exceed $10^3 M^{-1}$ liter. This binding site does not use neutral oxygen donors and is nonselective; magnesium binds almost equally. In line with the very adaptable stereochemistry of the Gla chain it also binds very rapidly, though there may be a subsequent slow conformation change in the rest of the protein. The selectivity of the action of calcium lies in its ability to bind to further groups after associating with the prothrombin Gla units. It cross-links the protein to negatively charged membranes, and this is one trigger for the blood-clotting cascade.

The calcium-binding site of the Gla-containing fragment undergoes considerable conformation change on binding the calcium ion (Pluck *et al.*, 1980). We note again that in this case, quite unlike that of phospholipase A_2, the sequence of the protein which binds calcium is highly charged. It is probable that it is disordered before calcium binding and binding is faster.

A rather similar situation occurs in the activation of trypsinogen to trypsin, and similar roles for calcium are found in activating blood clot formation, in complement activation, and in the binding of proteins to the exposed sugars of cell surfaces, e.g., lectin action (Table V).

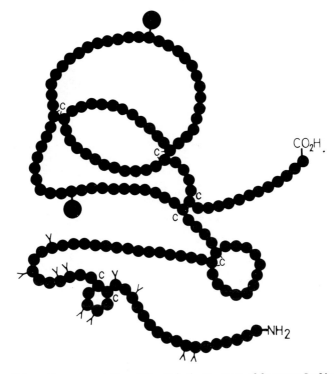

Fig. 2. Schematic representation of the Kringle structure of fragment I of bovine prothrombin (Esnouf, 1977). C, Cysteine residues, Y, Gla side chains.

2. Bridge Swivel

Some of the calcium in extracellular fluids is permanently bound into proteins. This calcium must have a binding constant of $\gtrsim 10^4$. A typical example is one site of fibrinogen which binds calcium through four carboxylate residues which are restricted in conformational freedom by an

TABLE V

Calcium-Requiring Extracellular Triggers

Protein	Calcium function
Prothrombin	Binds protein to membrane
Complement	Binds to components r and s
Lectin (concanavalin A)	Binds adjacent to Mn^{2+} site and far from sugar site; allows binding to cell surface
Fibrinogen	Binds to C-terminal

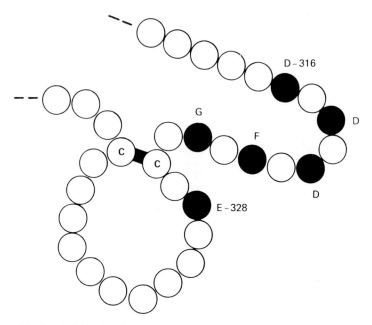

Fig. 3. Calcium-binding site of fibrinogen; schematic (Harrison, 1980).

S—S bridge [Fig. 3, and compare prothrombin (Fig. 2)]. The function of this calcium is not known, but we can make a suggestion. From model studies we know that the motions of calcium ligands on the surface of the calcium ion are rapid. Thus calcium can act as a link about which groups can swivel without dissociation, that is, a low-energy bridge swivel. Here calcium is a permanent part of the machinery and not a piece to be inserted. Another example is provided by α-lactalbumin.

It appears that calcium cross-links may have a number of different functions much as S—S bonds do. They can act (1) to provide thermal stability, (2) to give considerable rigidity, and (3) to allow limited (local) mobility while preventing greater segmental motion; i.e., they restrict the entropy of a polymer to a limited set of essential conformations. Such cross-links are found in polysaccharide and at lipid surfaces, as well as in proteins, and are part of the quaternary as well as the tertiary structure.

E. Calcium Bone and Tooth Proteins

There are two very different classes of bone and tooth proteins. The first contains Gla units and is found in bone (Hauschka and Gallop, 1977).

TABLE VI

Calcium-Binding Proteins and Molecules of Hard Structures

Protein	Function
Phosphoprotein (dentine)	To assist in nucleation of dentine
Gla proteins	To limit the crystal growth of apatite
Proline-rich proteins	Salivary gland excretion to protect teeth
(Phosphocitrate	Prevents crystallization of calcium phosphate)
(Trimetaphosphates	Prevents crystal growth of bone and shell)

It is thought to be an inhibitor of bone growth by absorption on the growing calcium phosphate crystals. The smallest form of this protein is a peptide of 40 amino acids, and it is unlikely to have a simple folded structure. We suspect that NMR studies will show that it is close to a random coil like many other highly charged peptides of this size.

We have studied another type of solid-state protein (Cookson *et al.*, 1980). It has a quite unusual sequence of phosphorylated serine and acidic amino acids. It has been shown that it is largely an extended polymer. It is known that this protein only occurs in tooth dentine, not in bone. The difference between these two types of protein is partly in the protein composition and partly related to the packing of the microcrystals of apatite $Ca_2(OH)(PO_4)$. It is reasonable to suppose that the proteins which have Gla or Ser-PO_4 are used differently to control the inorganic matrix. Probably they control the way in which the mineral is laid down both by positively assisting growth in certain directions and by inhibiting growth in others. Many small organic molecules are known which affect crystal growth (Table VI), and they have similarities with the bone and tooth proteins in that they have tightly packed anions on a molecular framework. No doubt similar principles apply to the growth of calcium carbonate crystals in shells (see later) but here polysaccharides are also involved.

III. CALCIUM ION BINDING TO LARGE PARTICLES, MEMBRANES, AND SURFACES

The fact that calcium ions are present in the environment at about 10^{-3} M (in soft water $10^{-5} M$) makes calcium useful in the organization of larger biological assemblies. The simplest cases are in the blood proteins, hemocyanin and erythrocruorin, and in viral coat proteins. However, calcium ions are very generally involved in the binding of anionic polymers to one another, in the adhesive strength of membranes, and in the attachment of cells to one another and to surfaces. Unfortunately here the proteins are not the only binding sites and the organization of extracellular structures from plant and bacterial cell walls through to the adhesion of animals to surfaces involves proteins, polysaccharides, and lipids. We cannot discuss the topic in full here, but we can give a general outline.

First, any polyanionic surface generates a negative potential ψ which assists binding above and beyond that established by direct association of the individual anions of the structure [see p in Eq. (1)]. We then distinguish layers of binding:

1. Direct binding (as discussed so far). Roughly speaking, association centers and constants for calcium are: SO_4^- or SO_3^- (very small); PO_4H^-, CO_2^- (10 M^{-1} liter); PO_4^{2-} (100 M^{-1} liter). However, all polyanions are based on hydroxylated polymers—lipids, proteins, polysaccharides, and some individual chelate sites give binding constants $>10^3\ M^{-1}$ liter on all surfaces.

2. The Gouy–Chapman diffuse layer which sets up a concentration of ions in the vicinity of the charged surface proportional to the charge. The layer extends from 5 to 50 Å from the surface.

3. A closely associated Stern layer, 5 Å thick, of *hydrated* ions in direct contact with the charged surface. This layer is very difficult to assess except using ions such as $[Co(NH_3)_6]^{3+}$ which cannot lose their first coordination sphere.

The binding layers are not independent, since the ions bound in one layer screen subsequent binding at all sites. We have an intractable problem. Even so it is very clear that the folding together, shaping, and attachment of biological membranes, of membranes to other membranes, of cell surfaces, and of cells to inert supports uses the calcium ion as a cement between mainly anionic polymers (Williams, 1979b). The effect can be a more or less permanent attachment or a loose association readily adjusted. We stress these points here, since the binding and function of calcium is dependent on a physical field effect (electrical here). Other field effects will become apparent later.

A. Polysaccharide–Calcium Interactions

Outside cells calcium can bind to the carboxylate and sulfate groups of many polysaccharides, many of which are also linked to proteins. A major function of the calcium is to stabilize the quarternary and not the tertiary structure so that the polysaccharides bind together in n-fold helices. The calcium ion cross-links through coordination to (1) anionic centers, (2) hydroxyl side chains of sugar, and (3) water molecules of hydration. The calcium ion is unique in its ability to stabilize particular structures. In the next few years we must expect a large development in the study of glycoproteins and the role of calcium in the control of extracellular protein–polysaccharide matrices, e.g., connective tissues, plant structures, and mucous fluids. These structures are not rigid, of course, but have mobility as a necessary part of their function. Already it is known that a large polysaccharide is involved in the coccolith formation of some unicellular organisms.

B. Lipid Binding

Many studies on the binding of calcium to lipids have been made. The following conclusions can be drawn (Hauser *et al.*, 1976): (1) Binding is weak to most individual head groups, log $K \lesssim 2.0$. (2) Binding to groups (two or three) of head groups in the membrane can be stronger, log $K \gtrsim 3.0$. (3) Fields (Williams, 1979b) increase this binding into the range log $K = 3.0$–5.0. (4) Selectivity rests on a combination of charge (usually phosphate oxygen) and neutral donors (sugar oxygen).

It is the opinion of the authors that statistical, fluctuating binding of calcium and magnesium ions to membranes in fields is inevitable and varies both with the nature of the head group and the field. In part this will cause changes in conformation of the head group and of the whole surface. The state of the membrane is then dependent on calcium and is in communication with the surrounding aqueous fluids (Williams, 1979b). In mitochondria calcium is extremely important, while in chloroplasts it is magnesium which fills the same interactive role with the membrane. The effects are quite different from the physical effects of changes in Gouy–Chapman double layers but are related to such changes. It is necessary, therefore, to observe the membrane *directly* by spectroscopic probes. While this has been done for Ln^{3+} (Hauser *et al.*, 1978), the study of calcium (or magnesium) binding to membranes is often too general to discover the effects of membrane binding separately from double-layer effects (Barber and Searle, 1979). Finally, since phosphorylation acts as a control over membrane structures, the precise anionic state and calcium binding of a membrane is related to metabolic activity.

IV. EXTRACELLULAR CALCIUM TRANSPORT

A. Proteins

Calcium needs to be mobilized between different compartments of a higher organism. There is then a need for calcium carrier proteins in fluids connecting calcium stores, e.g., in bone and through blood or lymph, but it is also necessary to transport the calcium across endothelial cells between waste liquids and the blood. The extracellular part of this transport will be described here, while the intracellular transport and transport through membranes are discussed in later sections of this chapter.

B. Calcium Concentration in Extracellular Fluids and Its Control

Hastings has summarized the advances in our knowledge of the state of calcium ions in extracellular fluids using serum as a specific example (Hastings, 1980). The salient point is that serum, like many other biological fluids, acts as a calcium buffer, and it is convenient to divide the total calcium pool into three pools in fast equilibrium: free calcium ions, calcium ions bound to small molecules, and calcium ions bound to proteins. In blood serum the total calcium is 2.5 mM, of which about 1.2 mM exists as free calcium ions, a small amount, about 0.1 mM, is bound to small molecules such as citrate, and about 1.2 mM is "nondiffusable" and has been described as calcium proteinate without specification as to its nature. These levels of calcium are sensitively related to the levels of precipitating anions, such as carbonate and phosphate, so that the bone matrix also acts as a slower buffer of calcium. The laying down or re-solution of bone (or shell) can then be manipulated by varying the concentration of calcium (or phosphate or carbonate) by a very small percentage, for example, ±5%. The precipitation equilibria operating between the buffer and the solids

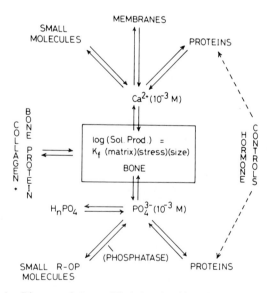

Fig. 4.   Diagram of the equilibria involved in calcium precipitation.

are under the control of three hormones, 1,25-hydroxylated vitamin D_3 (cholecalciferol), calcitonin, and parathyroid hormone. We then have the extremely and exquisitely sensitive system of a balanced set of reactions shown in Fig. 4. The enzyme alkaline phosphatase is also a key control of phosphate.

It might be thought that there is a simple single constant that can represent the activity of bone, a solubility product. This is not the case, since the crystals laid down are only about 100 Å across and not of a precisely determined size. In such circumstances the activity of the calcium and other ions in the crystallites is a function of crystal size and stress on the crystals (see later) and is dependent also on the exact nature of the collagen material which interacts with the crystallites (Williams, 1975). Thus bone formation has control exerted upon it not only from the side of the equilibria involving the hormones and their effects on free calcium, free phosphate, and free additional ions such as magnesium, fluoride, hydroxide, and carbonate (all of which are found in bone), but also by the state of the bone. Where the bone is under constant stress, i.e., a pressure, then locally it has a different activity in the absence of stress. Moreover there can be local changes in the proteins. Thus bone can be likened to an organic polymer, such as glycogen, in an organism involved in constant metabolic activity, degradation and synthesis, and interacts with a large variety of other processes within cells which require calcium or phosphate. The "bone" is then transferable from one place to another in the body, e.g., during pregnancy. This is a very quick overview of the thermodynamics of bone. The kinetics are equally intriguing. Without going into detail it is clear that the rates at which bone is laid down or redissolved are related to the presence of many inhibitors (or accelerators) such as magnesium ions and pyrophosphate. Thus in Fig. 4 we should include a variety of rate controls, but at this time few of these are well understood (see Gla proteins above).

Apart from the importance of calcium ions in precipitation reactions which generate solid matrices, calcium is involved in a host of activities associated with the functions of serum and other body fluids. It is an essential component in blood clotting, in complement reactions involving antibodies, and in digestion, in addition to its role as a messenger. Moreover, it is also involved in cell adhesion and cell movement on various substrates and in many interactions between different membranes. Calcium has a role in many diseases, e.g., in viral invasion of cells, and in the stability of bacterial membranes. The importance of calcium ion in fluids around cells cannot be overstated. In complex organisms the controlled buffering of this ion outside the cell (homeostasis) is equally as important   as the control of its concentration within the cell.

TABLE VII

Proteins Associated with Calcium Which Are Virtually in a Random Coil State

Mineralizing Proteins	Gla proteins, Phosphoproteins
Storage devices	Phosvitin, chromagranin A, calsequestrin

It is unfortunate that at the present time the exact nature of the calcium transport proteins in blood is not known. In certain organisms this is extremely important, as calcium is rapidly mobilized from stores to bone or shell formation, e.g., in the transformation of a tadpole to a frog. The transport protein must be recognized by selected cells. There are strong suggestions that phosphorylated proteins, like phosvitin of eggs, are involved, but until these are thoroughly studied we shall not know how they function. Note that phosvitin, like many of the other proteins associated with calcium, is highly mobile—virtually a random coil (Table VII). We include the storage protein of vesicles inside cells, as vesicles often have calcium concentrations as high as those of external fluids, >1 mM.

C. General Summary of Extracellular Calcium Proteins

While the range of activities of calcium in association with proteins is huge, certain general principles have emerged in the above descriptions.

1. No matter whether the protein as a whole is hydrophobic or hydrophilic, the calcium site is a hydrophilic loop or hole. We shall find that this is also true for internal calcium-binding proteins, though some enzyme sites may be exceptional.

2. Calcium-binding proteins can be highly ordered or almost disordered.

3. Selectivity of binding arises from multidentate chelation—usually more than three donor centers are involved. While invariably there is at least one carboxylate, there is usually at least one neutral oxygen donor site.

4. Calcium-binding constants are usually in the range of external calcium concentrations, approximately $\sim 10^3$ mM.

5. Calcium often cross-links proteins, but this is not a gross restriction on internal mobility.

6. Calcium can act as a cross-link between proteins, e.g., in viral coats, but quaternary cross-linkage is more usual in polysaccharides.

V. VESICLES: CALCIUM STORES

A vesicle is really extracellular in that it is separated from the cytoplasm of the cell by a membrane. Calcium is often accumulated in such vesicles ready for release as a trigger but also as a source of cations for the production of calcium carbonate or calcium phosphate. The stores are not just of free calcium ions. In the sarcoplasmic reticulum, for example, there is a protein, *calsequestrin* (Table VI) (Ostwald *et al.*, 1974), which binds about 50 calcium ions per monomer. It appears to lie very close to one surface of the membrane. Its purpose is not known, but its binding constant for calcium, $10^3-10^5 M^{-1}$ liter, and its locality suggest that it is a surface buffer. Release of calcium from the sarcoplasmic reticulum by nerve triggering is therefore followed immediately by recovery of the free calcium concentration in exactly the local region from which it was lost, by calcium from this buffer protein. The low binding constants for calcium will ensure a rapid response. In a sense calsequestrin also acts in a manner not unlike a vesicle store of free calcium, since it is also able to maintain the calcium input to the cell initially by maintaining the local concentration close to the input channel. We suspect that other calcium-storing devices are similar in nature. The composition of the protein is that of a highly negatively charged polymer, and we suspect that in the absence of a membrane it is effectively a protein without a tertiary fold. Such a protein is found in chromaffin granules, chromagrannin A, and it binds calcium kinetics of conformational change.

VI. INTRACELLULAR PROTEINS

While there are many functions for calcium outside cells, there appear to be a very limited number of uses for this cation in the cell where at rest it is only at a level of $10^{-8} M$. This is not to say that these uses are anything but extremely important, but they are largely associated with triggers and transport. We shall distinguish proteins which act as internal triggers, proteins which have structural calcium, calcium buffer and transport (aqueous) proteins, and calcium transport (membrane) proteins, pumps. It is unfortunate that, although quite a few are well characterized, the functions of many are not yet known.

A. Troponin C and Calmodulins (Triggers)

There now appear to be calcium trigger proteins in all cells (Means and Dedman, 1980). Some are associated with various contractile devices, for example, in muscle. These proteins are called troponins C and certain myosin chains. Elsewhere in cells the trigger proteins called calmodulins are known to switch on a diversity of major functions including (1) glycolysis, (2) dehydrogenation, (3) ion (including calcium) pumping, and (4) formation and hydrolysis of cAMP. Apart from these known activities there is reason to suppose that calcium-triggering proteins are involved in (5) activation of the release of stored chemicals (calcium, acetylcholine, epinephrine, etc.) from intracellular vesicles, (6) cell division, and (7) a variety of functions of filaments. Very many, if not all, of these activities are activated by calmodulins, but it does not follow that activation of each protein molecule is exactly the same or occurs at exactly the same calcium ion concentration, as we shall see.

Both troponins and calmodulins have been examined in detail by NMR (Levine *et al.*, 1977a, 1978; Seamon, 1979). Based on the observed sequence analogies with parvalbumin, the outline structure of the calcium-bound form of these proteins has been deduced on the basis of the known crystal structure of parvalbumin. The generalization has been most clearly described by Kretsinger (1980). We have shown (Levine *et al.*, 1977b; Evans *et al.*, 1980) that the calcium-free form of these proteins is less strongly structured and, indeed, it appears that roughly half of it is close in energy to a random coil. This result has also been found by microcalorimetry (Tskalkova and Privalov, 1980). All these proteins have a large proportion of hydrophilic, especially charged, amino acids, and in this respect they resemble many other control proteins, e.g., histones and proteins which control the function of ribosomes and membranes. We have suggested that enzymes and control proteins differ in their amino acid composition (Fig. 5), and that through a consequent difference in mobility from enzymes (especially enzyme active sites) this leads to a different relation among sequence, mobility within structure, and function (Williams, 1979a,b).

A feature of all these calcium-binding proteins is that they bind with constants in the range 10^5–10^8 M^{-1} liter and that this occurs very rapidly with on rates close to the diffusion control limit. As we have shown elsewhere (Williams, 1979a; Levine and Williams, 1981), this implies that the site of binding, which is 3 or 4 carboxylates plus 2 neutral oxygen donors on a continuous-sequence stretch of 12 amino acids (Fig. 6), has considerable mobility. Figure 7 illustrates the most extended form and the most contracted form as calcium binds. The uptake of calcium is linked to many other changes in the protein over a large part of its structure, and these changes have been recognized by the study of fragments (Evans *et al.*, 1980). Thus the act of binding calcium to an internal trigger protein can be compared with the binding of an external trigger, but is quite unlike the

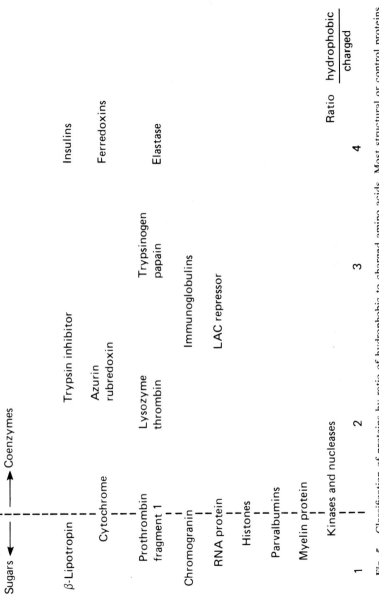

Fig. 5. Classification of proteins by ratio of hydrophobic to charged amino acids. Most structural or control proteins are on the left, while the less mobile proteins are on the right (Williams, 1979a,b).

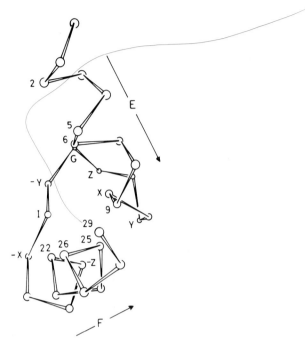

Fig. 6. Helix–loop–helix configuration of the calcium-binding site of parvalbumin (Kret-singer, 1980). The residues X, Y, Z, −X, −Y, and −Z provide the closely sequential calcium-binding side chains (cf. Table IV).

binding to an external enzyme in which little conformational change is induced (Levine *et al.*, 1977a).

We do not imply that the troponin and calmodulin proteins are random proteins in the absence of calcium. In fact the presence of order is easily shown by NMR (Evans *et al.*, 1980) and by comparing the binding constants of calcium to the proteins with those to protein fragments (Leavis *et al.*, 1978), (Table VIII). It is seen that the calcium binding is reduced by a factor of several powers of 10 in the fragment which is composed of a single hand. We do not believe, however, that the binding sites of the fragments are chemically different from those in the whole proteins, since the directly comparable NMR signals in the folded forms and fragments (calcium bound) are very similar. Thus the fold in the complete protein reduces the configurational entropy of the protein, and this makes a high binding constant, 10^5–10^8 M^{-1} liter, possible in the EE hands for the sequences found in troponins and calmodulins (Williams, 1977). This explains the fact that the discrimination in binding between calcium and magnesium remains at $10^3 : 10^4$ at the binding sites of calmodulin, troponin, and the fragments.

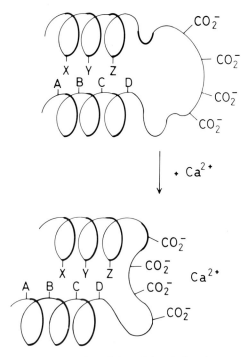

Fig. 7. Diagrammatic representation of the calcium-induced structural reorganization.

The nature of sites which give this level of discrimination has been described elsewhere (Williams, 1977; Levine and Williams, 1981). Note that the same Ca/Mg discriminatory factor of $10^3:10^4$ is present at the strong and weaker binding sites of the proteins. This has led to a descrip-

TABLE VIII

Calcium-Binding Constants in Peptides

Fragment	CO_2^-	Isolated peptide log K_b	Log_b intact protein
Parvalbumin[a]			
CD site	4	4	8
EF site	4	2	8
Troponin C[b]			
Site II	4	3.8	5.3
Site III	4	5.2	7.3
Site IV	4	4.2	7.3

[a] Derancourt et al., 1978; parvalbumin
[b] Leavis et al., 1978; troponin-C

tion of the strong sites as Ca^{2+}-Mg^{2+} sites, since *at physiological concentrations* of calcium and magnesium ions both can bind at these sites with constants of 10^8 M^{-1} liter (Ca^{2+}) and 10^2 or 10^3 M^{-1} liter (Mg^{2+}). It is then the weak sites which are free from magnesium and accept (rapidly) a pulse of calcium, while the strong sites may well be bound by magnesium and may not then exchange magnesium for calcium during the time course of a short pulse. However, we see immediately that these deductions depend on the precise concentration of magnesium and calcium both before and after a pulse, on the duration of the pulse, and on the value of K in an *organized* system of interacting proteins. We return to these points later.

Apart from the calcium-binding sites, the trigger proteins have other regions which bind them to the particular part of the cell organization which they activate. Again we have used NMR to find some of these regions (Evans and Levine, 1980). We note that they are usually relatively charged, hydrophilic loops (Fig. 8) which bind to charged hydrophilic loops of the proteins which they activate (e.g., troponin C activates troponin I) and that these regions of mutual interaction are mobile. In fact this is an essential feature of transmission. (A machine, artificial or biological, must have moving parts.) The interaction energy of a calmodulin molecule has the following states, assuming the calmodulin to be bound at its site in the calcium-bound and calcium-free states:

$$\Delta G(\text{calcium-free}) \rightleftarrows \Delta G(\text{calcium-bound})$$

where ΔG is a function of the binding to the particular acceptor. Since

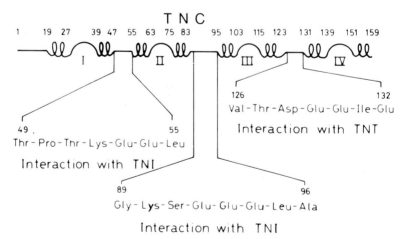

Fig. 8. Hinge regions of troponin C which form surfaces of contact with troponins I and T. The drug trifluoperazine interacts with residues 148–155 and 72–81. See also Fig. 10.

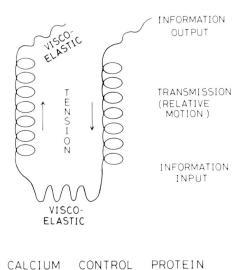

CALCIUM   CONTROL   PROTEIN

Fig. 9. Transmission of information upon calcium binding to a receptor protein. The helices are shown moving in the plane, but they actually roll around one another.

receptor binding and calcium binding are interactive, the exact mode and strength of binding of calcium by calmodulin is a function of the particular receptor concerned. If the receptor is part of an organized structure, then calcium binding and magnesium binding are a function of the precise character of the organization, e.g., the chemical, mechanical, and electrical fields being exerted at a particular time on that organization (Williams, 1979b). We see that the association of calcium sites with structures of the cell means that calcium binding, triggering, etc., can be functions of the cell as a whole, e.g., the temperature or the pressure to which it is subjected (Fig. 9). Furthermore, each component of the organization can be triggered at a different calcium ion concentration, even by the same calmodulin.

In this general context perhaps a useful way to look at a calmodulin is as follows. When calcium binds to this type of protein, whether it is an organized structure or not, it adjusts the shape of the protein, especially at certain points remote from the calcium. If these points are points of contact with an enzyme, the effect of binding the calcium is to relay conformation changes in calmodulins to enzymes. If it is asked Why not let the calcium act directly on the enzyme? then only one simple answer occurs to us. The calmodulin is used to spread the effect of calcium binding over a very large surface of contact, which will permit more precise simultaneous control of changes at a multitude of contact sites between the calmodulin and the enzyme.

B. Calcium Proteins *in Situ*

If the above description of an intracellular calcium-binding protein is correct, we can describe its properties *in situ* in the following way. Let us suppose that the calcium-binding constant at a site on the protein can vary steadily from, say, $\log K_{Ca}^1 = 5$ to $\log K_{Ca}^1 = 8$, where K^1 is the binding constant for a given constraint on the protein, i.e., a given strain accepted by the protein. We can partition the binding energy into the local binding for calcium minus the energy constraint imposed at and by the relay site R:

$$\log K_{Ca}^1 = \log K_{Ca}^0 - \mathscr{F}(\Delta S/RT)_R$$

Here we treat this as a restriction on the number of conformations and not their potential energy. In the equation the strength of calcium binding to the ligand is altered by a loss or gain of entropy at the constraining relay site, R. We immediately see that the function depends on temperature (and pressure). Now let us connect the relay site to another protein which also causes entropic effects at the relay site. $\log K_{Ca}^1$ now depends on the state of the second protein via its effect on ΔS. If the second protein has many conformational states (ΔS is open to variable tuning) and this second protein can be energized continuously, e.g., by variable force applied to it such as a tension, then $\log K_{Ca}^1$ will reflect this tension. Both proteins vary as a continuous function of pressure, temperature, or field applied to them, and all their constants also vary continuously. The reverse is also true, in the sense that the calcium-binding site is of variable entropy. We then have a continuous relationship between input and output and not an $R \rightleftarrows T$ allosteric switch. We shall call this type of motile protein *rhetatic*.

It is interesting to consider the way in which drugs (inhibitors) can act on such proteins. Figure 10 shows the way in which a phenothiazine drug binds to calmodulin (Klevit *et al.*, 1981). The drug binds at a site far from the calcium near the wrist of the EF hand and could perhaps act as a clamp on the hand or affect the relay site, although it binds at neither.

There is an immediate relationship between this description of $\log K'$ and that of the solubility products of bone, for the product is only constant in the absence of mechanical and/or electrical fields. We wish to stress that each individual bone or protein assembly responds in a continuous fashion to the stress. The contrast with an allosteric response is that in the latter the response can be continuous only in a statistical sense and within

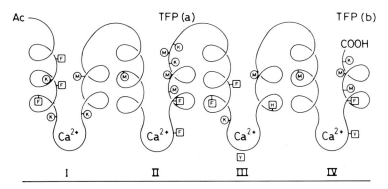

Fig. 10. Representation of the mode of binding of trifluoperazine to the two domains of calmodulin.

a large assembly of units. This allows an allosteric complement of proteins, e.g., hemoglobin, to act within the whole cell, while the rhetatic protein can act singly at the level of a local region of a cell.

C. Rates of Calcium Binding

The rates of binding of calcium to trigger proteins are very rapid, since the binding occurs in a series of steps (Eigen, 1963). The on rate is close to the diffusion control limit, 10^{10} sec^{-1}. The off rate is then 10^3–10^4 sec^{-1}. The fast off rate is essential for rapid trigger relaxation. Magnesium binding constants to these proteins are about 10^4. Since the on rate is $<10^6$ sec^{-1}, the off rate is slow, 10^2 sec^{-1}. This means that magnesium cannot be part of fast switches.

We can compare the rate of this switch with that of a two-state allosteric switch, which will be much slower, since the intermediates between the two states are of higher energy and yet they are necessary for the change. Again we can compare the process with the on–off rates of calcium to an enzyme site, e.g., phospholipase A$_2$, noting that the very limited mobility of such a site causes the activation-free energy of the binding step to be high, since several water molecules need to be replaced at once for the calcium to enter the preformed hole.

Before leaving the trigger proteins we state again our belief that many more such calcium proteins will be uncovered, e.g., S-100 of nerve cells.

Now, although the on rate is fast, the conformation change that follows is relatively slow, and we have shown that in free calmodulin it is of the order of 10^{-3} sec.

VII. CALCIUM-BINDING INTESTINAL PROTEINS

There is a protein which is related to troponins and calmodulins and is found in the epithelial cells of the intestine (Wasserman and Feher, 1977). It is likely that this protein is common to a wide range of cells required in the transport of calcium ions, e.g., in glands and possibly in bone cells. It may also be related to parvalbumin which is a binding protein of many muscle cells (Derancourt *et al.*, 1978) but which is of unknown function. The intestinal calcium-binding protein has properties, a sequence, mobility within structure, calcium-binding constants, and changes in mobility and structure with calcium binding all of which are reminiscent of calmodulins. An outline structure (Levine *et al.*, 1977b) is shown in Fig. 11. This structure has now been solved in detail by X-ray diffraction (Moffatt, personal communication), and this outline has been confirmed. In the intestinal cell it appears that it is present as a free mobile protein within the solution matrix. In the next section we shall propose a function for it. (Added in proof: The structure is now known, Szebenyi *et al.*, 1981.)

Possible Function of Intestinal Calcium-Binding Protein

It has now been proved that this protein is present in the cytoplasm of the cells of the intestinal surface (Taylor, 1980). It is highly probable that it

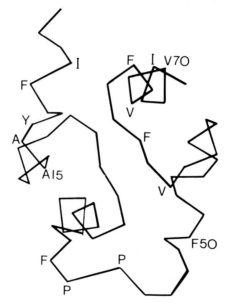

Fig. 11. Outline fold of the vitamin D-induced intestinal calcium-binding protein (cf. Fig. 6).

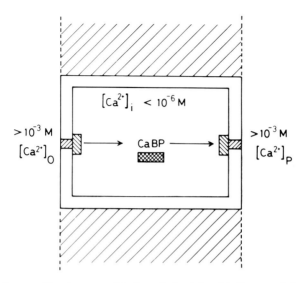

Fig. 12.   Proposed involvement of the intestinal calcium-binding protein in the translocation of calcium.

transports calcium from the external fluids into the lumen. The protein occurs in many other cell layers through which calcium transport is necessary, e.g., salt glands. The protein has two binding sites for calcium and undergoes a marked change in conformation on calcium binding (Levine *et al.*, 1977b). Our NMR studies, together with the sequence, show that it resembles calmodulin but that it is about one-half the size. The amount of the protein in a cell can exceed 3% of the total protein, so that the protein could reach approximately millimolar concentrations. We propose the following scheme for its function as an aqueous intracellular calcium transporter (i.e., an aqueous phase ionophore). The diffusion rate of free calcium across a cell at $10^{-7}\,M$ will be increased by the carrier ($10^{-3}\,M$) by a factor of $>10^2$. Moreover, transport of calcium in this form will not interfere with cell metabolism. If the protein can recognize calcium channel gates in the membrane, then it can transport calcium with no energy involvement, as in Fig. 12. Let the external calcium be able to enter a channel to the intestinal cell but unable to pass the gate unless the protein is bound in calcium-free form. On calcium binding to the protein the calcium protein changes conformation and enters the aqueous internal phase, leaving the gate closed. On the opposite side of the cell calcium-bound protein binds to the gated channel and can release calcium down the channel, provided that binding to the channel reduces calcium binding by $>10^3$. In this type of ferrying of calcium there is no energy requirement. We note that parvalbumins can be described in a similar way, except that they distribute calcium rapidly between intracellular sites.

VIII. CALCIUM PROTEINS IN MEMBRANES

The fact that calcium moves through membranes in specific channels means that there are proteins in the membrane designed to handle calcium, although they may not bind it, but not other ions. These proteins are connected to gates and pumps which control the calcium flow. We turn to these proteins now.

The calcium proteins of major interest are calcium ATPases. Such proteins must have a channel through a membrane which is highly selective for calcium against magnesium, sodium, and potassium. In addition there must be a head group which, by using ATP hydrolysis, drives calcium out of the channel into the external medium. The channels are blocked by ions of too high a charge, but the correct size, e.g., Ln^{3+}, but ions of a lower charge, e.g., Na^+, can perhaps use them poorly. This suggests that selectivity is based on principles related to those for the binding of calcium to phospholipase A_2. There is a fixed channel (hole) even in the absence of calcium which accepts and binds only an ion of the size and charge of calcium. A different size cannot enter, and a different charge either binds too tightly to allow relaxation and mobility; it blocks the channel (higher charge) or it binds too weakly (lower charge).

We must also envisage a mechanism of gating. The simplest mechanism is based on a supposed effect of phosphorylation and dephosphorylation of the protein, which causes a conformation change probably of a loop of the protein. This is an anion-binding effect comparable to the cation (calcium)-binding effect shown in Fig. 13, and it could be transmitted so as to alter geometry in a second loop which was in fact the binding site for calcium. Thus there is the sequence:

Conformation I	$\rightarrow$ conformation II $\rightarrow$	conformation III
Ca^{2+}-binding constant	PO_4^{3-}-bound	Ca^{2+}-binding constant
(in) log $K = 7$	Ca^{2+}-bound	(out) log $K = 3$

The principles are not different from those of the modulation of calcium binding, through changes in the environment of proteins accept that spatial significance arises through incorporation of the protein into a membrane.

It is not possible to give a picture of a channel which is more than an outline physical model. A model device is described by Edmonds (1980) for the sodium channel, which has many intriguing possibilities.

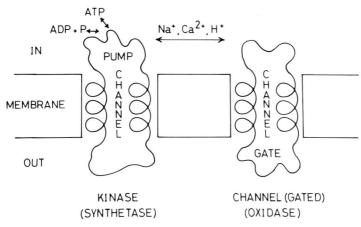

Fig. 13. Schematic membrane gating mechanism.

IX. SUMMARY: INTRACELLULAR CALCIUM

The role of calcium inside cells is not fully explored because of the difficulty of following the various binding centers of this cation when it is only at 10^{-7} M. From our knowledge of the physiological effects of small changes in calcium levels inside cells we suspect that this ion is concerned with the regulation of cell differentiation and division as well as with the major triggering devices which control intracellular chemical, mechanical, and electrical activity. These control proteins will demand variations on the theme of the calmodulins which can only be fully explored when the (hypothetical) proteins have been isolated, but we are suggesting that in the above we have given a general overview of the physical principles lying behind calcium control proteins in cells.

REFERENCES

Aguiar, A. (1979). Ph.D. Thesis, Oxford University.
Aguiar, A. (1981). To be published.
Aguiar, A., de Haas, G. H., Jansen, E. H. J. M., Slotboom, A. J., and Williams, R. J. P. (1979). Proton nuclear magnetic resonance pH titration studies of the histidines of pancreatic phospholipase A_2. *Eur. J. Biochem.* **100**, 511–518.
Barber, J., and Searle, G. F. W. (1979). Double layer theory and the effect of pH on cation-induced chlorophyll fluorescence. *FEBS Lett.* **103**, 241–245.
Blake, C. C. F., Grace, D. E. P., Johnson, L. N., Perkins, S. J., Phillips, D. C., Cassels, R., Dobson, C. M., Poulsen, F. M., and Williams, R. J. P. (1978). Physical and chemical properties of lysozyme in solution. *Ciba Found. Symp.* [N.S.] **60**, 137–185.

Campbell, I. D., and Dobson, C. M. (1979). The application of high resolution nuclear magnetic resonance to biological systems. *Methods Biochem. Anal.* **25**, 1–33.

Campbell, I. D., Dobson, C. M., and Williams, R. J. P. (1975). Assignment of proton NMR spectra of proteins. *Proc. R. Soc. London, Ser. A* **345**, 23–40.

Cookson, D. J., Levine, B. A., Williams, R. J. P., Jontell, M., Linde, A., and de Bernard, B. (1980). Cation binding by the rat incisor dentine phosphoprotein: A spectroscopic study. *Eur. J. Biochem.* **110**, 273–278.

Cotton, F. A., Bier, C. J., Day, V. W., Hagen, E. J., and Larsen, S. (1971). Some aspects of the structure of staphylococcal nuclease. *Cold Spring Harbor Symp. Quant. Biol.* **36**, 239–243.

Cummings, S., Enderby, J. E., and Howe, R. A. (1980). Ion hydration in aqueous $CaCl_2$ solutions. *J. Phys. C* **13**, 1–12.

Derancourt, J., Haiech, J., and Pechere, J. F. (1978). Binding of calcium by parvalbumin fragments. *Biochim. Biophys. Acta* **532**, 373–375.

Edmonds, D. T. (1980). Membrane ion channels and ionic hydration energies. *Proc. R. Soc. London, Ser. B* **211**, 51–62.

Eigen, M. (1963). Fast elementary steps in chemical reaction mechanisms. *Pure Appl. Chem.* **6**, 97–142.

Esnouf, M. P. (1977). Biochemistry of blood coagulation. *Br. Med. Bull.* **33**, 213–218.

Evans, J. S., and Levine, B. A. (1980). Protein-protein interaction sites in the calcium modulated troponin complex. *J. Inorg. Biochem.* **12**, 227–239.

Evans, J. S., Levine, B. A., Leavis, P. C., Gergely, J., Grabarek, Z., and Drabikowski, W. (1980). Proton magnetic resonance studies on proteolytic fragments of troponin-C: Structural homology with the native molecule. *Biochim. Biophys. Acta* **623**, 10–20.

Forsen, S., Thulin, E., Drakenberg, T., Krebs, J., and Seamon, K. (1980). A ^{113}Cd NMR study of calmodulin and its interaction with calcium, magnesium and trifluoperazine. *FEBS Lett.* **117**, 189–194.

Harrison, C. M. (1980). The importance of calcium binding for the function of human fibrinogen. In "Calcium-Binding Proteins and Calcium Function" (F. L. Siegel *et al.*, eds.) pp. 405–406. Elsevier/North-Holland Publ., Amsterdam.

Hastings, A. B. (1980). The state of calcium in the thirties. *Trends Biochem. Sci.* **5**, 84–85.

Hauschka, P. V., and Gallop, P. M. (1977). Calcium binding properties of osteocalcin, the γ-carboxyglutamate-containing protein of bone. In "Calcium-Binding Proteins and Calcium Function" (R. H. Wasserman, R. A. Corradino, E. Carafoli, R. H. Kretsinger, D. H. MacLennan, and F. L. Siegel, eds.), pp. 338–347. Elsevier/North-Holland Publ., Amsterdam and New York.

Hauser, H., Levine, B. A., and Williams, R. J. P. (1976). Interaction of ions with membranes. *Trends Biochem. Sci.* **1**, 278–281.

Hauser, H., Guyer, W., Levine, B. A., Skrabal, P., and Williams, R. J. P. (1978). Conformation of the polar group of lysophosphatidylcholine in water: Conformational changes induced by polyvalent cations. *Biochim. Biophys. Acta* **508**, 450–463.

Johansen, J. T., and Vallee, B. L. (1975). Environment and conformation dependent sensitivity of the arsanilazotyrosine-248 carboxypeptidase A chromaphore. *Biochemistry* **14**, 649–660.

Kannan, K. K., Notstrand, B., Fridborg, K., Lougren, S., Ohlsson, A. and Petef, M. (1975). Crystal structure of human erythrocyte carbonic anhydrase B. *Proc. Natl. Acad. Sci. U.S.A.* **72**, 51–55.

Klevit, R., Esnouf, P., Levine, B. A., and Williams, R. J. P. (1981). A study of calmodulin and its interaction with trifluoperazine by high resolution proton NMR spectroscopy. *FEBS Lett.* **123**, 25–29.

Kretsinger, R. H. (1980). Structure and evolution of calcium modulated proteins. *CRC Crit. Rev. Biochem.* **8**, 119–174.

Leavis, P. C., Rosenfeld, S. S., Gergely, J., Grabarek, Z., and Drabikowski, W. (1978). Proteolytic fragments of troponin-C: Localization of high and low affinity binding sites and interactions with troponin-I and troponin-T. *J. Biol. Chem.* **253**, 5452–5459.

Levine, B. A., and Williams, R. J. P. (1981). The chemistry of the calcium ion and its biological relevance. *In* "Calcium in Normal and Pathological Biological Systems" (L. J. Anghileri, ed.). CRC Press, Cleveland, Ohio (in press).

Levine, B. A., Mercola, D., Coffman, D., and Thornton, J. M. (1977a). Calcium binding by troponin-C: A proton magnetic resonance study. *J. Mol. Biol.* **115**, 743–760.

Levine, B. A., Williams, R. J. P., Fullmer, C. S., and Wasserman, R. H. (1977b). NMR studies of various calcium-binding proteins. *In* "Calcium-Binding and Proteins and Calcium Function" (R. H. Wasserman, R. A. Corradino, E. Carafoli, R. H. Kretsinger, D. H. MacLennan, and F. L. Siegel, eds.), pp. 29–37. Elsevier/North-Holland Publ., Amsterdam and New York.

Levine, B. A., Thornton, J. M., Fernandes, R., Mercola, D., and Kelly, C. M. (1978). Comparison of the calcium and magnesium induced structural changes of troponin-C: A protein magnetic resonance study. *Biochim. Biophys. Acta* **535**, 11–24.

Levine, B. A., Moore, G. R., Ratcliffe, R. G., and Williams, R. J. P. (1979). Nuclear magnetic resonance studies of the solution structure of proteins. *In* "Chemistry of Macromolecules" (R. E. Offord, ed.), Vol. 24, pp. 77–141. University Park Press, Baltimore, Maryland.

Means, A. R., and Dedman, J. R. (1980). Calmodulin—An intracellular calcium receptor. *Nature (London)* **285**, 73–77.

Moore, G. R., and Williams, R. J. P. (1980a). Nuclear magnetic resonance studies of ferrocytochrome-c. *Eur. J. Biochem.* **103**, 513–521.

Moore, G. R., and Williams, R. J. P. (1980b). The solution structures of tuna and horse cytochromes-c. *Eur. J. Biochem.* **103**, 533–541.

Noggle, J. H., and Schirmer, R. E. (1971). "The Nuclear Overhauser Effect." Academic Press, New York.

Ostwald, T. J., MacLennan, D. H., and Dorrington, K. J. (1974). Effects of cation binding on the conformation of calsequestrin and the high affinity calcium binding protein of the sarcoplasmic reticulum. *J. Biol. Chem.* **249**, 5867–5871.

Pluck, N., Esnouf, P., Israel, E. A., and Williams, R. J. P. (1980). The study of bovine prothrombin fragments by high resolution proton NMR. *In* "The Regulation of Coagulation" (P. Mann and A. Taylor, eds.), pp. 67–74. Elsevier/North-Holland Publ., Amsterdam and New York.

Seamon, K. (1979). Cation dependent conformations of brain calcium-dependent regulator protein detected by NMR. *Biochem. Biophys. Res. Commun.* **86**, 1256–1265.

Taylor, A. N. (1980). Immunocytochemical localization of the vitamin-D induced calcium binding protein. *In* "Calcium-Binding Proteins and Calcium Function" (F. L. Siegel *et al.*, eds.), pp. 393–400. Elsevier/North-Holland Publ., Amsterdam and New York.

Tsalkova, T. N., and Privalov, P. L. (1980). Stability of troponin-C. *Biochim. Biophys. Acta* **624**, 196–204.

Wasserman, R. H., and Feher, J. J. (1977). Vitamin-D dependent calcium binding proteins. *In* "Calcium-Binding Proteins and Calcium Function" (R. H. Wasserman, R. A. Corradino, E. Carafoli, R. H. Kretsinger, D. H. MacLennan, and F. L. Siegel, eds.), pp. 293–302. Elsevier/North-Holland Publ., Amsterdam and New York.

Wasserman, R. H., Corradino, R. A., Carafoli, E., Kretsinger, R. H., MacLennan, D. H., and Siegel, F. L., eds. (1977). "Calcium-Binding Proteins and Calcium Function." Elsevier/North-Holland Publ., Amsterdam and New York.

Williams, R. J. P. (1975). Phases and phase structure in biological systems. *Biochim. Biophys. Acta* **416**, 237–286.

Williams, R. J. P. (1977). Calcium chemistry and its relation to protein binding. *In* "Calcium-Binding Proteins and Calcium Function" (R. H. Wasserman, R. A. Corradino, E. Carafoli, R. H. Kretsinger, D. H. MacLennan, and F. L. Siegel, eds.), pp. 3–12. Elsevier/North-Holland Publ., Amsterdam and New York.

Williams, R. J. P. (1978). Energy states of proteins, enzymes and membranes. *Proc. R. Soc. London, Ser. B* **200**, 358–389.

Williams, R. J. P. (1979a). The conformational properties of proteins in solution. *Biol. Rev. Cambridge Philos. Soc.* **54**, 389–437.

Williams, R. J. P. (1979b). Cation and proton interactions with proteins and membranes. *Biochem. Soc. Trans.* **7**, 481–509.

Chapter 2

Mitochondrial Regulation of Intracellular Calcium

GARY FISKUM
ALBERT L. LEHNINGER

  I. Introduction and Scope 39
 II. Mechanisms, Kinetics, and Regulation of Mitochondrial
     Ca²⁺ Transport 41
     A. Ca²⁺ Influx 41
     B. Ca²⁺ Efflux 50
III. Evidence for Mitochondrial Regulation of Cellular Ca²⁺ 62
     A. Ca²⁺ Content of Isolated Mitochondria 62
     B. Mitochondrial Involvement in Cellular Ca²⁺ Fluxes 64
     C. Comparison of Mitochondrial Transport Kinetics with
        Those of Other Cellular Transport Processes 66
 IV. Steady-State Buffering of Free Ca²⁺ by Mitochondria 67
     References 71

I. INTRODUCTION AND SCOPE

The intracellular distribution of Ca^{2+} and, in particular, the cytosolic free Ca^{2+} concentration, have profound regulatory influences on eukaryotic cellular activities. These Ca^{2+}-sensitive functions include various aspects of metabolism, secretion, motility, and membrane transport (for reviews, see Rasmussen and Goodman, 1977; Cheung, 1980; and the other chapters in this book and in Volume I). Control of these activities by Ca^{2+} requires that an exceedingly low level of free Ca^{2+} in the cytosol (10^{-7}–$10^{-6} M$) be maintained in the face of a 10^3- to 10^4-fold greater concentration in the extracellular milieu, i.e., approximately $1.3 \times 10^{-3} M$. Maintenance of this Ca^{2+} gradient is accomplished by Ca^{2+} transport processes taking place in the membranes of mitochondria, endoplasmic reticulum, and the

CALCIUM AND CELL FUNCTION, VOL. II

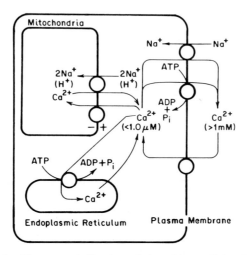

Fig. 1.   Transport-mediated regulation of intracellular Ca^{2+}.

plasmalemma, as shown in Fig. 1 (for reviews, see Borle, 1981; Carafoli and Crompton, 1978).

Ca^{2+} enters cells through the plasma membrane passively, down its electrochemical potential via either a nonspecific leak or by a specific Ca^{2+} pore or gate. The active extrusion of Ca^{2+} from cells occurs via two different transport mechanisms: (1) an ATP hydrolysis-dependent mechanism of Ca^{2+} efflux (Ca^{2+}-transporting ATPase) through the plasma membrane, present in most cells, and (2) a Na^+–Ca^{2+} antiport mechanism that directly couples the extrusion of Ca^{2+} with the influx of Na^+ down the electrochemical gradient of Na^+ across the plasma membrane of most excitable cells.

Maintenance of low cytosolic Ca^{2+} is also provided by uptake of Ca^{2+} from the cytosol into the cisternae of the sarcoplasmic or endoplasmic reticulum, also mediated by a Ca^{2+}-transporting ATPase. Efflux of Ca^{2+} from the endoplasmic reticulum back into the cytosol is poorly understood (Endo, 1977), even though it is particularly important in the triggering of muscle contraction. A third way in which cytosolic Ca^{2+} is maintained at a low level is by transport of Ca^{2+} into the mitochondrial matrix. In contrast to endoplasmic or sarcoplasmic reticulum and the plasmalemma, the mitochondrial inner membrane does not possess a specific Ca^{2+}-transporting ATPase. The mechanism of mitochondrial Ca^{2+} influx and efflux involves utilization of the electrochemical gradient of H^+ and other ions that are generated across the mitochondrial inner membrane by electron transport or ATP hydrolysis (Section II).

Knowledge of the relative contributions and characteristics of these different Ca^{2+} transport processes in maintaining and regulating the intracellular distribution of Ca^{2+} is essential in understanding how Ca^{2+} affects cellular activities under both normal and pathological conditions. Progress in this area has come both from the independent analysis of each of these membrane transport processes and from studies employing whole cells as well as reconstituted systems using combinations of Ca^{2+}-transporting subcellular organelles.

The aim of this chapter is to summarize what is currently known about mitochondrial Ca^{2+} transport and how it may participate in regulating the cytosolic free Ca^{2+} concentration. This includes assessment of the mechanisms and kinetics of mitochondrial Ca^{2+} transport, the factors that can regulate transport activities both *in vitro* and *in vivo*, and the evidence that supports the cooperation of mitochondrial and other cellular Ca^{2+} transport processes in the regulation of intracellular Ca^{2+}.

II. MECHANISMS, KINETICS, AND REGULATION OF MITOCHONDRIAL Ca^{2+} TRANSPORT

A vast literature has accumulated on this topic. It has been most comprehensively reviewed by Saris and Åkerman (1980), Bygrave (1978), and Lehninger *et al.* (1967); recent shorter reviews include those of Fiskum and Lehninger (1980), Nicholls and Crompton (1980), and Carafoli (1979).

A. Ca^{2+} Influx

1. Background

It was found almost 20 years ago that rat kidney mitochondria could accumulate and retain Ca^{2+} at the expense of energy provided either by respiration or ATP hydrolysis (Vasington and Murphy, 1961; DeLuca and Engström, 1961). Energy-dependent Ca^{2+} uptake has since been found to occur in mitochondria isolated from a variety of tissues from various families of vertebrates and invertebrates, but only a few fungal and plant species (Carafoli and Lehninger, 1971; Chen and Lehninger, 1973). Although most of the research on the properties of energy-dependent Ca^{2+} influx has been carried out on rat liver mitochondria, nearly all types of animal tissue mitochondria appear to have qualitatively similar properties.

2. Mechanism

It is now widely agreed that Ca^{2+} enters mitochondria electrophoretically in response to the negative-inside membrane potential developed

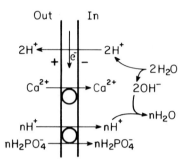

Fig. 2. Respiration-dependent mitochondrial uptake of Ca^{2+} via electrophoretic uniport and simultaneous uptake of phosphate via phosphate–H^+ symport.

across the inner membrane by electron transport (Fig. 2). A suggestion that Ca^{2+} uptake occurs via a Ca^{2+}–$anion^{1-}$ symport mechanism (Moyle and Mitchell, 1977) must be discounted, since several laboratories have found that Ca^{2+} is transported inward via a uniport process (Fig. 2) in which each Ca^{2+} ion carries two net positive charges that are electrically compensated for by extrusion of two H^+ from the matrix (see, e.g., Åkerman, 1978a; Nicholls, 1978; Vercesi *et al.,* 1978; Fiskum *et al.,* 1979; Deana *et al.,* 1979). This process has the effect of converting much of the transmembrane potential to a pH gradient, thus alkalinizing the mitochondrial matrix. Matrix alkalinization can have deleterious effects on the structure and permeability of the inner membrane and can inhibit further membrane potential-dependent Ca^{2+} influx. The profound rise in internal pH is normally prevented by the simultaneous influx of H^+ coupled to the influx of phosphate on the H^+–$H_2PO_4^-$ symporter (Coty and Pedersen, 1975), as shown in Fig. 2, or by the inward diffusion of undissociated lipophilic free acid forms of an anion such as acetate.

Although mitochondrial Ca^{2+} uptake is primarily energized by electron transport, if respiration is interrupted by ischemia or by a respiratory inhibitor such as cyanide, mitochondria can still accumulate and retain Ca^{2+} so long as ATP is available, since ATP hydrolysis by mitochondria also generates a transmembrane potential and thus can support Ca^{2+} uptake (Bielawski and Lehninger, 1966; Brand and Lehninger, 1975; Alexandre *et al.,* 1978), as shown in Fig. 3. Either respiration-dependent or ATP-dependent mitochondrial Ca^{2+} uptake is inhibited or blocked when the membrane potential is lowered or abolished. One common means by which this process (known as uncoupling) occurs is by an increase in the electrophoretic backflow (or leak) of H^+ into the matrix (Fig. 3). Any agent or activity that promotes H^+ backflow also inhibits Ca^{2+} uptake and

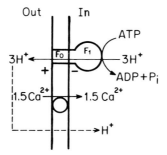

Fig. 3. ATP hydrolysis-dependent mitochondrial uptake of Ca^{2+}.

stimulates the release of Ca^{2+}. For instance, phospholipase affects mitochondria in this way by producing free fatty acids and lysophosphatides which increase the leakiness of the membrane to protons and other ions (Scarpa and Lindsay, 1972).

3. Kinetics

a. Rate of Influx and Affinity for Ca^{2+}. Some kinetic properties of mitochondrial Ca^{2+} uptake are listed in Table I. Values for K_m and V_{max} from different laboratories vary widely because of major differences in experimental conditions such as temperature, the composition of experimental media, and the methods of measurement. Reasonable values for K_m and V_{max} are 10 μM free Ca^{2+} and 500 nmoles Ca^{2+} min^{-1} mg^{-1} protein, respectively. It has recently been shown, however, that the maximal rate of mitochondrial Ca^{2+} uptake is limited only by the rate of respiration or ATP hydrolysis when these processes constitute the driving force (Bragadin *et al.*, 1979). Moreover, when an outward-directed K^+ diffusion potential was utilized as the driving force for mitochondrial Ca^{2+} uptake, a V_{max} of ~1000 nmoles min^{-1} mg^{-1} was obtained. This finding indicates that previous measurements of V_{max} and also of K_m for the Ca^{2+} uptake driven by respiration or ATP hydrolysis may not accurately define the true kinetics of Ca^{2+} translocation. There is general agreement that, under many conditions, including those likely to occur *in vivo*, sigmoid kinetics are observed for mitochondrial Ca^{2+} influx. A Hill coefficient of approximately 2 suggests that the translocator has two Ca^{2+}-binding sites.

b. Bivalent Metal Ion Specificity. Further evidence that mitochondrial Ca^{2+} uptake is mediated by a transport protein (or proteins) comes from studies on transport specificity (Table I). Several divalent cations (Sr^{2+}, Mn^{2+}, Ba^{2+}, Fe^{2+}, etc.) appear to be translocated by the same mechanism,

TABLE I

Some Kinetic Properties of Mitochondrial Ca^{2+} Uptake

V_{max}	>500 nmoles min^{-1} mg^{-1}	Limited by the rate of respiration or ATP hydrolysis
K_m	$\sim 10\ \mu M$	Uncertain because of underestimated V_{max}
Hill coefficient	2	Sigmoid kinetics under physiological conditions
Specificity	Ca^{2+}, Sr^{2+}, Mn^{2+}, Ba^{2+}, Fe^{2+}, and heavy-metal ions	Transport of Ca^{2+} and perhaps Mn^{2+} and Fe^{2+} are physiologically important
Inhibitors	Ruthenium red, La^{3+}	Act specifically on the Ca^{2+} influx system
	Mg^{2+}	Significant, noncompetitive inhibition at cytosolic concentrations

with only Sr^{2+} being transported at a rate similar to that for Ca^{2+}. However, the relatively slow uptake of Fe^{2+} via the Ca^{2+} transporter may be physiological, since the synthesis of heme, a relatively slow process, occurs within the mitochondrial matrix.

c. **Stimulation of Ca^{2+} Uptake by Phosphate.** The presence of phosphate at concentrations normally present in the cytosol increases the capacity of mitochondria to accumulate Ca^{2+}. As mentioned in Section II,A,2, uptake of phosphate during Ca^{2+} influx helps buffer the pH of the matrix and thereby assists in the maintenance of a high membrane potential. Equally important is the precipitation of insoluble calcium phosphate as electron-dense granules within the matrix (Greenawalt et al., 1964). This process greatly lowers the concentration of free Ca^{2+} within the matrix, as determined by the solubility product of calcium phosphate. Lowering of the internal free Ca^{2+} concentration decreases the transmembrane Ca^{2+} gradient and thus favors an uphill influx of Ca^{2+}. Both ATP (or ADP) and Mg^{2+} are known to assist in the precipitation of calcium phosphate in the matrix and accordingly also increase the capacity of mitochondria to sequester Ca^{2+}. In a medium with physiological concentrations of phosphate, ATP, and Mg^{2+}, respiration-energized mitochondria can accumulate more than 2000 nmoles Ca^{2+} mg^{-1} protein without loss of ability to catalyze oxidative phosphorylation on subsequent additions of ADP. As discussed in Section IV, the capacity of mitochondria to accumulate Ca^{2+} in amounts 100-fold greater than normally present in mitochondria *in situ* may be important in the maintenance of Ca^{2+} homeostasis by injured cells.

When mitochondria accumulate large quantities of calcium and phosphate, electron-dense granules of amorphous calcium phosphate are observed in the matrix (Greenawalt *et al.*, 1964). These appear to be present in the mitochondria of many animal tissues *in vivo*. They constitute a major fraction of the dry weight of mitochondria isolated from the hepatopancreas of the blue crab (*Callinectes sapidus*), which has been studied as a model of rapid biological calcification (Becker *et al.*, 1974). The amorphous nature of these deposits is intriguing, since $Ca^{2+}(PO_4)_2$ in the test tube undergoes spontaneous transition into crystalline hydroxyapatite in minutes at pH 7.4 and body temperature. It therefore appears likely that mitochondria contain a factor or factors that inhibit this process. Extracts of rat liver mitochondria contain such an inhibitor with properties resembling those of 3-phosphocitrate, an extremely potent inhibitor of hydroxyapatite crystal growth (Tew *et al.*, 1980).

d. Inhibitors of Ca^{2+} Influx. Mitochondrial Ca^{2+} uptake is inhibited noncompetitively by the mucopolysaccharide stain ruthenium red and competitively by various rare earth metal ions such as La^{3+}. These inhibitors are very potent ($K_i \sim 10^{-7} M$) and are believed to act specifically on the Ca^{2+} translocase. Mg^{2+} in millimolar concentrations, such as exist in the cell cytosol, significantly inhibits mitochondrial Ca^{2+} uptake. It has been suggested that this inhibition is a consequence of the ability of Mg^{2+} to screen negative charges on the membrane at or near the Ca^{2+}-binding sites (Åkerman *et al.*, 1977). These observations indicate that the Mg^{2+} concentration must be taken into account when assessing the physiological significance of data obtained from Ca^{2+} transport experiments using isolated mitochondria.

4. Isolation of the Translocase

There has been limited success in the isolation of the protein(s) responsible for mitochondrial Ca^{2+} uptake. Several high-affinity Ca^{2+}-binding glycoproteins have been isolated from mitochondria (Gomez-Puyou *et al.*, 1972; Sottocasa *et al.*, 1972; Carafoli *et al.*, 1978). Antibody raised against the Sottocasa glycoprotein, which occurs in the space between the outer and inner mitochondrial membranes, was found to inhibit Ca^{2+} uptake by rat liver mitochondria (Panfili *et al.*, 1976). Reconstitution of Ca^{2+} transport across artificial phospholipid bilayer systems by these glycoproteins has not been demonstrated. It is more likely that they are involved in the recognition and binding of Ca^{2+} in the intermembrane space rather than in its actual translocation across the membrane.

Recently the isolation of a Ca^{2+}-binding protein from the inner membrane of calf heart mitochondria has been described (Jeng and Shamoo, 1980a,b). This hydrophobic protein, called calciphorin, has a molecular weight of only $\sim$3000 and binds one Ca^{2+} ion with a dissociation constant of 5.2×10^{-6} M. Calciphorin-mediated translocation of Ca^{2+} through a bulk organic phase is inhibited by ruthenium red and La^{3+} and exhibits properties consistent with the activity of the electrophoretic Ca^{2+} carrier responsible for mitochondrial Ca^{2+} uptake. Further experiments demonstrating reconstitution of electrophoretic Ca^{2+} transport in artificial membranes will, however, be necessary to support this hypothesis.

5. Regulation

a. Experimental Analysis. Much research has been devoted to elucidating the biological factors that regulate both mitochondrial Ca^{2+} uptake and release, since they are keys to understanding how mitochondria may participate in control of the intracellular distribution of Ca^{2+} (Saris and Åkerman, 1980; Borle, 1981; Bygrave, 1978). However, in most such investigations it has been difficult to determine whether an observed change in Ca^{2+} transport activity is due to direct regulation of Ca^{2+} influx or efflux or whether an alteration of some other aspect of mitochondrial function has occurred that indirectly affects Ca^{2+} movements. Also, the great majority of the published experiments have been conducted under rather unphysiological conditions. In particular, the concentrations of the major cytosolic ions, such as Mg^{2+}, $H_2PO_4^-$, K^+, Na^+, ATP^{4-}, and ADP^{3-}, have generally not been maintained at levels approximating those present in the cytosol. The temperature at which such experiments are run is generally 25°C and occasionally lower, but rarely 37°C. This factor is very important because the mitochondrial membrane undergoes transitions at certain critical temperatures and transport systems frequently have very large temperature coefficients. Moreover, the level of Ca^{2+} used in most such experiments is 10–1000 times higher than the cytosolic concentration, which is less than 1 μM Ca^{2+}, and much greater than 10 nmoles Ca^{2+} mg^{-1} mitochondrial protein, approximately the normal endogenous content. Thus, the biological factors that have been found to alter mitochondrial Ca^{2+} transport *in vitro* must be considered only as *possible* physiological regulators unless their effects are also demonstrated *in situ* within the cell or are observed under conditions that mimic the intracellular environment.

A representative list of some of the factors that affect mitochondrial Ca^{2+} uptake measured *in vitro* is given in Table II. These can be consid-

TABLE II

Some Potential Regulators of Mitochondrial Ca^{2+} Uptake

Factor	Effect on uptake	Representative references
Cellular factors		
Tissue type	—	Carafoli and Lehninger, 1971; Reynafarje and Lehninger, 1973
Intracellular location of mitochondria	—	McMillin-Wood et al., 1980
Stage of cell development	—	Bygrave and Ash, 1977; Smith and Bygrave, 1978
Extracellular factors		
Mitogens	Increase	Carpentieri and Sordahl, 1980
Hormones		
Triiodothyronine	Increase	Herd, 1978
Adrenal glucocorticoids	Decrease	Kimura and Rasmussen, 1977
Glucagon	Increase	Friedmann et al., 1979; Taylor et al., 1980; Yamazaki et al., 1980
Catecholamines	Increase	Taylor et al., 1980
Intracellular factors		
Unknown second messengers	—	
H^+ concentration	Decrease	Reed and Bygrave, 1975; Studer and Borle, 1980

ered under three headings: (1) factors related to the specific cell type, function, and stage of development, (2) extracellular stimuli, and (3) intracellular stimuli.

b. Cellular Factors. Both K_m and V_{max} for mitochondrial Ca^{2+} uptake vary with the type of tissue. For instance, the apparent V_{max} for Ehrlich ascites or L1210 ascites tumor cell mitochondria (Reynafarje and Lehninger, 1973) is more than twice that of most other types of mitochondria. However, this may be due to differences in maximal rates of electron transport rather than differences in Ca^{2+} influx per se (Section II,A,3,a).

Differences in rates of Ca^{2+} influx are also observed among different populations of mitochondria isolated from the same tissue. For example, in heart, intramyofibrillar mitochondria exhibit an approximately 2-fold higher V_{max} and a 2.5-fold higher K_m for Ca^{2+} uptake than subsarcolemmal mitochondria (McMillin-Wood et al., 1980). Intracellular Ca^{2+} transport is likely to differ among tissues that perform different types of Ca^{2+}-

dependent activities, such as glycogenolysis, secretion, and contraction, or differ in specialized locations within a given cell type, such as at the nerve cell body compared to its presynaptic terminal.

Changes in mitochondrial Ca^{2+} uptake during development have also been observed. In mitochondria isolated from blowfly flight muscle, Ca^{2+} uptake is minimal 1 day prior to emergence of the adult fly but increases by more than 500% within the next 24 hr and then slowly decreases by 50% during the following 5 days (Smith and Bygrave, 1978). Mitochondria isolated from livers of rat fetuses taken 2–3 days before birth also exhibit very low rates of Ca^{2+} uptake compared to mitochondria from newborn rats (Bygrave and Ash, 1977). Possibly related to these effects is the observation that mitochondria from human lymphocytes stimulated to divide and develop with the mitogen phytohemagglutinin accumulate Ca^{2+} several times faster than mitochondria from control lymphocytes (Carpentieri and Sordahl, 1980). Further studies are necessary to determine whether or not such changes in mitochondrial Ca^{2+} transport are simply the result of an increase in maximal rates of coupled respiration and whether differences that occur during cell development are evident when transport is measured at physiological concentrations of Ca^{2+}.

c. **Extracellular Factors.** Many reports have indicated that different hormones influence mitochondrial Ca^{2+} uptake (Table II). It has been considered for many years that the stimulatory effect of thyroid hormones on cellular energy metabolism is mediated by a functional alteration of the mitochondria (Hoch, 1974). The direct addition of 0.1–1.0 μM triiodothyronine (T3) to mitochondria isolated from thyroidectomized rats has been shown to increase the V_{max} for Ca^{2+} uptake without affecting the K_m (Herd, 1978). However, other investigators found no difference in Ca^{2+} influx using mitochondria from rats that were euthyroid, thyroidectomized, or thyroidectomized and injected with T3 for 3 days, although changes in Ca^{2+} efflux rates were found (Greif et al., 1980).

The adrenal glucocorticoids also profoundly affect cellular energy metabolism as well as mitochondrial Ca^{2+} accumulation. Mitochondria isolated from the livers of rats injected with dexamethasone exhibit a lower rate and extent of respiration-dependent Ca^{2+} uptake than control mitochondria (Kimura and Rasmussen, 1977). Since no change in the K_m for uptake was observed and since the alterations were not evident when uptake was driven by ATP hydrolysis, it is likely that the effect of dexamethasone on Ca^{2+} transport was indirect.

Glucagon, when preinjected into rats, perfused through the liver, or added directly to suspensions of isolated hepatocytes, has been observed to stimulate the initial rate of mitochondrial Ca^{2+} uptake (Friedmann,

1980; Taylor *et al.*, 1980; Yamazaki *et al.*, 1980; but see Foden and Randall, 1978, for alternative results). Glucagon also increases the retention time of a given load of Ca^{2+}, as will be discussed later. Several reports demonstrating an increased respiratory capacity of mitochondria from glucagon-treated rats or hepatocytes (Halestrap, 1978; Yamazaki *et al.*, 1980; Siess and Wieland, 1980) indicate that this may be the mechanism by which glucagon stimulates the initial rate of Ca^{2+} influx (Section II,A,3,a).

There have also been reports that both insulin and catecholamines have effects similar to those of glucagon on mitochondrial Ca^{2+} uptake. The fact that insulin is active in this regard when preadministered to the animal (Dorman *et al.*, 1975; Hughes and Barritt, 1978) but not when perfused directly through the liver suggests that its action may be indirect, possibly the consequence of increased circulating levels of glucagon (Hughes and Barritt, 1978). Another intriguing study suggests the existence of a low-molecular-weight factor present in the liver cytosol of insulin-treated rats that can stimulate the rate and extent of mitochondrial Ca^{2+} uptake (Turakulov *et al.*, 1977). In contrast to the case of insulin, the catecholamines phenylephrine and epinephrine stimulated Ca^{2+} influx in mitochondria isolated after these hormones were perfused directly through the liver (Taylor *et al.*, 1980). The changes induced by epinephrine were blocked by a β- but not an α-adrenergic antagonist. Although in this study an increased rate of respiration, such as that seen with glucagon, was observed, no such respiratory stimulation by catecholamines was observed in isolated hepatocytes by other investigators (Siess and Wieland, 1980).

d. Intracellular Factors. The observation that several different extracellular stimuli can affect mitochondrial Ca^{2+} uptake suggests that one or more intracellular second messengers may be involved. With the exception of the single preliminary report for insulin (Turakalov *et al.*, 1977), the demonstration of an intracellular messenger that is active in altering Ca^{2+} influx by isolated mitochondria is lacking. Although cyclic AMP, a second messenger for intracellular alterations induced by glucagon and catecholamines, may effect release of mitochondrial Ca^{2+}, as discussed in Section II,B,5,b, there is no evidence that it affects mitochondrial Ca^{2+} influx per se.

Hydrogen ion concentration is an intracellular factor that could influence mitochondrial Ca^{2+} uptake under physiological conditions. In the presence of phosphate a decrease in pH of the experimental medium from 7.4 to 7.0 depresses the apparent V_{max} for rat kidney mitochondrial Ca^{2+} by 25% without affecting K_m (Studer and Borle, 1980). An increase in pH from 7.4 to 7.8 decreases the K_m by 25% without affecting the V_{max}. A similar variation in K_m with pH has also been observed with rat liver mitochondria (Reed and Bygrave, 1975). The physiological significance of these observations is demonstrated by the effect of pH on the steady-state exchange of mitochondrial Ca^{2+} with the suspending medium and the influence of pH on mitochondrial Ca^{2+} metabolism studied in intact cells. For a fixed concentration of extramitochondrial Ca^{2+} an increase in pH stimulates the rate of exchange and increases the level of mitochondrial Ca^{2+} (Studer and Borle, 1980). In intact kidney cells, alkalosis causes an increase in Ca^{2+} exchange between the cytosolic and mitochondrial compartments and an increase in the exchangeable mitochondrial pool of Ca^{2+} (Studer and Borle, 1979).

B. Ca^{2+} Efflux

1. Background

Release of accumulated mitochondrial Ca^{2+} can be induced by uncoupling agents (Section II,A,2) or by inhibition of electron transport. In this case Ca^{2+} efflux can occur via reversal of the Ca^{2+} uptake uniporter, since under these conditions the outward flux of Ca^{2+} proceeds down the Ca^{2+} concentration gradient unopposed by a positive-outside membrane potential. However, unlike the inhibition of Ca^{2+} uptake, ruthenium red only partially inhibits the uncoupler-induced release of Ca^{2+} from mitochondria (Sordahl, 1974; Puskin et al., 1976; Fiskum and Cockrell, 1978). The addition of ruthenium red alone to tightly coupled energized mitochondria causes a slow but significant release of Ca^{2+}. Reversal of the uniport carrier under these conditions is highly unlikely because of the presence of the high positive-outside membrane potential. Similarly, the energy-dependent uptake of Ca^{2+} into "inside-out" submitochondrial vesicles (Loyter et al., 1969; Pedersen and Coty, 1972; Lötscher et al., 1979), which have a positive-inside membrane potential, is not readily accounted for by the action of the electrophoretic Ca^{2+} influx system. These observations have been interpreted as evidence for the existence of a ruthenium red-insensitive mechanism of Ca^{2+} efflux that is distinctly different from the ruthenium red-sensitive Ca^{2+} influx system.

2. Mechanism

It has been postulated both on theoretical (Puskin et al., 1976) and experimental grounds (Fiskum and Cockrell, 1978; Åkerman, 1978b) that an electroneutral Ca^{2+}–$2H^+$ antiport carrier mediates ruthenium red-insensitive Ca^{2+} efflux from respiring mitochondria isolated from nonexcitable tissues (Fig. 4A). Several recent reports support this view (Fiskum

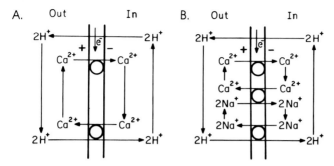

Fig. 4. Mitochondrial cycling of Ca^{2+} via independent influx and efflux processes. (A) Uniport-mediated influx accompanied by $Ca^{2+}-2H^+$ antiport-mediated efflux. (B) Uniport-mediated influx accompanied by $Ca^{2+}-2Na^+$ antiport-mediated efflux.

and Lehninger, 1979; Wehrle and Pedersen, 1979; Tsokos *et al.*, 1980). However, additional evidence for a specific electroneutral $Ca^{2+}-2H^+$ antiporter is needed. Release of Ca^{2+} from mitochondria isolated from excitable tissues, such as heart and brain, as well as brown fat mitochondria, appears to occur by a $Ca^{2+}-Na^+$ exchange process (Crompton *et al.*, 1976b) which recent evidence suggests occurs by electroneutral $Ca^{2+}-2Na^+$ antiport (Affolter and Carafoli, 1980; Fig. 4B).

As seen in Fig. 4A, steady-state Ca^{2+} influx–efflux cycling and maintenance of a steady-state distribution of Ca^{2+} and H^+ across the inner membrane of respiring mitochondria can be accounted for by electrophoretic Ca^{2+} influx in response to the negative-inside membrane potential and simultaneous Ca^{2+} efflux in an independent electroneutral Ca^2-2H^+ efflux system that responds to the pH gradient (Fiskum and Lehninger, 1979). This scheme also provides for cycling of protons via electron transport-dependent H^+ ejection and $Ca^{2+}-2H^+$ antiport. In excitable tissue mitochondria cycling of Ca^{2+} can occur via $Ca^{2+}-2Na^+$ exchange (Crompton and Heid, 1978), as shown in Fig. 4B. In this case Na^+ and H^+ cycling are coupled to Ca^{2+} influx–efflux by electroneutral release of Na^+ and uptake of H^+ via Na^+-H^+ antiport (Douglas and Cockrell, 1974).

3. Kinetics

If the Ca^{2+} influx and efflux systems possessed identical kinetic properties, there would be no net accumulation of Ca^{2+} by mitochondria. Moreover, futile cycling of both Ca^{2+} and H^+ would take place with dissipation of the membrane potential, in effect bringing about uncoupling of respiration and phosphorylation. However, it has been found that the V_{max} of Ca^{2+} efflux in energized rat liver mitochondria ($\sim 10-20$ nmoles min^{-1} mg^{-1}) is only about $\frac{1}{100}$ of the V_{max} for Ca^{2+} influx (Crompton *et al.*, 1978;

Fiskum and Lehninger, 1979). In fact, in the presence of physiological concentrations of phosphate, ATP, Mg^{2+}, and K^+, and with the normal endogenous level of mitochondrial Ca^{2+} (about 10 nmoles mg^{-1} protein), the rate of Ca^{2+} efflux is only about 0.5–5.0 nmoles min^{-1} mg^{-1}. Thus, under steady-state conditions the extramitochondrial (i.e., cytosolic) free Ca^{2+} concentration can be maintained at such a level that mitochondrial Ca^{2+} influx proceeds at the same slow rate as Ca^{2+} efflux (discussed further in Section IV). Since 6 Ca^{2+} are accumulated per atom of oxygen reduced when the mitochondria respire on an NAD-linked substrate such as pyruvate or β-hydroxybutyrate (Rossi and Lehninger, 1964), the influx–efflux cycling of Ca^{2+} across the mitochondrial membrane *in vivo* would consume only a very small fraction of the total respiratory energy, equivalent to the consumption of less than 1 ng-atom oxygen min^{-1} mg^{-1} of mitochondrial protein, compared to the total oxygen consumption rate of 100–200 ng-atoms min^{-1} mg^{-1} during maximal state-3 oxidative phosphorylation. Thus only 0.5–1% of the respiratory energy would be required to maintain a low cytosolic Ca^{2+} concentration in the steady state.

No rigorous kinetic analysis of mitochondrial Ca^{2+} efflux has been reported, primarily because of the extreme difficulty in determining the true concentration of free Ca^{2+} in the mitochondrial matrix. The matrix normally contains high concentrations of both phosphate and adenine nucleotides which can complex with Ca^{2+} and also participate in its precipitation in the matrix as a phosphate salt. Ca^{2+} efflux from rat liver mitochondria appears to be specific for Ca^{2+}; mitochondrial Mn^{2+} efflux is much slower (Gunter *et al.*, 1978). Ruthenium red-induced release of Ca^{2+} from mitochondria in the presence or absence of Na^+ is inhibited by lanthanide cations (Åkerman, 1978b), although the concentrations required are 1000-fold greater than those necessary to inhibit Ca^{2+} influx (Crompton *et al.*, 1979). The degree of inhibition of Ca^{2+} efflux by different lanthanide cations is a function of their ionic radii and is quite different from the lanthanide specificity in the inhibition of Ca^{2+} influx, a finding that further supports the view that different transport systems mediate mitochondrial Ca^{2+} uptake and release.

4. Isolation

Recently the partial purification of what appears to be a $Ca^{2+}-n H^+$ antiporter from bovine heart mitochondria has been reported (Dubinsky *et al.*, 1979). The solubilization of this factor, which is trypsin-sensitive, is enhanced by the presence of La^{3+}. $Ca^{2+}-n H^+$ exchange was reconstituted in liposomes and was found to be insensitive to ruthenium red (W. Dubinsky, personal communication). It exhibited a K_m for Ca^{2+} of approximately 0.4 mM. Isolation and reconstitution of $Ca^{2+}-Na^+$ antiport activity have not been reported.

5. Regulation

a. Experimental Analysis. As discussed earlier for Ca^{2+} uptake (Section II,A,5,a), assessing the physiological significance of alterations in Ca^{2+} efflux measured *in vitro* by isolated mitochondria is complicated by the large number of experimental parameters that affect mitochondrial energy coupling and ion transport. This is particularly true for Ca^{2+} efflux, since it is sensitive to many intramitochondrial factors that are difficult to measure and control at levels that approximate those within the cell. Few data have been reported on the regulation of Ca^{2+} efflux under steady-state conditions that resemble those present in intact cells (Section IV). Thus, the following factors that affect Ca^{2+} efflux under various conditions can be used only as clues to the biological factors that regulate mitochondrial Ca^{2+} transport and the cytosolic free Ca^{2+} concentration *in vivo*.

b. Intracellular Factors. Much research on the regulation of mitochondrial Ca^{2+} efflux has been concerned with intracellular factors whose levels can fluctuate in response to changes in metabolism and to extracellular stimuli. Some of the factors that have been observed to affect the release of mitochondrial Ca^{2+} are listed in Table III.

The amount of Ca^{2+} sequestered in the mitochondrial matrix has a significant influence on the rate of Ca^{2+} efflux. The accumulation of large quantities of Ca^{2+} can cause spontaneous release of Ca^{2+} (see, e.g., Rossi and Lehninger, 1964; Zoccarato *et al.*, 1981). However, even at Ca^{2+} loads where no spontaneous release occurs, the rate of Ca^{2+} efflux, as induced by the addition of ruthenium red, is dependent on the level of previously accumulated Ca^{2+} (Crompton *et al.*, 1976; Fiskum and Cockrell, 1978). In addition, recent evidence indicates that varying the level of mitochondrial Ca^{2+} within a very small and presumably physiological range of from 5 to 20 nmoles mg^{-1} protein significantly alters the rate of steady-state Ca^{2+} efflux, influx–efflux cycling, and the extramitochondrial free Ca^{2+} concentration in a cytosol-like medium (Becker *et al.*, 1980).

The spontaneous release of Ca^{2+} induced by large Ca^{2+} (and phosphate) loads is accompanied by respiratory uncoupling and has been interpreted as due to Ca^{2+}-induced "damage" to the mitochondrial inner membrane, resulting in leakiness to H^+ and a drop in the membrane potential (Rossi and Lehninger, 1964; Hunter *et al.*, 1976; Lötscher *et al.*, 1980). Many of the regulatory factors listed in Table III affect the amount of accumulated Ca^{2+} necessary to cause spontaneous release; they also affect the length of time a given load of Ca^{2+} is retained prior to spontaneous release. However, they may also influence Ca^{2+} efflux under conditions where no apparent damage has occurred (see, e.g., Roos *et al.*, 1980). Thus, depending on the experimental conditions, regulation of Ca^{2+} efflux can occur di-

TABLE III

Some Potential Intracellular Regulating Factors of Mitochondrial Ca^{2+} Efflux

Regulator	Effect on efflux	Representative references
Ca^{2+}	Increase	Lehninger et al., 1967; Fiskum and Cockrell, 1978; Crompton et al., 1976
Phosphate	Increase	Siliprandi et al., 1979; Coelho and Vercesi, 1980; Roos et al., 1980
ATP (ADP)	Decrease	Harris et al., 1979; Zoccarato et al., 1981
Phosphoenolpyruvate	Increase	Chudapongse and Haugaard, 1973; Peng et al., 1974
Pyrophosphate	Increase	Sordahl and Asimakis, 1978; Vercesi, 1981
Palmitoyl-CoA	Increase	Asimakis and Sordahl, 1977; Wolkowicz and Wood, 1979
Redox ratio of mitochondrial NAD (P)	Increase	Lehninger et al., 1978; Fiskum and Lehninger, 1979; Lötscher et al., 1980; Beatrice et al., 1980
H^+ concentration	Increase	Åkerman, 1978b; Tsokos et al., 1980
Na^+	Increase	Crompton et al., 1976, 1978; Al-Shaikhaly et al., 1979
cAMP	Increase	Borle, 1974; Badyshtov and Seredenin, 1977; Arshad and Holdsworth, 1980

rectly at the level of the putative Ca^{2+} efflux system, indirectly via effects on Ca^{2+}-induced nonspecific changes in membrane permeability, or via both mechanisms. Clarification of these factors is a major goal of current and future research on mitochondrial Ca^{2+} transport.

Although phosphate can potentiate the uptake of Ca^{2+} by mitochondria, as described in Section II,A,2, it can under some conditions promote the release of accumulated Ca^{2+} (Drahota et al., 1965; Coelho and Vercesi, 1980; Roos et al., 1980). These effects occur within the range of cytosolic phosphate concentrations (0.5–5.0 mM). Phosphate may stimulate the release of mitochondrial Ca^{2+} by inducing the release of mitochondrial Mg^{2+} and adenine nucleotides (Coelho and Vercesi, 1980; Zoccarato et al., 1981). These substances could be necessary for the stabilization of precipitated calcium phosphate in the matrix (Lehninger, 1970), and they may also directly influence the permeability of the membrane (Hunter *et al.*, 1976; Harris *et al.*, 1979; Panov *et al.*, 1980).

The presence of either Mg^{2+} or adenine nucleotides in the medium inhibits the phosphate-induced release of both substances in addition to inhibiting Ca^{2+} efflux (Rossi and Lehninger, 1964; Siliprandi *et al.*, 1978; Zoccarato *et al.*, 1981). Although Ca^{2+} efflux is inhibited by extramitochondrial ATP, ADP, or AMP, when interconversion of ATP and AMP to ADP via adenylate kinase is blocked only ADP is effective (Zoccarato *et al.*, 1981). The cytosolic concentration of free Mg^{2+} probably does not vary significantly and is, therefore, unlikely to be a physiological regulator of mitochondrial Ca^{2+} efflux. The levels of cellular adenine nucleotides vary, particularly in response to changes in energy metabolism (Krebs and Veech, 1969), and may thus be involved in the regulation of mitochondrial Ca^{2+} transport *in vivo*. This is supported by the recent report that the steady-state extramitochondrial free Ca^{2+} concentration is decreased from 1.6 to 0.7 μM when the Mg^{2+}-ATP concentration is increased from 0 to 3 mM (Becker, 1980).

Several biological factors in addition to phosphate stimulate mitochondrial Ca^{2+} efflux, in part as a result of adenine nucleotide efflux from the intramitochondrial pool. These include phosphoenolpyruvate (Chudapongse and Haugaard, 1973; Peng *et al.*, 1974; Sul *et al.*, 1976), pyrophosphate (Sordahl and Asimakis, 1978; Asimakis and Aprille, 1980; Vercesi; 1981), and palmitoyl-CoA (Asimakis and Sordahl, 1977; Wolkowicz and Wood, 1979). Net release of adenine nucleotides is apparently induced via an alteration in the activity of the well-characterized adenine nucleotide translocator which normally catalyzes a 1 : 1 exchange of external ADP for internal ATP (Klingenberg, 1980). These releasing agents are effective at physiological concentrations, and their intracellular levels are known to fluctuate significantly in response to changes in cellular metabolism. However, the stimulatory effects of these agents on Ca^{2+} efflux are strongly inhibited by extramitochondrial adenine nucleotides at levels normally present in the cytosol. Until it is demonstrated that phosphoenolypyruvate, pyrophosphate, and palmitoyl-CoA affect Ca^{2+} influx–efflux cycling in the presence of endogenous levels of mitochondrial Ca^{2+} and cytosolic concentrations of Mg^{2+} and adenine nucleotides, their role in the cellular regulation of mitochondrial Ca^{2+} transport will remain uncertain.

Another potential physiological regulator of mitochondrial Ca^{2+} efflux that has received much recent attention is the oxidation–reduction state of intramitochondrial pyridine nucleotides. This mode of regulation was initially suggested by the results of experiments employing intact Ehrlich ascites tumor cells (Landry and Lehninger, 1976). It was found that the net uptake of Ca^{2+} by these cells was much greater when they were respiring on succinate in the presence of rotenone to block electron transport chain-mediated oxidation of NADH and NAD-linked substrates than when rotenone was absent and the pyridine nucleotides were in a relatively oxidized state. It was subsequently shown that the net release of Ca^{2+} from respiring mitochondria isolated from rat liver, heart, or Ehrlich tumor cells could be induced by shifting the redox state of the mitochondrial pyridine nucleotides to a relatively oxidized steady state (Lehninger et al., 1978; Fiskum and Lehninger, 1979). This was accomplished by adding acetoacetate or oxaloacetate, oxidants of mitochondrial NADH via β-hydroxybutyrate, and malate dehydrogenases, respectively, to mitochondria suspended in a medium containing rotenone to prevent net electron flow from NADH to oxygen. When the redox steady state of the mitochondrial pyridine nucleotides was shifted to a more reduced condition by the addition of an appropriate substrate such as malate or β-hydroxybutyrate, reuptake of the released Ca^{2+} occurred. Experiments employing ruthenium red to block mitochondrial Ca^{2+} uptake indicated that the release of Ca^{2+} induced by an oxidized steady state of the pyridine nucleotides was due to the stimulation of Ca^{2+} efflux rather than the inhibition of Ca^{2+} influx.

Simultaneous measurements of movements of Ca^{2+}, H^+, and phosphate between mitochondria and the medium, as well as oxygen consumption and the redox state of the pyridine nucleotides, support the hypothesis that regulation of Ca^{2+} efflux is mediated by changes in the activity of the electroneutral $Ca^{2+}-2H^+$ efflux pathway (Fiskum and Lehninger, 1979). However, this conclusion must be tempered by the finding that medium Mg^{2+} and adenine nucleotides also inhibit the Ca^{2+} efflux promoted by the oxidized state of the pyridine nucleotides (Nicholls and Brand, 1980; Fiskum et al., 1980). As expected from studies employing other Ca^{2+}-releasing agents, it has recently been found that oxidation of intramitochondrial pyridine nucleotides also induces net release of both Mg^{2+} (Siliprandi et al., 1979; Coelho and Vercesi, 1980) and adenine nucleotides, even in the presence of ruthenium red to block possible respiratory uncoupling due to rapid Ca^{2+} efflux–influx cycling (D. R. Pfeiffer, personal communication; Vercesi, 1981).

Thus it is possible that all the aforementioned factors that stimulate mitochondrial Ca^{2+} efflux act indirectly by affecting the mitochondrial permeability to other substances, particularly Mg^{2+} and adenine nucleotides. One possible explanation for their mode of action is that they potentiate a Ca^{2+}-induced structural "transition" of the inner membrane, which activates hydrophilic channels that allow the nonspecific passive diffusion of a variety of ions including Ca^{2+} (Hunter et al., 1976; Haworth and Hunter, 1980). It has also been proposed that Ca^{2+}-releasing agents act in conjunction with intramitochondrial Ca^{2+} to stimulate mitochondrial phospholipase A_2, which then produces fatty acids and lysophosphatides known to increase the nonspecific permeability of the inner membrane (Beatrice *et al.*, 1980). It has also been suggested that fatty acids can specifically increase mitochondrial permeability to Ca^{2+} (Roman *et al.*, 1979). Although the local anesthetic nupercaine, an inhibitor of phospholipase, inhibits the stimulated release of Ca^{2+} (Beatrice *et al.*, 1980), it can have other more direct effects on membrane permeability. In fact, the inhibition by nupercaine of Ca^{2+} efflux stimulated by the oxidized steady state of mitochondrial pyridine nucleotides has been interpreted as evidence for the involvement of a specific, regulatable Ca^{2+} efflux translocase (Dawson *et al.*, 1979; Dawson and Fulton, 1980). In this context, it has recently been reported that NAD^+ binds specifically to a Ca^{2+}-binding glycoprotein isolated from rat liver mitochondria that may be involved in mitochondrial Ca^{2+} transport (Panfili *et al.*, 1980). An antibody against this glycoprotein inhibits Ca^{2+} efflux induced by shifting the redox state of the mitochondrial pyridine nucleotides to a more oxidized state with oxaloacetate.

Another candidate for a physiological regulator of mitochondrial Ca^{2+} efflux that was previously discussed with respect to Ca^{2+} influx (Section II,A,5,d) is the pH of the extramitochondrial milieu. A decrease in pH of from 7.4 to 6.8 has been demonstrated to induce net release of Ca^{2+} from rat liver mitochondria and has been interpreted as due to stimulation of a $Ca^{2+}-nH^+$ antiport-mediated Ca^{2+} efflux (Åkerman, 1978b). Release occurred at low levels of mitochondrial Ca^{2+} (10 nmoles mg^{-1} protein) and was not inhibited by ruthenium red or Mg^{2+}. An effect of adenine nucleotides on acid pulse-induced Ca^{2+} efflux has not been reported. A recent preliminary report has indicated that net release of mitochondrial Ca^{2+} can be induced by a decrease in the medium pH of as little as 0.1 pH unit (Tsokos *et al.*, 1980). Changes in pH of this magnitude have also been demonstrated to affect the steady-state extramitochondrial free Ca^{2+} concentration. One study reported that the steady-state free Ca^{2+} steadily increased from 0.35 to 0.75 μM with sequential decreases in the suspending medium pH of from 7.4 to 6.6 (Nicholls, 1978). This change is consistent with an increase in the pH-dependent Ca^{2+} efflux but can also be explained by a decrease in Ca^{2+} influx (Section II,A,5,d).

An assessment of the physiological role that pH has on regulating mitochondrial Ca^{2+} efflux will await the results from experiments performed under conditions that mimic those present in the cytosol. There is evidence that intracellular pH can affect the cytosolic free Ca^{2+} concentration; a lowering of the pH of the cytosol has been accompanied by an increase in the cytosolic Ca^{2+} concentration in barnacle muscle fibers (Lee and Ashley, 1978). Although these observations are consistent with the known effects of pH on mitochondrial Ca^{2+} transport, a causal relationship between the two has not been established.

As mentioned in Section II,B,2, Na^+ drives net Ca^{2+} efflux from energized mitochondria isolated from many excitable tissues via $2Na^+$–Ca^{2+} antiport. For heart mitochondria, the curve relating the Ca^{2+} efflux (in the presence of ruthenium red to block reuptake) to the sodium concentration is sigmoidal with an apparent K_{Na} of 8 mM (Crompton et al., 1977), which is close to the microelectrode-determined concentration of cytosolic Na^+ in heart of 6 mM (Lee and Fozzard, 1975). Na^+-induced mitochondrial Ca^{2+} efflux occurs at low endogenous levels of mitochondrial Ca^{2+} (Carafoli et al., 1974), although rather high concentrations of Na^+ may be required. The activity of this mode of Ca^{2+} efflux is apparently not accompanied by mitochondrial damage or uncoupling (Crompton et al., 1977) but may be affected by intramitochondrial adenine nucleotides (Harris, 1979).

Transient changes in sarcoplasmic Na^+ that occur in response to cellular depolarization are quite small (<0.1 mM) and are unlikely to affect sarcoplasmic Ca^{2+} via changes in the rate of mitochondrial Ca^{2+} efflux (Carafoli and Crompton, 1978). On the other hand, Crompton et al. (1976b) have proposed that mitochondrial Na^+–Ca^{2+} antiport may be involved in the positive ionotropic effect of cardiac glycosides which probably cause a sustained rise in the steady-state sarcoplasmic Na^+ concentration that could increase mitochondrial Ca^{2+} efflux, thereby increasing the sarcoplasmic free Ca^{2+} concentration. Further work along the lines of that of Brand and DeSelincourt (1980), who demonstrated a relationship between the extramitochondrial Na^+ concentration and the steady-state free Ca^{2+} concentration, will be necessary to understand the true physiological role of this Ca^{2+} transport system.

Cyclic AMP (cAMP) is another very important intracellular constituent that has been reported to stimulate release of Ca^{2+} from energized mitochondria isolated from rat kidney, liver, and heart (Borle, 1974) and adrenal medulla (Matlib and O'Brien, 1974). Net Ca^{2+} efflux was observed with mitochondria suspended in a cytosol-like medium when 10^{-7}–3×10^{-6} M cAMP was added (Borle, 1974). However, a substantial and sustained increase in the extramitochondrial Ca^{2+} concentration was obtained only when the mitochondrial Ca^{2+} content was quite high (several hundred nmoles per milligram of protein).

Because of the potential physiological importance of cAMP-regulated mitochondrial Ca^{2+} transport, many investigators have attempted to reproduce these results and have been unable to do so (Scarpa et al., 1976;

Borle, 1976; Schotland and Mela, 1977). However, there have been recent reports that demonstrate cAMP-induced Ca^{2+} efflux from heart and liver mitochondria under different conditions than those previously employed and at low levels of mitochondrial Ca^{2+} (Badyshtov and Seredenin, 1977; Arshad and Holdsworth, 1980). Clearly, more investigation is necessary to resolve this controversial issue.

c. Extracellular Factors. Although relatively little is known about the control of mitochondrial Ca^{2+} efflux by extracellular stimuli, it has attracted much attention because of its potential importance in the field of endocrinology. Interest has grown because of increasing evidence that the intracellular distribution of Ca^{2+} acts as a second messenger for many extracellular stimuli. In most cases a given signal is followed by an increase in cytosolic Ca^{2+}, which is often in part due to the release of Ca^{2+} from intracellular stores such as mitochondria. Although such changes may be brought about by hormonal, electrical, or other stimuli (e.g., mitogens), direct evidence for alterations in mitochondrial Ca^{2+} transport per se is limited to studies employing hormones, particularly glucagon and thyroxine.

Many reports indicate that glucagon, in addition to stimulating mitochondrial Ca^{2+} influx (Section II,A,5,c), leads to an inhibition of Ca^{2+} efflux. This has been demonstrated with mitochondria isolated from livers of rats injected with glucagon (Prpic *et al.*, 1978; Hughes and Barritt, 1978; Yamazaki *et al.*, 1980) and with livers directly perfused with glucagon (Taylor *et al.*, 1980). Most studies have simply indicated prolongation of Ca^{2+} retention by mitochondria from glucagon-treated animals, although actual inhibition of Ca^{2+} efflux (as induced by a pulse of EGTA) has also been demonstrated (Taylor *et al.*, 1980). The effect of glucagon on mitochondrial Ca^{2+} retention is rapid, being maximal in rat livers excised only 1 min after intravenous glucagon injection (Yamazaki *et al.*, 1980). The influence of glucagon on this and other mitochondrial activities has been shown to be blocked by cycloheximide (Hughes and Barritt, 1978) and puromycin (Prpic *et al.*, 1978), indicating an involvement of protein synthesis.

It appears unlikely that inhibition of mitochondrial Ca^{2+} efflux by glucagon is due to its aforementioned stimulatory effect on respiration (Section II,A,5,c). Inhibition of efflux is more likely due to the comparatively higher endogenous levels and better retention of adenine nucleotides found in mitochondria from glucagon-treated tissue (Prpic *et al.*, 1978; Taylor *et al.*, 1980). It has also recently been suggested by Prpic and Bygrave (1980) that glucagon inhibits Ca^{2+} efflux via its effect on the intramitochondrial NADPH redox state. These investigators reported a slower oxaloacetate-induced Ca^{2+} efflux (Section II,B,5,b) in mitochondria from glucagon-treated rats which correlated with relatively decreased oxidation of NADPH (but not NADH). A relatively reduced steady state of NADPH might also explain the increased retention of Ca^{2+} by mitochondria from glucagon-treated animals.

The physiological significance of the effects of glucagon on mitochondrial Ca^{2+} transport has yet to be established. The relatively slow Ca^{2+} efflux could be associated with depressed Ca^{2+} efflux–influx cycling and maintenance of a relatively low cytosolic steady-state free Ca^{2+} concentration. However, Brand and DeSelincourt (1980) have indicated that, with low loads of Ca^{2+}, mitochondria from the livers of glucagon-treated animals maintain the same steady-state extramitochondrial free Ca^{2+} concentration as control mitochondria. Further experimentation along these lines will be necessary to determine the influence of glucagon on mitochondrial Ca^{2+} buffering under physiologically realistic conditions.

Inhibitory effects on net Ca^{2+} efflux such as those observed with glucagon have also been observed with animals treated with insulin (Dorman *et al.*, 1975; Hughes and Barritt, 1978) and catecholamines (Taylor *et al.*, 1980). The effects of insulin when administered to the whole animal may actually be due to increased circulating levels of glucagon (Hughes and Barritt, 1978), whereas catecholamines appear to act directly, probably via increasing the endogenous content and retention of mitochondrial adenine nucleotides (Taylor *et al.*, 1980). Prolonged retention of Ca^{2+} by mitochondria from phenylephrine-perfused livers is, however, inconsistent with observations made with whole cells indicating α-adrenergic agonist-induced elevation of cytosolic Ca^{2+} due to net mitochondrial Ca^{2+} efflux (Blackmore *et al.*, 1979; Babcock *et al.*, 1979; Murphy *et al.*, 1980).

In contrast to the inhibitory influence of glucagon and catecholamines on release of Ca^{2+} from isolated mitochondria, adrenal glucocorticoids significantly shorten the retention time of accumulated Ca^{2+} (Kimura and Rasmussen, 1977). Available evidence indicates that alterations in adenine nucleotide translocation can also explain glucocorticoid-induced changes in mitochondrial Ca^{2+} efflux.

Triiodothyronine has also been demonstrated to affect the release of mitochondrial Ca^{2+} significantly, efflux release being stimulated by the direct addition of T3 to suspensions of mitochondria isolated from either the livers of thyroidectomized rats (Herd, 1978) or the hearts of euthyroid rats (Harris *et al.*, 1979). These studies suggest loss of intramitochondrial adenine nucleotides as the regulatory factor in T3-induced Ca^{2+} efflux. It has also recently been reported that the release of Ca^{2+}, which is stimulated by an oxidized redox state of pyridine nucleotides, is very slow in mitochondria isolated from the livers of either thyroidectomized or hypophysectomized rats (Greif *et al.*, 1980; Fiskum *et al.*, 1980). Treatment of these animals with either T3 for 3 days or growth hormone resulted in an elevation of Ca^{2+} efflux to a level comparable with that observed with control mitochondria. Such studies should be pursued further to determine if the effects of T3 and growth hormone on mitochondrial Ca^{2+} transport relate to the stimulation by these hormones of cellular energy metabolism as observed *in vivo*.

d. Cellular Factors. There is much evidence that Ca^{2+} efflux is both quantitatively and qualitatively different in mitochondria isolated from different types of tissue. This was originally suggested by reports demonstrating a lack of respiratory uncoupling of tumor mitochondria after accumulation and retention of levels of Ca^{2+} that are not retained and that totally uncouple oxidative phosphorylation in liver mitochondria (McIntyre and Bygrave, 1974; Thorne and Bygrave, 1974). A relatively high capacity for Ca^{2+} accumulation and retention has also been documented for carefully prepared heart mitochondria (Vercesi *et al.*, 1978). The massive capacity for Ca^{2+} sequestration by tumor mitochondria suggested among other things that Ca^{2+} efflux by these mitochondria might be relatively inactive. It was subsequently demonstrated that ruthenium red-insensitive Ca^{2+} efflux by Ehrlich ascites tumor mitochondria was less than 10% as active as efflux by rat liver mitochondria over a wide range of Ca^{2+} loads (Fiskum and Cockrell, 1978).

As previously discussed, mitochondrial Ca^{2+} efflux varies greatly depending on the experimental conditions and presence of regulatory factors. The concentration of Na^+ is one of the most influential of these parameters and has been shown to stimulate Ca^{2+} efflux to varying degrees in mitochondria isolated from different tissues. For instance, release of Ca^{2+} by mitochondria from tissues such as heart, brain, adrenal cortex, parotid gland, and brown fat is significantly stimulated by the presence of 5–25 mM Na^+ (Carafoli *et al.*, 1974; Crompton *et al.*, 1978; Al-Shaikhaly *et al.*, 1979). In contrast, Na^+ has little or no effect on Ca^{2+} efflux from mitochondria from liver, kidney, lung, and uterus muscle (Crompton *et al.*, 1978). Although this suggests that these mitochondria do not possess a Na^+–Ca^{2+} efflux antiport system (Section II,B,2), recent evidence indicates that this transport process is present in these tissues but that its maximal activity is only 10–20% of that in most electrically excitable tissues (Haworth *et al.*, 1980; Hughes *et al.*, 1980).

It is likely that the mechanisms and kinetics of regulated Ca^{2+} efflux vary considerably in mitochondria from different tissues, since intracellular Ca^{2+} affects a multitude of cell-specific processes. In addition to Na^+-stimulated Ca^{2+} efflux, that regulated by fatty acids (Roman *et al.*, 1979), palmitoyl-CoA (Wolkowicz and Wood, 1980), and the redox state of mitochondrial pyridine nucleotides (Coelho and Vercesi, 1980) has been reported to differ in mitochondria from liver, kidney, and heart. Ca^{2+} efflux may also vary in mitochondria present at different locations within the same cell (Wolkowicz and Wood, 1980).

The relevance of these observations to the regulation of cellular Ca^{2+} metabolism is as yet unknown. Many differences in mitochondrial Ca^{2+} efflux may not be apparent when experiments are performed under conditions that more closely resemble the intracellular environment (e.g., in the presence of adenine nucleotides). Thus, further investigations employing isolated mitochondria and whole cells will be required to establish firmly if and how extracellular stimuli modulate mitochondrial Ca^{2+} transport.

III. EVIDENCE FOR MITOCHONDRIAL REGULATION OF CELLULAR Ca^{2+}

Current understanding of the mechanisms, kinetics, and regulation of mitochondrial Ca^{2+} transport strongly supports the proposal that mitochondria play an important role in maintaining and regulating the intracellular distribution of Ca^{2+}, although the degree to which mitochondria participate in cellular Ca^{2+} homeostasis in different tissues has not been resolved. A detailed analysis of this issue is beyond the scope of this chapter; however, examples of the major lines of evidence that mitochondria assist in controlling intracellular Ca^{2+} will be presented. The reader is also referred to informative articles by Carafoli and Crompton (1978) and Borle (1981).

A. Ca^{2+} Content of Isolated Mitochondria

One of the first observations strongly suggesting a role of mitochondria in cellular Ca^{2+} homeostasis was that a large fraction of tissue Ca^{2+} was found in isolated mitochondria. Determination of the distribution of ^{40}Ca or ^{45}Ca among cell fractions isolated by ultracentrifugation indicates that, for various tissues, including liver, kidney, and ascites tumor cells, an average of approximately 40% of the cellular Ca^{2+} is found in the mitochondrial fraction (Borle, 1981). Isolated liver and kidney mitochondria contain an average of 0.75 mmole Ca^{2+} kg^{-1} wet tissue. This can be compared to the value determined from experiments which monitor the exchange of exogenous ^{45}Ca with intact tissue slices and isolated cells. From such measurements Borle and co-workers and others (e.g., Claret-Berthon *et al.*, 1977) have calculated the size and exchange rates of various intracellular pools of Ca^{2+}. The exchangeable mitochondrial pool so determined is less than 0.2 mmole kg^{-1} wet wt for liver and pancreas and ~0.3 mmole kg^{-1} for kidney, muscle, and the pituitary (Borle, 1981). This suggests that the level of Ca^{2+} found in isolated mitochondria is artificially high and/or that greater than 50% of mitochondrial Ca^{2+} present within cells is unexchangeable or very slowly exchangeable.

It has been argued that much of the Ca^{2+} found in isolated mitochondria is actually accumulated during the isolation procedure and is thus significantly higher than that present *in vivo*. This is certainly plausible, since endogenous oxidizable substrates as well as ATP could be utilized to drive the rapid accumulation of "contaminating" Ca^{2+} (from interstitial fluid, disrupted endoplasmic reticulum, etc.) by mitochondria present in the tissue homogenate. In an attempt to alleviate this problem, tissues have been homogenized in the presence of EGTA, to chelate contaminating Ca^{2+}, and ruthenium red, to block mitochondrial Ca^{2+} uptake. Although these procedures lower the endogenous content of mitochondrial Ca^{2+}, significant quantities of Ca^{2+} remain in mitochondria so isolated. Since it has been demonstrated that EGTA and ruthenium red can induce the net release of mitochondrial Ca^{2+}, these procedures may actually result in an underestimate of true endogenous mitochondrial Ca^{2+}. Thus, the presence of Ca^{2+} at levels of from 5 to 200 nmoles mg^{-1} protein in mitochondria isolated from different tissues in the presence of EGTA or ruthenium red strongly indicates that mitochondria contain a substantial fraction of total cellular Ca^{2+} *in vivo*. As will be discussed later, it is also significant that endogenous mitochondrial Ca^{2+} can be released upon addition of uncouplers, ionophores, etc., thus indicating that it is a readily mobilizable pool of Ca^{2+}.

The level of endogenous mitochondrial Ca^{2+} varies considerably depending on the tissue from which the mitochondria are isolated. This is actually to be expected, since the total mitochondrial content of different tissues is also highly variable. Thus, heart mitochondria contain 10–40 nmoles Ca^{2+} mg^{-1} protein and are present at 90 mg protein g^{-1} of tissue, whereas myometrium mitochondria contain up to 200 nmoles Ca^{2+} mg^{-1} protein but are only present at 12 mg protein g^{-1} of tissue. However, the tissue mitochondrial content is probably not the only determinant of endogenous mitochondrial Ca^{2+}; factors such as endogenous phosphate and adenine nucleotides are likely also to be important in this regard. In addition, recent reports indicate that hormonal stimuli can have significant effects. Ca^{2+} present in mitochondria isolated in the presence of EGTA from livers perfused with phenylephrine (Blackmore *et al.*, 1979) or hepatocytes treated with norepinephrine (Babcock *et al.*, 1979; Murphy *et al.*, 1980) has been reported to be 40–60% lower than that in mitochondria from control tissue. A good correlation also exists between the magnitude and time course of the decrease in mitochondrial Ca^{2+} and the increase in cytosolic Ca^{2+} and Ca^{2+}-sensitive activities, including glycogen phosphorylase and release of cellular glucose. These results strongly support an involvement of net mitochondrial Ca^{2+} efflux in the mechanism of action of α-adrenergic stimuli [but see Althaus-Salzmann *et al.* (1980) and Poggioli *et al.* (1980) for an alternative explanation].

B. Mitochondrial Involvement in Cellular Ca^{2+} Fluxes

Numerous studies have demonstrated that movements of Ca^{2+} in cells and tissue slices are dependent upon mitochondrial Ca^{2+} transport. These include correlations of mitochondrial respiration and ion transport with net uptake and release of cellular Ca^{2+}, steady-state exchange of extracellular Ca^{2+} with intracellular pools, and changes in the actual cytosolic free Ca^{2+} concentration. The following includes a few examples of each of these lines of investigation.

Energy-dependent net uptake of Ca^{2+} into intact viable cells such as hepatocytes (Kleineke and Stratman, 1974; Dubinsky and Cockrell, 1975; Krell *et al.*, 1979), Ehrlich ascites tumor cells (Cittadini *et al.*, 1973; Landry and Lehninger, 1976; Charlton and Wenner, 1978), and lymphocytes (Landry *et al.*, 1978) has been characterized with the aid of measurements employing ^{45}Ca, atomic absorption, Ca^{2+} indicator dyes, and Ca^{2+} electrodes. Uptake of Ca^{2+} is dependent on mitochondrial energy coupling, as indicated by its being severely inhibited by respiratory inhibitors, uncouplers, and mitochondrial ATPase inhibitors. Cellular Ca^{2+} influx is also associated with respiratory stimulation, as in mitochondrial Ca^{2+} uptake. Evidence that accumulated Ca^{2+} is actually being sequestered by mitochondria includes its inhibition by La^{3+} and ruthenium red, as well as the localization of accumulated Ca^{2+} in subsequently isolated mitochondria. Physiological factors which influence Ca^{2+} uptake and release by mitochondria *in vitro* also affect net uptake and release of Ca^{2+} by whole cells. Thus phosphate, adenine nucleotides, and the availability of respiratory substrates increase net cellular Ca^{2+} uptake (Landry and Lehninger, 1976; Landry *et al.*, 1978; Charlton and Wenner, 1978; Krell *et al.*, 1979), whereas an oxidized redox state of mitochondrial pyridine nucleotides decreases net uptake and retention of cellular Ca^{2+} and promotes net Ca^{2+} efflux (Landry and Lehninger, 1976; Krell *et al.*, 1979).

In addition to measuring net movements of cellular Ca^{2+}, kinetic analyses of ^{45}Ca exchange experiments have allowed the identification of different cellular Ca^{2+} pools (see, e.g., Borle, 1972, 1973; Claret–Berthon, *et al.*, 1977). With the aid of respiratory inhibitors it has been possible to identify mitochondria as a relatively large, slowly exchangeable pool of intracellular Ca^{2+}. The rate of exchange of Ca^{2+} by this pool is affected in ways similar to those observed for net Ca^{2+} movements in isolated mitochondria and whole cells. Also, of particular interest is the observation that physiological parameters, such as the intracellular pH and phosphate concentration, can significantly influence mitochondrial Ca^{2+} exchange *in vivo* (Studer and Borle, 1979; Borle, 1973). Certain hormones have also been demonstrated to affect the steady-state exchange of intracellular mitochondrial Ca^{2+}. For example, it has been reported that hyperparathyroidism increases the size and exchange rate of the mitochondrial Ca^{2+} pool in kidney slices (Borle and Clark, 1981). The results of ^{45}Ca exchange experiments have also provided support for a role of mitochondrial Ca^{2+} transport in hormone- and secretagogue-induced release of ions and macromolecules by exocrine tissue (Kanagasuntheram and Randle, 1976; Kondo and Schulz, 1976; Exton, 1980).

Compelling evidence for the presence and transport of mitochondrial Ca^{2+} *in vivo* has also come from direct measurements of the intracellular distribution of Ca^{2+} using fluorescent and luminescent Ca^{2+} indicators, as well as electron microprobe X-ray analysis.

Rose and Lowenstein (1975) have monitored the cytosolic free Ca^{2+} concentration of *Chironomus* salivary gland cells by recording the luminescence of microinjected aequorin with the aid of a microscope and an image intensifier coupled to a television camera. These studies have demonstrated a respiration-dependent restriction of the mobility of microinjected Ca^{2+}. Moreover, experiments employing injections of uncouplers and ruthenium red firmly established that mitochondrial Ca^{2+} uptake was responsible for maintaining the low cytosolic free Ca^{2+} concentration in these cells.

Chlortetracycline, a fluorescent probe of membrane-bound Ca^{2+}, has also been used to monitor qualitative changes in the intracellular distribution of Ca^{2+}. Studies using this technique have strongly suggested that the increase in cytosolic Ca^{2+} of thymocytes treated with mitogens (Mikkelsen and Schmidt-Ulrich, 1980) and hepatocytes treated with norepinephrine (Babcock *et al.*, 1979) is due to the net release of mitochondrially sequestered Ca^{2+}.

Further evidence that mitochondria actually store Ca^{2+} *in vivo* has resulted from electron microprobe measurements of total Ca^{2+} in different areas of thin sections of various types of cells. Although several reports have documented the presence of large quantities of Ca^{2+} in the electron-dense deposits often observed within mitochondria *in situ* (see, e.g., Berridge *et al.*, 1975; Parducz and Joo, 1976; Saetersdal *et al.*, 1977), large gradients of Ca^{2+} between mitochondria and the cytoplasm have not consistently been observed in some types of tissue (Gupta and Hall, 1978) such as vascular smooth muscle (Somlyo et al., 1979). Electron microprobe analysis has, however, left little doubt that mitochondria contain the majority of cellular Ca^{2+} when large quantities of Ca^{2+} enter the cell, such as occurs under many pathological conditions.

C. Comparison of Mitochondrial Transport Kinetics with Those of Other Cellular Transport Processes

Experimental evidence indicates that the maximal rate and capacity of Ca^{2+} accumulation by mitochondria isolated from a variety of tissues, e.g., liver, kidney, nerve, and smooth muscle, far exceeds those of endoplasmic reticulum or plasmalemma (Ash and Bygrave, 1977; Batra, 1973; Blaustein et al., 1980; Moore et al., 1974). Thus, there is little disagreement that mitochondria are the primary controllers of intracellular Ca^{2+} at high levels of total Ca^{2+}. However, serious objections have been raised against the view that mitochondrial Ca^{2+} transport is important in establishing and regulating the cytosolic free Ca^{2+} concentration of normal "resting" cells at 10^{-7}–10^{-6} M. These objections are based, in part, on the observations that the K_m for mitochondrial Ca^{2+} influx is significantly greater than 1 μM and that the rate of mitochondrial Ca^{2+} influx–efflux cycling at <1 μM is extremely low (see, e.g., Blaustein et al., 1980). However, a theoretical analysis of cellular Ca^{2+} transport kinetics, such as that reported by Borle (1973, 1981), indicates that such arguments may not be valid.

Even though the apparent affinity of the mitochondrial Ca^{2+} uptake process is low, relative to the Ca^{2+}-transporting ATPases of the endoplasmic reticulum and the plasmalemma, the actual rates of Ca^{2+} influx are comparatively so high that the steady-state flux of mitochondrial Ca^{2+} at cytosolic concentrations is equal to or greater than the flux of Ca^{2+} across other membranes. Borle (1981) has calculated that, for a K_{Ca} of 5 μM, a V_{max} of 400 nmoles mg^{-1} mitochondrial protein, a Hill coefficient of 2, and a cytosolic free Ca^{2+} concentration of 10^{-7} M, steady-state cycling of Ca^{2+} across the mitochondrial inner membrane would take place at 0.8 nmoles min^{-1} mg^{-1} mitochondrial protein. Considering the mitochondrial content of liver cells, this works out to be approximately 266×10^{-18} moles min^{-1} per cell. Similar treatment of the transport kinetics and content of liver plasmalemma and smooth endoplasmic reticulum yield values of 40 and 0.33×10^{-18} mole min^{-1} mg^{-1}, respectively. Thus total cellular mitochondrial Ca^{2+} cycling is on the same order of magnitude as that for the plasma membrane and more than two orders of magnitude greater than that for smooth endoplasmic reticulum. These differences become even larger at levels of cytosolic free Ca^{2+} above 10^{-7} M.

Obviously such a theoretical comparison of cellular Ca^{2+} transport processes is open to criticism because of the assumptions which must be made (Borle, 1981). For instance, the analysis would be much stronger if solid data were available on transport kinetics at $10^{-7}-10^{-6}\,M$ free Ca^{2+} under physiological conditions. It does, however, serve to illustrate that mitochondrial Ca^{2+} transport, even at very low Ca^{2+} concentrations, could be extremely important in the Ca^{2+} homeostasis of resting cells as well as those exposed to transient or chronic elevations of intracellular Ca^{2+}. The steady-state mitochondrial Ca^{2+}-buffering studies described in the next section provide even further evidence for this proposal.

IV. STEADY-STATE BUFFERING OF FREE Ca^{2+} BY MITOCHONDRIA

Most studies on isolated mitochondria have been concerned with the kinetics and bioenergetics of Ca^{2+} transport and have not yielded much information on the dynamics of cytosolic Ca^{2+} homeostasis per se. Cellular Ca^{2+} experiments have been useful in this regard but are restricted in the amount of detailed information they give concerning the roles the different cellular transport mechanisms play in establishing and regulating the intracellular distribution of Ca^{2+}. Further understanding of transport-mediated regulation of cellular Ca^{2+} has recently been achieved by investigations on subcellular Ca^{2+} transport with respect to maintenance of free Ca^{2+} concentrations at or near steady-state conditions. The results of these studies further support an important involvement of mitochondrial Ca^{2+} influx and efflux in the regulation of cytosolic Ca^{2+} and demonstrate how different transport systems can work together to maintain cellular Ca^{2+} homeostasis.

Table IV describes the reported free Ca^{2+} concentrations at which isolated energized mitochondria buffer experimental suspension media under steady-state conditions. These have been measured either spectrophotometrically with Ca^{2+}-sensitive dyes or with the aid of Ca^{2+}-sensitive electrodes. The recorded signals have been standardized against known values of free Ca^{2+} concentrations using Ca^{2+}–EGTA or Ca^{2+}–nitriloacetate buffers. Most of these values have been obtained with mitochondria in the presence of low (endogenous) levels of Ca^{2+}, thus being more physiologically realistic than data obtained from many of the previously discussed experiments.

An examination of the reported mitochondrial Ca^{2+} "set points" in Table IV indicates that, in general, mitochondria can maintain a medium free Ca^{2+} concentration at $<1.0\,\mu M$. Depending on the experimental con-

TABLE IV

Steady-State Free Ca^{2+} Concentrations of Mitochondrially Buffered Media

Tissue	Experimental medium	Free Ca^{2+} (μM)	References
Liver	High K^+, pH 7.0, 30°C	0.7	Brand and DeSelincourt, 1980
	High K^+		Nicholls, 1978
	pH 7.0, 22°C	0.5	
	pH 7.0, 37°C	0.8	
	pH 7.4, 30°C	0.3	
	pH 6.6, 30°C	0.8	
	pH 7.0, 30°C		
	Plus 0.1 mM Mg^{2+}	0.8	
	Plus 3.0 mM Mg^{2+}	1.6	
	High K^+, pH 7.0, 25°C	0.7	Becker, 1980
	Plus 1.0 mM Mg^{2+}	1.5	
	Plus 1.0 mM Mg^{2+}, 3.0 mM Mg-ATP	0.8	
	High K^+, pH 7.0, 25°C		Becker *et al.*, 1980
	Plus 1.0 mM Mg^{2+}, 3.0 mM Mg-ATP		
	17 nmoles total Ca^{2+} mg^{-1} protein	0.5	
	10 nmoles total Ca^{2+} mg^{-1} protein	0.3	
	High K^+, pH 7.4, 30°C	0.8	Hughes *et al.*, 1980
	Plus 30 mM Na^+	1.0	
	High Na^+, pH 7.0, 30°C	0.6	Nicholls and Brand, 1980
	High Na^+, pH 7.0, 30°C	1.0	Dawson and Fulton, 1980
	Plus 100 μM nupercaine	0.6	
Heart	High K^+, pH 7.0, 30°C	0.2	Brand and DeSelincourt, 1980
	Plus 5 mM Na^+	1.3	
Brain	High K^+, pH 7.0, 30°C, 0.2 mM ATP	0.3	Nicholls and Scott, 1980
	Plus 10 mM Na^+	0.4	

ditions, such steady-state levels can be maintained for 2 hr or more (Becker *et al.*, 1980). As originally proposed by Drahota *et al.* (1965), the steady-state extramitochondrial free Ca^{2+} is determined by the rate of mitochondrial Ca^{2+} influx–efflux cycling (Nicholls, 1978; Becker, 1980; Becker *et al.*, 1980). If a pulse of Ca^{2+} is added to a steady-state system, there is a rapid rise in the free Ca^{2+} concentration followed by a rapid return to the original steady-state concentration, reflecting a transient acceleration of Ca^{2+} uptake by energized mitochondria. A similar response is evoked by a pulse of the Ca^{2+} chelator EGTA, which lowers the free Ca^{2+} concentration. In this case, there is net release of Ca^{2+} from the mitochondria into the medium until the free Ca^{2+} concentration is restored to the original value. Thus the bidirectional buffering of the ambient free

Ca^{2+} by respiring mitochondria could provide the effective buffering of either a net increase or decrease in cytosolic Ca^{2+} elicited by changes in the flux of Ca^{2+} across other cellular membranes.

The mitochondrial set point for free Ca^{2+} is affected by many factors that have been shown to influence both Ca^{2+} influx and efflux and provides insight into what factors may regulate cytosolic Ca^{2+} *in vivo*. Mg^{2+}, which inhibits Ca^{2+} influx, accordingly increases the mitochondrial set point (Nicholls, 1978; Becker 1980). The inhibition of Ca^{2+} efflux by ATP is reflected in its lowering of the steady-state free Ca^{2+} concentration (Becker, 1980; Nicholls and Scott, 1980). These effects are also elicited by the local anesthetic nupercaine (Dawson and Fulton, 1980). As expected, Na^+ increases the Ca^{2+} set point for heart (Brand and DeSelincourt, 1980) and brain mitochondria (Nicholls and Scott, 1980) because of its stimulation of Ca^{2+} efflux. Other factors that affect steady-state buffering of extramitochondrial Ca^{2+} include temperature (increases free Ca^{2+}) and pH (acidity decreases the set point), as demonstrated by Nicholls (1978). Further investigations are necessary to determine the mode of their action.

One very important outcome of steady-state Ca^{2+} buffering studies has been the finding that mitochondria actually can maintain free Ca^{2+} at less than 1.0 μM under physiologically realistic conditions. With a medium whose ionic composition simulates that of liver cytosol with respect to K^+ (125 mM), Mg^{2+}-ATP (3 mM), phosphate (2 mM), and free Mg^{2+} (1 mM), rat liver mitochondria buffer free Ca^{2+} to about 0.5 (Becker *et al.*, 1980) or 0.8 μM (Becker, 1980). In the presence of ATP, these and other types of mitochondria can maintain such low set points even when loaded with Ca^{2+} at levels 10- to 20-fold greater than their endogenous content (Nicholls and Scott, 1980). The extremely high Ca^{2+} buffering capacity of mitochondria may allow them to maintain cytosolic Ca^{2+} at submicromolar levels even under pathological conditions where in many cases total cellular Ca^{2+} rises dramatically. Steady-state mitochondrial Ca^{2+} buffering utilizes negligible amounts of energy, since the steady-state rate of influx–efflux cycling at 10^{-6}–10^{-7} M free Ca^{2+} is extremely low. In the absence of ATP and Mg^{2+}, Nicholls (1978) indirectly determined the rate of mitochondrial Ca^{2+} cycling to be approximately 5 nmoles min^{-1} mg^{-1} protein at a set point of approximately 1 μM. Direct measurement of ruthenium red-insensitive Ca^{2+} efflux at a steady-state extramitochondrial free Ca^{2+} concentration of 0.5 μM in the presence of ATP and Mg^{2+} has indicated that under these conditions cycling may actually be <1.0 nmole min^{-1} mg^{-1} (G. Fiskum, unpublished results). As expected, an increase in the rate of influx–efflux cycling is associated with an increase in the mitochondrial Ca^{2+} set point.

The mitochondrial set point for free Ca^{2+} measured in the presence of a cytosol-like medium may still be significantly higher than the level at which mitochondria buffer *in vivo*. This was suggested by the results of Becker *et al.* (1980) which indicate that the set point is sensitive to changes in intramitochondrial Ca^{2+} below 15 nmoles mg^{-1} protein. When EGTA was used to lower the intramitochondrial Ca^{2+} content from an endogenous level of 17 nmoles mg^{-1} to 10 nmoles mg^{-1}, the steady-state extramitochondrial Ca^{2+} dropped from 0.5 to 0.3 μM. Since the level of rat liver mitochondrial Ca^{2+} *in situ* may actually be at least this low, this supports the proposal that mitochondria can maintain and regulate cytosolic free Ca^{2+} at $\sim 10^{-7}$ M. There is actually one preliminary report indicating buffering of free Ca^{2+} by isolated myometrium mitochondria at approximately 10^{-8} M (Batra, 1973), however, this must be verified.

The variation in the mitochondrial Ca^{2+} set point with the level of intramitochondrial Ca^{2+} at low Ca^{2+} loads strongly suggests that Ca^{2+} efflux regulates influx–efflux cycling under these conditions. This also indicates that the *intra*mitochondrial free Ca^{2+} concentration varies at low levels of total Ca^{2+}. Strong support for this comes from the work of Denton *et al.* (1980), McCormack and Denton (1980), and Hansford (1981), who have demonstrated regulation of intramitochondrial pyruvate dehydrogenase by alterations in mitochondrial Ca^{2+} influx and efflux at low levels of mitochondrial Ca^{2+}. The activities of the pyruvate dehydrogenase, 2-oxoglutarate dehydrogenase, and NAD-isocitrate dehydrogenase complexes are sensitive to Ca^{2+} in the 10^{-7}–10^{-5} M concentration range. Thus mitochondrial Ca^{2+} transport may, in addition to buffering cytosolic Ca^{2+}, serve as a regulator of the intramitochondrial free Ca^{2+} concentration, which in turn may regulate the activity of the tricarboxylic acid cycle.

The importance of the relationship among the level of mitochondrial Ca^{2+}, the rate of influx–efflux cycling, and the buffering of extramitochondrial free Ca^{2+} has been illustrated by studies assessing the contribution of both mitochondrial and endoplasmic reticulum Ca^{2+} transport in establishing steady-state free Ca^{2+} concentrations. Becker *et al.* (1980) reported that, after a free Ca^{2+} concentration of near 0.5 μM was attained by energized mitochondria suspended in a cytosol-like medium containing ATP, the subsequent addition of rat liver microsomes (mainly endoplasmic reticulum) resulted in a lowering of the steady-state level to approximately 0.2 μM. This was accompanied by a decrease in mitochondrial Ca^{2+} of about 7 nmoles mg^{-1}, similar to that observed when EGTA was utilized to lower the steady-state mitochondrially buffered free Ca^{2+} concentration. The reduction of free Ca^{2+} caused by the addition of microsomes in the presence of mitochondria was thus interpretted as a consequence of Ca^{2+} sequestration by endoplasmic reticulum followed by net release of mitochondrial Ca^{2+} to the point where mitochondrial Ca^{2+}

efflux became equivalent to the new, lower rate of influx. Therefore, the lower set point was due not only to microsomal Ca^{2+} uptake but also to a readjustment of mitochondrial Ca^{2+} influx–efflux cycling. This was not observed at elevated levels of mitochondrial Ca^{2+} (>30 nmoles mg^{-1}), where the set point was ~ 0.5 μM in the absence or presence of microsomes, since the relatively small capacity for Ca^{2+} uptake by rat liver microsomes (<10 nmoles mg^{-1}) precluded a lowering of mitochondrial Ca^{2+} below the level (~ 15 nmoles mg^{-1}) which appeared to be saturating for Ca^{2+} efflux under the particular experimental conditions.

Results very similar to those obtained with the reconstituted mitochondria and microsome system have also been obtained using suspensions of isolated hepatocytes whose plasma membranes were made permeable to ions, energy-yielding substrates, and other molecules by the inclusion of a very low concentration of digitonin in the medium (Becker *et al.*, 1980). This procedure allows measurements of mitochondrial and endoplasmic reticulum Ca^{2+} transport *in situ* within the cells with an extracellular electrode. It has also been utilized to measure the buffering of Ca^{2+} by mitochondria and smooth endoplasmic reticulum within presynaptic nerve terminals (Blaustein *et al.*, 1978, 1980). The results of these studies as well as those using digitonin-treated hepatocytes and mixtures of isolated organelles strongly suggest that Ca^{2+} transport processes of mitochondria and endoplasmic reticulum work together to maintain the cytosolic free Ca^{2+} concentration in the range 0.1–0.6 μM. This is also supported by the results of Murphy *et al.* (1980); these investigators measured the free extracellular Ca^{2+} concentration at the null point where no net flux of Ca^{2+} occurred between the cytosol and the extracellular milieu upon addition of digitonin to intact hepatocytes and found it to be approximately 0.2 μM. Since plasma membrane Ca^{2+} transport did not take place in the experiments of Becker *et al.* (1980) but was included in the determinations of Murphy *et al.* (1980), the plasma membrane Ca^{2+} transport systems of hepatocytes appear unlikely to be the sole determinants of the cytosolic free Ca^{2+} concentration. However, the most important conclusion to be drawn from these investigations is that, depending on the level of total cellular Ca^{2+}, the activity of any of the mitochondrial, endoplasmic reticulum, or plasmalemmal Ca^{2+} transport processes could serve to regulate the cytosolic free Ca^{2+} concentration.

REFERENCES

Affolter, H., and Carafoli, E. (1980). The Ca^{2+}-Na^{+} antiporter of heart mitochondria operates electroneutrally. *Biochem. Biophys. Res. Commun.* **95**, 193–196.
Åkerman, K. E. O. (1978a). Charge transfer during valinomycin-induced Ca^{2+} uptake in rat liver mitochondria. *FEBS Lett.* **93**, 293–296.

Åkerman, K. E. O. (1978b). Effect of pH and Ca^{2+} on the retention of Ca^{2+} by rat liver mitochondria. *Arch. Biochem. Biophys.* **189**, 256–262.

Åkerman, K. E. O., Wikström, M. K. F., and Saris, N. E. L. (1977). Effect of inhibitors on the sigmoidicity of the calcium ion transport kinetics in rat liver mitochondria. *Biochim. Biophys. Acta* **464**, 287–294.

Alexandre, A., Reynafarje, B., and Lehninger, A. L. (1978). Stoichiometry of vectorial H^+ movements coupled to electron transport and to ATP synthesis in mitochondria. *Proc. Natl. Acad. Sci. U.S.A.* **75**, 5296–5300.

Al-Shaikhaly, M. H. M., Nedergaard, J., and Cannon, B. (1979). Sodium-induced calcium release from mitochondria in borwn adipose tissue. *Proc. Natl. Acad. Sci. U.S.A.* **76**, 2350–2352.

Althaus-Salzmann, M., Carafoli, E., and Jakob, A. (1980). Ca^{2+}, K^+ redistributions and α-adrenergic activation of glycogenolysis in perfused rat livers. *Eur. J. Biochem.* **106**, 241–248.

Arshad, J. H., and Holdsworth, E. S. (1980). Stimulation of calcium efflux from rat liver mitochondria by adenosine 3', 5'-cyclic monophosphate. *J. Membr. Biol.* **57**, 207–212.

Ash, G. R., and Bygrave, F. L. (1977). Ruthenium red as a probe in assessing the potential of mitochondria to control intracellular calcium in liver. *FEBS Lett.* **78**, 166–168.

Asimakis, G. K., and Aprille, J. R. (1980). *In vitro* alteration of the size of the liver mitochondrial adenine nucleotide pool: Correlation with respiratory functions. *Arch. Biochem. Biophys* **203**, 307–316.

Asimakis, G. K., and Sordahl, L. A. (1977). Effects of atractyloside and palmitoyl coenzyme A on calcium transport in cardiac mitochondria. *Arch. Biochem. Biophys.* **179**, 200–210.

Babcock, D. F., Chen, J. J., Yip, B. P., and Lardy, H. A. (1979). Evidence for mitochondrial localization of the hormone-responsive pool of Ca^{2+} in isolated hepatocytes. *J. Biol. Chem.* **254**, 8117–8120.

Badyshtov, B. A., and Seredenin, S. B. (1977). Investigation of the mechanism of action of ouabain and cyclic AMP of the transport of calcium ions by rat heart mitochondria. *Bull. Exp. Biol. Med. (Engl. Transl.)* **83**, 155–158.

Batra, S. (1973). The role of mitochondrial calcium uptake in contraction and relaxation of the human myometrium. *Biochim. Biophys. Acta* **305**, 428–432.

Beatrice, M. C., Palmer, J. W., and Pfeiffer, D. R. (1980). The relationship between mitochondrial membrane permeability, membrane potential, and the retention of Ca^{2+} by mitochondria. *J. Biol. Chem.* **255**, 8663–8671.

Becker, G. L. (1980). Steady-state regulation of extramitochondrial Ca^{2+} by rat liver mitochondria: Effects of Mg^{2+} and ATP. *Biochim. Biophys. Acta* **591**, 234–239.

Becker, G. L., Chen, C.-H., Greenawalt, J. W., and Lehninger, A. L. (1974). Calcium phosphate granules in the hepatopancreas of the blue crab *Callinectes sapidus*. *J. Cell Biol.* **61**, 316–326.

Becker, G. L., Fiskum, G., and Lehninger, A. L. (1980). Regulation of free Ca^{2+} by liver mitochondria and endoplasmic reticulum. *J. Biol. Chem.* **255**, 9009–9012.

Berridge, M. J., Oschman, J. C., and Wall, B. J. (1975). Intracellular calcium reservoirs in *Calliphora* salivary glands. *In* "Calcium Transport in Contraction and Secretion" (E. Carafoli, F. Clementi, W. Drabikowski, and A. Margreth, eds.), pp. 131–138. North-Holland Publ., Amsterdam.

Bielawski, J., and Lehninger, A. L. (1966). Stoichiometric relationships in mitochondrial accumulation of calcium and phosphate supported by hydrolysis of adenosine triphosphate. *J. Biol. Chem.* **241**, 4316–4322.

Blackmore, P. F., Dehaye, J., and Exton, J. H. (1979). Studies on α-adrenergic activation of hepatic glucose output: The role of mitochondrial calcium release in α-adrenergic activation of phosphorylase in perfused rat liver. *J. Biol. Chem.* **254,** 6945–6950.

Blaustein, M. P., Ratzlaff, R. W., and Schweitzer, E. S. (1978). Calcium buffering in presynaptic nerve terminals. II. Kinetic properties of the nonmitochondrial Ca sequestration mechanism. *J. Gen. Physiol.* **72,** 43–66.

Blaustein, M. P., Ratzlaff, R. W., and Schweitzer, E. S. (1980). Control of intracellular calcium in presynaptic nerve terminals. *Fed. Proc., Fed. Am. Soc.'Exp. Biol.* **39,** 2790–2795.

Borle, A. B. (1972). Kinetic analysis of calcium movements in cell culture. V. Intracellular calcium distribution in kidney cells. *J. Membr. Biol.* **10,** 45–66.

Borle, A. B. (1973). Calcium metabolism at the cellular level. *Fed. Proc., Fed. Am. Soc. Exp. Biol.* **32,** 1944–1950.

Borle, A. B. (1974). Cyclic AMP stimulation of calcium efflux from kidney, liver, and heart mitochondria. *J. Membr. Biol.* **16,** 221–236.

Borle, A. B. (1976). On the problem of the release of mitochondrial calcium by cyclic AMP. *J. Membr. Biol.* **29,** 205–208.

Borle, A. B. (1981). Control, modulation and regulation of cell calcium. *Rev. Physiol., Biochem. Pharmacol.* **90,** 13–153.

Borle, A. B., and Clark, I. (1981). Effects of chronic and acute endogenous hyperparathyroidism on calcium distribution and transport in rat kidney. *Endocrinology* (in press).

Bragadin, M., Pozzan, T., and Azzone, G. F. (1979). Kinetics of Ca^{2+} carrier in rat liver mitochondria. *Biochemistry* **18,** 5972–5978.

Brand, M. D., and DeSelincourt, C. (1980). Effects of glucagon and Na^+ on the control of extramitochondrial free Ca^{2+} concentration by mitochondria from liver and heart. *Biochem. Biophys. Res. Commun.* **92,** 1377–1382.

Brand, M. D., and Lehninger, A. L. (1975). Superstoichiometric Ca^{2+} uptake supported by hydrolysis of endogenous ATP in rat liver mitochondria. *J. Biol. Chem.* **250,** 7958–7960.

Bygrave, F. L. (1978). Mitochondria and the control of intracellular calcium. *Biol. Rev. Cambridge Philos. Soc.* **53,** 43–79.

Bygrave, F. L., and Ash, G. R. (1977). Development of mitochondrial calcium transport activity in rat liver. *FEBS Lett.* **80,** 271–274.

Carafoli, E. (1979). The calcium cycle of mitochondria. *FEBS Lett.* **104,** 1–5.

Carafoli, E., and Crompton, M. (1978). The regulation of intracellular calcium. *Curr. Top. Membr. Transp.* **10,** 151–216.

Carafoli, E., and Lehninger, A. L. (1971). A survey of the interaction of calcium ions with mitochondria from different tissues and species. *Biochem. J.* **122,** 681–690.

Carafoli, E., Tiozzo, R., Lugli, G., Crovetti, F., and Kratzing, C. (1974). The release of calcium from heart mitochondria by sodium. *J. Mol. Cell. Cardiol.* **6,** 361–371.

Carafoli, E., Schwerzmann, K., Roos, I., and Crompton, M. (1978). Protein in mitochondrial calcium transport. *In* "Transport by Proteins" (G. Blauer and H. Sund, eds.), pp. 171–186. de Gruyter, Berlin.

Carpentieri, U., and Sordahl, L. A. (1980). Respiratory and calcium transport functions of mitochondria isolated from normal and transformed human lymphocytes. *Cancer Res.* **40,** 221–224.

Charlton, R. R., and Wenner, C. E. (1978). Calcium-ion transport by intact Ehrlich ascitestumor cells: Role of respiratory substrates, P_i and temperature. *Biochem. J.* **170,** 537–544.

Chen, C.-H., and Lehninger, A. L. (1973). Ca²⁺ transport activity in mitochondria from some plant tissues. *Arch. Biochem. Biophys.* **157**, 183–196.

Cheung, W. Y. (1980). Calmodulin plays a pivotal role in cellular regulation. *Science* **207**, 19–27.

Chudapongse, P., and Haugaard, N. (1973). The effect of phosphoenolpyruvate on calcium transport by mitochondria. *Biochim. Biophys. Acta* **307**, 599–606.

Cittadini, A., Scarpa, A., and Chance, B. (1973). Calcium transport in intact Ehrlich ascites tumor cells. *Biochim. Biophys. Acta* **291**, 246–259.

Claret-Berthon, B., Claret, M., and Mazet, J. L. (1977). Fluxes and distribution of calcium in rat liver cells: Kinetic analysis and identification of pools. *J. Physiol. (London)* **272**, 529–552.

Coelho, R. L. C., and Vercesi, A. E. (1980). Retention of Ca²⁺ by rat liver and rat heart mitochondria: Effect of phosphate, Mg²⁺, and NAD(P) redox state. *Arch. Biochem. Biophys.* **204**, 141–147.

Coty, W. A., and Pedersen, P. L. (1975). Phosphate transport in rat liver mitochondria: Kinetics, inhibitor sensitivity, energy requirements, and labelled components. *Mol. Cell. Biochem.* **9**, 109–124.

Crompton, M., and Heid, I. (1978). The cycling of calcium, sodium, and protons across the inner membrane of cardiac mitochondria. *Eur. J. Biochem.* **91**, 599–608.

Crompton, M., Capano, M., and Carafoli, E. (1976). The sodium-induced efflux of calcium from heart mitochondria: A possible mechanism for the regulation of mitochondrial calcium. *Eur. J. Biochem.* **69**, 453–462.

Crompton, M., Kunzi, M., and Carafoli, E. (1977). The calcium-induced and sodium-induced effluxes of calcium from heart mitochondria: Evidence for a sodium-calcium carrier. *Eur. J. Biochem.* **79**, 549–558.

Crompton, M., Moser, R., Ludi, H., and Carafoli, E. (1978). The interrelations between the transport of sodium and calcium in mitochondria of various mammalian tissues. *Eur. J. Biochem.* **82**, 25–31.

Crompton, M., Siegel, E., and Carafoli, E. (1979). Distinguishing the two calcium carriers of cardiac mitochondria. In "Function and Molecular Aspects of Biomembrane Transport" (E. Quagliariello *et al.*, eds.), pp. 171–174. Elsevier/North Holland Publ., Amsterdam.

Dawson, A. P., and Fulton, D. V. (1980). The action of nupercaine on calcium efflux from rat liver mitochondria. *Biochem. J.* **188**, 749–755.

Dawson, A. P., Selwyn, M. J., and Fulton, D. V. (1979). Inhibition of Ca²⁺ efflux from mitochondria by nupercaine and tetracaine. *Nature (London)* **277**, 484–486.

Deana, R., Arrabaca, J. D., Mathien-Shire, Y., and Chappell, J. B. (1979). The electric charge stoichiometry of calcium influx in rat liver mitochondria and the effect of inorganic phosphate. *FEBS Lett.* **106**, 231–234.

DeLuca, H. F., and Engström, G. W. (1961). Calcium uptake by rat kidney mitochondria. *Proc. Natl. Acad. Sci. U.S.A.* **47**, 1744–1750.

Denton, R. M., McCormack, J. G., and Edgell, N. J. (1980). Role of calcium ions in the regulation of intramitochondrial metabolism: Effects of Na⁺, Mg²⁺, and ruthenium red on the Ca²⁺-stimulated oxidation of oxoglutarate and on pyruvate dehydrogenase activity in intact rat heart mitochondria. *Biochem. J.* **190**, 107–117.

Dorman, D M Barritt. G. J., and Bygrave. F. L. (1975). Stimulation of hepatic mitochondrial calcium transport by elevated plasma insulin concentration. *Biochem. J.* **150**, 389–395.

Douglas, M. G., and Cockrell, R. S. (1974). Mitochondrial cation-hydrogen ion exchange. *J. Biol. Chem.* **249**, 5464–5471.

Drahota, Z., Carafoli, E., Rossi, C. S., Gamble, R. L., and Lehninger, A. L. (1965). The steady state maintenance of accumulated Ca^{++} in rat liver mitochondria. *J. Biol. Chem.* **240**, 2712–2720.

Dubinsky, W., Kandrach, A., and Racker, E. (1979). Resolution and reconstitution of a calcium transporter from bovine heart mitochondria. *In* "Membrane Bioenergetics" (C. P. Lee, G. Schatz, and L. Ernster, eds.), pp. 267–280. Addison-Wesley, Reading, Massachusetts.

Dubinsky, W. P., and Cockrell, R. S. (1975). Ca^{2+} transport across plasma and mitochondrial membranes of isolated hepatocytes. *FEBS Lett.* **59**, 39–43.

Endo, M. (1977). Calcium release from the sarcoplasmic reticulum. *Physiol. Rev.* **57**, 71–108.

Exton, J. H. (1980). Mechanisms involved in α-adrenergic phenomena: Role of calcium ions in actions of catecholamines in liver and other tissues. *Am. J. Physiol.* **238**, E3–E12.

Fiskum, G., and Cockrell, R. S. (1978). Ruthenium red sensitive and insensitive calcium transport in rat liver and Ehrlich ascites tumor cell mitochondria. *FEBS Lett.* **92**, 125–128.

Fiskum, G., and Lehninger, A. L. (1979). Regulated release of Ca^{2+} from respiring mitochondria by Ca^{2+}/2H^{+} antiport. *J. Biol. Chem.* **254**, 6236–6239.

Fiskum, G., and Lehninger, A. L. (1980). The mechanisms and regulation of mitochondrial Ca^{2+} transport. *Fed. Proc., Fed. Am. Soc. Exp. Biol.* **39**, 84–88.

Fiskum, G., Reynafarje, B., and Lehninger A. L. (1979). The electric charge stoichiometry of respiration-dependent Ca^{2+} uptake by mitochondria. *J. Biol. Chem.* **254**, 6288–6295.

Fiskum, G., Grief, R., Vercesi, A., and Lehninger, A. L. (1980). The bioenergetics of mitochondrial Ca^{2+} efflux and stimulation by triiodothyronine (T3). *Fed. Proc., Fed. Am. Soc. Exp. Biol.* **39**, 1706.

Foden, S., and Randle, P. J. (1978). Calcium metabolism in rat hepatocytes. *Biochem. J.* **170**, 615–625.

Friedmann, N. (1980). Studies on the mechanism of the glucagon elicited changes in mitochondrial Ca^{2+} uptake. *J. Supramol. Struct., Suppl.* **4**, 107.

Gomez-Puyou, A., de Gomez-Puyou, M. T., Becker, G., and Lehninger, A. L. (1972). An insoluble Ca^{2+}-binding factor from rat liver mitochondria. *Biochem. Biophys. Res. Commun.* **47**, 814–819.

Greenawalt, J. W., Rossi, C. S., and Lehninger, A. L. (1964). Effect of active accumulation of calcium and phosphate ions on structure of rat liver mitochondria. *J. Cell Biol.* **23**, 21–38.

Greif, R. L., Fiskum, G., Sloane, D. A., and Lehninger, A. L. (1980). Thyroid hormone and regulation of mitochondrial calcium ion efflux. *Proc. Am. Thyroid Assoc., 56th Meet.* Abstract T-23.

Gunter, T. E., Gunter, K. K., Puskin, J. S., and Russell, P. R. (1978). Efflux of Ca^{2+} and Mn^{2+} from rat liver mitochondria. *Biochemistry* **17**, 339–345.

Guptu, B. J., and Hall, T. A. (1978). Electron microprobe x-ray analysis of calcium. *Ann. N.Y. Acad. Sci.* **307**, 28–51.

Halestrap, A. P. (1978). Stimulation of the respiratory chain of rat liver mitochondria between cytochrome c$_1$ and cytochrome c by glucagon treatment of rats. *Biochem. J.* **172**, 399–405.

Hansford, R. G. (1981). Effect of micromolar concentrations of free Ca^{2+} ions on pyruvate dehydrogenase interconversion in intact rat heart mitochondria. *Biochem. J.* **194**, 721–732.

Harris, E. J. (1979). Modulation of Ca^{2+} efflux from heart mitochondria. *Biochem. J.* **178,** 673–680.

Harris, E. J., Al-Shaikhaly, M., and Baum, H. (1979). Stimulation of mitochondrial calcium ion efflux by thiol specific reagents and by thyroxine: The relationship to adenosine diphosphate retention and to mitochondrial permeability. *Biochem. J.* **182,** 455–464.

Haworth, R. A., and Hunter, D. R. (1980). Allosteric regulation of the Ca^{2+} activated hydrophilic channel of the mitochondrial inner membrane by nucleotides. *J. Membr. Biol.* **54,** 231–236.

Haworth, R. A., Hunter, D. R., and Berkoff, H. A. (1980). Na^+ releases Ca^{2+} from liver, kidney, and lung mitochondria. *FEBS Lett.* **110,** 216–218.

Herd, P. A. (1978). Thyroid hormone-divalent cation interactions: Effects of thyroid hormone on mitochondrial calcium metabolism. *Arch. Biochem. Biophys.* **188,** 220–225.

Hoch, F. L. (1974). Metabolic effects of thyroid hormones. *In* "Handbook of Physiology" (R. O. Greep, and E. B. Astwood, eds.), Sect. 7, Vol. 3, pp. 391–411. Am. Physiol. Soc., Washington, D.C.

Hughes, B. P., and Barritt, G. J. (1978). Effects of glucagon and $N^6,O^{2'}$-dibutyryladenosine $3':5'$-cyclic monophosphate on calcium transport in isolated rat liver mitochondria. *Biochem. J.* **176,** 295–304.

Hughes, B. P., Blackmore, P. F., and Exton, J. H. (1980). Exploration of the role of sodium in the α-adrenergic regulation of hepatic glucogenolysis. *FEBS Lett.* **121,** 260–264.

Hunter, D. R., Haworth, R. A., and Southard, J. H. (1976). Relationship between configuration, function, and permeability in calcium-treated mitochondria. *J. Biol. Chem.* **251,** 5069–5077.

Jeng, A. Y., and Shamoo, A. E. (1980a). Isolation of a Ca^{2+} carrier from calf heart inner mitochondrial membrane. *J. Biol. Chem.* **255,** 6897–6903.

Jeng, A. Y., and Shamoo, A. E. (1980b). The electrophoretic properties of a Ca^{2+} carrier isolated from calf heart inner mitochondrial membrane. *J. Biol. Chem.* **255,** 6904–6912.

Kanagasuntheram, P., and Randle, P. J. (1976). Calcium metabolism and amylase release in rat parotid acinar cells. *Biochem. J.* **160,** 547–564.

Kimura, S., and Rasmussen, H. (1977). Adrenal glucocorticoids, adenine nucleotide translocation, and mitochondrial calcium accumulation. *J. Biol. Chem.* **252,** 1217–1225.

Kleineke, H., and Stratman, F. W. (1974). Calcium transport in isolated rat hepatocytes. *FEBS Lett.* **43,** 75–80.

Klingenberg, M. (1980). The ADP-ATP translocation in mitochondria, a membrane potential controlled transport. *J. Membr. Biol.* **56,** 97–105.

Kondo, S., and Schulz, I. (1976). Ca^{++} fluxes in isolated cells of rat pancreas: Effect of secretagogues and different Ca^{++} concentrations. *J. Membr. Biol.* **29,** 185–203.

Krebs, H. A., and Veech, R. L. (1969). Equilibrium relations between pyridine nucleotides and adenine nucleotides and their role in the regulation of metabolic processes. *Adv. Enzyme Regul.* **7,** 397–416.

Krell, H., Baur, H., and Pfaff, E. (1979). Transient ^{45}Ca uptake and release in isolated rat liver cells during recovery from deenergized states. *Eur. J. Biochem.* **101,** 349–364.

Landry, Y., and Lehninger, A. L. (1976). Transport of calcium ions by Ehrlich ascites tumor cells. *Biochem. J.* **158,** 427–438.

Landry, Y., Vincent-Viry, M., and Jodin, C. (1978). Energy requirement for calcium uptake by thymus lymphocytes. *FEBS Lett.* **88**, 305–308.

Lee, C. O., and Fozzard, H. G. (1975). Activities of potassium and sodium ions in rabbit heart muscle. *J. Gen. Physiol.* **65**, 695–708.

Lee, T. J., and Ashley, C. C. (1978). Increase in free Ca^{2+} in muscle after exposure to CO_2. *Nature (London)* **275**, 236–238.

Lehninger, A. L. (1970). Mitochondria and calcium ion transport. *Biochem. J.* **119**, 129–138.

Lehninger, A. L., Carafoli, E., and Rossi, C. S. (1967). Energy-linked ion movements in mitochondrial systems. *Adv. Enzymol.* **29**, 259–320.

Lehninger, A. L., Vercesi, A., and Bababunmi, E. A. (1978). Regulation of Ca^{2+} release from mitochondria by the oxidation-reduction state of pyridine nucleotides. *Proc. Natl. Acad. Sci. U.S.A.* **75**, 1690–1694.

Lötscher, H.-R., Schwerzmann, K., and Carafoli, E. (1979). The transport of Ca^{2+} in a purified population of inside-out vesicles from rat liver mitochondria. *FEBS Lett.* **99**, 194–198.

Lötscher, H.-R., Winterhalter, K. H., Carafoli, E., and Richter, C. (1980). Hydroperoxide-induced loss of pyridine nucleotides and release of calcium from rat liver mitochondria. *J. Biol. Chem.* **255**, 9325–9330.

Loyter, A., Christiansen, R. O., Steensland, H., Saltzgaber, J., and Racker, E. (1969). Energy-linked ion translocation in submitochondrial particles. 1. Ca^{++} accumulation in submitochondrial particles. *J. Biol. Chem.* **244**, 4422–4427.

McCormack, J. G., and Denton, R. M. (1980). Role of calcium ions in the regulation of intramitochondrial metabolism: Properties of the Ca^{2+}-sensitive dehydrogenases within intact uncoupled mitochondria from the white and brown adipose tissue of the rat. *Biochem. J.* **190**, 95–105.

McIntyre, H. J., and Bygrave, F. L. (1974). Retention of calcium by mitochondria isolated from Ehrlich ascites tumor cells. *Arch. Biochem. Biophys.* **165**, 744–748.

McMillin-Wood, J., Wolkowicz, P. E., Chu, A., Tate, C. A., Goldstein, M. A., and Entman, M. L. (1980). Calcium uptake by two preparations of mitochondria from heart. *Biochim. Biophys. Acta* **591**, 251–265.

Matlib, A., and O'Brien, J. P. (1974). Adenosine 3′ : 5′-cyclic monophosphate stimulation of calcium efflux. *Biochem. Soc. Trans.* **2**, 997–1000.

Mikkelsen, R. B., and Schmidt-Ulrich, R. (1980). Concanavalin A induces the release of intracellular Ca^{2+} in intact rabbit thymocytes. *J. Biol. Chem.* **255**, 5177–5183.

Moore, L., Fitzpatrick, D. F., Chen, T. S., and Landon, E. J. (1974). Calcium pump activity of the renal plasma membrane and renal microsomes. *Biochim. Biophys. Acta* **345**, 405–418.

Moyle, J., and Mitchell, P. (1977). The lanthanide-sensitive calcium phosphate porter of rat liver mitochondria. *FEBS Lett.* **77**, 136–140.

Murphy, E., Coll, K., Rich, T. L., and Williamson, J. R. (1980). Hormonal effects on calcium homeostasis in isolated hepatocytes. *J. Biol. Chem.* **255**, 6600–6608.

Nicholls, D. G. (1978). The regulation of extramitochondrial free calcium ion concentration by rat liver mitochondria. *Biochem. J.* **176**, 463–474.

Nicholls, D. G., and Brand, M. D. (1980). The nature of the calcium ion efflux induced in rat liver mitochondria by the oxidation of endogenous nicotinamide nucleotides. *Biochem. J.* **188**, 113–118.

Nicholls, D. G., and Crompton, M. (1980). Mitochondrial calcium transport. *FEBS Lett.* **111**, 261–268.

Nicholls, D. G., and Scott, I. D. (1980). The regulation of brain mitochondrial calcium-ion transport. *Biochem. J.* **186**, 833–839.

Panfili, E., Sandri, G., Sottocasa, G. L., Linazzi, G., Liut, G., and Graziosi, G. (1976). Specific inhibition of mitochondrial Ca^{2+} transport by antibodies directed to the Ca^{2+}-binding glycoprotein. *Nature (London)* **264**, 185–186.

Panfili, E., Sottocasa, G., Sandri, G., and Liut, G. (1980). The Ca^{2+}-binding glycoprotein as the site of metabolic regulation of mitochondrial Ca^{2+} movements. *Eur. J. Biochem.* **105**, 205–210.

Panov, A., Filippova, S., and Lyakhovich, V. (1980). Adenine nucleotide translocase as a site of regulation of ADP and of the rat liver mitochondria permeability of H^+ and K^+ ions. *Arch. Biochem. Biophys.* **199**, 420–426.

Parducz, A., and Joo, F. (1976). Visualization of stimulated nerve endings by preferential calcium accumulation of mitochondria. *J. Cell Biol.* **69**, 513–517.

Pedersen, P. L., and Coty, W. (1972). Energy-dependent accumulation of calcium and phosphate by purified inner membrane vesicles of rat liver mitochondria. *J. Biol. Chem.* **247**, 3107–3113.

Peng, C. F., Price, D. W., Bhuvaneswaran, C., and Wadkins, C. L. (1974). Factors that influence phosphoenolpyruvate-induced calcium efflux from rat liver mitochondria. *Biochem. Biophys. Res. Commun.* **56**, 134–140.

Poggioli, J., Berthon, B., and Claret, M. (1980). Calcium movements in *in situ* mitochondria following activation of α-adrenergic receptors in rat liver cells. *FEBS Lett.* **115**, 243–246.

Prpic, V., and Bygrave, F. L. (1980). On the inter-relationship between glucagon action, the oxidation-reduction state of pyridine nucleotides, and calcium retention by rat liver mitochondria. *J. Biol. Chem.* **255**, 6193–6199.

Prpic, V., Spencer, T. L., and Bygrave, F. L. (1978). Stable enhancement of calcium retention in mitochondria isolated from rat liver after administration of glucagon to the intact animal. *Biochem. J.* **176**, 705–714.

Puskin, J. S., Gunter, T. E., Gunter, K. K., and Russell, P. R. (1976). Evidence for more than one Ca^{2+} transport mechanism in mitochondria. *Biochemistry* **15**, 3834–3842.

Rasmussen, H., and Goodman, B. P. (1977). Relationship between calcium and cyclic nucleotides in cell activation. *Physiol. Rev.* **57**, 422–509.

Reed, K. C., and Bygrave, F. L. (1975). A kinetic study of mitochondrial calcium transport. *Eur. J. Biochem.* **55**, 497–504.

Reynafarje, B., and Lehninger, A. L. (1973). Ca^{2+} transport by mitochondria from L1210 mouse ascites tumor cells. *Proc. Natl. Acad. Sci. U.S.A.* **70**, 1744–1748.

Roman, I., Gmaj, P., Nowicka, C., and Angielski, S. (1979). Regulation of Ca^{2+} efflux from kidney and liver mitochondria by unsaturated fatty acids and Na^+ ions. *Eur. J. Biochem.* **102**, 615–623.

Roos, I., Crompton, M., and Carafoli, E. (1980). The role of inorganic phosphate in the release of Ca^{2+} from rat liver mitochondria. *Eur. J. Biochem.* **110**, 319–325.

Rose, B., and Lowenstein, W. R. (1975). Calcium ion distribution in cytoplasm visualized by aequorin: Diffusion in cytosol restricted by energized sequestering. *Science* **190**, 1204–1206.

Rossi, C. S., and Lehninger, A. L. (1964). Stoichiometry of respiratory stimulation, accumulation of Ca^{++} and phosphate, and oxidative phosphorylation in rat liver mitochondria. *J. Biol. Chem.* **239**, 3971–3980.

Saetersdal, T. S., Myklebust, R., and Berg Justesen, N. P. (1977). Calcium containing particles in mitochondria of heart muscle cells as shown by cryo-ultramicrotomy and x-ray microanalysis. *Cell Tissue Res.* **182**, 17–31.

Saris, N., and Akerman, K. E. O. (1980). Uptake and release of bivalent cations in mitochondria. *Curr. Top. Bioenerg.* **10**, 104–179.

Scarpa, A., and Lindsay, J. G. (1972). Maintenance of energy-linked functions in rat liver mitochondria aged in the presence of nupercaine. *Eur. J. Biochem.* **27**, 401–407.

Scarpa, A., Malmström, K., Chiesei, M., and Carafoli, E. (1976). On the problem of the release of mitochondrial calcium by cyclic AMP. *J. Membr. Biol.* **29**, 205.

Schotland, J., and MeLa, L. (1977). Role of cyclic nucleotides in the regulation of mitochondrial calcium uptake and efflux kinetics. *Biochem. Biophys. Res. Commun.* **75**, 920–924.

Siess, E. A., and Wieland, O. H. (1980). Early kinetics of glucagon action in isolated hepatocytes at the mitochondrial level. *Eur. J. Biochem.* **110**, 203–210.

Siliprandi, D., Toninello, F., Zoccarato, F., Rugolo, M., and Siliprandi, N. (1978). Efflux of magnesium and potassium ions from liver mitochondria induced by inorganic phosphate and by diamide. *J. Bioenerg. Biomembr.* **10**, 1–10.

Siliprandi, N., Rugolo, M., Siliprandi, D., Toninello, A., and Zoccarato, F. (1979). Induction of Mg^{2+} efflux and Ca^2 cycling in rat liver mitochondria by inorganic phosphate and oxidizing agents. *In* "Membrane Bioenergetics" (C. P. Lee, G. Schatz, and L. Ernster, eds.), pp. 533–545. Addison-Wesley, Reading, Massachusetts.

Smith, R. L., and Bygrave, F. L. (1978). Enrichment of ruthenium red-sensitive Ca^{2+} transport in a population of heavy mitochondrial isolated from flight-muscle of *Lucilia cuprina. FEBS Lett.* **95**, 303–306.

Somlyo, A. P., Somlyo, A. V., and Shuman, H. (1979). Electron probe analysis of vascular smooth muscle: Composition of mitochondria, nuclei, and cytoplasm. *J. Cell Biol.* **81**, 316–335.

Sordahl, L. A. (1974). Effects of magnesium, ruthenium red and the antibiotic ionophore A-23187 on initial rates of calcium uptake and release by heart mitochondria. *Arch. Biochem. Biophys.* **167**, 104–115.

Sordahl, L. A., and Asimakis, G. K. (1978). Calcium retention and release in heart mitochondria. *In* "The Proton and Calcium Pumps" (G. F. Azzone, M. Avron, J. C. Metcalfe, E. Quagliariello, and N. Siliprandi, eds.), pp. 273–282. Elsevier/North-Holland Publ., Amsterdam.

Sottocasa, G., Sandri, G., Panfili, E., Gazzotti, P., Vasington, F. D., and Carafoli, E. (1972). Isolation of a soluble Ca^{2+} binding glucoprotein from ox liver mitochondria. *Biochem. Biophys. Res. Commun.* **47**, 808–813.

Studer, R. K., and Borle, A. B. (1979). Effect of pH on the calcium metabolism of isolated rat kidney cells. *J. Membr. Biol.* **48**, 325–341.

Studer, R. K., and Borle, A. B. (1980). The effect of hydrogen ion on the kinetics of calcium transport by rat kidney mitochondria. *Arch. Biochem. Biophys.* **203**, 707–718.

Sul, H. S., Shrago, E., and Shug, A. L. (1976). Relationship of phosphoenolpyruvate transport, acyl coenzyme A inhibition of adenine nucleotide translocase and calcium ion efflux in guinea pig heart mitochondria. *Arch. Biochem. Biophys.* **172**, 230–237.

Taylor, W. M., Prpic, V., Exton, J. H., and Bygrave, F. L. (1980). Stable changes to calcium fluxes in mitochondria isolated from rat livers perfused with α-adrenergic agonists and with glucagon. *Biochem. J.* **188**, 443–450.

Tew, W. P., Mahle, C., Benavides, J., Howard, J. E., and Lehninger, A. L. (1980). Synthesis and characterization of phosphocitric acid, a potent inhibitor of hydroxyapatite crystal growth. *Biochemistry* **19**, 1983–1988.

Thorne, R. F. W., and Bygrave, F. L. (1974). Calcium does not uncouple oxidative phosphorylation in tightly-coupled mitochondria from Ehrlich ascites tumor cells. *Nature (London)* **248**, 348–351.

Tsokos, J., Cornwell, T. F., and Vlasuk, G. (1980). Ca^{2+} efflux from liver mitochondria induced by a decrease in extramitochondrial pH. *FEBS Lett.* **119**, 297–300.

Turakulov, Y. K., Gainutdinov, M. K., Lavina, I. I., and Akhmatov, M. S. (1977). Insulin-dependent cytoplasmic regulator of the transport of Ca^{2+} ions in liver mitochondria. *Dokl. Akad. Nauk SSSR* **234**, 1471–1473.

Vasington, F. D., and Murphy, J. V. (1961). Active binding of calcium by mitochondria. *Fed. Proc., Fed. Am. Soc. Exp. Biol.* **20**, 146.

Vercesi, A. (1981). Sequence of events in Ca^{2+} efflux induced by the oxidized steady state of mitochondrial NAD(P). *Fed. Proc., Fed. Am. Soc. Exp. Biol.* **40**, 1781.

Vercesi, A., Reynafarje, B., and Lehninger, A. L. (1978). Stoichiometry of H^+ ejection and Ca^{2+} uptake coupled to electron transport in rat heart mitochondria. *J. Biol. Chem.* **253**, 6379–6385.

Wehrle, J. P., and Pedersen, P. L. (1979). Phosphate transport in rat liver mitochondria. *J. Biol. Chem.* **254**, 7269–7275.

Wolkowicz, P. E., and Wood, J. M. (1979). Effect of malonyl-CoA on calcium uptake and pyridine nucleotide redox state in rat liver mitochondria. *FEBS Lett.* **10**, 63–66.

Wolkowicz, P. E., and Wood, J. M. (1980). Dissociation between mitochondrial calcium ion release and pyridine nucleotide oxidation. *J. Biol. Chem.* **255**, 10348–10353.

Yamazaki, R. K., Mickey, D. L., and Story, M. (1980). Rapid action of glucagon on hepatic mitochondrial calcium metabolism and respiratory rates. *Biochim. Biophys. Acta* **592**, 1–12.

Zoccarato, F., Ruglo, M., Siliprandi, D., and Siliprandi, N. (1981). Correlated adenine nucleotides, Mg^{2+} and Ca^{2+} effluxes induced in rat liver mitochondria by external Ca^{2+} and phosphate. *Eur. J. Biochem.* **114**, 195–199.

Chapter 3

Calcium Movement and Regulation in Presynaptic Nerve Terminals

CATHERINE F. MCGRAW
DANIEL A. NACHSHEN
MORDECAI P. BLAUSTEIN

I. Introduction	81
II. Calcium Content and Intraterminal Ca²⁺ Distribution	82
A. Total Ca²⁺ in Presynaptic Nerve Terminals	82
B. Distribution of Ca²⁺ in Nerve Terminals	83
C. Intraterminal Ca²⁺ Sequestration: Evidence of Low $[Ca^{2+}]_{in}$	84
D. The Morphology of Ca²⁺-Storing Organelles	89
III. Ca²⁺ Entry Mechanisms	92
A. Ca²⁺-Dependent Transmitter Release and Ca²⁺ Channels in Squid and Frog Nerve Endings	92
B. Heterogeneity of Ca²⁺ Channels	93
C. "Fast" and "Slow" Ca²⁺ Channels in Synaptosomes	95
D. Na⁺–Ca²⁺ Exchange	99
IV. Ca²⁺ Efflux from Nerve Terminals	100
Na⁺–Ca²⁺ Exchange and Other Extrusion Mechanisms	100
V. Ca²⁺ Movements and Transmitter Release	101
VI. Summary	102
References	103

I. INTRODUCTION

A number of physiological processes are critically dependent upon the level of ionized intracellular calcium ($[Ca^{2+}]_{in}$). How intracellular Ca²⁺ mediates such diverse neuronal activities as neurotransmitter release

CALCIUM AND CELL FUNCTION, VOL. II
ISBN 0-12-171402-0

(e.g., Katz, 1968), regulation of membrane permeability (e.g., Meech, 1978; Eckert and Tillotson, 1978; Krnjevic et al., 1975), microtubule polymerization (Weisenberg, 1972; Schliwa, 1976; Dedman et al., 1979), and axonal transport (Hammerschlag, 1980; Ochs, 1980) is not completely understood. Nevertheless, it is clear that the spatial and temporal distribution of $[Ca^{2+}]_{in}$ must be precisely controlled so that these activities can be regulated independently.

This chapter reviews the mechanisms of Ca^{2+} entry and exit at nerve terminals, and the intracellular mechanisms that contribute to the regulation of $[Ca^{2+}]_{in}$.

II. CALCIUM CONTENT AND INTRATERMINAL Ca^{2+} DISTRIBUTION

A. Total Ca^{2+} at Presynaptic Nerve Terminals

Intracellular Ca^{2+} is unevenly distributed throughout presynaptic nerve terminals. It is present in bound (sequestered), complexed, and free (ionized) forms that may or may not be readily exchangeable. Ionized Ca^{2+} is only a very small fraction (probably $<0.1\%$, at rest) of total intraterminal Ca^{2+}; thus, Ca^{2+} influx produces physiologically significant increases in $[Ca^{2+}]_{in}$ despite a minute increase in the total Ca. Also, redistribution of Ca^{2+} within the terminals during or immediately after periods of activity may alter $[Ca^{2+}]_{in}$ substantially, with no significant changes in the total Ca^{2+}. Nonetheless determination of the distribution of the total cell Ca^{2+} within the terminals, at steady state (rest) and under various experimental conditions, may provide clues as to which components are involved in the regulation of $[Ca^{2+}]_{in}$.

The concentration of total cell Ca still cannot be directly measured in situ. However, Ca^{2+} content can be measured with X-ray microprobe methods in quick-frozen tissue samples (Ornberg and Reese, 1980); as will be discussed below, X-ray microprobe methods can be used to determine the subcellular localization of Ca^{2+}. Standard atomic absorption methods can also be used to measure net Ca^{2+} content in cellular or subcellular neuronal preparations such as squid giant nerve fiber axoplasm (e.g., Hodgkin and Keynes, 1957; Requena et al., 1979) and pinched-off nerve endings (synaptosomes) from mammalian brain (Blaustein et al., 1978a; Schweitzer and Blaustein, 1980).

Synaptosomes are functionally "intact" in that they retain many of the properties of in situ presynaptic nerve terminals (Bradford, 1975; Blaustein et al., 1977). These preparations are therefore very useful for analyzing the Ca sequestering and buffering systems of nerve terminals of the central nervous system.

The total Ca content of freshly isolated synaptosomes, prepared from rat brain and homogenized in sucrose with 0.2 mM [ethyleneglycol-bis(β-aminoethyl ether)-N,N'-tetracetic acid] (EGTA), is about 5–8 μmoles g^{-1} synaptosome protein (Blaustein et al., 1978a; Schweitzer and Blaustein, 1980). This determination may, however, be an overestimate if a substantial amount of Ca^{2+} enters the terminals when they depolarize during mincing and homogenization.

B. Distribution of Ca^{2+} in Nerve Terminals

The distribution of (net) Ca^{2+} in subcellular compartments has been determined by comparing measurements obtained from intact terminals and from osmotically lysed terminals incubated in the presence of the Ca^{2+} ionophore A23187 and/or the mitochondrial uncoupler, carbonyl cyanide p-trifluoromethoxyphenylhydrazone (FCCP). The FCCP releases Ca^{2+} from mitochondria selectively, while the A23187 releases Ca^{2+} from other membrane-bounded stores as well (Blaustein et al., 1978c). Data obtained from lysed synaptosomes incubated with one or both of these agents indicate that approximately 1.2 μmoles Ca^{2+} g^{-1} synaptosome protein is in the mitochondria, and about 3.0 μmoles Ca^{2+} g^{-1} protein is in nonmitochondrial membrane-bounded compartments (Schweitzer and Blaustein, 1980). The difference in the total calcium content between intact terminals and osmotically lysed terminals incubated without FCCP or A23187 is about 1.5 μmoles Ca^{2+} g^{-1} protein; this is an upper limit for the amount of Ca^{2+} bound to cytoplasmic proteins (Schweitzer and Blaustein, 1980). The latter value is, however, too large, because it includes the Ca^{2+} released from intraterminal organelles during osmotic lysis.

The total Ca^{2+} concentrations detected in synaptosomes only approximate the actual Ca^{2+} content of presynaptic terminals in situ. However, these measurements, and measurements of the total Ca^{2+} under different experimental conditions (e.g., in the presence of various Ca^{2+} loads), enable us to estimate the net flux of Ca^{2+} across the plasmalemma and within the subcellular compartments. The net flux data, in combination with measurements of ionized calcium (see below), provide useful information about the buffering capacity of intraterminal organelles (Schweitzer and Blaustein, 1980).

The physiological [Ca^{2+}]$_{in}$ in the cytoplasm of presynaptic terminals, as well as in nerve axons and cell bodies, is on the order of 10^{-7} M at rest (DiPolo et al., 1976; Christofferson and Simonsen, 1977; Schweitzer and Blaustein, 1980; Alvarez-Leefmans et al., 1981). The extracellular Ca concentration is in the millimolar range (cf. Blaustein, 1974). This concentration gradient, together with a large, negative resting potential, sets up a very steep electrochemical gradient for Ca^{2+} across the plasmalemma of nerve terminals. Following an action potential and an increase in the Ca^{2+} conductance of the plasmalemma (Katz and Miledi, 1969b; Nachshen and Blaustein, 1980), sufficient Ca^{2+} could enter the terminals to raise $[Ca^{2+}]_{in}$ to $10^{-6} M$, or even perhaps even $10^{-5} M$. The increase in $[Ca^{2+}]_{in}$ triggers a burst of transmitter release (Miledi, 1973; Llinas and Nicholson, 1975), but the transmitter release evoked by the action potential is transient and declines within in about 1–2 msec. This decline is probably the result of an abrupt fall in $[Ca^{2+}]_{in}$ (Katz and Miledi, 1968). However, it is unlikely that all (or most) of the Ca^{2+} that enters during depolarization can be extruded from the terminals within 1–2 msec. Thus, intracellular Ca^{2+} buffering mechanisms must bind and/or sequester the excess Ca^{2+} until it can be extruded.

Some Ca^{2+} may bind to vesicle membranes or plasma membranes, but much of it probably diffuses away (Parsegian, 1977) from the active zone (sites of vesicle exocytosis). Some Ca^{2+} may bind passively to proteins in the cytoplasm with a high affinity for Ca^{2+} (Wolff et al., 1977), similar to those in squid axoplasm described by Baker and Schlaepfer (1978). This diffusion and binding of Ca^{2+}, analogous to the buffering of cytoplasmic Ca^{2+} in muscle by parvalbumins (Somlyo et al., 1981), could account for the rapid decline in evoked transmitter release that occurs when nerve terminals are repolarized.

These Ca^{2+}-binding molecules probably do not have sufficient capacity to buffer all the Ca^{2+} that enters during a train of action potentials. Therefore, intraterminal organelles with a large Ca^{2+} storage capacity, which actively sequester Ca^{2+}, are probably responsible for the somewhat slower buffering that is required under these circumstances. The stored Ca^{2+} is then gradually released from these organelles and extruded across the plasmalemma. Several intraterminal organelles have been shown to sequester Ca^{2+} by ATP-dependent mechanisms and may thus be important in Ca^{2+} buffering. Likely candidates for this function include mitochondria, smooth endoplasmic reticulum (SER), synaptic vesicles, coated vesicles, and microvesicles. The contribution to Ca buffering made by each of these organelles is discussed below.

C. Intraterminal Ca^{2+} Sequestration: Evidence of Low $[Ca^{2+}]_{in}$

1. Ca^{2+} Sequestration in Mitochondria

It has been known for some time (cf. Slater and Cleland, 1953) that most mitochondria, including those from neuronal tissue (Lazarewicz et al., 1974), are able to accumulate Ca^{2+}. Data from numerous studies have shown that the Ca^{2+} transport system is located in the inner mitochondrial membrane and is dependent on energy derived either from electron transport or from ATP hydrolysis (for reviews, see Lehninger, 1970; Bygrave, 1978).

In the past, several lines of indirect evidence have led to the speculation that mitochondria play a major role in the control of $[Ca^{2+}]_{in}$. For example, Drahota et al. (1965) have claimed that isolated liver mitochondria have a "high" affinity for Ca^{2+} ($K_{Ca} \simeq 2 \mu M$) and can reduce $[Ca^{2+}]$ in incubation media to below 0.2 μM in the presence of ATP and Mg. Alnaes and Rahamimoff (1975) have estimated that mitochondria make up 6–7% of the nerve terminal volume at the frog neuromuscular junction. Other ultrastructural studies show that Ca^{2+} granules are accumulated in the mitochondrial matrix after exposure to Ca^{2+}-containing media (Greenawalt et al., 1964). In addition, biochemical studies on Ca^{2+} uptake by microsomal and mitochondrial fractions from brain preparations indicate that the mitochondria can accumulate more Ca^{2+} than other fractions (Lazarewicz et al., 1974). Accumulation of Ca^{2+} by mitochondria takes precedence over oxidative phosphorylation in the presence of large Ca^{2+} loads. Physiological evidence indicates that mitochondrial inhibitors cause an increase in spontaneous transmitter release, presumably by inducing a release of sequestered Ca (Kraatz and Trautwein, 1957; Alnaes and Rahaminoff, 1975).

There is general agreement that mitochondria can accumulate substantial amounts of Ca (Fig. 1a); but it is doubtful that they actually accumulate substantial amounts of Ca^{2+} under normal physiological conditions. The systems that buffer intracellular Ca^{2+} in vivo must have a very high affinity for Ca^{2+} and a very rapid initial rate of uptake. Many experiments with rat liver mitochondria indicate that these requirements are not satisfied in the presence of physiological concentrations of Mg^{2+} ($\sim 10^{-3}$ M) and Ca^{2+} ($< 10^{-6} M$) (e.g., Vinogradov and Scarpa, 1973; Hutson et al., 1976). Data from studies on neuronal mitochondria show similar results: Although mitochondria from rat brain and squid axon can accumulate Ca from media with near-physiological Ca^{2+} concentrations, they do so at a very slow rate (Rahamimoff et al., 1976; Brinley et al., 1978).

Two types of experiments are particularly striking in this regard. Data on brain mitochondria from our laboratory (Blaustein et al., 1978a; Schweitzer and Blaustein, 1980), and on liver mitochondria from Lehninger's laboratory (Becker et al., 1980), show that (1) mitochondria are unable to sequester Ca when $[Ca^{2+}]$ in the media falls below about 0.5–1 μM, and (2) intracellular organelles other than mitochondria buffer $[Ca^{2+}]$ to much lower levels. Furthermore, when intact synaptosomes are depolarized to induce the uptake of small Ca^{2+} loads, very little of the accumulated Ca^{2+} is found in mitochondria: Nearly all the accumulated

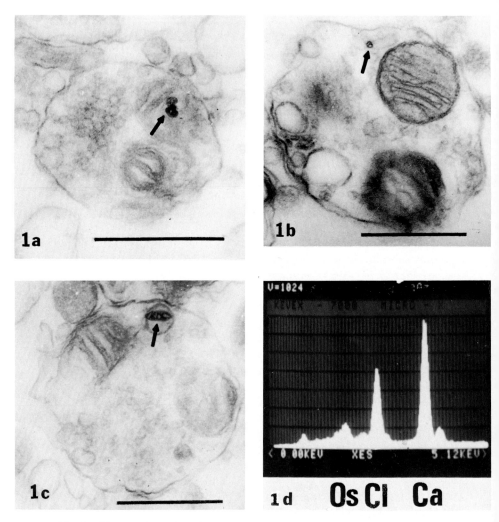

Fig. 1. Electron-dense deposits (arrows) are located in: a mitochondrion (a), a SER cisternal profile (b), and a large vesicular profile (perhaps a swollen segment of SER) (c). The synaptosomes were all treated with saponin and were incubated with Ca^{2+}, ATP, and potassium oxalate. The bars indicate 0.5 μm in all three micrographs. (d) Electron probe analysis of the electron-dense deposit in the vesicular profile (c) confirms the presence of Ca^{2+}. (From McGraw *et al.*, 1980b, with permission.)

Ca^{2+} is sequestered in other membrane-bounded organelles (Blaustein *et al.*, 1978c, 1980a,b). It is difficult to escape the conclusion that mitochondria *do not* play a central role in intracellular Ca^{2+} buffering under normal physiological conditions (also see Brinley, 1980).

2. Ca²⁺ Sequestration in Smooth Endoplasmic Reticulum

Numerous investigators demonstrated nonmitochondrial, ATP-dependent, Ca^{2+} sequestration in microsomal preparations from brain (e.g., Otsuka *et al.,* 1965; Yoshida *et al.,* 1966; Robinson and Lust, 1968; Diamond and Goldberg, 1971; Trotta and DeMeis, 1975) and peripheral nerves (e.g., Lieberman *et al.,* 1967; Perkins and Wright, 1969; Eroglu and Keen, 1977). These microsomal fractions were composed of heterogeneous vesicles and membrane fragments of neuronal and glial origin, so that identification of the Ca^{2+} sequestering component was not possible. As mentioned in the previous section, synaptosomes isolated from rat brains also contain a nonmitochondrial component that accumulates Ca^{2+} in the presence of Mg and ATP (Blaustein *et al.,* 1978b,c). This Ca^{2+} uptake is insensitive to mitochondrial poisons such as oligomycin, sodium azide, 2,4-dinitrophenol, ruthenium red, and FCCP; however, it is sensitive to trypsin and to the Ca^{2+} ionophore A23187. In addition, previously bound Ca^{2+} is rapidly released by A23187, whereas the Ca^{2+} chelator, EGTA, does not cause Ca^{2+} release. These data are consistent with the idea that the Ca^{2+} is sequestered into a membrane-bounded organelle and is not merely bound to exposed membranes.

We have attempted to study this membrane-bounded organelle by assaying subfractions of lysed synaptosomes for ATP-dependent Ca^{2+} uptake activity (Blaustein *et al.,* 1978b). Neither the synaptic vesicles nor the mitochondrial fractions appear to be associated with the Ca^{2+} uptake activity measured at low Ca^{2+}. Also, the apparent half-saturation constant for Ca^{2+} (K_{Ca}), determined by examining the effect of varying ionized Ca concentrations in the incubation media on ^{45}Ca uptake, is about 0.4 μM (Blaustein *et al.,* 1978c). This value is much lower than the K_{Ca} for Ca^{2+} uptake of the mitochondrial or synaptic vesicular systems (see below).

What is the identity of the nonmitochondrial Ca^{2+} sequestering organelle? Smooth endoplasmic reticulum is a likely candidate (Robinson and Lust, 1968; Lieberman, 1971). Indeed, data from a variety of preparations (e.g., Hales *et al.,* 1974; Eroglu and Keen, 1977), including squid axons (Henkart *et al.,* 1978; Henkart, 1980) and presynaptic motor nerve terminals (cf. Ornberg and Reese, 1980), indicate that SER sequesters Ca^{2+}. Recent morphological studies on synaptosome preparations (see Section II.D) have established that intraterminal SER stores Ca^{2+} in the presence of ATP, Mg^{2+}, and potassium oxalate (McGraw *et al.,* 1980a,b). Calcium uptake by SER is prevented by A23187 but not by mitochondrial poisons. Thus, SER most likely corresponds to the previously characterized nonmitochondrial component of Ca^{2+} sequestration.

There are striking similiarities between the nonmitochondrial Ca^{2+} uptake system in nerve terminals and the Ca^{2+} sequestration system of skeletal muscle sarcoplasmic reticulum (SR) (cf. Blaustein *et al.*, 1978c). Both systems (1) require ATP and Mg^{2+}, (2) have a high affinity for Ca^{2+}, (3) are sensitive to A23187, (4) are insensitive to mitochondrial inhibitors, and (5) are enhanced by anions such as oxalate. Both systems also appear to be mediated by a Mg^{2+}-dependent Ca^{2+}-ATPase with a stoichiometry of 2 moles of Ca^{2+} transported per mole of ATP hydrolyzed (cf. Blaustein *et al.*, 1978c). In view of the well-demonstrated role of SR in the regulation of skeletal muscle Ca^{2+} (e.g., MacLennan and Holland, 1975) these similarities support, albeit indirectly, the speculation that SER is the primary organelle responsible for intracellular Ca^{2+} buffering; SR may simply be a particularly well-developed example of Ca^{2+}-sequestering SER.

3. Ca^{2+} Storage in Synaptic Vesicles

Other organelles have also been reported to play a role in Ca^{2+} buffering in presynaptic nerve terminals; These include (1) the neurotransmitter-storing synaptic vesicles (Bohan *et al.*, 1973; Pappas and Rose, 1976; Schmidt *et al.*, 1976 Israel *et al.*, 1980; Michaelson *et al.*, 1980), (2) the coated vesicles (Blitz *et al.*, 1977) implicated in synaptic vesicle recycling, and (3) the microvesicles present in neurosecretory nerve terminals (Nordmann and Chevalier, 1980).

Morphological observations on synaptic vesicles fixed in the presence of high concentrations of Ca^{2+} show a Ca^{2+}-binding site on the synaptic vesicle membrane: This appears as an easily recognized electron-dense spot on the vesicle surface (Bohan *et al.*, 1973; Pappas and Rose, 1976; Politoff *et al.*, 1974). In several studies, synaptic vesicle fractions from vertebrate brains have been assayed for ATP-dependent Ca^{2+} uptake activity, but relatively little activity has been observed (e.g., DeRobertis *et al.*, 1967; Diamond and Goldberg, 1971; Blaustein *et al.*, 1978b). However, in two recent studies, Mg-ATP-dependent Ca uptake activity was observed in synaptic vesicles isolated from the *Torpedo* electric organ (Michaelson *et al.*, 1980 Israel *et al.*, 1980). This activity is sensitive to A23187 and to osmotic lysis, but it is not enhanced by the precipitating anion, oxalate. The distribution of the vesicular markers, and of membrane-bounded acetylcholine and ATP is identical to the distribution of Ca^{2+} uptake activity; this suggests that the Ca^{2+} uptake activity is not due to microsomal contaminants. These cholinergic synaptic vesicles also have a Ca^{2+}-dependent ATPase. However, the affinity of the Ca^{2+} transport system for Ca^{2+} is low ($K_{Ca} \geq 50 \ \mu M$), and Ca transport itself is apparently not tightly coupled to ATP hydrolysis. Thus, the role that these synaptic vesicles play in buffering transient increases in Ca^{2+} is likely to be very limited.

Fractions isolated from calf brain, consisting of a mixture of coated and smooth vesicles, were shown to contain a Mg^{2+}-dependent Ca^{2+}-ATPase activity that was half-saturated at 0.8 μM Ca^{2+} (Blitz et al., 1977). This activity, however, was only apparent when the coat material was removed by urea treatment and the resulting aggregate was dissolved in Triton X-100. In the presence of Mg^{2+}, ATP, and 20 μM or 200 μM Ca^{2+}, the coated vesicle fraction displayed a Ca^{2+} uptake that was enhanced by the addition of potassium oxalate. The role these vesicles may play in buffering $[Ca^{2+}]_{in}$ is difficult to assess, because information on the kinetics of the Ca^{2+} uptake process under physiological conditions is not available.

It has recently been established that a microvesicle-containing fraction from neurohypophyseal nerve terminals sequesters Ca^{2+} by an ATP-dependent mechanism (Nordmann and Chevalier, 1980). This uptake system has a high affinity for Ca^{2+}, with an apparent K_{Ca} (0.4–0.6 μM) comparable to that of the SER of synaptosomes. Morphological studies have provided evidence that Ca^{2+} is present in microvesicles of neurosecretory terminals (Shaw and Morris, 1980). The microvesicles appear to move toward the "active zones" of the terminals during depolarization. This has led to the suggestion that the microvesicles play an important role in buffering the increase in $[Ca^{2+}]_{in}$ that occurs during activity—by accumulating Ca^{2+} and perhaps by then releasing this Ca during the process of exocytosis. However, two considerations appear to render this hypothesis unlikely. One is that the Ca^{2+} accumulated by nerve terminals during periods of activity can all be extruded during a subsequent rest period when exocytosis is curtailed (cf. Blaustein and Ector, 1976). Second, vesicle recycling may actually be associated with Ca entry: The vesicle membrane pinched off from the plasmalemma is likely to engulf a small volume of extracellular fluid that will then be carried into the terminal (cf. Nordmann et al., 1974).

D. The Morphology of Ca^{2+}-Storing Organelles

As outlined in the previous sections, the Ca^{2+}-sequestering activities of intraterminal nonmitochondrial organelles have been well characterized by biochemical and physiological techniques. How do such findings correlate with ultrastructural studies on these organelles?

The identification of Ca sequestration or Ca^{2+}-binding sites is a difficult task: Calcium is a diffusible ion that can be lost from, or translocated within, the tissue during conventional fixation processes for electron microscopy. A variety of cytochemical methods have been used to localize Ca^{2+}. The methods involve the addition to the fixation medium of either (1) Ca^{2+} or other electron-dense cations that bind to Ca^{2+}-binding sites, or

(2) a Ca^{2+}-precipitating agent such as potassium oxalate, potassium pyroantimonate, or naphthylhydroxymethylamine. The latter agents form electron-dense deposits that can be visualized by electron microscopy at Ca^{2+} sequestration sites. The major drawback of these methods is that the fixation time is slow in comparison to the rate of ion movement in the tissue. This can cause serious errors in the identification of Ca^{2+}-storing organelles. Other problems are (1) nonspecific binding to membranes, (2) precipitation of ions other than Ca^{2+} (Simson and Spicer, 1975), and (3) artifactual precipitation caused by the fixation method itself (cf. Gray and Paula-Barbosa, 1974). These problems can be overcome, however. Calcium translocation and loss can be minimized by the use of quick-frozen samples of fresh tissue. Subsequent freeze-substitution of acetone for water causes the ions to precipitate *in situ* (Henkart *et al.*, 1978; Ornberg and Reese, 1980). After the tissue is embedded in plastic, thin sections can be analyzed by electron probe to determine the subcellular distribution of Ca^{2+}. Correlative studies, using electron probe analysis to determine the elemental content of precipitates, can confirm the presence of Ca^{2+}, thereby eliminating some false interpretations.

Some organelles that sequester Ca^{2+} are permeable to anions such as oxalate; this anion can then be used to precipitate sequestered Ca^{2+} and hold it *in situ* (cf. Makinose and Hasselbach, 1965). In fact, if the stored Ca^{2+} is precipitated, the free Ca^{2+} concentration within the organelles will be reduced and more Ca^{2+} can be transported in and trapped. This effect is observed in synaptosomes with disrupted plasma membranes (Blaustein *et al.*, 1978a), and we have exploited it to localize the Ca^{2+}-sequestering sites in the terminals (McGraw *et al.*, 1980b). Saponin was used to make the plasma membranes leaky, because it causes only limited distortion of the synaptosome structure and spares the cholesterol-poor membranes of intraterminal SER and mitochondria.

After treatment with saponin, synaptosomes were incubated in the presence of Ca^{2+}, ATP, and potassium oxalate. Following incubation, the tissue samples were fixed and prepared for electron microscopy and electron probe analysis by conventional fixation and embedding procedures (McGraw *et al.*, 1980b). Ultrastructural analysis of these synaptosomes revealed the presence of large, dense deposits within mitochondrial and large vesicular and cisternal profiles of SER (Fig. 1a–c). Electron probe analysis of the calcium oxalate deposits provided positive identification of the Ca^{2+} (Fig. 1d). When mitochondrial poisons were added to the incubation media, mitochondrial deposits were rarely observed, even though Ca^{2+} deposits were present in the SER. In synaptosomes incubated with A23187, no deposits were found in any intraterminal organelles. These findings correlate well with our biochemical observations on the nonmitochondrial Ca^{2+} uptake system (see above). The evidence suggests that SER is, indeed, the nonmitochondrial Ca^{2+}-sequestering organelle we have characterized physiologically. However, it must be recognized that all structures identified as SER by morphological criteria may not be a functionally homogeneous group of structures.

Serial sections of intraterminal SER demonstrate that these organelles are large, flattened sacs or cisterns; they often have thin, tubular branches that may resemble synaptic vesicles in cross section (McGraw *et al.*, 1980b). Profiles of SER are present throughout the terminals.and are commonly observed adjacent to mitochondria. This juxtaposition of SER to the mitochondria may have important physiological consequences: since SER has a higher affinity for Ca^{2+}, it may rapidly buffer the Ca^{2+} in the immediate vicinity of the mitochondria. This would lower the free Ca^{2+} so that the mitochondria would use the energy derived from electron transport not to accumulate Ca but rather to phosphorylate ADP. Thus SER may protect the energy-producing machinery of the nerve terminal.

Buffering of $[Ca^{2+}]_{in}$ by Intraterminal Organelles

We have presented evidence that Ca^{2+} is stored both in mitochondria and in nonmitochondrial organelles (including SER) of presynaptic nerve terminals. To what extent do these organelles actually buffer $[Ca^{2+}]_{in}$? In other words, how much does $[Ca^{2+}]_{in}$ actually change in response to a given load of Ca^{2+}? This has been determined in squid giant axons with antipyrylazo III (a Ca^{2+}-sensitive metallochrome) and aequorin (a Ca^{2+}-sensitive photoprotein): $[Ca^{2+}]_{in}$ increases by only about $0.2–0.6$ nM per 1000 nM Ca^{2+} load (Brinley, 1978 and 1980). Brinley's data also indicate that mitochondria are *not* the organelles responsible for most of this buffering, because Ca^{2+} buffering is unimpaired when the axons are treated with the mitochondrial uncoupler FCCP.

Two types of studies have been carried out on synaptosomes to obtain information about Ca^{2+} distribution and Ca^{2+} buffering in mammalian central neurons. When intact synaptosomes are given a small load of Ca^{2+} (by depolarizing in solutions with a low $[Ca^{2+}]$), most of the accumulated Ca^{2+} is sequestered in a pool that is releaseable by the Ca^{2+} ionophore A23187 but not by FCCP (Blaustein *et al.*, 1980b). This is an indication that mitochondria do not serve as the primary intracellular Ca^{2+} buffering sites. With much larger (perhaps pathological) Ca^{2+} loads, however, intraterminal mitochondria sequester substantial amounts of Ca^{2+} (Blaustein *et al.*, 1978a,c).

In another type of experiment, Ca^{2+} buffering by intraterminal organelles was determined with arsenazo III, in osmotically shocked synaptosomes (Schweitzer and Blaustein, 1980). In these experiments, too, small

Ca^{2+} loads were well-buffered, even in the presence of mitochondrial poisons, to levels (200–500 nM) below that at which the mitochondria appear to come into play (~500–1000 nM or higher). Nevertheless, the mitochondria have a much larger capacity to sequester Ca^{2+} (~30 μmoles Ca^{2+} g^{-1} synaptosome protein) than the other intraterminal organelles (probably SER, ~3–5 μmoles Ca^{2+} g^{-1} synaptosome protein).

Our main conclusion is that the short-term regulation of $[Ca^{2+}]_{in}$ in resting neurons (including mammalian presynaptic nerve terminals) is mainly under the control of Ca^{2+} pumps in the SER: The Ca^{2+} that enters the neurons during activity is rapidly buffered by cytoplasmic proteins and then sequestered by SER until it can be extruded from the cells (see Section IV). Both the SER Ca^{2+} pumps and those in the plasmalemma (see Section IV) help to maintain $[Ca^{2+}]_{in}$ at about 100–300 nM at rest. Only with very large (perhaps pathological) Ca^{2+} loads do the mitochondria participate in Ca^{2+} buffering.

III. Ca^{2+} ENTRY MECHANISMS

A. Ca^{2+}-Dependent Transmitter Release and Ca^{2+} Channels in Squid and Frog Nerve Endings

The role of Ca^{2+} in transmitter release has been most extensively studied in two preparations: the squid giant synapse and the frog neuromuscular junction. In addition to Ca^{2+} dependence, transmitter release from these two types of nerve endings shares many common features:

1. Release is voltage-regulated and can proceed in the absence of Na^+-dependent action potentials if the terminals are directly depolarized with a current-passing electrode (Katz and Miledi, 1967a,c,d; Llinas *et al.*, 1976).

2. The release evoked in this manner is unaffected by tetrodotoxin (TTX), a potent blocker of Na^+ channels, and is not abolished by agents that block K^+ channels, such as tetramethylammonium ions (TEA) (Katz and Miledi, 1967b,c,d).

3. The Ca^{2+}-dependent release is, however, greatly diminished by a variety of polyvalent cations, including Mg^{2+} (Del Castillo and Engbaek, 1954; Takeuchi and Takeuchi, 1962), Mn^{2+} and Co^{2+} (Katz and Miledi, 1969a; Meiri and Rahaminoff, 1972), and La^{3+} (Miledi, 1971; Heuser and Miledi, 1971). These metal ions diminish release at concentrations that have little effect on the Na^+-dependent nerve action potential.

4. The heavy alkaline earth cations, Sr^{2+} and, to some extent, Ba^{2+}, can substitute for Ca^{2+} in sustaining release (Miledi, 1966; Katz and Miledi, 1969a; Silinsky, 1978).

An impressive array of observations, including the aforementioned, have led to widespread acceptance of the "calcium hypothesis" of excitation–secretion coupling at presynaptic nerve terminals (cf. Katz and Miledi, 1967d). According to this hypothesis, the Ca^{2+} conductance of the nerve terminal increases as a consequence of depolarization (normally the result of a nerve action potential). Calcium then enters the terminal, moving down the steep electrochemical gradient for Ca^{2+} (see preceding section). The consequent rise in $[Ca^{2+}]_{in}$ triggers exocytosis by an as yet unknown mechanism. An important question that follows is: to what extent do these characteristics of Ca^{2+}-dependent release reflect the properties of the membrane's Ca^{2+} conductance?

The squid giant synapse has proven particularly useful in providing an answer: its large size allows microelectrodes to be inserted into the presynaptic fibers for electrical recording and for injecting Ca^{2+}-sensitive probes such as aequorin. Experimental results show that many release properties are, in fact, determined by Ca^{2+} entry via the Ca^{2+} channels of the presynaptic terminal membrane. These Ca^{2+} channels, like the Ca^{2+} channels in many other preparations—both vertebrate and invertebrate (cf. Reuter, 1973; Hagiwara, 1975)—are voltage-regulated, insensitive to TTX and TEA, permeable to Sr^{2+} and Ba^{2+}, and blocked by Mg^{2+}, transition metals, and lanthanides (Katz and Miledi, 1967d, 1969a; Llinas *et al.*, 1976).

B. Heterogeneity of Ca^{2+} Channels

Despite the substantial similarities among Ca^{2+} channels in various types of cells, there may also be significant differences. Some of these differences are perhaps artifactual because of the difficulties involved in measuring the electrical current carried by Ca^{2+}. The Ca^{2+} current is often relatively small in magnitude by comparison with the Na and K currents in the same cell. The K^+ currents are particularly troublesome because the time course of K^+ channel activation may be slow, and comparable to that of the Ca^{2+} channel (Katz and Miledi, 1967d). Moreover, some K^+ channels are Ca^{2+}-activated (Meech, 1972, 1978): thus, properties of the Ca^{2+} channels may, in some cases, be masked by those of the K^+ channels (see Adams and Gage, 1979, 1980, for discussions of this problem). Nonetheless, even when these problems are taken into account, differences among various divalent cation channels are apparent; for example:

1. Most Ca^{2+} channels are permeable to Sr^{2+} and Ba^{2+}, in addition to Ca^{2+} (see Reuter, 1973). However, the relative magnitudes of the maximal inward currents (I_{max}) carried by each of the alkaline earth cations, at high concentrations, varies in different tissue preparations. In barnacle muscle, Hagiwara (1975) reported I_{max} (Sr^{2+}) > I_{max} (Ba^{2+}) > I_{max} (Ca^{2+}), whereas in molluscan neurons, Akaike *et al.* (1978b) observed that I_{max} (Ba^{2+}) = I_{max} (Sr^{2+}) > I_{max} (Ca^{2+}). It is unclear to what extent this difference reflects tissue and species variation as opposed to differences in experimental method, ionic conditions, etc.

2. The Ca^{2+}–Na^+ selectivity ratio differs in various preparations. In mealworm larvae, for example, Co^{2+}-sensitive action potentials can be elicited in the presence of TTX, either in Ca^{2+}-free media containing Na^+ or in Na^+-free media containing Ca^{2+} (Yamamoto and Washio, 1979). Since Co^{2+} is believed to be a specific Ca^{2+} channel blocker (e.g., Hagiwara, 1975), the implication of this finding is that both Na^+ and Ca^{2+} pass through a common Ca^{2+} channel. Other Ca^{2+} channels appear far more selective and cannot support regenerative spike activity in Ca^{2+}-free Na^+-containing media (eg., Llinas *et al.*, 1976).

3. Mn^{2+} passes through some Ca^{2+} channels (Hagiwara and Miyazaki, 1977; Fukuda and Kawa, 1977; Anderson, 1979) but blocks other Ca^{2+} channels (Katz and Miledi, 1968; Hagiwara and Takahashi, 1967).

4. In the presence of a maintained depolarizing potential, some Ca^{2+} channels turn off, or "inactivate," while others do not (cf. Llinas *et al.*, 1976; Akaike *et al.*, 1978b). Among the channels that inactivate, the reported time constants of inactivation range, in various preparations, from several milliseconds to tens of milliseconds (cf. Akaike *et al.*, 1978b; Adams and Gage, 1979). There is evidence that cardiac Purkinje fibers have two types of Ca^{2+} channels with different inactivation characteristics and with different susceptibilities to blockage by Mn^{2+} (Seigelbaum, 1978). Some central neurons (in the inferior olive: Yarom and Llinas, 1979; and in the cerebellum: Llinas and Sugimori, 1980) also have two Ca^{2+} conductances, with different inactivation characteristics.

5. Calcium channels from different preparations have different sensitivities to the organic Ca^{2+} antagonists D-600 and verapamil. These drugs reduce Ca^{2+} conductance in mammalian cardiac and smooth muscle at concentrations of $10^{-6} M$ or less (reviewed by Fleckenstein, 1977). This contrasts with the much higher concentrations (10^{-6}–$10^{-5} M$) needed to block neuronal Ca^{2+} channels (Hauesler, 1972; Van der Kloot and Kita, 1975; Gotgilf and Magazanik, 1977; Nachshen and Blaustein, 1979).

6. Although many polyvalent cations effectively block Ca^{2+} channels, the relative blocking efficiency differs in various cells. For example, in *Helix* neurons, the order of inhibitory efficacy is Ni^{2+} > La^{3+} > Co^{2+}

(Akaike *et al.*, 1978a), whereas in *Balanus* (barnacle) muscle, the sequence is $La^{3+} > Co^{2+} > Ni^{2+}$ (Hagiwara, 1975).

Because of the apparent heterogeneity of Ca^{2+} channels, it is important to consider Ca^{2+} channel properties in vertebrate, as well as in invertebrate presynaptic nerve terminal preparations. The vertebrate neuromuscular junction, although an excellent system for studying Ca^{2+}-dependent transmitter release, is of limited use in unraveling the specific properties of Ca^{2+} conductance. The small size of the terminals makes them inaccessible to presynaptic probes for Ca. Also, the relationships between $[Ca^{2+}]_{out}$ and $[Ca^{2+}]_{in}$ and between $[Ca^{2+}]_{in}$ and transmitter release are unknown. The many parallels between Ca^{2+}-dependent release at the neuromuscular junction and at the squid synapse certainly suggest that similar Ca^{2+} channels are involved, but this cannot be tested directly.

Even less is known about Ca^{2+} channels in presynaptic terminals of the vertebrate central nervous system (CNS). As with the neuromuscular junction, the similarities between Ca^{2+}-dependent release at central synapses and at the squid giant synapse implicate voltage-regulated Ca^{2+} channels, but the nerve terminal Ca^{2+} conductance has not been studied electrophysiologically.

C. "Fast" and "Slow" Ca^{2+} Channels in Synaptosomes

An alternative approach, used to study Ca^{2+} channels in CNS presynaptic nerve terminals, involves direct measurement of ^{45}Ca entry into synaptosomes. Synaptosomes release transmitter in a Ca^{2+}-dependent fashion (Blaustein, 1975; Drapeau and Blaustein, 1981) and regulate Ca^{2+} entry by voltage-sensitive mechanisms (Blaustein, 1975; Nachshen and Blaustein, 1980).

Many early studies on Ca^{2+} channels in synaptosomes are difficult to interpret because precautions were not taken to ensure membrane integrity. Synaptosomes were often prepared in a hyperosmotic sucose gradient and then resuspended in solutions of normal or low ionic strength and osmolarity; this may have subjected them to hypoosomotic shock and rupture of the plasma membranes. Another problem has been that measurements of undirectional Ca^{2+} influx are rarely made. Researchers most often measure net entry, but Ca^{2+} entering the synaptosomes is avidly sequestered (see above). Thus, the measurements may reflect the properties of the intraterminal organelles that accumulate Ca^{2+}, instead of the properties of the presynaptic membrane. If, however, proper handling precautions are taken, and if flux measurements are made at short times

(Nachshen and Blaustein, 1980), tracer studies on synaptosomes can yield useful information about Ca^{2+} entry mechanisms.

Tracer studies are particularly helpful in characterizing the selectivity properties of Ca^{2+} channels, because movement of Ca^{2+} is measured directly. As already mentioned, Ca^{2+} currents are often masked by K^+ currents that activate slowly and that must be eliminated. This is usually accomplished by pharmacological agents (e.g., TEA) or by replacement of internal K^+ with impermeant cations; but the possible effects of these manipulations on the Ca^{2+} channels must then be considered. Furthermore, tracer studies permit simultaneous measurement of several ion fluxes when more than one ion species passes through the channel (e.g., Sr and Ba).

In order to study the properties of Ca^{2+} channels in vertebrate CNS presynaptic terminals, we have measured ^{45}Ca influx in synaptosomes. When synaptosomes are incubated in a standard physiological salt solution, the "resting" Ca^{2+} influx is about $0.1-0.2$ nmoles sec^{-1} mg^{-1} protein (see legend to Fig. 2). Influx is increased when the synaptosomes are incubated in K^+-rich depolarizing solutions (Blaustein and Goldring, 1975), and the influx is graded with increasing levels of external K^+. For example, Ca^{2+} influx increases about 2- to 3-fold when one-quarter of the external NaCl (normally 145 mM) is replaced by KCl, and $5-10$-fold when one-half the NaCl is replaced by KCl.

The time course of the K^+-stimulated Ca^{2+} influx is biphasic (Fig. 2). There is a brief period of rapid Ca^{2+} accumulation, the fast phase, which is complete within $1-3$ sec. Subsequently, influx proceeds at a reduced rate, the slow phase, lasting for $10-20$ sec. This slow phase is obscured as a backflux of ^{45}Ca from the synaptosomes becomes significant (after 10 sec).

The two phases of K^+-stimulated Ca^{2+} influx, fast and slow, are probably mediated by separate channels, since their properties are different:

1. The fast phase is blocked by low concentrations of lanthanides ($K_I <$ 0.5 μM) and Mn^{2+} ($K_I \simeq 70$ μM). In contrast, blockage of the slow phase requires substantially higher concentrations of La^{3+} ($K_I > 50$ μM) and Mn^{2+} ($K_I \simeq 300$ μM).

2. The fast phase is abolished if the synaptosomes are preincubated for several seconds in depolarizing solutions before the influx is measured. However, predepolarization has no effect on the slow phase of Ca^{2+} influx.

3. The selectivity properties of the two phases for the divalent cations Ca^{2+}, Sr^{2+}, and Ba^{2+}, are different (Nachshen, 1981). Permeability ratios were determined for these ions by measuring the K^+-stimulated tracer

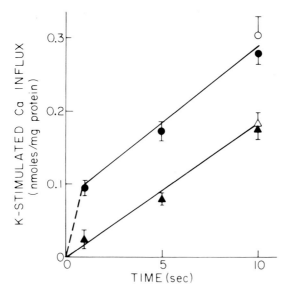

Fig. 2. The time course of K^+-stimulated Ca^{2+} uptake in synaptosomes. ^{45}Ca uptake was measured in low-K and in K-rich solutions; the difference is referred to as K-stimulated Ca^{2+} uptake. Prior to incubation with ^{45}Ca, the synaptosomes were preincubated for 15 sec in low-K^+ (●) or high-K^+ (▲) solution. Low-K^+ solutions contained the following millimolar concentrations: NaCl, 145; KCl, 5; $CaCl_2$, 0.02; $MgCl_2$, 0.5; glucose, 10; Tris-HEPES (buffered to pH 7.4 at 30°C), 10. The high-K^+ solutions were of similar composition, except that half of the NaCl was replaced by KCl, giving the following millimolar concentrations: NaCl, 72.5; KCl, 77.5. Open symbols indicate that the ^{45}Ca-containing media also included 10 μM TTX.

influxes of the three ionic species. During the fast phase, the relative $P_{Ca}/P_{Sr}/P_{Ba}$ permeabilities were 6 : 3 : 1; during the slow phase, the $P_{Ca}/P_{Sr}/P_{Ba}$ ratios were 3 : 1.5 : 1.

Although these data show that the fast and slow channels are distinct, they also have several similarities:

1. K-stimulated akaline earth cation influxes during both fast and slow phases saturate with increasing concentrations of Ca^{2+}, Sr^{2+}, and Ba^{2+} (Nachshen, 1981).

2. Both channels are blocked by H^+, Mg^{2+}, and Ni^{2+} with approximately equal K_I values for the two channels (Nachshen and Blaustein, 1980).

3. Both phases are highly selective for Ca^{2+} over Na^+ (Krueger and Nachshen, 1979) and are not affected by TTX (Nachshen and Blaustein, 1980; Fig. 2).

4. Both phases are blocked by the organic Ca^{2+} antagonists verapamil and D-600, but only at relatively high concentrations (50–100 μM; Nachshen and Blaustein, 1979).

Three observations indicate that the fast Ca^{2+} channels are voltage-regulated. First, as mentioned above, the Ca^{2+} entry is promoted by K^+-rich (depolarizing) solutions and is graded with the external K concentration. Second, the fast Ca^{2+} channels inactivate when the synaptosomes are incubated in depolarizing solutions. Finally, when the Mg^{2+} concentration is increased (>5 mM), the relationship between the external K^+ concentration and Ca^{2+} influx is shifted in the hyperpolarizing direction; i.e., a higher K^+ concentration (greater depolarization) is required to open a given fraction of the Ca^{2+} channels (Nachshen and Blaustein, 1981). This is evidence that surface charges (cf. Frankenhaeuser and Hodgkin, 1957; Muller and Finkelstein, 1974) influence a voltage-regulating mechanism that opens the Ca^{2+} channels.

The slow phase of Ca^{2+} influx does not inactivate. However, the slow channels also appear to be voltage-regulated because they are activated by a variety of depolarizing agents such as veratridine and gramicidin D, as well as by elevated external K (Nachshen and Blaustein, 1980). The Na^+–Ca^{2+} exchange mechanism that has been described in synaptosomes may be voltage-sensitive (see below), and depolarization would be expected to enhance Ca^{2+} entry by this mechanism. However, available evidence (Nachshen and Blaustein, 1980; also see below) indicates that the slow phase of Ca^{2+} influx, described above, does not involve the Na^+–Ca^{2+} exchange mechanism.

The physiological significance of the fast and slow Ca^{2+} channels has not yet been established. It is possible that the fast entry mechanism modulates primarily the phasic release of transmitter following an action potential. The slow entry mechanism may then modulate transmitter release following a more long-lasting depolarization of presynaptic nerve endings (for example, as a consequence of repetitive stimulation or of presynaptic depolarization at axo-axonic synapses). However, there is preliminary evidence (Drapeau and Blaustein, 1981) for fast and slow phases of Ca^{2+}-dependent transmitter release from synaptosomes.

Because of the heterogeneous nature of the synaptosome preparations we use, it is not known whether fast and slow channels coexist in the same terminals. Preliminary data indicate that both classes of channels are present in constant proportions in synaptosome preparations obtained from several specific brain regions.

D. Na^+-Ca^{2+} Exchange

One particular type of Ca^{2+} channel, the Na^+-Ca^{2+} carrier, has been studied in a large variety of tissues (for reviews, see Blaustein *et al.*, 1974; Requena and Mullins, 1979; Sulakhe and St. Louis, 1980; Blaustein and Nelson, 1981), including synaptosomes (Blaustein and Oborn, 1975; Blaustein and Ector, 1976). This carrier exchanges one Ca^{2+} for each three or more Na^+ countertransported (cf. Blaustein and Nelson 1981). Since net charge is moved by the carrier, the transport can be voltage-regulated (e.g., Blaustein *et al.*, 1974; Mullins and Brinley, 1975).

Energy to drive the exchange is provided by the Na^+ electrochemical gradient across the plasmalemma. The exchange appears to be symmetric in that Ca^{2+} can be moved in either direction across the plasma membrane, depending upon the prevailing Na^+ gradient. The Na^+ gradient in this way maintains the large electrochemical gradient for Ca^{2+} across the plasmalemma that has been observed in many types of cells at rest. But when the Na^+ gradient is reduced (e.g., by inhibiting Na-K pump activity and increasing the intracellular Na^+ concentration or by depolarizing the cells), net Ca^{2+} influx may ensue.

Internal Na^+-dependent Ca^{2+} influx and external Na^+-dependent Ca^{2+} efflux have been studied in synaptosomes (e.g., Blaustein and Oborn, 1975; Blaustein and Ector, 1976). In brief, Ca^{2+} influx is increased when external Na^+ is reduced and/or internal Na^+ is increased. Conversely, Ca^{2+} efflux from ^{45}Ca-loaded synaptosomes is reduced when external Na^+ is lowered.

The stoichiometry of the internal Na^+-dependent Ca^{2+} influx has not been precisely determined in synaptosomes, but the available data are consistent with at least a $2Na^+$-$1Ca^{2+}$ exchange (cf. Blaustein and Ector, 1976). If, as in other tissues, the carrier is voltage-regulated, this would have important consequences for nerve terminal function: during the nerve action potential, Ca^{2+} would enter the terminal via the Na^+-Ca^{2+} carriers, as well as via the voltage-regulated fast and slow Ca^{2+} channels. Also, Na^+ accumulation in nerve endings, after repetitive action potential activity, might enhance Ca^{2+} entry and increase the resting $[Ca^{2+}]_{in}$. Given a Na^+ channel site density at the synaptosomes of 20 μm^{-2} (Krueger *et al.*, 1979) and a single Na^+ channel conductance of 8 pS, Na^+ in a spherical terminal of 0.75 μm diameter could increase by 0.5–1 mM following a single action potential. Tetanic stimulation could raise $[Na^+]_{in}$ to levels that would significantly accelerate Ca^{2+} influx: In synaptosomes ^{45}Ca uptake increases by ≈ 0.3 nmole mg^{-1} protein as $[Na^+]_{in}$ is raised from 30 to 70 mM (Blaustein and Oborn, 1975).

Several lines of evidence indicate that the mechanism for Na^+-dependent Ca^{2+} influx in synaptosomes is distinct from the mechanisms for the fast and slow phases of K^+-stimulated Ca^{2+} influx:

1. The K_{Ca} value (i.e., apparent half-saturation for external Ca) for the Na^+–Ca^{2+} carrier, in solutions containing 72.5 mM NaCl, is ≈ 1.5 mM; K_{Ca} values for the fast and slow entry mechanisms are ≈ 0.2–0.4 mM.

2. The external Na^+-dependent Ca^{2+} efflux (the reverse mode of the Na^+–Ca^{2+} exchanger; Blaustein and Ector, 1976) is only slightly blocked by Mn^{2+} (2–5 mM); under similar conditions K^+-stimulated Ca^{2+} influx is reduced by at least 90%.

3. The internal Na^+-dependent Ca^{2+} influx is significantly stimulated when the external Na^+ concentration is halved; this manipulation does not cause enhanced uptake of Sr^{2+} or Ba^{2+}. The K^+-stimulated Ca^{2+} channel pathways can, however, mediate the influx of Sr^{2+} and Ba^{2+}, as well as Ca^{2+} (see Section III,C).

IV. Ca^{2+} EFFLUX FROM NERVE TERMINALS

Na^+–Ca^{2+} Exchange and Other Extrusion Mechanisms

In order for Ca^{2+} to act as a second messenger in the nerve terminal, the normal (resting) $[Ca^{2+}]_{in}$ must be maintained at very low levels (on the order of 10^{-7} M; see below). Also, as $[Ca^{2+}]_{in}$ rises, mitochondria may be expected to cease phosphorylation and begin to accumulate Ca^{2+} (Rossi and Lehninger, 1964; Drahota et al., 1965). Finally, although intracellular organelles are extremely effective in buffering internal Ca^{2+} (Section II), excess Ca^{2+} must eventually be extruded across the plasma membranes. Thus, the importance of efficient Ca^{2+} extrusion mechanisms becomes evident.

One primary mechanism for Ca extrusion is the Na^+–Ca^{2+} exchanger described in the preceding section. However, even though the Na electrochemical gradient may provide the energy required for Ca^{2+} extrusion via this mechanism, there is still some question about the possible direct role of ATP in Ca^{2+} extrusion. Two roles have been suggested: one is an effect on Na^+–Ca^{2+} exchange, and the second involves an independent Ca^{2+}–ATPase transport system such as the one observed in red blood cells (cf. Schatzmann and Burgin, 1978) and perhaps in squid axons (DiPolo and Beauge, 1979; but see Nelson and Blaustein, 1981).

In squid axons (e.g., DiPolo, 1974; Blaustein, 1977) and barnacle muscle fibers, ATP appears to alter the kinetic properties of the Na^+–Ca^{2+}

exchange system. But there is as yet no information about whether ATP directly influences the properties of the Na^+–Ca^{2+} exchange system in mammalian nerve terminals.

Some workers have favored the view that a second, independent Ca^{2+} transport system, involving a Ca^{2+}-dependent ATPase, may also play a role in Ca^{2+} extrusion. It has recently been observed that membrane vesicles, derived from rat brain synaptosomes, can exchange Na for Ca in the presence of a Na^+ gradient (Rahaminoff and Spanier, 1979; Gill *et al.,* 1981). Synaptic membrane vesicles can also exchange Ca^{2+} for H^+ and exhibit ATP-dependent Ca^{2+} uptake. The interpretation of these observations is complicated by the fact that some intraterminal organelles sequester Ca^{2+} at the expense of ATP hydrolysis (Section II). Thus, further work is required to show conclusively that the synaptic membrane vesicles are, indeed, derived from the nerve terminal plasma membrane rather than from Ca^{2+}-transporting organelles.

V. Ca^{2+} MOVEMENTS AND TRANSMITTER RELEASE

Voltage-dependent (K^+-stimulated) Ca^{2+} influx in synaptosomes during the initial 1 sec of uptake amounts to about 2 nmoles mg^{-1} protein when the extracellular solution contains 1–2 mM Ca^{2+} and 77 mM K^+ (Nachshen and Blaustein, 1980). If all the Ca^{2+} enters as a net flow of current, the Ca^{2+} current density corresponds to 1.6 μA cm^{-2}. This determination is based on three assumptions (cf. Blaustein, 1975): (1) that the terminals are spheres with diameters of about 0.75 μm, (2) that 80% of the occluded water space (4.5 μl mg^{-1} total protein) represents the internal volume of osmotically intact synaptosomes, and (3) that the Ca^{2+} channel distribution is homogeneous in the plasma membranes of all the terminals.

In fact, only about one-third of the osmotically intact synaptosomes may be functional (Fried and Blaustein, 1978). Moreover, the initial rate of Ca^{2+} influx is probably underestimated, because most of the Ca^{2+} channels are inactivated by the end of the 1-sec incubation. Thus, actual Ca^{2+} current density may be as much as 10 times greater than the calculated value (or, on the order of 16 μA/cm^2). This compares favorably with a Ca^{2+} current density of 35 μA/cm^2 at the squid giant synapse (Llinas, 1977).

Using the aforementioned assumptions, it can be shown that the Ca^{2+} influx that occurs during a 1-msec depolarizing pulse (about the duration of an action potential) may be sufficient to increase $[Ca^{2+}]_{in}$ from the resting level of about 10^{-7} M to about 10^{-6} M. The relationship between $[Ca^{2+}]_{in}$ and transmitter release has not yet been determined in nerve terminals; but, interestingly enough, 10^{-6} M is the level of intracellular Ca^{2+} that gives half-maximal release of catecholamine from leaky bovine adrenal medullary cells (Baker and Knight, 1978; Baker et al., 1980).

As emphasized above, Ca^{2+} plays a crucial role in many secretory mechanisms. Furthermore, there appear to be important parallels between depolarization–neurotransmitter release coupling at nerve terminals and depolarization–secretion coupling in many other types of secretory cells (e.g., Douglas, 1968; Moriarty, 1978; Putney et al., 1977; Rubin, 1970; Schulz and Stolze, 1980; Williams, 1980). However, special features of nerve terminals such as a large surface/volume ratio and a large Ca^{2+} buffering capacity, are likely to have important implications for the dynamics of the Ca^{2+} entry–transmitter release process. For instance, the data of Llinas et al. (1976) show a linear relationship between quantal transmitter release and the inward Ca^{2+} current (I_{Ca}) in the squid giant synapse under voltage clamp. These data seem to rule out "cooperative" models of transmitter release in which n Ca^{2+} ions (where $n > 1$) must bind to an internal release site to trigger the release of a single quantum of transmitter (Dodge and Rahamimoff, 1967; Hubbard et al., 1968a). There is also no evidence of cooperativity between Ca^{2+} ions in voltage-sensitive Ca^{2+} channels (e.g., Hagiwara, 1975; Akaike et al., 1978b; Nachshen and Blaustein, 1980). Thus, the reality of a cooperative binding site that regulates Ca-dependent transmitter release is questionable. Nevertheless, apparent cooperativity has been observed in numerous studies on the relationship between $[Ca^{2+}]_{out}$ and transmitter release at a variety of synapses (e.g., Dodge and Rahamimoff, 1967; Hubbard et al., 1968a,b; Katz and Miledi, 1970). One possibility is that the relationship between Ca^{2+} entry and $[Ca^{2+}]_{in}$, as a result of intraterminal Ca^{2+} buffering, gives rise to apparent cooperativity (Nachshen and Drapeau, 1981).

VI. SUMMARY

We cannot fully understand the dynamic aspects of calcium's role in chemical neurotransmission without determining how it enters nerve terminals, how it is handled in the cytoplasm, and how it is extruded across the plasma membrane. These aspects of Ca^{2+} metabolism have been reviewed in this chapter. It should be readily apparent that the dynamic features of Ca^{2+} metabolism (e.g., reduced Ca entry because of Ca^{2+} channel inactivation, and the kinetics of intraterminal Ca^{2+} buffering by smooth endoplasmic reticulum), may contribute to the synaptic transfer function. It is also conceivable that long-term modulation of synaptic transfer may be, in part, the result of changes in systems that regulate Ca^{2+} metabolism at presynaptic nerve terminals.

ACKNOWLEDGMENTS

We thank Ms. T. Patterson for preparing the typescript. This work was supported by NIH research grants NS-16106 (to MPB) and NS-16461 (to DAN). C.F.M. was supported by NIH training grant IT32-NS07071.

REFERENCES

Adams, D. J., and Gage, P. W. (1979). Characteristics of sodium and calcium conductance changes produced by membrane depolarization in an *Aplysia* neurone. *J. Physiol. (London)* **289**, 143–161.

Adams, D. J., and Gage, P. W. (1980). Divalent ion currents and the delayed potassium conductance in an *Aplysia* neurone. *J. Physiol. (London)* **304**, 297–313.

Akaike, N., Fishman, H. M., Lee, K. S., Moore, L. E., and Brown, A. M. (1978a). The units of calcium conductance in *Helix* neurones. *Nature (London)* **274**, 379–382.

Akaike, N., Lee, K. S., and Brown, A. M. (1978b). The calcium current of *Helix* neuron. *J. Gen. Physiol.* **71**, 509–531.

Alnaes, E., and Rahamimoff, R. (1974). Dual action of praeseodymium (Pr^{3+}) on transmitter release at the frog neuromuscular synapse. *Nature (London)* **247**, 478–479.

Alnaes, E., and Rahamimoff, R. (1975). On the role of mitochondria in transmitter release from motor nerve terminals. *J. Physiol. (London)* **248**, 285–306.

Alvarez-Leefmans, F. J., Rink, T. J., and Tsien, R. Y. (1981). Free Ca^{2+} in neurones of *Helix aspersa* measured with ion-selective microelectrodes. *J. Physiol. (London)* (in press).

Anderson M. (1979). Mn^{2+} ions pass through Ca^{2+} channels in myoepithelial cells. *J. Exp. Biol.* **82**, 227–238.

Baker, P. F., and Knight, D. E. (1978). Calcium-dependent exocytosis in bovine adrenal medullary cells with leaky plasma membranes. *Nature (London)* **276**, 620–622.

Baker, P. F., and Knight, D. E. (1980). Gaining access to the site of exocytosis in bovine adrenal medullary cells. *J. Physiol. (Paris)* **76**, 497–504.

Baker, P. F., and Schlaepfer, W. (1978). Uptake and binding of calcium by axoplasm isolated from giant axons of *Loligo* and *Myxicola*. *J. Physiol. (London)* **276**, 103–125.

Baker, P. F., Knight, D. E., and Whitaker, M. J. (1980). The relation between ionized calcium and cortical granule exocytosis in eggs of the sea urchin *Echinus esculentus*. *Proc. R. Soc. London, Ser. B* **207**, 149–161.

Becker, G. L., Fiskum, G., and Lehninger, A. L. (1980). Regulation of free Ca^{2+} by liver mitochondria and endoplasmic reticulum. *J. Biol. Chem.* **255**, 9009–9012.

Birks, R. J., and Cohen, M. W. (1968). The influence of internal sodium on the behavior of motor nerve endings. *Proc. R. Soc. London, Ser. B* **170**, 401–421.

Blaustein, M. P. (1974). The interrelationship between sodium and calcium fluxes across cell membranes. *Rev. Physiol., Biochem. Pharmacol.* **70**, 33–82.

Blaustein, M. P. (1975). Effects of potassium, veratridine and scorpion venom on calcium accumulation and transmitter release by nerve terminals *in vitro*. *J. Physiol. (London)* **247**, 617–655.

Blaustein, M. P. (1977). Effects of internal and external cations and of ATP on sodium-calcium exchange and calcium-calcium exchange in squid axons. *Biophys. J.* **20**, 79–111.

Blaustein, M. P., and Ector, A. C. (1976). Carrier-mediated sodium-dependent and calcium-dependent calcium efflux from pinched-off presynaptic nerve terminals (synaptosomes) *in vitro*. *Biochim. Biophys. Acta* **419**, 295–308.

Blaustein, M. P., and Goldring, J. M. (1975). Membrane potentials in pinched-off presynaptic nerve terminals monitored with a fluorescent probe: Evidence that synaptosomes have potassium diffusion potentials. *J. Physiol.* (*London*) **247**, 589–615.

Blaustein, M. P., and Nelson, M. T. (1981). Sodium-calcium exchange: Its role in the regulation of cell calcium. *In* "Calcium Transport across Biological Membranes (E. Carafoli, ed.). Academic Press, New York (in press).

Blaustein, M. P., and Oborn, C. J. (1975). The influence of sodium on calcium fluxes in pinched-off nerve terminals *in vitro*. *J. Physiol.* (*London*) **247**, 657–686

Blaustein, M. P., Russell, J. M., and DeWeer, P. (1974). Calcium efflux from internally dialyzed squid axons: The influence of external and internal cations. *J. Supramol. Struct.* **2**, 558–581.

Blaustein, M. P., Kendrick, N. C., Fried, R. C., and Ratzlaff, R. W. (1977). Calcium metabolism at the mammalian presynaptic nerve terminal: Lessons from the synaptosome. *Soc. Neurosci. Symp.* **2**, 172–194.

Blaustein, M. P., Ratzlaff, R. W., and Kendrick, N. C. (1978a). The regulation of intracellular calcium in presynaptic nerve terminals. *Ann. N.Y. Acad. Sci.* **307**, 195–212.

Blaustein, M. P., Ratzlaff, R. W., Kendrick, N. C., and Schweitzer, E. S. (1978b). Calcium buffering in presynaptic nerve terminals. I. Evidence for involvement of a nonmitochondrial ATP-dependent sequestration mechanism. *J. Gen. Physiol.* **72**, 15–41.

Blaustein, M. P., Ratzlaff, R. W., and Schweitzer, E. S. (1978c). Calcium buffering in presynaptic nerve terminals. II. Kinetic properties of the nonmitochondrial Ca sequestration mechanism. *J. Gen. Physiol.* **72**, 43–66.

Blaustein, M. P., McGraw, C. F., Somlyo, A. V., and Schweitzer, E. S. (1980a). How is the cytoplasmic calcium concentration controlled in nerve terminals? *J. Physiol.* (*Paris*) **76**, 459–470.

Blaustein, M. P., Ratzlaff, R. W., and Schweitzer, E. S. (1980b). Control of intracellular calcium in presynaptic nerve terminals. *Fed. Proc., Fed. Am. Soc. Exp. Biol.* **39**, 2790–2795.

Blitz, A. L., Fine, R. E., and Toselli, P. A. (1977). Evidence that coated vesicles from brain are calcium-sequestering organelles resembling sarcoplasmic reticulum. *J. Cell Biol.* **75**, 135–147.

Bohan, T. B., Boyne, A. F., Guth, P. S., Narayan, Y., and Williams, T. H. (1973). Electron-dense particles in cholinergic synaptic vesicles. *Nature* (*London*) **244**, 32–33.

Bradford, H. F. (1975). Isolated nerve terminals as an *in vitro* preparation for the study of dynamic aspects of transmitter metabolism and release. *In* "Handbook of Psychopharmacology" (L. L. Iverson, S. D. Iverson, and S. H. Snyder, eds.), Vol. 1, pp. 191–252. Plenum, New York.

Brinley, F. J., Jr. (1978). Calcium buffering in squid axons. *Annu. Rev. Biophys. Bioeng.* **7**, 363–392.

Brinley, F. J., Jr. (1980). Regulation of intracellular calcium in squid axons. *Fed. Proc., Fed. Am. Soc. Exp. Biol.* **39**, 2778–2782.

Brinley, F. J., Jr., Tiffert, T., and Scarpa, A. (1978). Mitochondria and other calcium buffers of squid axon studied *in situ*. *J. Gen. Physiol.* **72**, 101–127.

Bygrave, F. L. (1978). Mitochondria and the control of intracellular calcium. *Biol. Rev. Cambridge Philos. Soc.* **53**, 43–79.

Christoffersen, G. R. J., and Simonsen, L. (1977). Ca^{++} sensitive microelectrode: Intracellular steady state measurement in nerve cell. *Acta Physiol. Scand.* **101**, 492–494.

Dedman, J. R., Brinkley, B. R., and Means, A. R. (1979). Regulation of microfilaments and microtubules by calcium and cyclic AMP. *Adv. Cyclic Nucleotide Res.* **11**, 131–174.

Del Castillo, J., and Engbaek, L. (1954). The nature of neuromuscular block produced by magnesium. *J. Physiol. (London)* **124**, 370–384.

DeRobertis, E., Rodriquez de Lores Arnaiz, G., Alberici, M., Butcher, R. W., and Sutherland, E. W. (1967). Subcellular distribution of adenyl cyclase and cyclic phosphodiesterase in rat brain cortex. *J. Biol. Chem.* **242**, 3487–3493.

Diamond, I., and Goldberg, A. L. (1971). Uptake and release of ^{45}Ca by brain microsomes, synaptosomes and synaptic vesicles. *J. Neurochem.* **18**, 1419–1431.

DiPolo, R. (1974). Effect of ATP on the calcium efflux in dialyzed squid giant axons. *J. Gen. Physiol.* **54**, 503–517.

DiPolo, R., and Beauge, L. (1979). Physiological role of ATP-driven calcium pump in squid axons. *Nature (London)* **278**, 271–273.

DiPolo, R., Requena, J., Brinley, R. J., Jr., Mullins, L. J., Scarpa, A., and Tiffert, (1976). Ionized calcium concentrations in squid axons. *J. Gen. Physiol.* **67**, 433–467.

Dodge, F. A., and Rahamimoff, R. (1967). Co-operative action of calcium ions in transmitter release at the neuromuscular junction. *J. Physiol. (London)* **193**, 419–432.

Douglas, W. W. (1968). Stimulus-secretion coupling: The concept and clues from chromaffin and other cells. *Br. J. Pharmacol. Chemother.* **30**, 612–619.

Drahota, Z., Carafoli, E., Rossi, C. S., Gamble, J. L., and Lehninger, A. L. (1965). The steady state maintenance of accumulated Ca^{++} in rat liver mitochondria. *J. Biol. Chem.* **240**, 2712–2720.

Drapeau, P., and Blaustein, M. P. (1981). Biphasic dopamine release from striatal synaptosomes parallels Ca entry. *Neurosci. Abstr.* **7**, 441.

Eckert, R., and Tillotson, O. (1978). Potassium activation associated with intraneuronal free calcium. *Science* **200**, 437–440.

Eroglu, L., and Keen, P. (1977). Active uptake of Ca^{45} by a microsomal fraction prepared from rat dorsal roots. *J. Neurochem.* **29**, 905–909.

Fleckenstein, A. (1977). Specific pharmacology of calcium in myocardium, cardiac pacemakers, and vascular smooth muscle. *Annu. Rev. Pharmacol. Toxicol.* **17**, 149–166.

Frankenhaeuser, B., and Hodgkin, A. L. (1957). The action of calcium on the electrical properties of squid axons. *J. Physiol. (London)* **137**, 218–244.

Fried, R. C., and Blaustein, M. P. (1978). Retrieval and recycling of synaptic vesicle membrane in pinched-off nerve terminals (synaptosomes). *J. Cell Biol.* **78**, 685–700.

Fukuda, J., and Kawa, K. (1977). Permeation of manganese, cadmium, zinc, and beryllium through calcium channels of an insect muscle membrane. *Science* **196**, 309–311.

Gill, D. L., Grollman, E. F., and Kohn, L. D. (1981). Calcium transport mechanisms in membrane vesicles from guinea pig brain synaptosomes. *J. Biol. Chem.* **256**, 184–192.

Gotgilf, M., and Magazanik, L. G. (1977). Action of calcium channels blocking agents (verapamil, D-600 and manganese ions) on transmitter release from motor nerve endings of frog muscle. *Neurophysiology (USSR)* **9**, 415–421.

Gray, E. G., and Paula-Barbosa, M. (1974). Dense particles within synaptic vesicles fixed with acid-aldehyde. *J. Neurocytol.* **3**, 487–496.

Greenawalt, J. W., Rossi, C. S., and Lehninger, A. C. (1964). Effect of active accumulation of calcium and phosphate ions on the structure of rat liver mitochondria. *J. Cell Biol.* **23**, 21–38.

Hagiwara, S. (1975). Ca-dependent action potentials. *Membranes* **3**, 359–382.

Hagiwara, S., and Miyazaki, S. (1977). Ca and Na spikes in egg cell membrane. *In* "Cellular Neurobiology" (Z. Hall, R. Kelly, and C. F. Fox, eds.), pp. 147–158. Alan R. Liss, Inc., New York.

Hagiwara, S., and Takahashi, K. (1967). Surface density of calcium ions and calcium spikes in the barnacle muscle fiber membrane. *J. Gen Physiol.* **50**, 583–601.

Hales, C. N., Luzio, J. P., Chandler, J. A., and Herman, L. (1974). Localization of calcium in the smooth endoplasmic reticulum of rat isolated fat cells. *J Cell Sci.* **15**, 1–15.

Hammerschlag, R. (1980). The role of calcium in the initiation of fast axonal transport. *Fed. Proc., Fed. Am. Soc. Exp. Biol.* **39**, 2809–2814.

Haeusler, G. (1972). Differential effects of verapamil on excitation-contraction coupling in smooth muscle and on excitation-secretion coupling in adrenergic nerve terminals. *J. Pharmacol. Exp. Ther.* **180**, 672–682.

Henkart, M. (1980). Identification and function of intracellular calcium stores in axons and cell bodies of neurons. *Fed. Proc., Fed. Am. Soc. Exp. Biol.* **39**, 2783–2789.

Henkart, M. P., Reese, T. S., and Brinley, F. J., Jr. (1978). Oxalate produces precipitates in endoplasmic reticulum of Ca-loaded squid axons. *Science* **202**, 1300–1303.

Heuser, J., and Miledi, R. (1971). Effect of lanthanum ions on function and structure of frog neuromuscular junctions. *Proc. R. Soc. London, Ser. B* **179**, 247–260.

Hodgkin, A. L., and Keynes, R. D. (1957). Movements of labelled calcium in squid giant axons. *J. Physiol. (London)* **138**, 253–281.

Hubbard, J. I., Jones, S. F., and Landau, E. M. (1968a). On the mechanism by which calcium and magnesium affect the spontaneous release of transmitter from mammalian motor nerve terminals. *J. Physiol. (London)* **194**, 355–370.

Hubbard, J. I., Jones, S. F., and Landau, E. M. (1968b). On the mechanism by which calcium and magnesium affect the release of transmitter by nerve impulses. *J. Physiol. (London)* **196**, 75–86.

Hutson, S. M., Pfeiffer, D. R., and Lardy, H. A. (1976). Effect of cations and anions on the steady state kinetics of energy-dependent Ca^{2+} transport in rat liver mitochondria. *J. Biol. Chem.* **251**, 5251–5258.

Israel, M., Manaranche, R., Marsal, J., Meunier, F. M., Morel, N., Frachon, P., and Lesbats, B. (1980). ATP-dependent calcium uptake by cholinergic synaptic vesicles isolated from *Torpedo* electric organ. *J. Membr. Biol.* **54**, 90–102.

Katz, B. (1968). "The Release of Neural Transmitter Substances." Thomas, Springfield, Illinois.

Katz, B., and Miledi, R. (1967a). Modification of transmitter release by electrical interference with motor nerve endings. *Proc. R. Soc. London, Ser. B* **167**, 1–7.

Katz, B., and Miledi, R. (1967b). Tetrodotoxin and neuromuscular transmission. *Proc. R. Soc. London, Ser. B* **167**, 8–22.

Katz, B., and Miledi, R. (1967c). The release of acetylcholine from nerve endings by graded electrical pulses. *Proc. R. Soc. London, Ser. B* **167**, 23–38.

Katz, B., and Miledi, R. (1967d). A study of synaptic transmission in the absence of nerve impulses. *J. Physiol. (London)* **192**, 407–436.

Katz, B., and Miledi, R. (1968). The role of calcium in neuromuscular facilitation. *J. Physiol. (London)* **195**, 481–492.

Katz, B., and Miledi, R. (1969a). The effect of divalent cations on transmission in the squid giant synapse. *Pubbl. Stn. Zool. Napoli* **37**, 303–310.

Katz, B., and Miledi, R. (1969b). Tetrodoxin-resistant electric activity in presynaptic terminals. *J. Physiol. (London)* **203**, 459–487.

Katz, B., and Miledi, R. (1970). Further study of the role of calcium in synaptic transmission. *J. Physiol. (London)* **207**, 789–801.

Kraatz, H. G., and Trautwein, W. (1957). Die Wirkung von 2,4-Dinitrophenol (DNP) auf die neuromuskulare Erregung Subertragung. *Arch. Exp. Pathol. Pharmakol.* **231**, 419–439.

Krnjevic, K., Puil, E., and Werman, R. (1975). Evidence for Ca^{2+}-activated K^+ conductance in cat spinal motoneurons from intracellular EGTA injections. *Can. J. Physiol. Pharmacol.* **53**, 1214–1218.

Krueger, B. K., and Nachshen, D. A. (1979). Selectivity of Na^+ and Ca^{++} channels in synaptosomes. *Fed. Proc., Fed. Am. Soc. Exp. Biol.* **39**, 2038 (abstr.).

Krueger, B. K., Ratzlaff, R. W., Stricharz, G. R., and Blaustein, M. P. (1979). Saxitoxin binding to synaptosomes, membranes and solubilized binding sites from rat brain. *J. Membr. Biol.* **80**, 287–310.

Lazarewicz, J. W., Haljamae, H., and Hamberger, A. (1974). Calcium metabolism in isolated brain cells and subcellular fractions. *J. Neurochem.* **22**, 33–45.

Lehninger, A. L. (1970). Mitochondria and calcium ion transport. *Biochem. J.* **119**, 129–138.

Lieberman, A. (1971). Microtubule-associated smooth endoplasmic reticulum in frog's brain. *Z. Zellforsch. Mikrosk. Anat.* **116**, 564–577.

Lieberman, E. M., Palmer, R. F., and Collins, G. H. (1967). Calcium ion uptake by crustacean peripheral nerve subcellular particles. *Exp. Cell Res.* **46**, 412–418.

Llinas, R. (1977). Calcium and transmitter release in squid synapse. *Soc. Neurosci. Symp.* **2**, 139–160.

Llinas, R., and Nicholson, C. (1975). Calcium role in depolarization-secretion coupling: An aequorin study in squid giant synapse. *Proc. Natl. Acad. Sci. U.S.A.* **72**, 187–190.

Llinas, R., and Sugimori, M. (1980). Electrophysiological properties of *in vitro* Purkinje cell dendrites in mammalian cerebellar slices. *J. Physiol. (London)* **305**, 197–213.

Llinas, R., Blinks, J. R., and Nicholson, C. (1972). Calcium transient in presynaptic terminal of squid giant synapse: Detection with aequorin. *Science* **176**, 1127–1129.

Llinas, R., Steinberg, I. Z., and Walton, K. (1976). Presynaptic calcium currents and their relation to synaptic transmission: Voltage clamp study in squid giant synapse and theoretical model for the calcium gate. *Proc. Natl. Acad. Sci. U.S.A.* **73**, 2918–2922.

McGraw, C. F., Somlyo, A. V., and Blaustein, M. P. (1980a). Probing for calcium at presynaptic nerve terminals. *Fed. Proc., Fed. Am. Soc. Exp. Biol.* **39**, 2796–2801.

McGraw, C. F., Somlyo, A. V., and Blaustein, M. P. (1980b). Localization of calcium in presynaptic nerve terminals: An ultrastructural and electron microprobe analysis. *J. Cell Biol.* **85**, 228–241.

MacLennan, D. H., and Holland, P. C. (1975). Calcium transport in sarcoplasmic reticulum. *Annu. Rev. Biophys. Bioeng.* **4**, 377–404.

Makinose, M., and Hasselbach, W. (1965). Der Einfluss von Oxalat auf den Calcium-Transport isolierter Vesikel des sarkoplasmatischen Reticulum. *Biochem. Z.* **343**, 360–382.

Meech, R. W. (1972). Intracellular calcium injection causes increased potassium conductance in *Aplysia* nerve cells. *Comp. Biochem. Physiol. A* **42A**, 493–499.

Meech, R. W. (1978). Calcium-dependent potassium activation in nervous tissues. *Annu. Rev. Biophys. Bioeng.* **7**, 1–18.

Meiri, U., and Rahamimoff, R. (1972). Neuromuscular transmission: Inhibition by manganese ions. *Science* **176**, 308–309.

Michaelson, D. M., Ophir, I., and Angel, I. (1980). ATP-stimulated Ca^{2+} transport into cholinergic *Torpedo* synaptic vesicles. *J. Neurochem.* **35**, 116–124.

Miledi, R. (1966). Strontium as a substitute for calcium in the process of transmitter release at the neuromuscular junction. *Nature (London)* **212**, 1233–1234.

Miledi, R. (1971). Lanthanum ions abolish the "calcium response" of nerve terminals. *Nature (London)* **229**, 410–411.

Miledi, R. (1973). Transmitter release induced by the injection of calcium ions into nerve terminals. *Proc. R. Soc. London, Ser. B* **183**, 421–425.

Moriarty, C. M. (1978). Role of calcium in the regulation of adenohypophyseal hormone release. *Life Sci.* **23**, 185–194.

Muller, R. U., and Finkelstein, A. (1974). The electrostatic basis of Mg^{++} inhibition of transmitter release. *Proc. Natl. Acad. Sci. U.S.A.* **71**, 923–926.

Mullins, L. J., and Brinley, F. J., Jr. (1975). The sensitivity of calcium efflux from squid axons to changes in membrane potential. *J. Gen. Physiol.* **65**, 135–152.

Nachshen, D. A. (1981). Influx of Ca, Sr, and Ba, through Ca channels in synaptosomes. *Biophys. J.* **33** (2, pt. 2), 145a.

Nachshen, D. A., and Blaustein, M. P. (1979). The effects of some organic "calcium antagonists" on calcium influx in presynaptic nerve terminals. *Mol. Pharmacol.* **16**, 579–586.

Nachshen, D. A., and Blaustein, M. P. (1980). Some properties of potassium-stimulated calcium influx in presynaptic nerve endings. *J. Gen. Physiol.* **76**, 709–728.

Nachshen, D. A., and Blaustein, M. P. (1981). In preparation.

Nachshen, D. A., and Drapeau, P. (1981). In preparation.

Nelson, M. T., and Blaustein, M. P. (1981). Effects of ATP and vanadate on calcium efflux from barnacle muscle fibres. *Nature (London)* **289**, 314–316.

Nordmann, J. J., and Chevalier, J. (1980). The role of microvesicles in buffering $[Ca^{2+}]_i$ in the neurohypophysis. *Nature (London)* **287**, 54–56.

Nordmann, J. J., Dreifuss, J. J., Baker, P. F., Ravazzola, M., Malaisse-Lagae, F., and Orci, L. (1974). Secretion-dependent uptake of extracellular fluid by the rat neurohypophysis. *Nature (London)* **250**, 155–157.

Ochs, S. (1980). Calcium requirement for axoplasmic transport and the role of the perineurial sheath. *In* "Nerve Repair and Regeneration: Its Clinical and Experimental Basis" (D. L. Jewett and H. R. McCarroll, eds.), pp. 77–88. Mosby, St. Louis, Missouri.

Ornberg, R. L., and Reese, T. S. (1980). A freeze-substitution method for localizing divalent cations: Examples from secretory systems. *Fed. Proc., Fed. Am. Soc. Exp. Biol.* **39**, 2802–2808.

Otsuka, M., Otsuki, I., and Ebashi, S. (1965). ATP-dependent Ca binding of brain microsomes. *J. Biochem. (Tokyo)* **58**, 188–190.

Pappas, G. D., and Rose, S. (1976). Localization of calcium deposits in the frog neuromuscular junction at rest and following stimulation. *Brain Res.* **103**, 362–365.

Parsegian, V. A. (1977). Diffusion of calcium and its influence on vesicle-membrane interactions. *Soc. Neurosci. Symp.* **2**, 161–171.

Perkins, M. S., and Wright, E. B. (1969). The crustacean axon. I. Metabolic properties: ATPase activity, calcium binding, and bioelectric correlations. *J. Neurophysiol.* **32**, 930–947.

Politoff, A. L., Rose S., and Pappas G. D. (1974). The calcium binding sites of synaptic vesicles of the frog sartorius neuromuscular junction. *J. Cell Biol.* **61**, 818–823.

Putney, J. W., Jr., Weiss, S. J., Leslie, B. A., and Marier, S. H. (1977). Is calcium the final mediator of exocytosis in the rat parotid gland. *J. Pharmacol. Exp. Ther.* **203**, 144–155.

Rahamimoff, H., and Spanier, R. (1979). Sodium-dependent calcium uptake in membrane vesicles derived from rat brain synaptosomes. *FEBS Lett.* **104,** 111–114.

Rahamimoff, R., Erulkar, S. D., Alnaes, E., Meiri, H., Rotshenker, S., and Rahamimoff, H. (1976). Modulation of transmitter release by calcium ions and nerve impulses. *Cold Spring Harbor Symp. Quant. Biol.* **40,** 107–116.

Requena, J., and Mullins, L. J. (1979). Calcium movement in nerves. *Q. Rev. Biophys.* **12,** 371–460.

Requena, J., Mullins, L. J., and Brinley, F. J., Jr. (1979). Calcium content and net fluxes in squid axons. *J. Gen. Physiol.* **73,** 327–342.

Reuter, H. (1973). Divalent cations as charge carriers in excitable membranes. *Prog. Biophys. Mol. Biol.* **26,** 1–43.

Robinson, J. D., and Lust, W. D. (1968). Adenosine triphosphate-dependent calcium accumulation by brain microsomes. *Arch. Biochem. Biophys.* **126,** 286–294.

Rossi, C. S., and Lehninger, A. L. (1964). Stoichiometry of respiratory stimulation, accumulation of Ca^{++} and phosphate, and oxidative phosphorylation in rat liver mitochondria. *J. Biol. Chem.* **239,** 3971–3980.

Rubin, R. P. (1970). The role of calcium in the release of neurotransmitter substances and hormones. *Pharmacol. Rev.* **22,** 389–428.

Schatzmann, H. J., and Burgin, H. (1978). Calcium in human red blood cells. *Ann. N.Y. Acad. Sci.* **307,** 125–146.

Schliwa, M. (1976). The role of divalent cations in the regulation of microtubule assembly. *In vivo* studies on microtubules of the heliozoan axopadium using the ionphore A-23187. *J. Cell Biol.* **70,** 527–540.

Schmidt, R., Zimmerman, H., and Whitaker, V. P. (1976). Quantitative analysis of the metal ion control of cholinergic synaptic vesicles by atomic absorption spectrophotometry and changes induced by electrical stimulation. *Exp. Brain Res.* **24,** 19–20.

Schulz, I., and Stolze, H. H. (1980). The exocrine pancreas: The role of secretogogues, cyclic nucleotides and calcium in enzyme secretion. *Annu. Rev. Physiol.* **42,** 127–156.

Schweitzer, E., and Blaustein, M. P. (1980). Calcium buffering in presynaptic nerve terminals: Free calcium levels measured with arsenaro III. *Biochim. Biophys. Acta* **600,** 912–921.

Seigelbaum, S. (1978). Calcium sensitive currents in cardiac Purkinje Fibers. Ph.D. Thesis, Yale University, New Haven, Connecticut.

Shaw, F. D., and Morris, J. F. (1980). Calcium localization in the rat neurohypophysis. *Nature (London)* **287,** 56–58.

Silinsky, E. M. (1978). On the role of barium in supporting the asynchronous release of acetylcholine quanta by motor impulses. *J. Physiol. (London)* **274,** 157–171.

Simson, J. A., and Spicer, S. S. (1975). Selective subcellular localization of cations with variants of the potassium (pyro) antimonate technique. *J. Histochem. Cytochem.* **23,** 575–598.

Slater, E. C., and Cleland, K. W. (1953). The effects of calcium on the respiratory and phosphorylative activities of heart-muscle sarcosomes. *Biochem. J.* **55,** 566–80.

Somlyo, A. V., Gonzalez-Serratos, H., Shuman, H., McClellan, G., and Somlyo, A. P. (1981). Calcium release and ionic changes in the sarcoplasmic reticulum of tetanized muscle: an electron-probe study. *J. Cell Biol.* **90,** 577–594.

Sulakhe, P. V., and St. Louis, P. J. (1980). Passive and active calcium fluxes across plasma membranes. *Prog. Biophys. Mol. Biol.* **35,** 135–195.

Takeuchi, A., and Takeuchi, N. (1962). Electrical changes in pre- and postsynaptic axons of the giant synapse of *Loligo*. *J. Gen. Physiol.* **45**, 1181–1193.

Trotta, E. E., and DeMeis, L. (1975). ATP-dependent calcium accumulation in brain microsomes: Enhancement by phosphate and oxalate. *Biochim. Biophys. Acta* **394**, 239–247.

Van der Kloot, W., and Kita, H. (1975). The effects of the "calcium-antagonist" verapamil on muscle action potentials in the frog and crayfish and on neuromuscular transmission in the crayfish. *Comp. Biochem. Physiol.* **50**, 121–125.

Vinogradov, A., and Scarpa, A. (1973). The initial velocities of calcium uptake by rat liver mitochondria. *J. Biol. Chem.* **248**, 5527–5531.

Weisenberg, R. C. (1972). Microtubule formation *in vitro* in solution containing low calcium concentrations. *Science* **177**, 1104–1105.

Williams, J. A. (1980). Regulation of pancreatic acinar cell function by intracellular calcium. *Am. J. Physiol.* **238**, G269–G279.

Wolff, D. J., Pairier, P. G., Brostrom, C. O., and Brostrom, M. A. (1977). Divalent cation binding properties of bovine brain Ca^{2+}-dependent regulator protein. *J. Biol. Chem.* **252**, 4108–4117.

Yamamoto, D., and Washio, H. (1979). Permeation of sodium through calcium channels of an insect muscle membrane. *Can. J. Physiol. Pharmacol.* **57**, 220–222.

Yarom, Y., and Llinas, R. (1979). Electrophysiological properties of mammalian inferior olive neuron in *in vitro* brain stem slices and *in vitro* whole brain stem. *Soc. Neurosci. Abstr.* **5**, 109.

Yoshida, H., Kadota, K., and Fujisawa, H. (1966). Adenosine triphosphate dependent calcium binding of microsomes and nerve endings. *Nature (London)* **212**, 291–292.

Chapter 4

Calmodulin and Calcium-Binding Proteins: Evolutionary Diversification of Structure and Function

JACQUES G. DEMAILLE

I.	Introduction	111
II.	Calcium Ions as Second Messengers	112
III.	The Evolution of the Ca^{2+}-Binding Protein Family	113
	A. The Evolutionary Tree	113
	B. The Ancestral One-Domain Polypeptide	116
	C. Evolutionary Rates and Evolution of the Calcium-Binding Loops	118
IV.	Parvalbumin as the Prototype of Suppressor Molecules	120
V.	Calmodulin as the Prototype of Sensor Molecules	124
	A. Ion-Binding Constants of Calmodulin	124
	B. Sequence of Ion Binding to Calmodulin	124
	C. Specificity of Calmodulin in the Activation of Target Enzymes	129
	D. The Role of the Trimethyllysine Residue	133
VI.	Sequential Activation–Deactivation of Ca^{2+}-Dependent Enzymes	135
VII.	Conclusion	138
	References	139

I. INTRODUCTION

In spite of extreme variations in their environment, metazoans are able to maintain homeostasis or a steady state. To this end, cells constantly receive, code, and analyze information carried by molecules called mes-

CALCIUM AND CELL FUNCTION, VOL. II

sengers. The number of first messengers, which carry information from cell to cell, has increased during evolution with the complexity of organisms. They include hormones, neurotransmitters, and neuropeptides. In contrast, very few second or intracellular messenger systems have been characterized so far in eukaryotic cells. Among them, cyclic adenosine $3',5'$-monophosphate (cAMP) and Ca^{2+} have been found to play a prominent role (Rasmussen, 1970; Robison *et al.*, 1971; Rasmussen and Goodman, 1977), whereas the function of the guanylate cyclase–cyclic guanosine $3',5'$-monophosphate (cGMP) system remains poorly understood because of the small number of specific substrates of cGMP-dependent protein kinases.

A major feature of second messengers is the apparent simplicity of their mechanism of action. In all eukaryotic cells, cAMP appears to act only through binding to the regulatory subunit of cAMP-dependent protein kinases, thereby leading to the dissociation of free, active catalytic subunits (Rubin and Rosen, 1975; Rosen *et al.*, 1977; Cohen, 1978; Krebs and Beavo, 1979). In turn, the catalytic subunit phosphorylates a number of target proteins, which are present or absent in differentiated cells, thereby explaining the various responses of cells to an increase in cAMP levels (e.g., relaxation of smooth muscle, inotropic and chronotropic effects in cardiac muscle, secretion of the parotid salivary gland, and steroidogenesis of the adrenal cortex).

The picture is somewhat different as regards the other fundamental second messenger, namely, Ca^{2+}. These ions act only through binding to intracellular calcium-binding proteins that are all derived from a common ancestor. The major difference between the Ca^{2+} and cAMP pathways is the number of Ca^{2+}-binding proteins, some of which mediate the triggering effect of Ca^{2+}, whereas others are involved in suppression of the Ca^{2+} signal.

Exhaustive reviews of the field of Ca^{2+}-binding proteins have been recently published (see, e.g., Kretsinger, 1980). This chapter will focus on the evolution of the family of intracellular Ca^{2+}-binding proteins with respect to their involvement in transduction of the calcium signal or its suppression.

II. CALCIUM IONS AS SECOND MESSENGERS

Cells are able to respond to various stimuli (e.g., nerve impulses or certain hormones) by a transient, localized increase in the free Ca^{2+} concentration. For a compound to be called a second messenger, the following elements must be present:

1. A generator of second messengers coupled to the external signal.

2. A sensor capable of binding and releasing the messenger molecule; when bound to the messenger, the sensor molecule is transconformed and the information is propagated to other macromolecules, thereby triggering a physiological response such as contraction or secretion.

3. A suppressor that removes the messenger from the cytosolic compartment either because it will also bind Ca^{2+} after a kinetic delay or because Ca^{2+} are transferred through a membrane at the expense of ATP. In eukaryotes, the free Ca^{2+} level in the cytosol is about 250 nM, whereas the extracellular Ca^{2+} level is about 2 mM. The cell also contains Ca^{2+} stores in the endoplasmic (or sarcoplasmic) reticulum in which the pCa can reach values of 2. Upon stimulation, the cytosolic free Ca^{2+} concentration increases, e.g., from 0.26 μM in the quiescent cardiac ventricle to up to 10 μM during contracture (Marban et al., 1980). The generator is described as a Ca^{2+} channel that opens upon stimulation to let Ca^{2+} ions diffuse along the concentration gradient.

The sensors are members of the Ca^{2+}-binding protein family. The best characterized are troponin C, the calcium switch of sarcomeric muscles (Perry, 1979), and calmodulin, the ubiquitous and multifunctional Ca^{2+}-binding protein (Cheung, 1980; Means and Dedman, 1980; Klee et al., 1980). The suppressor molecules are either membrane-bound Ca^{2+}, Mg^{2+}-ATPases, which extrude Ca^{2+} from the cytosolic space, or soluble calcium-binding proteins, such as the parvalbumins of fast skeletal muscle and nervous tissue (Baron et al., 1975), which remove Ca^{2+} from the sensor molecule faster than the Ca^{2+} pumps could by themselves.

We will focus on the Ca^{2+}-binding proteins that can serve as sensors or as suppressors of the calcium signal.

III. THE EVOLUTION OF THE Ca^{2+}-BINDING PROTEIN FAMILY

The variety of functions of intracellular Ca^{2+}-binding proteins is paralleled by their various structures and ion-binding properties. They differ in the number of domains (two, three, or four), in the absence or presence of either Ca^{2+}-specific or Ca^{2+}–Mg^{2+} sites, and in the inability or ability to bind other macromolecules, either permanently or reversibly (Table I).

A. The Evolutionary Tree

The genealogical history of the family was reconstructed by using the maximum parsimony method (Goodman and Pechère, 1977; Goodman et

TABLE I

Structural and Functional Features of the Members of the Calcium-Binding Protein Family[a]

Protein	Domains	Ca²⁺-specific sites[c]	Ca^{2+}–Mg^{2+} high-affinity sites[d]	Domains that bind no divalent metal	Protein–protein interaction	Site of interaction
Pa	II, III, IV	None	III, IV	II	None	—
RLC	I, II, III, IV	I	None	II, III, IV	Myosin heavy chain (permanent) Kinase and phosphatase	Unknown Ser-15
ALK-LC	I, II, III, IV	None	None	I, II, III, IV	Myosin heavy chain (permanent)	Unknown
STNC	I, II, III, IV	I, II	III, IV	None	Troponin T Troponin I (permanent, enhanced by Ca²⁺)	Domain IV Domain II (46–77) and III (89–120)
HTNC	I, II, III, IV	II	III, IV	I	Troponin T, troponin I, (permanent, enhanced by Ca²⁺)	Probably similar to those of STNC
CaM	I, II, III, IV	I, II, III, IV	None	None	CaM-dependent enzymes (only in the presence of Ca²⁺) Other subunits of phosphorylase kinase (permanent, enhanced by Ca²⁺?)	Domains II (72–80) and III (by homology with STNC)
ICaBP	I, II	I, II	None	None	Unknown	

[a] Adapted from Goodman et al. (1979), with permission of Springer-Verlag.

[b] Pa, Parvalbumin; RLC, the phosphorylatable regulatory light chain of myosin; ALK-LC, the alkali light chains of myosin; STNC, skeletal troponin C; HTNC, heart troponin C; CaM, calmodulin; ICaBP, intestinal calcium-binding protein.

[c] Specific for Ca²⁺ at millimolar Mg²⁺, i.e., pK_d Ca²⁺ < 7, pK_d Mg²⁺ < 3.

[d] pK_d Ca²⁺ > 7, pK_d Mg²⁺ > 3.

al., 1979; Demaille *et al.*, 1980). The evolutionary tree, depicted in Fig. 1, includes 13 parvalbumins, 3 myosin regulatory light chains, 5 troponins C, 3 calmodulins, 2 myosin alkali light chains, 2 intestinal Ca^{2+}-binding proteins, and 1 component of brain S-100 protein.

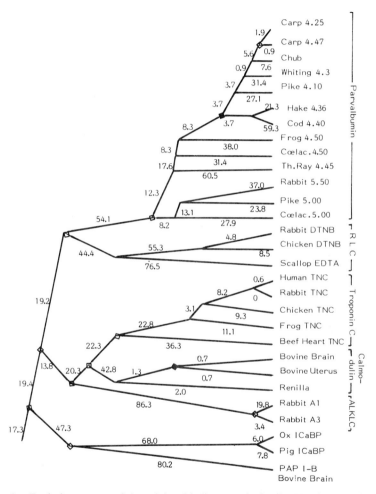

Fig. 1. Evolutionary tree of the calcium-binding protein family. Numbers on the links are nucleotide replacements per 100 codons. Where the augmentation algorithm for superimposed mutations increased the number of nucleotide replacements on a link the augmented value is shown. Symbol for an obvious gene duplication is ◇. Wherever there is only circumstancial evidence for gene duplication, ◆ is used. Reprinted from Demaille *et al.* (1980), with permission of Pergamon Press.

In agreement with observations from Collins (1974) and with the model proposed by Weeds and McLachlan (1974), Ca^{2+}-binding proteins appear to have evolved through tandem duplications in a precursor gene that caused a doubling and then a quadrupling in size of a primordial one-domain polypeptide ~40 residues long, consisting of two α helices on each side of a 12-residue ion-binding loop (Moews and Kretsinger, 1975). The final four-domain (I,II,III,IV) protein had domain I genetically closer to III, and II closer to IV.

The seven major lineages appear to be grouped into three major subfamilies. One of them includes the intestinal calcium-binding protein and brain S-100 protein, the functions of which are as yet unknown. Another subfamily is composed of myosin regulatory light chains and parvalbumins, which exhibit high-affinity sites for Ca^{2+} and Mg^{2+}. Finally the last subfamily includes myosin alkali light chains, the function of which is still unknown since they have lost their ion-binding properties and interact permanently with myosin heavy chains, and the typical sensor proteins, troponin C and calmodulin.

Troponin C and calmodulin appear to be closely related, as could be anticipated from their close primary structure homology (Collins *et al.*, 1977; Watterson *et al.*, 1980b) and from calmodulin's ability to replace troponin C in the activation of actomyosin ATPase (Amphlett *et al.*, 1976).

B. The Ancestral One-Domain Polypeptide

The ancestral one-domain polypeptide was reconstructed from individual domain ancestors and ancestors of domains I + III and II + IV (Fig. 2) and was found, after chemical synthesis, to exhibit Ca^{2+}-binding properties (Maximov *et al.*, 1978). The presence of acidic clusters in both the N- and C-terminal (residues 1–9 and 37–40) is remarkable. In particular, the sequence TEEEL (residues 3–7) has been strongly conserved and is found in domain III of calmodulin and troponin C in the form SEEEI or SEEEL. It has been proposed that such an acidic cluster may be involved in the binding of troponin C to troponin I (Leavis *et al.*, 1978), even though calmodulin binding to its target proteins rather involves a hydrophobic patch at the surface of the Ca^{2+}-loaded molecule (La Porte *et al.*, 1980). Also, the ancestral polypeptide contains methionyl residues at the C-terminus of the second α helix. Oxidation of such methionines in calmodulin domain II was found to alter calmodulin Ca^{2+}-binding properties and to abolish its interaction with phosphodiesterase and troponin I (Walsh and Stevens, 1978).

TOP.

Fig. 2. Reconstruction of the primordial one-domain polypeptide. The top reconstruction, obtained from the evolutionary tree in which intestinal calcium-binding protein was aligned with domain I and II sequences, represents the root of this tree, i.e., the place at which the domain I–domain III branch converges with the domain II–domain IV branch. The bottom reconstruction was obtained from an alternative, not quite as parsimonious tree (costing 1140 instead of 1137 nucleotide replacements, the cost of the tree for the top reconstruction). In this alternative tree intestinal calcium-binding protein was aligned with domain III and IV sequences, and the calmodulin lineage in each of the four domain branches converged on the root. Again the primordial sequence was taken from the root where domain I–domain III and domain II–domain IV branches converged. Reprinted from Goodman *et al.* (1979).

C. Evolutionary Rates and Evolution of the Calcium-Binding Loops

1. Nonuniformity of Evolutionary Rates

As shown in Table II, the evolution of troponin C was much faster during earlier evolutionary periods as compared to later ones. The deceleration of the evolutionary rate is particularly obvious in the last 75 million years, when it amounted to only 0.4 nucleotide replacements per 100 codons and per 10^8 years. Similar nonuniformity was observed for parvalbumins and myosin regulatory light chains (Goodman *et al.*, 1979). The slowest rate of evolution was observed for calmodulin, with only 0.3 nucleotide replacements per 100 codons and per 10^8 years.

2. Evolution of the Middomain Positions

The very slow rate of evolution of calmodulin suggests that it is probably very close to the ancestral four-domain protein. In line with this suggestion, the distribution of calmodulin is ubiquitous in all eukaryotic cells examined so far. It is therefore likely that the ancestral four-domain protein exhibited Ca^{2+}-specific sites similar to those of calmodulin, i.e., sites that are not saturated by Mg^{2+} in the resting cell. It was of interest to see whether evolution of the sites toward either the loss of ion-binding capacity or the acquisition of high-affinity $Ca^{2+}-Mg^{2+}$ sites, such as those found in parvalbumins (Haiech *et al.*, 1979), could be followed. To this

TABLE II

Rates of Nucleotide Replacements in Known Evolutionary Periods of Different Members of the Family of Intracellular Calcium-Binding Proteins[a]

	Age[b] (MyrBP)	Nucleotide replacements per 100 codons per 10^8 years
Calmodulin		
Cinidaria-vertebrate ancestor to *Renilla* calmodulin	700–0	0.3
Cinidaria-vertebrate ancestor to bovine calmodulin	700–0	0.3
Skeletal muscle troponin C		
Tetrapod ancestor to frog STNC	300–0	3.3
Tetrapod to amniote STNC ancestor	340–300	7.8
Amniote ancestor to chicken STNC	300–0	3.1
Amniote to eutherian STNC ancestor	300–75	3.6
Eutherian to human and rabbit STNCs	75–0	0.4 (0.0–0.8)

[a] Reprinted from Demaille *et al.* (1980).
[b] MyrBP, Million year before present; STNC, skeletal troponin C.

end, the number of nucleotide replacements in the middomain positions, i.e., the 12 residues of the loop, was compared to the number of nucleotide replacements at other positions, which are approximately twice as numerous. For a similar rate of evolution in both middomain and other positions, the number of nucleotide replacements at the middomain positions would therefore be expected to be approximately one-half that found at other positions. Table III shows that this is the case for lineages after they lost Ca^{2+}-binding activity or those which have conserved Ca^{2+}-specific sites. In contrast, lineages which are either losing their Ca^{2+}-binding activity or acquiring high-affinity $Ca^{2+}-Mg^{2+}$ sites (such as parvalbumin and troponin C domains III and IV) exhibit a number of nucleotide replacements almost as high as that found at the other positions. Finally, once $Ca^{2+}-Mg^{2+}$ sites have been acquired, the middomain positions evolve much slower than other positions as a result of a strong stabilization. It is therefore possible to describe the history of calcium-binding properties solely on the basis of evolutionary study.

TABLE III

Effects of Evolutionary Changes in Calcium-Binding Activity on the Proportion of Nucleotide Replacements Found in Middomain Positions to That Found in Other Domain Positions[a]

Links representing	No. of nucleotide replacements at each type of position		Average no. of nucleotide replacements per position[b]	
	Middomain	Other	Middomain	Other
Early ancestral lineages in which middomains were low-affinity Ca^{2+} sites	22	92[c]	1.83	3.83
Later lineages in which middomains were still low-affinity Ca^{2+} sites	29	68	2.42	2.83
Lineages losing Ca^{2+}-binding activity	57	64[c]	4.75	2.67
Evolutionary transition of middomains to high-affinity $Ca^{2+}-Mg^{2+}$ sites	18	21[d]	1.50	0.88
Later lineages after they acquired the high-affinity $Ca^{2+}-Mg^{2+}$ sites	52	244[c]	4.33	10.17
Lineages after they lost Ca^{2+}-binding activity	92	200	7.67	8.33

[a] Reprinted from Goodman et al. (1979), with permission of Springer-Verlag.
[b] On the average each domain consists of 36 positions, 12 middomain positions, and 24 other domain positions.
[c] Significantly different from group 6 at the 1% level.
[d] Significantly different from group 6 at the 5% level.

In contrast, the prediction of the type of site, e.g., Ca^{2+}-specific versus high-affinity $Ca^{2+}-Mg^{2+}$ sites, is not possible from a simple inspection of the middomain primary structure. Ca^{2+} specificity was suggested by Potter *et al.* (1977b) to be associated with the presence of a glycyl residue between the Y and Z ligating positions as defined by the three-dimensional structure of carp parvalbumin (Moews and Kretsinger, 1975). However, such a feature is present at the high-affinity $Ca^{2+}-Mg^{2+}$ site of parvalbumin domain IV (Haiech *et al.*, 1979). Domain–domain interaction is obviously involved in the binding properties of individual domains, as shown by the large decrease in the affinity for Ca^{2+} upon limited proteolysis and separation of parvalbumin domains (Derancourt *et al.*, 1978).

Finally, there is indirect evidence that the affinity for Ca^{2+} of calmodulin, for instance, is modified upon binding of calmodulin to target proteins, such as phosphodiesterase (Crouch and Klee, 1980; Klee and Haiech, 1980).

IV. PARVALBUMIN AS THE PROTOTYPE OF SUPPRESSOR MOLECULES

Parvalbumins are low-molecular-weight (MW 12,000) acidic proteins initially isolated from lower vertebrate (fish and amphibian) muscles (Pechère *et al.*, 1973) and subsequently found in higher vertebrates as well (Lehky *et al.*, 1974; Pechère, 1974). Their tissue distribution seems restricted to tissues which exhibit fast responses to external stimuli, namely, fast-twitch skeletal muscle and nervous tissue (Baron *et al.*, 1975). In this respect, it is of interest to note that parvalbumin synthesis in chick embryo leg muscles is a late event that occurs at about the time of hatching (Fig. 3). This synthesis is coordinated with that of myosin light chain kinase (Le Peuch *et al.*, 1979) and of sarcoplasmic reticulum Ca^{2+}, Mg^{2+}-ATPase (Martonosi *et al.*, 1977). In contrast, the triggering protein calmodulin is synthesized much earlier, more than a week before hatching.

Parvalbumin was shown to behave as a relaxing factor when added to myofibrils (Pechère *et al.*, 1977). This resolved the following apparent paradox: How can muscle contract in the presence of 1 mM parvalbumin, which in fish white muscle is 10 times more abundant than troponin C and exhibits a higher affinity for Ca^{2+} by about one order of magnitude.

Parvalbumin will not prevent contraction if and only if it binds Ca^{2+} after a delay. The rationale for such a kinetic delay is the existence of two high-affinity $Ca^{2+}-Mg^{2+}$ sites in domains III and IV of parvalbumins, as shown in Table IV (Haiech *et al.*, 1979). Since the affinity for Mg^{2+} is quite

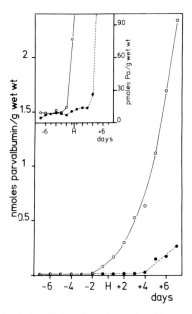

Fig. 3.   Parvalbumin levels in chick embryo leg (○) and breast (●) muscles, determined by radioimmunoassay. Inset: The ordinate has been enlarged by a factor of 10 to better illustrate the beginning of the synthesis; H, hatching. The precision of the assay is 0.1%. Reprinted from Le Peuch *et al.* (1979).

high, from 16 to 30 μM, parvalbumins are saturated by Mg^{2+} in the resting cell, where the free Mg^{2+} concentration has been estimated to be 0.44–0.6 mM by ^{31}P nuclear magnetic resonance (NMR) studies (Gupta and Yushok, 1980; Gupta and Moore, 1980). Ca^{2+} ions will only bind to Ca^{2+}–Mg^{2+} sites after dissociation of Mg^{2+}. Finally, parvalbumin returns to the Mg^{2+}-loaded form when Ca^{2+} are pumped out by the sarcoplasmic reticulum Ca^{2+}, Mg^{2+}-ATPase, as shown *in vitro* by the removal of Ca^{2+} from parvalbumins by sarcoplasmic reticulum vesicles (Gillis and Gerday, 1977).

The general scheme of the Ca^{2+} cycle in fast-twitch skeletal muscle is illustrated in Fig. 4 (Haiech *et al.*, 1979). Confirmation of this scheme was provided by two independent approaches. First, a simulation was made of the distribution of Ca^{2+} among the Ca^{2+}-specific or regulatory sites of troponin C, i.e., domains I and II (Potter and Gergely, 1975; Leavis *et al.*, 1978), and the Ca^{2+}–Mg^{2+} sites of troponin C and parvalbumin, i.e., domains III and IV of both proteins. It was shown (Gillis, 1980) that the regulatory sites of troponin C were 97% saturated immediately after a Ca^{2+} pulse, while binding of Ca^{2+} to the Ca^{2+}–Mg^{2+} sites proceeded

TABLE IV

Magnesium and Calcium Binding to Parvalbumins Studied by Equilibrium and Flow Dialysis[a,b]

		Rabbit Pa pI = 5.55	Frog Pa pI = 4.88	Frog Pa pI = 4.50	Hake Pa pI = 4.36
Mg^{2+} binding					
	K_d (μM)	16	21	27	30
	ν	2.3	1.9	1.9	1.4[c]
Ca^{2+} binding in the presence of Mg^{2+}					
No Mg^{2+}	ν	ND	ND	ND	1.0 ± 0.2
11.25 mM	K_d (μM)	—	—	—	1.5
	ν	—	—	—	1.1
22.5 mM	K_d (μM)	—	—	—	2.2
	ν	—	—	—	1.3
50 mM	K_d (μM)	15	11	6	6.1[d]
	ν	2.0	1.7	1.5	1.5
100 mM	K_d (μM)	38	31	10	17
	ν	1.7	1.9	2.0	1.3
112.5 mM	K_d (μM)	ND	ND	ND	9
	ν	—	—	—	1.1
150 mM	K_d (μM)	55	48	14	16
	ν	2.0	1.3	2.0	1.3[d]
Calculated	$K_{dCa^{2+}}$ (nM)	6.6	7.8	2.2	3.2

[a] Reprinted from Haiech *et al.* (1979), with permission of the American Chemical Society.

[b] ND, Not determined; PA, parvalbumin.

[c] Determination of the number of sites at saturating Mg^{2+} pointed to $\nu = 1.0 \pm 0.34$ ($\overline{X} \pm$ SD, $n = 4$).

[d] Experiments performed in the presence of 1 mM DTT indicated K_d values of 4.0 μM (1.4 sites) and 9.7 μM (1.5 sites) in the presence of 50 and 150 mM Mg^{2+}, respectively.

slowly. The transfer of Ca^{2+} from the former to the latter sites occurs in about 200 msec, i.e., on a time scale comparable to a twitch duration.

A second confirmation came from proton NMR experiments (Birdsall *et al.*, 1979) performed in the Mg^{2+}- and Ca^{2+}-loaded parvalbumin and showing different conformational states in the Ca^{2+}- or Mg^{2+}-bound species. These differences occur for the signals assigned to carboxylate residues and for resonance deriving from phenylalanine side chains and ring-shifted methyl groups. Changes in the upfield aromatic spectral region corresponding to the side chains contributing to the hydrophobic core indicate a different protein folding. Also the absence of Mg^{2+}-induced perturbation of the ring-shifted methyl group signal assigned to Leu-63 is remarkable.

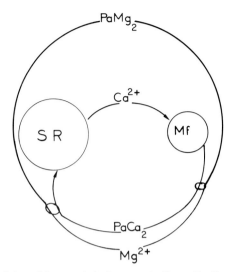

Fig. 4. Scheme of the calcium cycle in fast-muscle fibers. Pa, Parvalbumin; Mf, myofi-bril; SR, sarcoplasmic reticulum. Parvalbumin binds two Mg^{2+} in resting muscle (pCa 8). Ca^{2+} released from SR binds first to the Ca^{2+}-specific sites of troponin C, since they are not occupied by Mg^{2+} (contraction). As soon as Mg^{2+} dissociates from parvalbumin, parvalbu-min removes Ca^{2+} from troponin C (relaxation), and in turn SR removes calcium from parvalbumin, which then binds two Mg^{2+}. Reprinted from Haiech *et al.* (1979).

The different conformational states of parvalbumin in the presence of Ca^{2+} or Mg^{2+} therefore permitted a study of the rate of Mg^{2+} exchange upon Ca^{2+} addition to the Mg^{2+}-loaded protein. Slow exchange effects were observed for phenylalanine side chain signals characteristic of res-idues in the hydrophobic core. Also, the upfield methyl signal titrated downfield, confirming the reorganization of hydrophobic contacts upon Ca^{2+} binding. The first-order constant for the structural rearrangement, hence for the Mg^{2+}–Ca^{2+} exchange is ≤ 100 sec^{-1}, compatible with the relaxing function proposed for parvalbumins.

A number of questions remain unanswered. As indicated by Gillis (1980), parvalbumins can only play a role if the ratio of Ca^{2+}–Mg^{2+} sites to Ca^{2+}-specific sites exceeds 5. This is the case for lower vertebrates that are poikilotherms; they may need parvalbumin, because the rate of Ca^{2+} uptake by the pump is assumed to be too slow at low temperature (Gillis, 1980). In higher vertebrates, parvalbumin and troponin C concentrations are in the same range and the evolutionary advantage of maintaining parvalbumin synthesis is as yet unknown. For instance, chick breast mus-cle contains very little parvalbumin (Blum *et al.*, 1977), which is synthe-sized later at a rate slower than in leg muscle (Le Peuch *et al.*, 1979), even though breast muscle is a typical fast muscle (Fig. 3). Cardiac muscle contains undetectable amounts of parvalbumin in rabbit, while the heart of a small mammal, the shrew, contains significant amounts of a parvalbumin-like protein, presumably related to the fast heartbeat, hence the need for rapid relaxation (Le Peuch *et al.*, 1978). Its level is lower than that of troponin C, again raising some doubt as to its efficacy in accelerating relaxation.

V. CALMODULIN AS THE PROTOTYPE OF SENSOR MOLECULES

A calcium-binding protein can mediate Ca^{2+} effects within the cell if, and only if, it binds Ca^{2+} reversibly between pCa 5 and 7 and if its sites are not occupied by Mg^{2+} with physiological concentrations of free K^+ and Mg^{2+} (150 and 0.6 mM, respectively). Also, it must undergo a conformational change upon Ca^{2+} binding and propagate this change to interacting macromolecules. It is therefore of interest to study the ion-binding properties and the site of interaction of a typical sensor. Calmodulin was chosen because it is multifunctional and can effectively replace other sensor molecules, troponin C (Amphlett *et al.*, 1976), and leiotonin C (Ebashi, 1980).

A. Ion-Binding Constants of Calmodulin

There is general agreement, from both primary structure (Watterson *et al.*, 1980b) and Ca^{2+}-binding studies, on the presence of four ion-binding sites in calmodulin. A considerable number of discrepancies are found, however, in the literature with respect to the affinity for Ca^{2+} and for Mg^{2+} of these sites (Teo and Wang, 1973; Lin *et al.*, 1974; Watterson *et al.*, 1976; Wolff *et al.*, 1977; Dedman *et al.*, 1977b; Potter *et al.*, 1977b; Jarrett and Kyte, 1979; Crouch and Klee, 1980). This has prompted a reexamination of the Ca^{2+}-binding parameters of calmodulin and of the effects of Mg^{2+} and K^+ on these parameters (Haiech *et al.*, 1981). The results indicate that K^+ and Mg^{2+} compete with Ca^{2+} at each site. Dissociation constants for Ca^{2+} range from 67 nM to 0.9 μM, for Mg^{2+} from 70 μM to 0.27 mM, and for K^+ from 1.5 to 10.6 mM (Table V). The values for each site suggest that Ca^{2+} binding is sequential and ordered, in agreement with NMR studies (Seamon, 1980).

B. Sequence of Ion Binding to Calmodulin

A first approach to the sequence of ion binding to the individual domains stems from the assumption that the affinity of a given site for K^+ is

TABLE V

Cation Dissociation Constants of Ram Testis Calmodulin[a]

Ion	Site 1	Site 2	Site 3	Site 4
K^+ (mM)	3.7	10.6	8.7	1.5
Mg^{2+} (μM)	70	270	100	90
Ca^{2+} (nM)	67	170	600	900

[a] Sites are listed in order of decreasing affinity for Ca^{2+}, with no reference to their position in the primary structure.

dependent on the number of carboxylate groups. This leads to one of the following possible sequences: II $\rightarrow$ I $\rightarrow$ III $\rightarrow$ IV or II $\rightarrow$ III $\rightarrow$ I $\rightarrow$ IV (Haiech et al., 1981).

A more direct approach took advantage of the presence of only two tyrosyl residues in the primary structure of mammalian calmodulin. One is a ligand of Ca^{2+} in the ion-binding loop of domain III (Tyr-99), while the other, Tyr-138, is next to a Ca^{2+} ligand in the ion-binding loop of domain IV (Watterson et al., 1980b). This prompted a study of the tyrosine fluorescence quantum yield upon excitation at 275 nm when Ca^{2+} were added to the metal-free protein (Kilhoffer et al., 1980a). Figure 5 clearly shows an increase in the tyrosine fluorescence quantum yield from 0.027

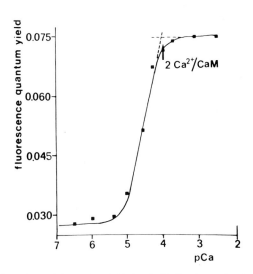

Fig. 5. Titration of metal-free calmodulin by Ca^{2+}. Conformational changes induced by successive additions of Ca^{2+} to 45 μM ram calmodulin in 100 mM Tris buffer (pH 7.6) were followed. The fluorescence quantum yield was plotted against pCa ($= -\log$ [Ca^{2+}]). Reprinted from Kilhoffer et al. (1980a).

to 0.08, which reaches a plateau for two Ca^{2+} bound per mole calmodulin. This indicates that Ca^{2+} binding to the two higher-affinity sites induces a major conformational change, in agreement with circular dichroic (CD) and ultraviolet (UV) difference spectroscopic studies (Klee, 1977; Crouch and Klee, 1980) and NMR studies (Seamon, 1980). Calcium binding affects the emission characteristics of both tyrosine residues. A detailed fluorescence study (Kilhoffer et al., 1981) shows that, in the absence of calcium, the tyrosyl residues are located in different microenvironments which are affected by Ca^{2+} binding for both Tyr-99 and Tyr-138. Mg^{2+} only partially mimic Ca^{2+} effects, especially in the presence of a physiological concentration of 150 mM K^+. The competition between K^+ and Mg^{2+} (Section V,A) results in small effects of Mg^{2+} added to the Ca^{2+}-free protein. Ca^{2+}-binding sites of calmodulin therefore behave quite differently from the Ca^{2+}–Mg^{2+} sites of domains III and IV of troponin C and parvalbumin.

The two higher-affinity sites were identified by using Tb^{3+} which competes with Ca^{2+} (Kilhoffer et al., 1980a). Tb^{3+} became luminescent when bound to a protein close to a tyrosine phenol group upon excitation of the latter at 275 nm. Upon titration of metal-free mammalian calmodulin with Tb^{3+} (Fig. 6), the tyrosine fluorescence quantum yield increased to a sharp maximum at 2 moles Tb^{3+} bound per mole calmodulin. Additional Tb^{3+} resulted in a strong quenching of tyrosine fluorescence. No significant increase in Tb^{3+} luminescence was observed up to 2 moles Tb^{3+} mole^{-1} calmodulin, followed by a 400-fold enhancement of Tb^{3+} luminescence upon subsequent Tb^{3+} addition. Tb^{3+} effects on both fluorescence and luminescence were reversed upon addition of excess Ca^{2+}. These experiments clearly identify the two higher-affinity sites as domains which lack tyrosine, namely, domains I and II. Therefore, the major conformational change which alters the environment of Tyr-99 and Tyr-138 in domains III and IV is a global transconformation of the molecule that occurs upon saturation of domains I and II.

The ion-binding sequence is likely to be II → I → III → IV. Confirmation of site III being saturated immediately after the higher-affinity sites I and II came from a similar study performed on octopus calmodulin (Kilhoffer et al., 1980b). Octopus calmodulin indeed contains a single tyrosyl residue, as shown by the UV absorption spectrum (Fig. 7) and the amino acid analysis (Molla et al., 1981).

Assignment of the tyrosine residue to a position homologous to that of Tyr-138 of mammalian calmodulin in the C-terminal tryptic peptide was obtained through tryptic digestion and separation of the peptides by high-performance liquid chromatography on a μBondapak phenyl column (Fig. 8). Only one major peptide absorbed at 275 nm and was eluted at 20.5 min, close to one of the tyrosine-containing peptides of mammalian cal-

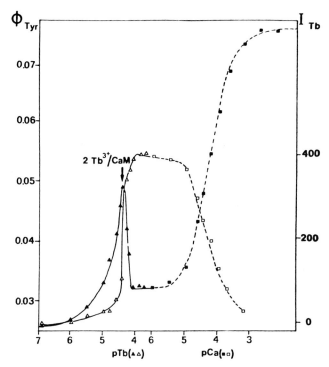

Fig. 6. Terbium binding to calmodulin. Solid lines indicate the fluorescence quantum yield (ϕ_{Tyr}) of calmodulin (▲) and the concomitant variation of terbium luminescence at 545 nm: I_{Tb} (△) as a function of added Tb^{3+} (pTb = −log [Tb^{3+}]). Ca^{2+}-induced Tb^{3+} removal is represented by dashed lines. ■, Tyrosine quantum yield; □, terbium luminescence. Ram calmodulin concentration was 22 μM in 100 mM Tris buffer (pH 6.9). The same effects were observed by using a 20 mM MOPS buffer (pH 6.5). When Tb^{3+} was >3 × 10^{-4} M, a Tb^{3+}-induced precipitation of calmodulin occurred that prohibited the spectroscopic study. Reprinted from Kilhoffer *et al.* (1980a).

modulin (20.9 min) (Kilhoffer *et al.*, 1980b). When compared to peptides 91–106 and 127–148 of bovine brain calmodulin (Watterson *et al.*, 1980b), the octopus tyrosine-containing peptide was obviously homologous in length and composition to the C-terminal peptide of mammalian calmodulin, pointing to a tyrosine position homologous to the domain IV Tyr-138 (Table VI). Upon Ca^{2+} binding to the metal-free protein, octopus calmodulin showed an increase in the tyrosine fluorescence quantum yield similar to that described for the mammalian protein. In contrast, addition of Tb^{3+} to metal-free octopus calmodulin (Fig. 9) resulted in a continuous increase in the tyrosine fluorescence quantum yield to a plateau at about 4 moles Tb^{3+} mole^{-1} protein, without the strong quenching induced by the third

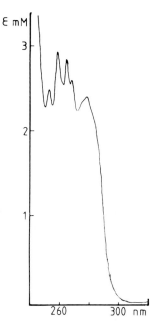

Fig. 7. Ultraviolet absorption spectrum of octopus calmodulin. The protein was dissolved in 20 μM CaCl$_2$, 0.27 M NaCl, 1 mM magnesium acetate, 20 mM imidazole buffer, pH 6.5. The extinction coefficient was obtained by amino acid analysis of an aliquot of the sample. Reprinted from Kilhoffer *et al.* (1980b).

Tb^{3+} bound to mammalian calmodulin. This indicates per se that Tyr-99 is responsible for such quenching and that domain III is saturated immediately after domains I and II.

Confirmation of this binding sequence was obtained through an examination of Tb^{3+} luminescence (Kilhoffer *et al.*, 1980b). Its increase upon Tb^{3+} addition to metal-free octopus calmodulin was insignificant up to 3 moles Tb^{3+} mole^{-1} protein, followed by only a 30-fold enhancement reached for 5 equiv Tb^{3+}. This confirms that the third saturated site is domain III and that domain IV exhibits a much lower affinity, as suspected from the Ca^{2+}-binding isotherms in the presence of 150 mM K$^+$ (Haiech *et al.*, 1981).

The sequence of ion binding to calmodulin is therefore firmly established as (I,II) → III → IV and, most likely, II → I → III → IV. This is at variance with that for troponin C, in which sites III and IV are the Ca^{2+}-Mg^{2+} high-affinity sites (Potter and Gergely, 1975; Leavis *et al.*, 1978; Nagy and Gergely, 1979; Leavis *et al.*, 1980). However, it must be kept in mind that sites III and IV of troponin C are saturated by Mg^{2+} in the

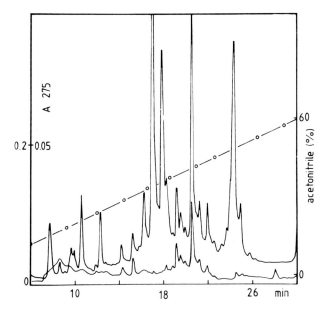

Fig. 8. High-performance liquid chromatography of an octopus calmodulin tryptic digest (10 nmoles) on a μBondapak phenyl column. Upper trace: absorbance of the eluate at 220 nm, 0.4 absorbance units full scale. Lower trace: absorbance of the eluate at 275 nm, 0.1 absorbance units full scale. Lower and upper traces were obtained on different and consecutive runs. In the illustrated part of the run (6–30 min), the acetonitrile gradient (O) was from 12 to 60%. Reprinted from Kilhoffer *et al.* (1980b).

resting cell and must behave like the relaxing sites of parvalbumin (Section IV). Finally, the Ca^{2+} signal will be received by sites I and II in both proteins, calmodulin and troponin C, under physiological conditions.

C. Specificity of Calmodulin in the Activation of Target Enzymes

Since all intracellular Ca^{2+}-binding proteins are evolutionarily related (Section III), it is of interest to determine which properties have been either lost or acquired in evolution, assuming that calmodulin is closest to the four-domain ancestral protein. Calmodulin was thus compared, on the one hand, to parvalbumins, which have lost the ability to interact with other macromolecules, lost domain I and the ion-binding capacity of domain II, and acquired relaxing sites ($Ca^{2+}–Mg^{2+}$ sites) in domains III and IV and, on the other hand, to troponin C, which is very closely related to calmodulin except for its relaxing sites in domains III and IV and for the absence of the trimethyllysine found at position 115 of the calmodulin peptide chain (Watterson *et al.*, 1980b). With respect to calmodulin,

TABLE VI

Amino Acid Composition of Tyrosine-Containing Tryptic Peptides from Octopus and Bovine Brain Calmodulins[a,b]

| Residue | Octopus calmodulin peptide eluted at 20.5 min | | Bovine brain calmodulin | |
	Found	Integer	Peptide 127–148	Peptide 91–106
Asx	4.0	4	4	3
Thr	1.0	1	1	0
Ser	1.1	1	0	1
Glx	4.0	4	5	1
Gly	2.0	2	2	2
Ala	1.7	2	2	2
Val	1.9	2	2	1
Met	1.5	1–2	2	0
Ile	1.1	1	1	1
Leu	0.3	0	0	1
Tyr	0.8	1	1	1
Phe	0.8	1	1	1
Lys	0.9	1	1	1
Arg	0	0	0	1
Total		21–22	22	16

[a] Reprinted from Kilhoffer *et al.* (1980b).
[b] Expressed as residues per mole after 24-hr hydrolysis. Values for bovine brain calmodulin peptides are from Watterson *et al.* (1980b).

troponin C and parvalbumin can be considered the closest and the most distant offspring, respectively, of the Ca^{2+}-binding protein family.

Apparently, calmodulin can replace other triggering proteins (troponin C and leiotinin C) as stated earlier (Section V). In contrast, there are considerable discrepancies in the literature with respect to the activation of calmodulin-dependent enzymes by other calcium-binding proteins. Parvalbumin was reported to be able (Potter *et al.*, 1977a) or unable (Le Donne and Coffee, 1979) to activate cyclic nucleotide phosphodiesterase. The same enzyme was reported to be activated by a 600-fold molar excess of troponin C over calmodulin (Dedman *et al.*, 1977a), whereas no such activation was found even at 1000-fold excess troponin C by others (Stevens *et al.*, 1976).

Convincing evidence for the activation of glycogen phosphorylase *b* kinase by troponin C and calmodulin was recently provided by Cohen *et al.* (1979). Troponin C binds to the same site where the second calmodulin

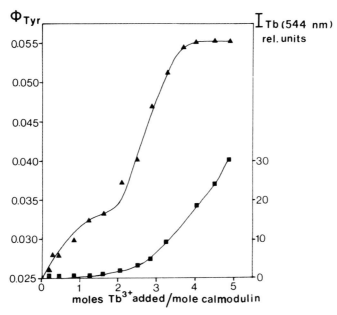

Fig. 9. Titration of octopus calmodulin by Tb^{3+} ions. Tb^{3+} ($TbCl_3$) was added to 21.7 μM calmodulin in 20 mM MOPS buffer, pH 6.9, and was assumed to be protein-bound. ϕ_{Tyr} is the tyrosine fluorescence quantum yield (▲). I_{Tb} is the variation in terbium luminescence (■) at 544 nm, expressed in relative units, upon excitation at the same wavelength (280 nm). Reprinted from Kilhoffer *et al.* (1980b).

molecule (which is dissociated in the absence of Ca^{2+}) binds to the $\alpha\beta\gamma\delta$ complex and is 200 times less effective than calmodulin.

Finally, in relation to myosin light chain kinase, troponin C was reported to be 10% as-effective as calmodulin in activation of the skeletal enzyme (Nairn and Perry, 1979). In contrast, addition of up to 10 μM parvalbumin or troponin C did not, in our hands (Walsh *et al.*, 1980), result in the activation of cardiac, skeletal, or smooth muscle myosin light chain kinase (Fig. 10). Such discrepancies prompted a reexamination of the activation of myosin light chain kinase by calmodulin and troponin C (Walsh *et al.*, 1981). As shown in Fig. 11, a 3125-fold molar excess of troponin C over calmodulin is required to achieve the same half-maximal activation. The possibility of calmodulin contaminating the troponin C preparation was ruled out in view of the lower V_{max} obtained with troponin C. It is therefore unlikely that troponin C, which is only 16 times more abundant than calmodulin in skeletal muscle (Cohen *et al.*, 1979), plays a role *in vivo* in the activation of myosin light chain kinase. This is further confirmed by experiments in which troponin C was added in the presence

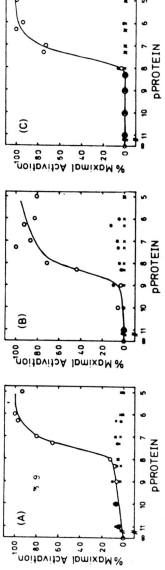

Fig. 10. Effect of increasing concentrations of bovine brain calmodulin (○), and frog pI 4.88 parvalbumin (x) on calcium-dependent myosin light chain kinases isolated from bovine cardiac muscle (A), rabbit skeletal muscle (B), and turkey gizzard (C). Assays of kinase activity were performed in the presence of saturating amounts of Ca^{2+} (0.1 mM). pProtein = $-\log[\text{protein}]$. Reprinted from Walsh *et al.* (1980).

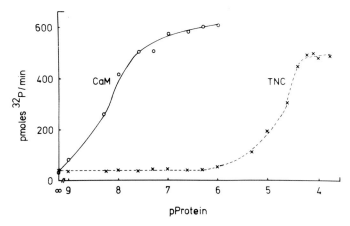

Fig. 11. Activation of skeletal myofibrillar myosin light chain kinase by calmodulin and troponin C. The rate of incorporation of ^{32}P into regulatory light chain of rabbit skeletal muscle myosin, catalyzed by canine skeletal myofibrillar myosin light chain kinase (2.5×10^{-7} M), was measured as a function of the concentration of calmodulin (O) or toponin C (X) under standard conditions in the presence of 0.1 mM $CaCl_2$. Reprinted from Walsh *et al.* (1981).

of excess Ca^{2+} to the kinase submaximally activated by calmodulin (Walsh *et al.*, 1981). Figure 12 shows that the addition of troponin C results in inhibition of the kinase, probably through competititon with calmodulin, as suggested by the incomplete reversal of the inhibition upon addition of excess calmodulin. The above experiments leave no doubt as to the inability of troponin C to activate skeletal myosin light chain kinase *in vivo*. There must be distinct features in the troponin C molecule that account for the loss of its capacity to activate at least some of the calmodulin targets. One of the structural differences frequently mentioned in the literature as responsible for such functional differences is the absence of trimethyllysine in the troponin chain. An investigation of the role of this residue in the activation by calmodulin of calmodulin-dependent myosin light chain kinase was therefore conducted.

D. The Role of the Trimethyllysine Residue

Mammalian and avian calmodulins contain 1 mole of trimethyllysine at position 115 (Watterson *et al.*, 1980a,b). Invertebrate calmodulins were also reported to contain one such residue, whether isolated from *Renilla reniformis* (Jones *et al.*, 1979) or from sea anemone and scallop (Yazawa *et al.*, 1980). However, octopus calmodulin, isolated from the whole animal,

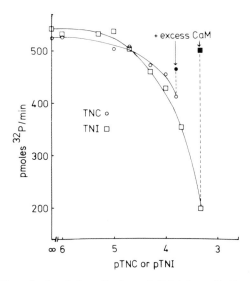

Fig. 12. Inhibition of calmodulin activation of skeletal myofibrillar myosin light chain kinase by troponin I and troponin C. The rate of incorporation of ^{32}P into the regulatory light chain of rabbit skeletal muscle myosin, catalyzed by skeletal myofibrillar myosin light chain kinase ($2.5 \times 10^{-7} M$), in the presence of 0.1 mM Ca^{2+} and at a limiting concentration (0.1 μM) of calmodulin, was measured as a function of the concentration of troponin I ($\square$) or troponin C ($\bigcirc$) under standard conditions. The initial velocity measured in the presence of a saturating concentration of calmodulin ($1.3 \times 10^{-5} M$) and in the absence of either troponin I or troponin C was 560 pmoles ^{32}P min^{-1}. Dashed lines lead to solid symbols which indicate the myosin light chain kinase activity observed when a saturating concentration of calmodulin ($1.3 \times 10^{-5} M$) was included in the incubation mixture at the corresponding troponin concentration. Reprinted from Walsh *et al.* (1981).

was found to contain only 0.1 mole each of mono-, di-, and trimethyllysine (Molla *et al.*, 1981). In this respect, it is similar to barley and fungi calmodulins (Grand *et al.*, 1980) which lack trimethyllysine and are less effective in activating myosin light chain kinase and phosphodiesterase by at least one order of magnitude. Octopus calmodulin was found to be at least as effective as ram testis calmodulin in activating both skeletal and smooth muscle myosin light chain kinases. In fact, the apparent K_d was 6 and 8 nM for the octopus calmodulin activation of skeletal and turkey gizzard enzymes versus 11 and 40 nM for ram testis protein (Fig. 13). It is obvious therefore that trimethyllysine is not required for myosin light chain kinase activation. In other words, the absence of trimethyllysine does not account for the inability of troponin C to activate this enzyme.

Studies from Perry (1979), Perry *et al.* (1979), and Cohen *et al.* (1978, 1979) suggest that calmodulin and troponin C may be biheaded proteins.

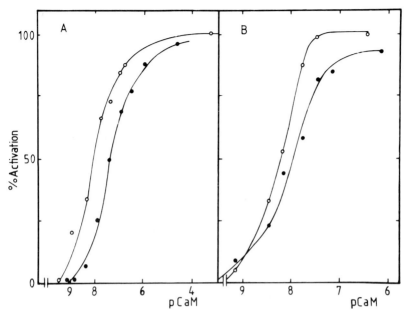

Fig. 13. Activation of turkey gizzard (A) and canine skeletal (B) myosin light chain kinases by octopus (○) and ram testis (●) calmodulins. Assays were performed as described by Walsh *et al.* (1980). Enzyme concentrations were 0.7 nM for the turkey gizzard enzyme and 4.8 μg ml^{-1} for the partially purified skeletal enzyme.

For instance, the whole troponin complex activates phosphorylase kinase even better than troponin C does. Also, the tightly bound calmodulin (δ subunit) of phosphorylase kinase activates another calmodulin-dependent enzyme, myosin light chain kinase. It is therefore possible, although there is at this point no firm experimental support for this hypothesis, that calmodulin can bind to two different molecules, e.g., a target enzyme on the one hand, and troponin I or an inhibitory protein such as calcineurin (Klee *et al.*, 1979a) on the other hand. If this were true, trimethyllysine might perhaps be involved in the interaction with inhibitory proteins.

VI. SEQUENTIAL ACTIVATION–DEACTIVATION OF Ca^{2+}-DEPENDENT ENZYMES

The minimal kinetic scheme for the activation of Ca^{2+}–calmodulin-dependent enzymes (Huang *et al.*, 1980) is depicted in Fig. 14. The final product, i.e., the *active* Ca^{2+}–calmodulin–enzyme ternary complex, is

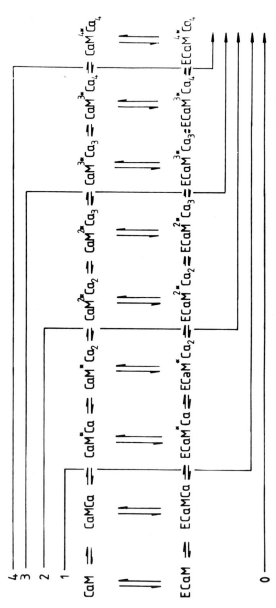

Fig. 14. The minimal kinetic scheme of activation of calmodulin-dependent enzymes. E, Activatable enzyme; CaM, calmodulin; CaM^{n*}, the transformed calmodulin after binding of the nth Ca^{2+} ion; 0–4, preferential kinetic pathways defining the hypothetical enzyme subsets E_0 to E_{IV}, respectively.

likely to contain at least 3 moles Ca^{2+} bound per mole calmodulin, as shown for instance for phosphodiesterase (Crouch and Klee, 1980) and skeletal myosin light chain kinase (Blumenthal and Stull, 1980). However, the pathway to reach such a final activated stage may be, at least in theory, quite different from enzyme to enzyme. For instance, phosphorylase kinase, which contains tightly bound calmodulin as its δ subunit (Cohen *et al.*, 1978), will follow pathway 0 and falls into the enzyme subset E_0, where 0 indicates that calmodulin binds to the enzyme even in the absence of Ca^{2+}. For other enzymes, such as phosphodiesterase and myosin light chain kinase, calmodulin binds to the enzyme only in the presence of Ca^{2+}. Since Ca^{2+} binding to calmodulin is sequential and ordered (Klee and Haiech, 1980) and induces sequential conformational changes in the protein (Section V,B), enzymes may perhaps bind calmodulin when it has bound one, two, three, or four Ca^{2+}, defining enzyme subsets E_I to E_{IV}, respectively. It can be assumed that the rate of the steps from the Ca^{2+}-free ternary complex to the active Ca^{2+}-loaded complex is not limiting when compared to the rate of Ca^{2+} binding and Ca^{2+}-induced transconformation of free calmodulin. If this assumption holds true, the activation pathways will tend to be slower and slower from 0 to 4.

A computer simulation of the kinetics of activation of enzyme subsets I to IV shows that this is indeed the case (Haiech and Demaille, 1981). The sequential binding of Ca^{2+} to calmodulin therefore generates the possibility of sequential enzyme activation.

Sequential deactivation of calmodulin-dependent enzymes may involve calmodulin-binding proteins or inhibitory proteins such as calcineurin (Klee *et al.*, 1979a; Klee and Haiech, 1980). Calcineurin is able to inhibit calmodulin-dependent enzymes because it binds calmodulin in a Ca^{2+}-dependent manner, and it does not exhibit any known enzymatic activity (Wang and Desai, 1977; Klee and Krinks, 1978; Sharma *et al.*, 1979; Wallace *et al.*, 1979). Its low-molecular-weight B subunit binds Ca^{2+} with high affinity (Klee *et al.*, 1979a) and exhibits some similarity to parvalbumin (Klee and Haiech, 1980). This may suggest that its binding sites are perhaps high-affinity Ca^{2+}-Mg^{2+}(or relaxing) sites.

Calcineurin has been shown to copurify with phosphodiesterase (Klee *et al.*, 1979b). On the other hand, calmodulin may well be a biheaded protein interacting with both a target enzyme and an inhibitory protein (Section V,D). There is therefore a distinct possibility that these proteins are associated within the cell in long-lived supramolecular assemblies, tentatively called "calcisomes" (Haiech and Demaille, 1981), that are localized in close vicinity to the membrane Ca^{2+} channels and Ca^{2+} pumps where the machinery responsible for short-term buffering of Ca^{2+} ions was shown to be present (Tillotson and Gorman, 1980). Maps of temporal and spatial distribution of Ca^{2+} within living cells (Taylor *et al.*, 1980) point to a restricted diffusion of Ca^{2+} in the cytosol, leading to transient and localized increases in Ca^{2+} concentration. The calcisome superstructure would permit the fast switch-off of Ca^{2+} effects if, as remains to be experimentally established, there is within the calcisome a migration of Ca^{2+} ions from the calmodulin sites to the high-affinity $Ca^{2+}-Mg^{2+}$ sites of calcineurin B after the kinetic delay introduced by Mg^{2+} dissociation (Haiech and Demaille, 1981). The calcisome cycle would then include an active step (active enzyme–calcium/calmodulin–magnesium/calcineurin), a refractory step (inactive enzyme–calmodulin–calcium/calcineurin), and an inactive step after removal of Ca^{2+} by the pumps and calmodulin dissociation. Sequential activation–deactivation of calmodulin-dependent enzymes provides the possibility of preventing futile cycles following cell stimulation.

Finally, since several calmodulin-dependent enzymes were shown to be phosphorylated by cAMP-dependent protein kinases, e.g., phosphorylase kinase (Cohen, 1973), smooth muscle myosin light chain kinase (Adelstein *et al.*, 1978), and phosphodiesterase (Sharma *et al.*, 1980), it is likely that cAMP-dependent protein kinase(s) and phosphoprotein phosphatase(s) are associated with calcisome components. The supramolecular assembly then becomes the quantum of the concerted regulation by calcium ions and cyclic nucleotides.

VII. CONCLUSION

Calcium ions represent the only ionic signal in eukaryotic cells. First shown to be involved in muscle contraction (Ebashi, 1972), Ca^{2+} are now known to control a number of physiological events, e.g., contraction, secretion, glycogenolysis, microtubule assembly–disassembly, and synaptic transmission (Means and Dedman, 1980; Cheung, 1980). Whereas the cAMP signal is mediated in eukaryotes by a single class of proteins, cAMP-dependent protein kinases, a number of Ca^{2+}-binding proteins evolved from a primordial one-domain ancestor which probably exhibited a Ca^{2+}-specific triggering site. A gain in affinity and in specificity was brought about by tandem duplications leading to a four-domain ancestral protein that was probably very close to calmodulin. The kinetic improvements needed for the function of fast tissues, either sarcomeric muscle or nervous tissue, came from the evolution of the ancestor molecule either toward simply high-affinity $Ca^{2+}-Mg^{2+}$ relaxing sites, such as domains III and IV of troponin C, or toward specialized proteins that do not interact with other macromolecules and contain only such relaxing sites, such as parvalbumins. Finally, triggering and relaxing sites may reside in the same molecule (troponin C) or in distinct entities of a supramolecular complex (calmodulin and calcineurin B) or in noninteracting proteins (e.g., parvalbumins and troponin C in fast-twitch muscle). Such possibilities provide the kinetic flexibility required for modulation of the Ca^{2+} signal in higher organisms.

ACKNOWLEDGMENTS

This chapter is dedicated to the memory of the late Jean-François Pechère. His major contributions to the study of parvalbumins helped open the field of intracellular calcium-binding proteins.

The author is deeply indebted to Dr. J. Goodman, to Dr. Claude B. Klee, and to Dr. Jacques Haiech for stimulating discussions.

This work was supported in part by grants from CNRS (ATP Modulation de l'action des hormones au niveau cellulaire), DGRST (ACC Biologie et Fonction du Myocarde and Parois artérielles), INSERM (CRL 78.4.086.1, 79.4.151.3, 80.5.032, ATP 63.78.95 Fibre musculaire cardiaque), MDA, NATO grant 1688, and Foundation pour la Recherche Médicale Française. The expert editorial assistance of Ms. Denise Waeckerlé is also gratefully acknowledged.

REFERENCES

Adelstein, R. S., Conti, M. A., Hathaway, D. R., and Klee, C. B. (1978). Phosphorylation of smooth muscle myosin light chain kinase by the catalytic subunit of adenosine $3':5'$-monophosphate-dependent protein-kinase. *J. Biol. Chem.* **253**, 8347–8350.

Amphlett, G. W., Vanaman, T. C., and Perry, S. V. (1976). Effect of troponin C-like protein from bovine brain (brain modulator protein) on the Mg^{2+}-stimulated ATPase of skeletal muscle actomyosin. *FEBS Lett.* **72**, 163–168.

Baron, G., Demaille, J., and Dutruge, E. (1975). The distribution of parvalbumins in muscle and in other tissues. *FEBS Lett.* **56**, 156–160.

Birdsall, W. J., Levine, B. A., Williams, R. J. P., Demaille, J. G., Haiech, J., and Pechère, J. F. (1979). Calcium and magnesium binding by parvalbumin: A proton magnetic resonance study. *Biochimie* **61**, 741–750.

Blum, H. E., Lehky, P., Kohler, L., Stein, E., and Fischer, E. H. (1977). Comparative properties of vertebrate parvalbumins. *J. Biol. Chem.* **252**, 2834–2838.

Blumenthal, D. K., and Stull, J. T. (1980). Activation of purified myosin light chain kinase from skeletal muscle. *Fed. Proc., Fed. Am. Soc. Exp. Biol.* **39**, 2707a.

Cheung, W. Y. (1980). Calmodulin plays a pivotal role in cellular regulation. *Science* **207**, 19–27.

Cohen, P. (1973). The subunit structure of rabbit skeletal muscle phosphorylase kinase, and the molecular basis of its activation reactions. *Eur. J. Biochem.* **34**, 1–14.

Cohen, P. (1978). The role of cyclic AMP-dependent protein kinase in the regulation of glycogen metabolism in mammalian skeletal muscle. *Curr. Top. Cell. Regul.* **14**, 118–192.

Cohen, P., Burchell, A., Foulkes, J. G., Cohen, P. T. W., Vanaman, T. C., and Nairn, A. C. (1978). Identification of the Ca^{2+}-dependent modulator protein as the fourth subunit of rabbit skeletal muscle phosphorylase kinase. *FEBS Lett.* **92**, 287–293.

Cohen, P., Picton, C., and Klee, C. B. (1979). Activation of phosphorylase kinase from rabbit skeletal muscle by calmodulin and troponin. *FEBS Lett.* **104**, 25–30.

Collins, J. H. (1974). Homology of myosin light chains, troponin-C and parvalbumins deduced from comparison of their amino acid sequences. *Biochem. Biophys. Res. Commun.* **58**, 301–308.

Collins, J. H., Greaser, M. L., Potter, J. D., and Horn, M. J. (1977). Determination of the amino acid sequence of troponin C from rabbit skeletal muscle. *J. Biol. Chem.* **252**, 6356–6362.

Crouch, T. H., and Klee, C. B. (1980). Positive cooperative binding of calcium to bovine brain calmodulin. *Biochemistry* **19**, 3692–3698.

Dedman, J. R., Potter, J. D., and Means, A. R. (1977a). Biological cross-reactivity of rat testis phosphodiesterase-activator protein and rabbit skeletal muscle troponin C. *J. Biol. Chem.* **252**, 2437–2440.

Dedman, J. R., Potter, J. D., Jackson, R. L., Johnson, J. D., and Means, A. R. (1977b). Physico-chemical properties of rat testis Ca^{2+}-dependent regulator protein of cyclic nucleotide phosphodiesterase: Relationship of Ca^{2+}-binding, conformational changes and phosphodiesterase activity. *J. Biol. Chem.* **252**, 8415–8422.

Demaille, J. G., Haiech, J., and Goodman, M. (1980). Calcium-binding proteins: From diversification of structure to diversification of function. *Protides Biol. Fluids* **28**, 95–98.

Derancourt, J., Haiech, J., and Pechère, J. F. (1978). Binding of calcium by parvalbumin fragments. *Biochim. Biophys. Acta* **532**, 373–375.

Ebashi, S. (1972). Calcium ions and muscle contraction. *Nature* (*London*) **240**, 217–218.

Ebashi, S. (1980). Regulation of muscle contraction. *Proc. R. Soc. London, Ser. B* **207**, 259–286.

Gillis, J. M. (1980). The biological significance of muscle parvalbumins. *In* "Calcium-Binding Proteins: Structure and Function" (F. L. Siegel *et al.*, eds.). Elsevier/North-Holland Publ., Amsterdam and New York, pp. 309–311.

Gillis, J. M., and Gerday, C. (1977). Calcium movements between myofibrils, parvalbumins and sarcoplasmic reticulum in muscle. *In* "Calcium-Binding Proteins and Calcium Function" (R. H. Wasserman, R. A. Corradino, E. Carafoli, R. H. Kretsinger, D. H. MacLennan, and F. L. Siegel, eds.), pp. 193–196. Elsevier/North Holland Publ., Amsterdam and New York.

Goodman, M., and Pechère, J. F. (1977). The evolution of muscular parvalbumins investigated by the maximum parsimony method. *J. Mol. Evol.* **9**, 131–158.

Goodman, M., Pechère, J. F., Haiech, J., and Demaille, J. G. (1979). Evolutionary diversification of structure and function in the family of intracellular calcium-binding proteins. *J. Mol. Evol.* **13**, 331–352.

Grand, R. J. A., Nairn, A. C., and Perry, S. V. (1980). The preparation of calmodulins from barley (*Hordeum* sp.) and basidiomycete fungi. *Biochem. J.* **185**, 755–760.

Gupta, R. K., and Moore, R. D. (1980). ^{31}P NMR studies of intracellular free Mg^{2+} in intact frog skeletal muscle. *J. Biol. Chem.* **255**, 3987–3993.

Gupta, R. K., and Yushok, W. D. (1980). Non-invasive ^{31}P NMR probes of free Mg^{2+}, Mg ATP and Mg ADP in intact Ehrlich ascites tumor cells. *Proc. Natl. Acad. Sci. U. S. A.* **77**, 2487–2491.

Haiech, J., and Demaille, J. G. (1981). Supramolecular organization of regulatory proteins into calcisomes: A model of the concerted regulation by calcium ions and cyclic adenosine 3':5' monophosphate in eukaryotic cells. *Int. Symp. Metab. Interconversion Enzymes* , pp. 303–313 Springer-Verlag, Berlin.

Haiech, J., Derancourt, J., Pechère, J. F., and Demaille, J. G. (1979). Magnesium and calcium binding to parvalbumins: Evidence for differences between parvalbumins and an explanation of their relaxing function. *Biochemistry* **18,** 2752–2758.

Haiech, J., Klee, C. B., and Demaille, J. G. (1981). Effects of cations on the affinity of calmodulin for calcium-ordered binding of calcium ions allows the specific activation of calmodulin-stimulated enzymes. *Biochemistry* **20,** 3890–3897.

Huang, C. Y., Chau, V., Chock, P. B., Sharma, R. K., and Wang, J. H. (1980). On the mechanism of Ca^{2+} activation of calmodulin-regulated cyclic nucleotide phosphodiesterase. *Fed. Proc., Fed. Am. Soc. Exp. Biol.* **39,** 288a.

Jarrett, H. W., and Kyte, J. (1979). Human erythrocyte calmodulin: Further chemical characterization and the site of its interaction with the membrane. *J. Biol. Chem.* **254,** 8237–8244.

Jones, H. P., Matthews, J. C., and Cormier, M. J. (1979). Isolation and characterization of Ca^{2+}-dependent modulator protein from the marine invertebrate *Renilla reniformis. Biochemistry* **18,** 55–60.

Kilhoffer, M. C., Demaille, J. G., and Gerard, D. (1980a). Terbium as luminescent probe of calmodulin binding sites: Domains I and II contain the high affinity sites. *FEBS Lett.* **116,** 269–272.

Kilhoffer, M. C., Gerard, D., and Demaille, J. G. (1980b). Terbium binding to octopus calmodulin provides the complete sequence of ion binding. *FEBS Lett.* **120,** 99–103.

Kilhoffer, M. C., Demaille, J. G., and Gérard, D. (1981). Tyrosine fluorescence of ram testis and octopus calmodulins. Effects of calcium, magnesium and ionic strength. *Biochemistry* **20,** 4407–4414.

Klee, C. B. (1977). Conformational transition accompanying the binding of Ca^{2+} to the protein of 3',5'-cyclic adenosine monophosphate phosphodiesterase. *Biochemistry* **16,** 1017–1024.

Klee, C. B., and Haiech, J. (1980). Concerted role of calmodulin and calcineurin in calcium regulation. *Ann. N. Y. Acad. Sci.* **356,** 43–54.

Klee, C. B., and Krinks, M. H. (1978). Purification of cyclic 3',5'-nucleotide phosphodiesterase inhibitory protein by affinity chromatography on activator protein coupled to Sepharose. *Biochemistry* **17,** 120–126.

Klee, C. B., Crouch, T. H., and Krinks, M. H. (1979a). Calcineurin: A calcium- and calmodulin-binding protein of the nervous sytem. *Proc. Natl. Acad. Sci. U. S. A.* **76,** 6270–6273.

Klee, C. B., Crouch, T. H., and Krinks, M. H. (1979b). Subunit structure and catalytic properties of bovine brain Ca^{2+}-dependent cyclic nucleotide phosphodiesterase. *Biochemistry* **18,** 722–729.

Klee, C. B., Crouch, T. H., and Richman, P. G. (1980). Calmodulin. *Annu. Rev. Biochem.* **49,** 489–515.

Krebs, E. G., and Beavo, J. A. (1979). Phosphorylation-dephosphorylation of enzymes. *Annu. Rev. Biochem.* **48,** 923–959.

Kretsinger, R. H. (1980). Structure and evolution of calcium-modulated proteins. *CRC Crit. Rev. Biochem.* **8,** 119–174.

La Porte, D. C., Wierman, B. M., and Storm, D. R. (1980). Calcium-induced exposure of a hydrophobic surface on calmodulin. *Biochemistry* **19,** 3814–3819.

Leavis, P. C., Rosenfeld, S. S., Gergely, J., Grabarek, Z., and Drabikowski, W. (1978). Proteolytic fragments of troponin C. Localization of high and low-affinity Ca^{2+}-

binding sites and interactions with troponin I and troponin T. *J. Biol. Chem.* **253**, 5452–5459.

Leavis, P. C., Nagy, B., Lehrer, S. S., Bialkowska, H., and Gergely, J. (1980). Terbium binding to troponin C: Binding stoichiometry and structural changes induced in the protein. *Arch. Biochem. Biophys.* **200**, 17–21.

Le Donne, N. C., Jr., and Coffee, C. J. (1979). Inability of parvalbumin to function as a calcium-dependent activator of cyclic nucleotide phosphodiesterase activity. *J. Biol. Chem.* **254**, 4317–4320.

Lehky, P., Blum, H. E., Stein, E. A., and Fischer, E. H. (1974). Isolation and characterization of parvalbumins from the skeletal muscle of higher vertebrates. *J. Biol. Chem.* **249**, 4332–4334

Le Peuch, C. J., Demaille, J. G., and Pechère, J. F. (1978). Radioelectrophoresis—A specific microassay for parvalbumins: Application to muscle biopsies from man and other vertebrates. *Biochim. Biophys. Acta* **537**, 153–159.

Le Peuch, C. J., Ferraz, C., Walsh, M. P., Demaille, J. G., and Fischer, E. H. (1979). Calcium and cyclic nucleotide dependent regulatory mechanisms during development of chick embryo skeletal muscle. *Biochemistry* **18**, 5267–5273.

Lin, Y. M., Liu, Y. P., and Cheung, W. Y. (1974). Cyclic 3':5'-nucleotide phosphodiesterase: Purification, characterization and active form of the protein activator from bovine brain. *J. Biol. Chem.* **249**, 4943–4954.

Marban, E., Rink, T. J., Tsien, R. W., and Tsien, R. Y. (1980). Free calcium in heart muscle at rest and during contraction measured with Ca^{2+}-sensitive microelectrodes. *Nature (London)* **286**, 845–850.

Martonosi, A., Roufa, D., Boland, R., Ryees, E., and Tillack, T. W. (1977). Development of sarcoplasmic reticulum in cultured chicken muscle. *J. Biol. Chem.* **252**, 318–332.

Maximov, E. E., Zapevalova, N. P., and Mitin, Yu. V. (1978). On the calcium-binding ability of the synthetic evolutionary ancestor of calcium-binding proteins. *FEBS Lett.* **88**, 80–82.

Means, A. R., and Dedman, J. R. (1980). Calmodulin—An intracellular calcium receptor. *Nature (London)* **285**, 73–77.

Moews, P. C., and Kretsinger, R. H. (1975). Refinement of the structure of carp muscle calcium-binding parvalbumin by model building and difference Fourier analysis. *J. Mol. Biol.* **91**, 201–228.

Molla, A., Kilhoffer, M. C., Ferraz, C., Audemard, E., Walsh, M. P., and Demaille, J. G. (1981). Octopus calmodulin: The trimethyllysyl residue is not required for myosin light chain kinase activation. *J. Biol. Chem.* **256**, 15–18.

Nagy, B., and Gergely, J. (1979). Extent and localization of conformational changes in troponin C caused by calcium binding: Spectral studies in the presence and absence of 6*M* urea. *J. Biol. Chem.* **254**, 12732–12737.

Nairn, A. C., and Perry, S. V. (1979). Calmodulin and myosin light chain kinase of rabbit fast skeletal muscle. *Biochem. J.* **179**, 89–97.

Pechère, J. F. (1974). Isolement d'une parvalbumine du muscle de lapin. *C. R. Hebd. Seances Acad. Sci., Ser. D* **278**, 2577–2579.

Pechère, J. F., Capony, J. P., and Demaille, J. (1973). Evolutionary aspects of the structure of muscular parvalbumins. *Syst. Zool.* **22**, 533–548.

Pechère, J. F., Derancourt, J., and Haiech, J. (1977). The participation of parvalbumins in the activation-relaxation cycle of vertebrate fast skeletal muscle. *FEBS Lett.* **75**, 111–114.

Perry, S. V. (1979). The regulation of contractile activity in muscle. *Biochem. Soc. Trans.* **7**, 593–617.

Perry, S. V., Grand, R. J. A., Nairn, A. C., Vanaman, T. C., and Wall, C. M. (1979). Calcium-binding proteins and the regulation of contractile activity. *Biochem. Soc. Trans.* **7**, 619–622.

Potter, J. D., and Gergely, J. (1975). The calcium and magnesium binding sites on troponin and their role in the regulation of myofibrillar adenosine triphosphatase. *J. Biol. Chem.* **250**, 4628–4633.

Potter, J. D., Dedman, J. R., and Means, E. R. (1977a). Ca^{2+}-dependent regulation of cyclic AMP phosphodiesterase by parvalbumin. *J. Biol. Chem.* **252**, 5609–5611.

Potter, J. D., Johnson, J. D., Dedman, J. R., Schreiber, W. E., Mandel, F., Jackson, R. L., and Means, A. R. (1977b). Calcium binding proteins: Relationship of binding, structure, conformation and biological function. *In* "Calcium-Binding Proteins and Calcium Function" (R. H. Wasserman, R. A. Corradino, E. Carafoli, R. H. Kretsinger, D. H. MacLennan, and F. L. Siegel, eds.), pp. 239–250. Elsevier/North-Holland Publ., Amsterdam and New York.

Rasmussen, H. (1970). Cell communication, calcium ion, and cyclic adenosine monophosphate. *Science* **170**, 405–412.

Rasmussen, H., and Goodman, D. B. P. (1977). Relationships between calcium and cyclic nucleotides in cell activation. *Physiol. Rev.* **57**, 421–509.

Robison, G. A., Butcher, R. W., and Sutherland, E. W. (1971). "Cyclic AMP." Academic Press, New York.

Rosen, O. M., Rangel-Aldao, R., and Erlichman, J. (1977). Soluble cyclic AMP-dependent protein-kinases: Review of the enzyme isolated from bovine cardiac muscle. *Curr. Top. Cell. Regul.* **12**, 39–74.

Rubin, C. S., and Rosen, O. M. (1975). Protein phosphorylation. *Annu. Rev. Biochem.* **44**, 831–887.

Seamon, K. B. (1980). Calcium- and magnesium-dependent conformational states of calmodulin as determined by nuclear magnetic resonance. *Biochemistry* **19**, 207–215.

Sharma, R. K., Desai, R., Waisman, D. M., and Wang, J. H. (1979). Purification and subunit structure of bovine brain modulator binding protein. *J. Biol. Chem.* **254**, 4276–4282.

Sharma, R. K., Wang, T. H., Wirch, E., and Wang, J. H. (1980). Purification and properties of bovine brain calmodulin-dependent cyclic nucleotide phosphodiesterase. *J. Biol. Chem.* **255**, 5916–5923.

Stevens, F. C., Walsh, M., Ho, H. C., Teo, T. S., and Wang, J. H. (1976). Comparison of calcium binding proteins: Bovine heart and bovine brain protein activators of cyclic nucleotide phosphodiesterase and rabbit skeletal muscle troponin C. *J. Biol. Chem.* **251**, 4495–4500.

Taylor, D. L., Blinks, J. R., and Reynolds, G. (1980). Contractile basis of ameboid movement. VIII. Aequorin luminescence during ameboid movement, endocytosis and capping. *J. Cell Biol.* **86**, 599–607.

Teo, T. S., and Wang, J. H. (1973). Mechanism of activation of a cyclic adenosine 3':5'-monophosphate phosphodiesterase from bovine heart by calcium ions: Identification of the protein activator as a Ca^{2+} binding protein. *J. Biol. Chem.* **248**, 5950–5955.

Tillotson, D., and Gorman, A. L. F. (1980). Non-uniform Ca^{2+}-buffer distribution in a nerve cell body. *Nature (London)* **286**, 816–817.

Wallace, R. W., Lynch, T. J., Tallant, E. A., and Cheung, W. Y. (1979). Purification and characterization of an inhibitor protein of brain adenylate cyclase and cyclic nucleotide phosphodiesterase. *J. Biol. Chem.* **254**, 377–382.

Walsh, M., and Stevens, F. C. (1978). Chemical modification studies on the Ca^{2+}-dependent protein modulator: The role of methionine residues in the activation of cyclic nucleotide phosphodiesterase. *Biochemistry* **17**, 3924–3933.

Walsh, M. P., Vallet, B., Cavadore, J. C., and Demaille, J. G. (1980). Homologous calcium-binding proteins in the activation of skeletal, cardiac and smooth muscle myosin light chain kinases. *J. Biol. Chem.* **255**, 335–337.

Walsh, M. P., and Guilleux, J. C. (1981). Calcium and cyclic AMP-dependent regulation of myofibrillar calmodulin-dependent myosin light chain kinases from cardiac and skeletal muscles. *Adv. Cyclic Nucleotide Res.* **14**, 375–390.

Wang, J. H., and Desai, R. (1977). Modulator binding protein: Bovine brain protein exhibiting the Ca^{2+}-dependent association with the protein modulator of cyclic nucleotide phosphodiesterase. *J. Biol. Chem.* **252**, 4175–4184.

Watterson, D. M., Harrelson, W. G., Jr., Keller, P. M., Sharief, F., and Vanaman, T. C. (1976). Structural similarities between the Ca^{2+}-dependent regulatory proteins of 3':5'-cyclic nucleotide phosphodiesterase and actomyosin ATPase. *J. Biol. Chem.* **251**, 4501–4513.

Watterson, D. M., Mendel, P. A., and Vanaman, T. C. (1980a). Comparison of calcium-modulated proteins from vertebrate brains. *Biochemistry* **19**, 2672–2676.

Watterson, D. M., Sharief, F., and Vanaman, T. C. (1980b). The complete amino acid sequence of the Ca^{2+}-dependent modulator protein (calmodulin) of bovine brain. *J. Biol. Chem.* **255**, 962–975.

Weeds, A. G., and McLachlan, A. D. (1974). Structural homology of myosin alkali light chains, troponin C and carp calcium binding protein. *Nature (London)* **252**, 646–649.

Wolff, D. J., Brostrom, M. A., and Brostrom, C. O. (1977). Divalent cation binding sites of CDR and their role in the regulation of brain cyclic nucleotide metabolism. *In* "Calcium Binding Proteins and Calcium Function" (R. H. Wasserman *et al.*, eds.) pp. 97–106. North Holland, New York.

Yazawa, M., Sakuma, M., and Yagi, K. (1980). Calmodulins from muscles of marine invertebrates, scallop and sea anemone: Comparison with calmodulins from rabbit skeletal muscle and pig brain. *J. Biochem. (Tokyo)* **87**, 1313–1320.

Chapter 5

Troponin

JAMES D. POTTER
J. DAVID JOHNSON

I.	Introduction	145
II.	Ca²⁺ Binding to Troponin and the Regulation of Muscle Contraction	147
	A. Ca²⁺ Binding to Tn, TnI · TnC, and TnC	147
	B. Relationship of Ca²⁺ Binding to Tn and the Activation of Myofibrillar ATPase	150
III.	Thin-Filament Protein Interactions in the Regulation of Muscle Contraction	151
IV.	Structure and Ca²⁺-Induced Structural Changes in Troponin	155
	A. X-Ray	156
	B. Microcalorimetry	157
	C. Proton Magnetic Resonance	157
	D. Circular Dichroism and Intrinsic Fluorescence	159
	E. Extrinsic Probes of Ca²⁺-Induced Structural Changes in Troponin C	160
V.	Propagation of the Ca²⁺-Induced Structural Changes in Troponin C to Thin-Filament Proteins	163
VI.	Rates of Ca²⁺ Exchange and Structural Changes in Troponin C	164
VII.	Rates of Ca²⁺ Exchange in Troponin	167
VIII.	Conclusions	168
	References	169

I. INTRODUCTION

The initial event in the activation (switching on) of muscle contraction is the binding of Ca²⁺ to the regulatory proteins present in the thin actin-containing filaments, which then allows myosin and actin to interact and contraction to occur. During the past two decades much has been learned

CALCIUM AND CELL FUNCTION, VOL. II

about the nature of the proteins which confer this Ca^{2+} requirement on the actomyosin system, and a brief review of our results and those of others on some of the properties of the skeletal and cardiac troponin systems will be presented here.

It was known from the work of Weber and Herz (1963) that the hydrolysis of ATP by myofibrils required a free Ca^{2+} concentration in the micromolar range, although the protein components which conferred this Ca^{2+} requirement were not known. Ebashi (1963) and Ebashi and Ebashi (1964) first showed in the early 1960s that a protein called "native tropomyosin" was required to confer this Ca^{2+} sensitivity on a Ca^{2+}-insensitive actomyosin preparation. It was subsequently shown by Ebashi et al. (1968) that native tropomyosin consisted of two proteins, tropomyosin and a new protein named troponin, both of which were required to confer Ca^{2+} sensitivity on a purified actomyosin preparation. These studies also demonstrated that the troponin component was the Ca^{2+} receptor of native tropomyosin. Subsequent to this, Hartshorne et al. (1968) found that troponin could be further separated into two components which were called troponin A and troponin B, both of which, in addition to tropomyosin, were required to confer Ca^{2+} sensitivity on synthetic actomyosin (purified actin and myosin). Schaub and Perry (1969) using sulfoethyl–Sephadex chromatography in the presence of 6 M urea were also able to separate troponin into two components. With the advent of sodium dodecyl sulfate (SDS) gels in the late 1960s and the use of DEAE-Sephadex chromatography in the presence of 6 M urea, Greaser and Gergely (1971, 1973) were able to separate Tn* into three subunits which they called troponin T (TnT, the tropomyosin binding subunit), troponin I (TnI, the actomyosin ATPase inhibitory subunit), and troponin C (TnC, the Ca^{2+}-binding subunit). All three subunits (reconstituted to form Tn), in addition to tropomyosin (Tm), were required to reconstitute native tropomyosin activity.

Various studies have shown that thin filaments contain 1 mole of Tn · Tm complex per 7 moles of actin (Ebashi et al., 1969; Potter, 1974) and are consistent with the amounts of Tn and Tm required to regulate synthetic actomyosin systems maximally (Greaser and Gergely, 1971, 1973). Thus, Ca^{2+} binding to TnC in the Tn complex probably results in structural changes in the Tn subunits, Tm, and perhaps actin, which result in

---

* Tn, the native complex of TnT, TnI, and TnC; S, skeletal; C, cardiac; STnT, skeletal troponin T (MW 30,503, Pearlstone et al., 1976); CTnT, cardiac TnT (MW ~38,000); STnI, skeletal troponin I (MW 20,864, Wilkinson and Grand, 1975); CTnI, cardiac TnI (MW 23,550 Wilkinson and Grand, 1978); STnC, skeletal troponin C, (MW 17,965, Collins et al., 1977); CTnC, cardiac TnC (MW 18,459, Van eerd and Takahashi, 1976).

the activation of muscle contraction. We will concentrate here on a discussion of the Ca^{2+}-binding properties of Tn, of the interaction of Tn, Tm, and actin, and of the structural changes which occur in TnC and Tn upon binding Ca^{2+}.

II. Ca^{2+} BINDING TO TROPONIN AND THE REGULATION OF MUSCLE CONTRACTION

A. Ca^{2+} Binding to Tn, TnI·TnC, and TnC

Many different techniques were used to measure the Ca^{2+}-binding properties of Tn and TnC when these proteins were first isolated (Greaser and Gergely, 1973; Fuchs and Briggs, 1968; Fuchs, 1972; Hartshorne and Pyun, 1971; Drabikowski and Barylko, 1971; Ebashi et al., 1968), yet there was no clear agreement from these studies as to the number of Ca^{2+}-binding sites or to their affinity for Ca^{2+}. The major difficulty that hampered all these studies was the difficulty in obtaining and regulating the free Ca^{2+} concentrations at low enough levels (nanomolar or less) to measure Ca^{2+} binding. Most of the studies were also carried out in the presence of concentrations of Mg^{2+} that were millimolar or less, which, as will be seen, complicated the resolution of one of the classes of sites on Tn and TnC. The use of metal chelators [ethyleneglycol bis(β-aminoethyl ether)-N,N'-tetraacetic acid (EGTA), ethylenediaminetetraacetic acid (EDTA)] in combination with equilibrium dialysis in the presence and absence of Mg^{2+} (Potter and Gergely, 1975) made it possible to resolve the discrepancies found in the earlier work.

The Ca^{2+}-binding parameters determined for the skeletal (S) and cardiac (C) proteins are listed in Table I. STnC and CTnC contain two (Table I) high-affinity Ca^{2+}-binding sites ($K_{Ca} \cong 10^7 \, M^{-1}$) that also bind Mg^{2+} competitively ($K_{Mg} \cong 10^3 \, M^{-1}$) and have been called the Ca^{2+}–Mg^{2+} sites (Potter and Gergely, 1975; Holroyde et al., 1980). STnC also contains two sites (Table I) of lower Ca^{2+} affinity ($K_{Ca} \cong 10^5 \, M^{-1}$) which are essentially specific for Ca^{2+} at physiological Mg^{2+} concentrations (millimolar) and have been called the Ca^{2+}-specific sites (Potter and Gergely, 1975). CTnC, in contrast to STnC, contains only one Ca^{2+}-specific site (Table I) similar in affinity to the two Ca^{2+}-specific sites in STnC (Johnson et al., 1980a; Holroyde et al., 1980). The existence of only three Ca^{2+}-binding sites on CTnC was predicted by Van eerd and Takahasi (1976) when they compared the sequences of CTnC with the sequence and structure of parvalbumin (Kretsinger and Nockolds, 1973) and the sequence of STnC (Collins et al., 1977) and discovered that two key coordinating aspartic acid

TABLE I

Ca^{2+}- and Mg^{2+}-Binding Properties of Skeletal and Cardiac Troponin Subunits and Complexes[a]

	Type of site	N	Metal	K_a (M^{-1})[b]	Predicted k_{off} (sec^{-1})	Experimental k_{off} (sec^{-1})
STnC	Ca^{2+}-specific	2	Ca^{2+}	3.2×10^5	312	~300 (a)
			Mg^{2+}	$(200)^c$	—	—
Ca^{2+}–Mg^{2+}		2	Ca^{2+}	2.1×10^7	4.8	1 (a), 3–4 (b)
			Mg^{2+}	4.0×10^3	25	8 (a)
S(TnI · TnC)	Ca^{2+}-specific	2	Ca^{2+}	3.5×10^6	29	—
			Mg^{2+}	$(200)^c$	—	—
	Ca^{2+}–Mg^{2+}	2	Ca^{2+}	2.2×10^8	0.45	—
			Mg^{2+}	4.0×10^4	2.5	—
STn	Ca^{2+}-specific	2	Ca^{2+}	4.9×10^6	20	20–25 (b)
			Mg^{2+}	$(200)^c$	—	—
	Ca^{2+}–Mg^{2+}	2	Ca^{2+}	5.3×10^8	0.2	0.3–0.6 (b)
			Mg^{2+}	4.0×10^4	2.5	4.7 (e)
CTnC	Ca^{2+}-specific	1	Ca^{2+}	2.5×10^5, 4.5×10^5 (c)	220–400	~300 (d)
			Mg^{2+}	$(200)^c$	—	—
	Ca^{2+}–Mg^{2+}	2	Ca^{2+}	1.4×10^7	7.2	—
			Mg^{2+}	7.0×10^2, 3.5×10^3 (c)	28–140	—
C(TnI · TnC)	Ca^{2+}-specific	1	Ca^{2+}	1×10^6	100	—
			Mg^{2+}	$(200)^c$	—	—
	Ca^{2+}–Mg^{2+}	2	Ca^{2+}	3.2×10^8	31	—
			Mg^{2+}	3.0×10^3	33	—
CTn	Ca^{2+}-specific	1	Ca^{2+}	2.5×10^6, 3×10^6 (c)	33–40	15–30 (f)
			Mg^{2+}	$(200)^c$	—	—
	Ca^{2+}–Mg^{2+}	2	Ca^{2+}	3.7×10^8	0.3	—
			Mg^{2+}	3.0×10^3, 2×10^4 (f)	5–33	—

[a] The predicted k_{off} rates were determined as k_{on}/K_A, where K_A is the equilibrium binding constant of metal for a particular site and k_{on} is the diffusion-limited on rate of the metal, assumed to be $\sim 10^8$ M^{-1} sec^{-1} for Ca^{2+} and $\sim 10^5$ M^{-1} sec^{-1} for Mg^{2+} after Eigen (1963). The experimentally determined k_{off} rates were obtained from fluorescence stopped-flow studies of the Ca^{2+} off reactions of various labeled Tn's and TnC's as follows: (a) Johnson et al., 1979; (b) Johnson et al., 1980b, 1981a; (c) Johnson et al., 1980a; (d) Johnson et al., 1978b; (e) Johnson et al., 1981b; (f) J. D. Johnson and J. D. Potter, unpublished studies with CTn$_{IA}$. These off rates with CTn$_{IA}$ are dependent on the state of phosphorylation of CTnI (Robertson et al., 1982; Solaro et al., 1981). In some cases where k_{off} (experimental) is $< k_{off}$ (predicted), a slower structural change which occurs subsequent to Ca^{2+} removal may be involved. In most cases the two values for

residues in region I were substituted for by a leucine and an alanine which could not contribute to Ca^{2+} coordination. Similar conclusions about CTnC have been drawn by Leavis and Kraft (1978) based on indirect binding measurements. Thus CTnC has essentially the same Ca^{2+}-binding properties as STnC except that it lacks the Ca^{2+}-specific site present in region I (Fig. 2) of STnC (Holroyde et al., 1980).

Ca^{2+} binding to intact Tn is quite different from binding to TnC and, in general, the Ca^{2+} affinity of all four sites is increased by an order of magnitude in the absence of Mg^{2+} (Table I). The two Ca^{2+}–Mg^{2+} sites in STn have a Ca^{2+} affinity of $\sim 10^8 M^{-1}$ and bind Mg^{2+} with an affinity which is also increased by an order of magnitude ($K_{Mg} \cong 10^4 M^{-1}$). The apparent affinity of these sites for Ca^{2+} is reduced in the presence of Mg^{2+} by the factor $1 + K_{Mg}[Mg^{2+}]$, such that at a free Mg^{2+} concentration of 3 mM this would be 121, resulting in an apparent Ca^{2+}-binding constant for these sites of $4.4 \times 10^6 M^{-1}$ for STn (Table I). Essentially the same results (Table I) were obtained with CTn (Holroyde et al., 1980).

The 1 : 1 molar ratio complexes of STnC and STnI [S(TnI · TnC)] and of CTnC and CTnI [C(TnI · TnC)] have essentially the same Ca^{2+}-binding properties as Tn (Table I). This clearly indicates a specific interaction between TnI and TnC, which probably occurs in the whole Tn complex. It also follows that, if TnI affects the Ca^{2+}-binding properties (structure) of TnC, Ca^{2+} binding to TnC probably alters the structure of TnI (evidence that this is true is presented later in the chapter). It is also interesting that the S(TnI · TnC) complex can replace Tn in regulating acto · Tm · myosin ATPase activity at a S(TnI · TnC)/actin molar ratio of 1 (Perry et al., 1972). The TnT subunit probably anchors the TnI · TnC complex to Tm at a specific position, thus allowing Tn to regulate the function of seven actins.

Although the Ca^{2+}-specific sites were originally thought to be specific for Ca^{2+}, recent evidence suggests that Mg^{2+} binding to sites distinct from the Ca^{2+}-specific sites affects their affinity. The evidence for this comes from studying the Ca^{2+} dependence of dansylaziridine-labeled STnC (STnC$_{DANZ}$; see Section IV,E) and STnT · STnI · STnC$_{DANZ}$ (STn$_{DANZ}$) fluorescence at different Mg^{2+} concentrations. It had been shown previously (Johnson et al., 1978a) that the Ca^{2+}-dependent increase in the fluorescence of STnC$_{DANZ}$ was due primarily to Ca^{2+} binding at the Ca^{2+}-

---

k_{off} are approximately equal, indicating that the fluorescence changes accurately report the rates of Ca^{2+} removal.

[b] The affinity of site(s) for Mg^{2+} distinct from the Ca^{2+}-specific sites which affect their Ca^{2+}-binding properties noncompetitively is shown in parentheses.

[c] Assumed to be the same as in STnC and STn.

specific sites, therefore, making $STnC_{DANZ}$ useful in assessing the Ca^{2+} affinity of these sites. The results of these studies (Potter *et al.*, 1980) show that the apparent K_{Mg} for producing shifts in the Ca^{2+} affinity of these sites is approximately 200 M^{-1} in both $STnC_{DANZ}$ and STn_{DANZ}. Because these shifts appear to be saturable by Mg^{2+} and because both TnC and Tn show the same apparent K_{Mg}, we have concluded that the effect is noncompetitive; i.e., Mg^{2+} does not bind directly to the Ca^{2+}-specific sites. Even if there were competitive binding at these sites, the affinity for Mg^{2+} would be low enough so that at physiological concentrations of Mg^{2+} these sites would still be specific for Ca^{2+}.

B. Relationship of Ca^{2+} Binding to Tn and to the Activation of Myofibrillar ATPase

Previous studies have suggested that only the Ca^{2+}-specific sites of STn's and CTn's are involved in the activation of myofibrillar ATPase. This finding has been based on static as well as kinetic measurements.

The initial studies (Potter and Gergely, 1975) were carried out by varying the free Mg^{2+} concentration and then examining the Ca^{2+} dependence of myofibrillar ATPase at these different Mg^{2+} concentrations. Since no shift was observed in the Ca^{2+} dependence of myofibrillar ATPase, which would have been expected if the $Ca^{2+}-Mg^{2+}$ sites were involved, it was concluded that only the Ca^{2+}-specific sites participated in regulation. Bremel and Weber (1972) had also presented evidence that all four Ca^{2+}-binding sites on Tn must be occupied before actomyosin ATPase activation could occur, which also supported this concept. It should be pointed out that there is still some question about the effect of Mg^{2+} in these systems, and we are currently investigating this problem.

Recent studies on cardiac myofibrillar ATPase (Holroyde *et al.*, 1980) as a function of pCa showed that activation of cardiac myofibrillar ATPase occurs over the same $[Ca^{2+}]$ range as binding to the single Ca^{2+}-specific site in CTn. Thus, in both the cardiac and skeletal systems, it appears that regulation occurs only upon Ca^{2+} binding to the Ca^{2+}-specific regulatory site(s) and not upon Ca^{2+} binding to the $Ca^{2+}-Mg^{2+}$-type sites.

Additional evidence comes from studying the kinetics of Ca^{2+} and Mg^{2+} exchange with the various Ca^{2+}-binding sites on Tn. Our initial studies (Johnson *et al.*, 1979) showed that the rate of dissociation of Ca^{2+} (Table I) from the $Ca^{2+}-Mg^{2+}$ sites of STnC was much slower ($\sim 1-5$ sec^{-1}) than that from the Ca^{2+}-specific sites (~ 300 sec^{-1}), again suggesting that only the Ca^{2+}-specific sites are involved in regulation. Recent work on Tn (Table I) (Johnson *et al.*, 1980b, 1981a; Potter *et al.*, 1980) has shown that the Ca^{2+} off rate from the $Ca^{2+}-Mg^{2+}$ sites is even slower than in TnC (0.6

sec^{-1}), while the off rate of Ca^{2+} from the Ca^{2+}-specific sites remains quite rapid ($\sim$23 sec^{-1}), as we had predicted (Johnson *et al.*, 1979). In addition, the off rate for Mg^{2+} is also very slow ($\sim$5 sec^{-1}) (Table I). Since the free Mg^{2+} in muscle is in the millimolar range, the Ca^{2+}–Mg^{2+} sites would be partially saturated with Mg^{2+} when the muscle is relaxed. This Mg^{2+} would have to dissociate before Ca^{2+} could completely fill these sites, thus reducing the on rate of Ca^{2+} (which is diffusion-limited to both classes of sites) to these sites to essentially the off rate of Mg^{2+}. Thus, not only would Ca^{2+} dissociate slowly from the Ca^{2+}–Mg^{2+} sites, but the association of Ca^{2+} would also be very slow, making it unlikely that these sites play a regulatory role in muscle contraction (Johnson *et al.*, 1980b, 1981a; Potter *et al.*, 1980; Robertson *et al.*, 1981). In contrast, the Ca^{2+}-specific sites bind and dissociate Ca^{2+} rapidly without interference from Mg^{2+} and thus are able to exchange Ca^{2+} rapidly enough to be involved in regulation. We have recently shown that this also holds true for the single Ca^{2+}-specific site of CTn based on stopped-flow measurements (Johnson *et al.*, 1981b) and in CTn as a function of its phosphorylated state (Robertson *et al.*, 1982). Thus it appears that, in both the cardiac and skeletal systems, the Ca^{2+}-specific sites are clearly the regulatory sites and that the Ca^{2+}–Mg^{2+} sites probably play a structural role since they always have either Ca^{2+} or Mg^{2+} bound to them depending on recent past muscle activity.

The role phosphorylation of CTn plays in modulating Ca^{2+} binding to the Ca^{2+}-specific site of CTn will not be discussed here, since we have recently reviewed these studies (Potter *et al.*, 1981; Solaro *et al.*, 1981). All the Ca^{2+}-binding measurements discussed so far have been carried out on Tn's *in vitro* and may not reflect their true Ca^{2+}-binding properties in the thin filament, as suggested by several laboratories (Bremel and Weber, 1972; Fuchs, 1977a,b, 1978; Fuchs and Bayuk, 1976). We are currently investigating the possibility that thin-filament interactions influence Tn Ca^{2+} binding using fluorescently labeled Tn's and direct binding experiments.

III. THIN-FILAMENT PROTEIN INTERACTIONS IN THE REGULATION OF MUSCLE CONTRACTION

Another area of prime interest is the interaction of all the thin-filament proteins (actin, Tm, Tn) and the influence of Ca^{2+} on these interactions. The skeletal muscle thin-filament system has been studied very thoroughly (Hitchcock *et al.*, 1973; Margossian and Cohen, 1973; Potter and Gergely, 1974; Hitchcock, 1975b), and the results of all our studies on the skeletal and cardiac proteins are listed in Table II.

TABLE II

Skeletal and Cardiac Troponin Subunit Interactions with Tropomyosin and Actin

Interaction		Skeletal	Cardiac
TnT + Tm	$\longrightarrow$ TnT · Tm	+	+
TnT + TnI	$\longrightarrow$ TnT · TnI	+	+
Tm · TnT + TnI	$\longrightarrow$ Tm · TnT · TnI	+	+
TnI + A	$\longrightarrow$ TnI · A	+	+
TnI + A · Tm	$\longrightarrow$ TnI · A · Tm	+	+
TnI + TnC	$\xrightarrow{+\,or\,-\,Ca^{2+}}$ CI	+	+
TnC + TnT	$\underset{-Ca^{2+}}{\overset{+Ca^{2+}}{\rightleftharpoons}}$ CT	+[a]	+[a]
CI + A (+ or − Ca^{2+})	$\longrightarrow\!\!\times\!\!\longrightarrow$,	−	−
CI + A · Tm	$\underset{+Ca^{2+}}{\overset{-Ca^{2+}}{\rightleftharpoons}}$ CI · A · Tm	+	+
TnI + Tm	$\longrightarrow\!\!\times\!\!\longrightarrow$	−	−
CI + Tm (+ or − Ca^{2+})	$\longrightarrow\!\!\times\!\!\longrightarrow$	−	−

[a] Occurs only at low ionic strengths in the presence of high Ca^{2+} (0.1 mM) with low affinity.

These interactions are, of course, of considerable interest, since Ca^{2+} binding to Tn is only the first step in the activation of muscle contraction and results in an alteration of these interactions which then allows myosin and actin to interact.

Starting with TnT, one of its primary interactions is with Tm (Greaser and Gergely, 1973), which can be demonstrated either by viscosity or sedimentation measurements to have a stoichiometry of 1:1 (Table II). The other primary interaction of TnT is with TnI and occurs only when the sulfhydryl groups on TnI are reduced (Table II). This was first reported by Horwitz et al. (1979) in the skeletal system, and we have shown that this is also true in the cardiac system (Table II). This has recently been shown by Hincke et al. (1979) as well. Again, the stoichiometry is 1:1 as measured by gel exclusion chromatography. Although TnT is essentially excluded on Sephadex G-200, TnI is not and thus any interaction between the two proteins can be detected by the presence of TnI in the excluded TnT peak. This is also observed when TnT is complexed with Tm (Table II). These measurements are made at fairly high ionic strengths because of the low solubility of both these proteins in the absence of TnC. Again there clearly is a strong stoichiometric interaction between these two proteins, suggesting that this is a vital interaction not previously thought to be of importance in the interaction of the Tn subunits, although several studies clearly suggested and gave evidence for a direct interaction between TnT and TnI (Hartshorne *et al.*, 1968; Hitchcock, 1975a; Katayama, 1979).

This is because it was previously thought that the primary interactions were between TnT and TnC and between TnC and TnI. Our recent experiments with cardiac subunits suggest that the interaction between CTnT and CTnC are weak and rather nonspecific and only occur in the presence of high Ca^{2+} (0.1 mM) at low ionic strengths (they are abolished even in the presence of Ca^{2+} above 0.3 M KCl). The reaction (Table II) appears to be nonspecific, because even at high added CTnC/CTnT molar ratios (6 : 1) a stoichiometry of only ~0.5 mole CTnC bound per CTnT could be obtained. Our current feeling is that TnC probably does not interact with TnT in the Tn complex (Fig. 1), and this is supported by the recent reactivity studies of Hitchcock, Zimmerman, and Smalley (1981), although it is still possible that this interaction may occur in the presence of all three subunits and Ca^{2+}. Thus the primary interactions of TnT are with Tm and with TnI, and the primary interaction of TnC is with TnI. Another important point about CTnT and STnT is their highly asymmetric shape that has been determined by gel exclusion chromatography (Stokes radius, 7.8 nm for CTnT and 4.1 nm for STnT) and from fluorescence and sedimentation studies (Prendergast and Potter, 1979). These studies indicate that the length of CTnT is approximately 10 nm. This has also been shown by Ohtsuki (1975, 1979) using immunoelectron microscopy, and by Horwitz *et al.* (1979) using gel filtration techniques. This strongly suggests that, since TnT covers a considerable portion of Tm, it may play an important role in affecting the structure of Tm (Fig. 1).

TnI forms a stoichiometric complex with TnC (Potter and Gergely, 1974), which is stable in the presence or absence of Ca^{2+} in nondenaturing solvents (Table II). Thus, in contrast to earlier ideas, the Tn complex is probably assembled via TnT · TnI · TnC (Fig. 1) rather than via TnT · TnC · TnI. There are probably two sites on STnC which interact with STnI (Weeks and Perry, 1977), and recent studies by Leavis *et al.* (1978) suggest that these are in region II between residues 46 and 77 and in region III between residues 89 and 120. Hitchcock (1981), using lysine reactivity studies, has come to a similar conclusion with the exception that the binding domain near 89–120 may be larger than estimated by Leavis *et al.* (1978) because of the reduced reactivity of Lys-136 and Lys-140. There are two regions of TnI which interact with TnC. One is located in the N-terminus between residues 1 and 47, and the other between residues 96 and 117, as suggested by the experiments of Syska *et al.* (1976). Thus, there appear to be two regions on each protein which are available for interaction with the other protein. In this regard, it is known that TnI will

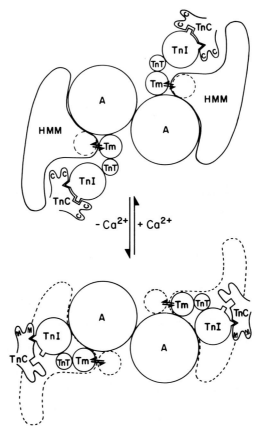

Fig. 1. Proposed model for the regulation of actomyosin interaction by Tn, Tm, and Ca^{2+}. The upper part of the figure illustrates all the thin-filament protein interactions in the presence of Ca^{2+} (activated state) with Tm in the nonblocking position. The lower part of the figure illustrates the relaxed state ($-Ca^{2+}$) where Ca^{2+} is removed from the Ca^{2+}-specific sites (Mg^{2+} has replaced Ca^{2+} at the Ca^{2+}–Mg^{2+} sites) and Tm occupies the blocking position.

bind to TnC independently of the Ca^{2+} concentration in the native state (Potter and Gergely, 1974), but under denaturing conditions (6 M urea) TnI will only bind to TnC in the presence of Ca^{2+} (Syska *et al.*, 1974). These results also suggest that there are two sites of interaction between TnC and TnI. One site is sensitive to Ca^{2+}, while the other site is Ca^{2+}-insensitive but sensitive to denaturants. In Fig. 1 we have illustrated only the Ca^{2+}-dependent change in interaction between TnC and TnI. At the present time there is little evidence supporting a Ca^{2+}-dependent interaction between TnC and TnT.

We have covered so far all the Tn subunit–subunit interactions and the interactions of the Tn subunits with Tm (only TnT interacts). We will now consider the interactions of the Tn subunits with actin. TnI is the only subunit (Table II) that appears to have a specific interaction with actin and is responsible for the well-known inhibition of actomyosin ATPase by TnI, whereas TnI, TnC, and TnI · TnC do not bind to Tm (with or without Ca^{2+}). The TnI · TnC complex, however, binds to actin · Tm in the absence of Ca^{2+} but only weakly in the presence of Ca^{2+} (Potter and Gergely, 1974; Hitchcock, 1975b). This is by far the most significant Ca^{2+}-dependent interaction among all the thin-filament proteins and probably is the key Ca^{2+}-dependent event which brings about the activation of muscle contraction (Fig. 1). That this interaction occurs in the Tn complex in thin filaments is supported by the recent cross-linking studies of Sutoh and Matsuzaki (1980).

Figure 1 incorporates all the information contained in Table II and also shows the newly proposed position of Tm and heavy meromyosin (HMM) in relation to the actin helix as suggested by Taylor and Amos (1981). In Fig. 1, Tm assumes two positions, one sterically blocking the binding of HMM (relaxed, $-Ca^{2+}$ form), and the other occupying a nonblocking position (activated, $+Ca^{2+}$ form). In the model, TnT binds to both Tm and TnI, and this binding is independent of the Ca^{2+} concentration (Fig. 1); TnC binds only to TnI, and the strength of this interaction is increased by Ca^{2+} (Fig. 1). This binding of Ca^{2+} (presumably to the Ca^{2+}-specific sites) also brings about the dissociation of TnI from actin, which allows Tm to move from the "blocking" to the "nonblocking" position, allowing myosin and actin to interact and contraction to occur.

IV. STRUCTURE AND Ca^{2+}-INDUCED STRUCTURAL CHANGES IN TROPONIN

Troponin C isolated from both rabbit skeletal and bovine cardiac muscle has a molecular weight near 18,000. Its structure is represented schematically in Fig. 2. Collins first recognized the four regions of sequence homology in TnC that correspond to the four Ca^{2+}-binding sites in skeletal TnC (Collins *et al.*, 1973). Each Ca^{2+}-binding site is flanked by two regions of α-helicity. Sites I and II represent the Ca^{2+}-specific regulatory sites, while sites III and IV are the high-affinity Ca^{2+}–Mg^{2+} sites. Although STnC and CTnC share many regions of sequence homology, Van eerd and Takahashi (1976) have suggested that site I has sufficient replacements in its Ca^{2+}-coordinating amino acids that it can no longer

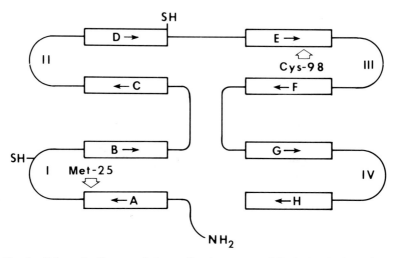

Fig. 2. Schematic diagram of the predicted structure of TnC. The horizontal arrows represent the direction of the polypeptide chain. The helical regions flanking each Ca^{2+}-binding site are labeled A to H. Sites I and II are Ca^{2+}-specific sites, and III and IV are the Ca^{2+}–Mg^{2+} sites. Vertical arrows indicate the location of Met-25 and Cys-98 in STnC. The two cysteine residues 35 and 84 of CTnC are designated by SH. Site I is the alleged mutated Ca^{2+}-specific site of CTnC.

bind Ca^{2+}. Thus CTnC has two Ca^{2+}–Mg^{2+} sites like STnC, one Ca^{2+}-specific regulatory site and one mutated low-affinity Ca^{2+}-binding site.

Since fluorescent probes located on Cys-98 of STnC responded to Ca^{2+} binding to the Ca^{2+}–Mg^{2+} sites and probes located at Met-25 responded to Ca^{2+} binding to the Ca^{2+}-specific sites, it seemed that the site assignment in Fig. 2 was correct (Potter *et al.*, 1976, 1977a; Johnson *et al.*, 1978a). Further, early evidence that the mutant site (site I) in CTnC was a Ca^{2+}-specific site (Potter *et al.*, 1977a; Leavis and Kraft, 1978) indicated that site I in STnC was a Ca^{2+}-specific site. Recently studies on proteolytic fragments of STnC by Leavis *et al.* (1978) seem to confirm that sites I and II are the Ca^{2+}-specific sites and sites III and IV are the Ca^{2+}–Mg^{2+} sites.

The actual amino acid residues involved in coordinating Ca^{2+} binding to each site and each class of site have been discussed in detail elsewhere (Potter *et al.*, 1977a). A variety of techniques have been utilized to determine the structure and Ca^{2+}-induced structural changes in troponin C.

A. X-Ray

The total three-dimensional structure of troponin C can only be determined by high-resolution x-ray crystallographic analysis, and these studies have been hampered by a lack of high-quality stable TnC crystals. Recently, Strasburg *et al.* (1980) have found crystals of chicken TnC containing one molecule per unit cell with diffractions extending to a resolution of better than 2.2 Å. These crystals should be suitable for a full high-resolution three-dimensional structural determination and should provide investigators with the first ''true'' structure of troponin C.

It should be mentioned that the three-dimensional model of TnC predicted by Kretsinger and Barry (1975) (by comparisons of sequence homology of parvalbumin and STnC) has been very useful in our structural studies. It will be of great interest to compare this predicted structure with the actual three-dimensional x-ray crystallography-determined structure.

B. Microcalorimetry

Potter *et al.* (1977b) have used microcalorimetry to measure the enthalpies of Ca^{2+} binding to the various classes of sites on STnC. While the enthalpy of Ca^{2+} binding to each site was ~ -7.7 kcal mole^{-1}, the entropy of Ca^{2+} binding to the $Ca^{2+}-Mg^{2+}$ sites was higher (14.7 eu) than that of Ca^{2+} binding to the Ca^{2+}-specific sites (8.0 eu). Thus the binding of Ca^{2+} to all four sites is driven by both enthalpy and entropy. If the entropy associated with Ca^{2+} binding to each class of sites results from conformational changes subsequent to Ca^{2+} binding, this would indicate that Ca^{2+} binding to the $Ca^{2+}-Mg^{2+}$ sites produces larger structural changes than Ca^{2+} binding to the Ca^{2+}-specific sites. Recently Yamada (1978) conducted microcalorimetry studies on Ca^{2+} binding to TnC in the presence of Mg^{2+}. They also found positive changes in entropy with Ca^{2+} binding to TnC. Mg^{2+} produces a substantial decrease in the apparent Ca^{2+} affinity of the $Ca^{2+}-Mg^{2+}$ sites of TnC (Table I), making it difficult to separate the $Ca^{2+}-Mg^{2+}$ from the Ca^{2+}-specific sites. Yamada's value for the enthalpy change with Ca^{2+} binding to the $Ca^{2+}-Mg^{2+}$ sites was less negative than that determined by Potter *et al.* (1977b). In the presence of Mg^{2+}, Ca^{2+} will displace Mg^{2+} from the $Ca^{2+}-Mg^{2+}$ sites, and the first enthalpy changes he observed could simply have been the differences in enthalpy between Ca^{2+} binding and Mg^{2+} binding. These changes in enthalpy would be less negative than the changes produced by direct Ca^{2+} binding.

C. Proton Magnetic Resonance

Proton magnetic resonance studies on TnC as a function of Ca^{2+} allow a determination of the general types of changes in structure that occur with

Ca^{2+} occupancy of the different classes of Ca^{2+}-binding sites. These studies have been conducted independently by Seamon et al. (1977) and by Levine et al. (1976, 1977).

The structure of apo-TnC has a definite conformation and is not a completely random coil. There is evidence for separate domains of preformed structure which themselves undergo structural changes with Ca^{2+} binding.

In the first stages of Ca^{2+} binding to the $Ca^{2+}-Mg^{2+}$ sites, a broadening of the resonances of phenylalanine, glutamate, and aspartate residues occurs (Levine et al., 1977). This change does not represent a major structural change but merely a stabilization of the constraints on these side chains occurring on a preformed structure. Both groups agree that occupancy of the $Ca^{2+}-Mg^{2+}$ sites results in very large structural changes. These take the form of the structuring of a hydrophobic core of aliphatic side chains and aromatic residues. In particular, phenylalanine residues are involved, with large structural changes also occurring about leucine, isoleucine, and valine residues. There is a general rearrangement of the polypeptide backbone, with a substantial increase in interactions among hydrophobic groups and evidence of unfolding–refolding transitions. It is clear that Ca^{2+} binding to the $Ca^{2+}-Mg^{2+}$ sites directs and stabilizes the folding or formation of much of the structure of TnC, including possible α-helix formation and the structuring of a hydrophobic core within the protein.

With Ca^{2+} binding to the Ca^{2+}-specific sites more subtle structural changes were observed. While there is some tightening of the polypeptide backbone, more localized structural changes predominate. The general picture was one of a more compact conformation with a higher degree of stability or stabilized residues. In the early stages of occupancy of the Ca^{2+}-specific sites, the environments of valine, leucine, and isoleucine residues were stabilized along with some neighboring phenylalanine residue side chains. These more subtle structural changes which occur with occupancy of the Ca^{2+}-specific sites also involved a series of complicated and ordered changes including alterations in glutamate and aspartate residues and changes in the structure of a group of hydrophobic residues which are very near one another. Finally, there is evidence of changes on the surface of the protein.

Ca^{2+} binding to the $Ca^{2+}-Mg^{2+}$ sites, therefore, results in large structural changes and rearrangement of the polypeptide backbone. Mg^{2+} binding to these sites produces almost identical structural changes in TnC (Seamon et al., 1977). Occupancy of the $Ca^{2+}-Mg^{2+}$ sites by Ca^{2+} or Mg^{2+} and the associated structural changes increase the thermal stability of TnC, consistent with the idea that the $Ca^{2+}-Mg^{2+}$ sites are simply "structural sites." Although more subtle, the ordered and more localized structural changes occurring with Ca^{2+} binding to the Ca^{2+}-specific sites presumably serve as the trigger in the regulation of muscle contraction.

D. Circular Dichroism and Intrinsic Fluorescence

Ca^{2+} produces large changes in the far ultraviolet (UV) circular dichroic (CD) spectra of both CTnC and STnC. Spectral minima occur near 222 and 208 nm suggestive of α-helical structure in TnC (Kawasaki and Van eerd, 1972; Van eerd and Kawasaki, 1972). While all reports agree that Ca^{2+} binding to TnC produces rather large increases in its α-helical content (from ~34% α-helix in apo-TnC to ~52% α-helix in Ca^{2+}-saturated TnC) and increases in tyrosine fluorescence, the Ca^{2+} dependence of these increases in α-helix and intrinsic fluorescence are controversial. The discrepancies reported in the literature probably arise from failure to determine accurately and compensate for contaminating Ca^{2+}, failure to use sufficient EGTA or EDTA to remove endogenous Ca^{2+}, the use of different Ca^{2+}–EGTA binding constants, and failure to compensate for the pH changes resulting from the deprotonation of EGTA with Ca^{2+} binding. As a result, contradictory results appear in the literature (McCubbin and Kay, 1973; McCubbin et al., 1974; Murray and Kay, 1972; Hincke et al., 1978).

Using the same Ca^{2+}–EGTA buffering system as used by Potter and Gergely (1975) to determine the Ca^{2+}-binding parameters for TnC, we have determined the Ca^{2+} dependence of the increase in $[\theta]$ 222 and intrinsic tyrosine fluorescence. We have found (Johnson and Potter, 1978) that approximately 62% of the total helical change and 81% of the total increase (60%) in tyrosine fluorescence occurs with Ca^{2+} binding to the Ca^{2+}–Mg^{2+} sites. The remainder of these spectral changes occur with Ca^{2+} binding to the Ca^{2+}-specific regulatory sites. Mg^{2+} could produce most of the spectral changes produced by Ca^{2+} binding to the Ca^{2+}–Mg^{2+} sites. These CD and intrinsic fluorescence results have been substantiated by Hincke et al. (1978) and Iio and Kondo (1980), respectively.

Thus, in agreement with the proton magnetic resonance (PMR) studies, large structural rearrangements involving a tightening of the polypeptide backbone and, in particular, increases in α-helicity, occur with occupancy of the Ca^{2+}–Mg^{2+} sites by either Ca^{2+} or Mg^{2+}. A more subtle compacting of the structure and increases in α-helicity occur with Ca^{2+} binding to the Ca^{2+}-specific sites of TnC. TnC undergoes very similar increases in α-helicity. Approximately 78% of the total change in α-helicity occurs with Ca^{2+} binding to the Ca^{2+}–Mg^{2+} sites, while the remainder of the increase occurs with Ca^{2+} binding to the Ca^{2+}-specific sites (Johnson et al., 1980a). The total Ca^{2+}-induced change in α-helicity for CTnC is from ~36 to ~48%.

E. Extrinsic Probes of Ca^{2+}-Induced Structural Changes in Troponin C

With Ca^{2+} binding to the Ca^{2+}–Mg^{2+} sites of STnC, the reactivity of its single cysteine residue is reduced (Potter *et al.*, 1976). Cys-98 lies on the Ca^{2+}–Mg^{2+} "side" of STnC, and all reported labels on Cys-98 respond primarily to Ca^{2+} binding to the Ca^{2+}–Mg^{2+} sites. Cys-98 labeled with a fluorinated sulfhydryl reagent (Seamon *et al.*, 1977) or with a sulfhydryl-directed spin label (Potter *et al.*, 1976) reports structural changes occurring with Ca^{2+} or Mg^{2+} binding to the Ca^{2+}–Mg^{2+} sites. Fluorescent probe molecules (which are more environmentally sensitive) directed toward Cys-98 show fluorescence changes with Ca^{2+} binding to the Ca^{2+}–Mg^{2+} sites and smaller fluorescence changes with Ca^{2+} occupancy of the Ca^{2+}-specific sites. For example, when Cys-98 is labeled with 2-(4'-iodoacetamidoanilino)naphthalene-6-sulfonic acid (IAANS), it undergoes a ~50% decrease in fluorescence with Ca^{2+} (or Mg^{2+}) binding to the Ca^{2+}–Mg^{2+} sites and a ~10% increase in fluorescence with Ca^{2+} binding to the Ca^{2+}-specific sites (J. D. Johnson and J. D. Potter, unpublished observations). Similarly, Potter *et al.* (1976) have labeled Cys-98 with dansyl-Cys-(Hg) and observed a 40% fluorescence increase with Ca^{2+} binding to the Ca^{2+}–Mg^{2+} sites followed by a 35% increase in fluorescence with Ca^{2+} binding to the Ca^{2+}-specific sites.

Since the Ca^{2+}–Mg^{2+} sites are probably structural sites and are always occupied by Ca^{2+} or Mg^{2+} (with each divalent producing similar large structural changes), we directed our efforts toward labeling both skeletal and cardiac TnC on the Ca^{2+}-specific side of each protein in an attempt to monitor the more subtle structural changes which occur with Ca^{2+} binding to the Ca^{2+}-specific regulatory sites of TnC. We have labeled STnC at Met-25 (Fig. 2) with the fluorescent probe molecule dansylaziridine to form TnC_{DANZ}. TnC_{DANZ} undergoes a small 15–20% fluorescence decrease with Ca^{2+} binding to the Ca^{2+}–Mg^{2+} sites and a large two fold or greater fluorescence increase with Ca^{2+} binding to the Ca^{2+}-specific sites (Johnson *et al.*, 1978a; Figure 3). In the presence of 3 mM Mg^{2+}, only the large Ca^{2+}-specific fluorescence increase is observed. This provides us with an accurate and convenient means of monitoring specifically Ca^{2+} binding to the Ca^{2+}-specific regulatory sites of TnC. TnC_{DANZ} still behaves like the native protein in terms of its Ca^{2+}-binding properties, its Ca^{2+}-induced increases in α-helicity, and its ability to complex with TnI and TnT. Complex formation of TnC_{DANZ} with TnI or TnI and TnT together increases the affinity of the Ca^{2+}-specific sites from ~5 × 10^5 to ~3 × 10^6 M^{-1}, as determined from the midpoint of these fluorescence changes. This is consistent with the increase in Ca^{2+} affinity of the Ca^{2+}-specific sites of TnC that

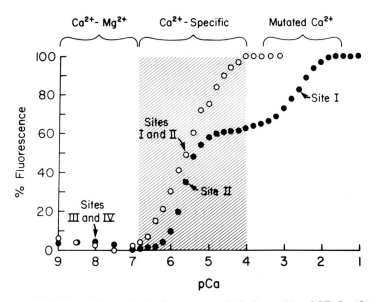

Fig. 3. Ca²⁺-induced changes in the fluorescence of STnC$_{DANZ}$ (○) and CTnC$_{IA}$ (●). The approximate pCa ranges where the Ca²⁺–Mg²⁺ sites, the Ca²⁺-specific sites, and the mutated Ca²⁺ site (of CTnC) would bind Ca²⁺ are as indicated. Note that Ca²⁺ binding to sites I and II of both STnC$_{DANZ}$ and CTnC$_{IA}$ produces a ~2.1-fold fluorescence increase but that the affinity of site I (the mutated site) in CTnC is greatly reduced. Labeled proteins were prepared and Ca²⁺ titrations were conducted as described in Johnson *et al.* (1978a) for STnC$_{DANZ}$ and in Johnson *et al.* (1980a) for CTnC$_{IA}$.

occurs with TnI binding, as determined by direct binding measurements (Table I).

The reactivity of dansylaziridine with Met-25 of TnC and the fluorescence of the noncovalent hydrophobic site-directed fluorescent probe molecule 1-anilinonaphthalene-8-sulfonate (ANS) increase with Ca²⁺ binding to the Ca²⁺-specific sites of TnC. Met-25 lies directly above a cluster of four phenylalanine residues (residues 19, 23, 26, and 75) which form a hydrophobic pocket in the predicted structure (Kretsenger and Barry, 1975) of TnC. We feel that with Ca²⁺ binding to the Ca²⁺-specific sites of TnC, additional increases in α-helicity in the A, B, and C, D helical sections of TnC (Fig. 2) form or expose a hydrophobic site comprised of the side chains of these phenylalanine residues. The formation of this site, with Ca²⁺ binding to the regulatory sites of TnC, would explain the greater reactivity of dansylaziridine with Met-25, the greater fluorescence of TnC$_{DANZ}$, and the higher fluorescence of ANS when Ca²⁺ occupies the Ca²⁺-specific sites of TnC. The role of the formation of this hydrophobic site in TnC is as yet unknown. Studies of Leavis *et al.* (1978) have shown, however, that in region II, residues 46–77 of TnC may be involved in TnI binding. It is tempting to correlate these structural changes near Met-25 with the observed alterations in phenylalanine resonance and structural changes in a cluster of hydrophobic residues which have been reported to occur (using PMR) as Ca^{2+} binds to the Ca^{2+}-specific sites of STnC (Levine *et al.*, 1977). These structural changes may play a role in propagation of the Ca^{2+} signal from TnC to other thin-filament proteins to alter the actomyosin interaction and regulate muscle contraction.

From studies on sequence homology, CTnC should have two Ca^{2+}–Mg^{2+} sites, one Ca^{2+}-specific site and a mutated Ca^{2+}-specific site (Van eerd and Takahashi, 1976). While the Ca^{2+}–Mg^{2+} sites of CTnC appeared to be very similar to their sister sites in STnC, the true affinity of Ca^{2+}-specific site II had been underestimated by direct Ca^{2+}-binding measurements because of contributions from mutated site I (for discussion, see Johnson *et al.*, 1980a; Holroyde *et al.*, 1980). CTnC is particularly amenable to fluorescence labeling because it has Cys-35 located off the A, B helices at mutated site I and Cys-84 located off the D helices of Ca^{2+}-specific site II. These cysteine residues were approximately equally labeled with the sulfhydryl-specific fluorescence probe molecule IAANS to form $CTnC_{IA}$. This fluorescent CTnC behaves as native TnC in terms of its Ca^{2+}-binding properties, Ca^{2+}-induced helical changes, and complex formation with other troponin subunits. $CTnC_{IA}$ undergoes a small fluorescence decrease with Ca^{2+} (or Mg^{2+}) binding to the Ca^{2+}–Mg^{2+} sites on the other side of the molecule, and a 2.1-fold biphasic increase in fluorescence with further Ca^{2+} binding (Fig. 3). Approximately 62% of this Ca^{2+}-specific fluorescence increase occurs with Ca^{2+} binding to Ca^{2+}-specific site II with $K_{Ca} \sim 4.5 \times 10^5\ M^{-1}$. The remaining 38% of the increase in $CTnC_{IA}$ fluorescence occurs with Ca^{2+} binding to a site(s) with $K_{Ca} \sim 5 \times 10^2\ M^{-1}$ (Johnson *et al.*, 1980a). This latter 38% fluorescence increase may occur with Ca^{2+} binding to mutated Ca^{2+}-specific site I of CTnC, since one label is near that site on Cys-35 and it is expected to have a greatly reduced Ca^{2+} affinity. Both sites responsible for these fluorescence increases are Ca^{2+}-specific, since the increases occur even in the presence of 30 mM Mg^{2+}. Thus, with fluorescent probes located near sites I and II of CTnC we can detect and characterize Ca^{2+} binding to both the single Ca^{2+}-specific regulatory site (site II) and the mutated Ca^{2+}-specific site (site I) of CTnC. This fluorescence study clearly pointed out, for the first time, the existence of a single Ca^{2+}-specific regulatory site in CTnC with similar affinity for the two Ca^{2+}-specific regulatory sites in STnC. This observation has been verified by direct Ca^{2+}-binding studies (Holroyde *et al.*, 1980).

When $CTnC_{IA}$ was complexed with TnI or reconstituted to whole troponin, it reported Ca^{2+} binding to a single Ca^{2+}-specific site ($K_{Ca} \sim 3 \times 10^6 M^{-1}$) with a ~25% fluorescence decrease (Johnson et al., 1980a). Direct binding studies verified this increase in Ca^{2+} affinity of the single Ca^{2+}-specific site of CTnC with complex formation (Holroyde et al., 1980). With these finding CTnC and CTn now seem much more similar to STnC and STn in terms of their Ca^{2+}-binding properties than when it was believed that CTnC and CTn lacked a high-affinity Ca^{2+}-specific site.

V. PROPAGATION OF THE Ca^{2+}-INDUCED STRUCTURAL CHANGES IN TROPONIN C TO THIN-FILAMENT PROTEINS

Fundamental to current concepts on the mechanism of muscle contraction is the propagation of the Ca^{2+}-induced structural changes in TnC throughout other thin-filament proteins to regulate the interaction of actin: with myosin, and contraction. In an effort to monitor and map the propagation of the Ca^{2+} signal from TnC throughout thin-filament proteins, we have labeled troponin I at a single cysteine residue with the fluorescent probe molecule IAANS and reconstituted it with TnT and TnC to form a fluorescent whole troponin, Tn_{IA}. Tn_{IA} undergoes a 25–30% fluorescence decrease with Ca^{2+} binding to the high-affinity Ca^{2+}–Mg^{2+} sites and a large 2.1-fold fluorescent decrease with Ca^{2+} binding to the Ca^{2+}-specific sites of whole Tn (Johnson et al., 1980b, 1981a). In the presence of 3 mM Mg^{2+}, only a large (2.1-fold) fluorescence decrease is observed, and it is half-maximal at pCa 6.5. Thus, Ca^{2+}-binding to the Ca^{2+}–Mg^{2+} sites and the Ca^{2+}-specific sites of TnC propagate structural changes into this region of TnI. In the presence of physiological concentrations of Mg^{2+}, only the structural (fluorescence) changes produced by Ca^{2+} binding to the Ca^{2+}-specific regulatory sites are observed. These structural changes in TnI which occur with occupancy of the Ca^{2+}-specific sites on TnC take place rapidly enough (see following section) to be involved in the chain of Ca^{2+}-dependent structural changes which begin on the Ca^{2+}-specific side of TnC and are propagated into TnI to alter the TnI–actin complex. This would allow Tm to move deeper into the superhelical grooves of F-actin and expose the myosin-binding site on actin, allowing actomyosin interaction and muscle contraction to occur (Fig. 1). Consistent with the general idea that the Ca^{2+} signal is propagated into other thin-filament proteins, Ohnishi et al. (1975) detected Ca^{2+}-induced structural changes in TnI labeled with a nitroxide spin label when TnI was complexed with TnC. Tonomura et al. (1969) used spin labels attached to Tm or actin. They detected Ca^{2+}-induced structural changes in Tm only in the presence of Tn, and Ca^{2+}-induced changes in actin were observed only when both Tn and Tm were present. Ohyashiki *et al.* (1976) used fluorescently labeled Tm and fluorescence polarization to show that Tn strengthened the interaction between Tm and F-actin and that Ca^{2+} reversed this effect. These studies indicate that the effects of Ca^{2+}-binding to TnC are felt even in Tm and F-actin.

Recently, Miki (1979) has observed fluorescence energy transfer from ϵ-ADP bound to F-actin to 7-chloro-4-nitrobenzo-2-oxa-1,3-diazole (NBD) bound to TnC. The efficiency of this energy transfer was decreased by the addition of Ca^{2+}. Although alternative interpretations are possible, this suggests that Ca^{2+} binding to TnC alters the relative distance (or orientation) between TnC and F-actin.

It would be of great interest to determine the Ca^{2+} dependence of the above effects to see if Ca^{2+} binding to the Ca^{2+}–Mg^{2+} sites or to the Ca^{2+}-specific side of TnC is involved. We predict that the regulatory sites are responsible. It is clear, however, that Ca^{2+} binding to TnC produces structural changes which are transmitted throughout the thin filament. These structural changes in thin-filament proteins may serve to alter the interactions of these proteins as represented schematically in Fig. 1.

VI. RATES OF Ca^{2+} EXCHANGE AND STRUCTURAL CHANGES IN TROPONIN C

We have used the small fluorescence decrease occurring with Ca^{2+} or Mg^{2+} binding to the Ca^{2+}–Mg^{2+} sites and the large fluorescence increase occurring with Ca^{2+} binding to the Ca^{2+}-specific sites of $STnC_{DANZ}$ and stopped-flow fluorometry to determine Ca^{2+} exchange rates with each class of site in STnC (Johnson *et al.*, 1979). In summary, we have found that Ca^{2+} binds to both the Ca^{2+}–Mg^{2+} sites and the Ca^{2+}-specific sites of TnC to produce their respective fluorescence changes within the 2.4-msec mixing time of the instrument. This suggests that Ca^{2+} binding and the structural changes reported by TnC_{DANZ} occur very rapidly, consistent with the idea that Ca^{2+} binding to each class of site on TnC is diffusion-controlled and occurs at $\sim 10^{8} \ M^{-1} \ sec^{-1}$.

Ca^{2+} removal from the Ca^{2+}-specific sites of TnC_{DANZ} produced the expected fluorescence decrease with a $\tau_{1/2}$ of 2–3 msec, at a rate of 230–346 sec^{-1}. Assuming that this was the rate of Ca^{2+} removal from the Ca^{2+}-specific sites proved valid, since the equilibrium binding constant of Ca^{2+} to these sites was approximately equal to the ratio of the Ca^{2+} on rate (assumed to be diffusion controlled, $1 \times 10^{8} \ M^{-1} \ sec^{-1}$) and the observed

Ca^{2+} off rate (~300 sec^{-1}) (Table I). Thus Ca^{2+} exchange with the Ca^{2+}-specific sites of STnC occurs very rapidly and is reported accurately by TnC$_{DANZ}$ fluorescence.

Removal of Ca^{2+} from the higher-affinity Ca^{2+}–Mg^{2+} sites produced the expected small increase in fluorescence with a $\tau_{1/2}$ of ~700 msec or at a rate of ~1 sec^{-1}. We suggested that this resulted from a somewhat slower structural change which occurs subsequent to Ca^{2+} binding to the Ca^{2+}–Mg^{2+} sites. From the known affinity of the Ca^{2+}–Mg^{2+} sites ($2 \times 10^7 M^{-1}$) and assuming a diffusion-controlled on rate ($1 \times 10^8 M^{-1}$ sec^{-1}), we predicted an off rate of Ca^{2+} from the high-affinity Ca^{2+}–Mg^{2+} sites of ~4.8 sec^{-1} (Table I; Johnson et $al.$, 1979). It is perhaps not surprising that the structural changes occurring at Met-25 (on the Ca^{2+}-specific side of the molecule) are somewhat slower than the actual rate of Ca^{2+} removal from the Ca^{2+}–Mg^{2+} sites, since these structural changes must be propagated across the protein to Met-25. Recently, we have labeled Cys-98 with IAANS and observed that it undergoes a 50% fluorescence decrease with Ca^{2+} binding to the Ca^{2+}–Mg^{2+} sites. Removal of Ca^{2+} from these sites with EGTA (or EDTA) results in a reversal of this fluorescence decrease at a rate of ~3 sec^{-1} (Johnson et $al.$, 1981a). Thus a label on the Ca^{2+}–Mg^{2+} side of TnC reports more accurately the actual Ca^{2+} off rates from these sites. Mg^{2+} removal from the Ca^{2+}–Mg^{2+} sites of TnC is also quite slow, ~8 sec^{-1} (Johnson et $al.$, 1979). In conclusion, we see that Ca^{2+} exchange with the Ca^{2+}-specific regulatory sites of STnC is very rapid, while Ca^{2+} exchange with the higher-affinity Ca^{2+}–Mg^{2+} sites is far too slow (3–4 sec^{-1}) to be involved directly in the regulation of muscle contraction. Our results using TnC$_{DANZ}$ fluorescence are in good agreement with the PMR studies of Levine et $al.$ (1977). They determined exchange rates of ≤10 sec^{-1} and 100–1000 sec^{-1} from the structural changes occurring with Ca^{2+} exchange with the Ca^{2+}–Mg^{2+} and Ca^{2+}-specific regulatory sites, respectively.

One very interesting finding from our study on TnC$_{DANZ}$ fluorescence was the existence of a transient state of the protein where the Ca^{2+}-specific sites of TnC were transiently occupied by Ca^{2+} during a Ca^{2+} flux, before a Ca^{2+}–EGTA equilibrium could be reached (Johnson et $al.$, 1979). This phenomenon occurred when TnC$_{DANZ}$ in 2 mM EGTA was rapidly mixed with 0.8 mM CaCl$_2$. At equilibrium the free Ca^{2+} should be $\sim1 \times 10^{-7} M$ and only the Ca^{2+}–Mg^{2+} sites would be occupied. Figure 4 shows a stopped-flow trace of this reaction. Within the mixing time of the instrument a highly fluorescent state occurred which had the same fluorescence intensity as TnC$_{DANZ}$ with the Ca^{2+}-specific sites occupied. This highly fluorescent state decayed back to baseline fluorescence with a $\tau_{1/2}$ of 2–3 msec, the same rate we observed for Ca^{2+} removal from the

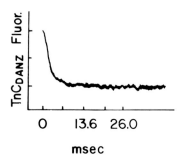

Fig. 4. The production of a highly fluorescent transient state in $STnC_{DANZ}$. A fluorescence stopped-flow trace of the reaction of $STnC_{DANZ}$ in 2 mM EGTA with 0.8 mM Ca^{2+} as described in Johnson *et al.* (1979). The fluorescence decays from a relative value of 2.0 to a value of 0.9 with a half-life of 2–3 msec or at a rate of 230–346 sec^{-1}. The highly fluorescent state is produced by the transient occupancy of the Ca^{2+}-specific regulatory sites of STnC during a rapid Ca^{2+} transient.

Ca^{2+}-specific sites of TnC_{DANZ}. We attributed this transient state to diffusion-limited Ca^{2+} binding to the Ca^{2+}-specific sites during the mixing time of the instrument. During the mixing time, EGTA also bound Ca^{2+} to reach an equilibrium Ca^{2+} concentration of, in this case, $1 \times 10^{-7} M$. This was insufficient Ca^{2+} to occupy the Ca^{2+}-specific sites, so that what we observed kinetically was the decay of the highly fluorescent state at the rate at which Ca^{2+} was removed from the Ca^{2+}-specific regulatory sites. This represents a transient occupancy of the regulatory sites by calcium (Johnson *et al.*, 1978a). This phenomena may be very important physiologically, since it shows that, during a very rapid Ca^{2+} transient, Ca^{2+} exchanges with the Ca^{2+}-specific regulatory sites, even though the free Ca^{2+} concentration at equilibrium is insufficient to support the occupancy of these sites.

This clearly points out the virtue of using lower-affinity Ca^{2+}-specific sites for regulating muscle contraction. These sites will bind and exchange Ca^{2+} extremely rapidly during a Ca^{2+} transient such as those postulated to occur in muscle. Ca^{2+}–Mg^{2+} sites on the other hand cannot exchange Ca^{2+} as rapidly and, in addition, they bind Mg^{2+} (which has a slow off rate itself) which would have to be removed (at a rate of ~5 sec^{-1} in whole Tn) before they could bind Ca^{2+} during a Ca^{2+} transient.

Very recently, Iio and Kondo (1980) repeated some of our fluorescence stopped-flow studies of TnC_{DANZ} in an effort to reconcile our results with their various results, obtained from studying the rates of tyrosine fluorescence changes in TnC. The difficulty in using tyrosine fluorescence is that their stopped-flow studies were analyzed assuming that Ca^{2+} binding to the Ca^{2+}–Mg^{2+} sites is responsible for the full increase in tyrosine fluorescence (Iio *et al.,* 1976), while in fact Ca^{2+} binding to both classes of sites results in increases in tyrosine fluorescence (Johnson *et al.,* 1978a) as verified by Iio and Kondo (1980). In their most recent study, Iio and Kondo (1980) verified our fluorescence and fluorescence stopped-flow studies on TnC_{DANZ} with one minor exception. When they mixed TnC_{DANZ} in EGTA with Ca^{2+} to pCa ~6.0, they observed the formation and decay of the highly fluorescent transient state as we reported (Fig. 3) and they reiterated our explanation of this phenomena. They suggested, however, that in this large fluorescence decrease, there was an additional very small fluorescence decrease which occurred with Ca^{2+} binding to the $Ca^{2+}-Mg^{2+}$ sites. Our data do not contain this small fluorescence decrease. The small decrease they saw probably resulted from a small degree of photobleaching of their sample which was at ~10 times the TnC_{DANZ} concentration we used. Other than this, they have confirmed our fluorescence stopped-flow studies with $STnC_{DANZ}$.

VII. RATES OF Ca^{2+} EXCHANGE IN TROPONIN

It must be remembered that, in whole Tn, the affinity of both the $Ca^{2+}-Mg^{2+}$ sites and the Ca^{2+}-specific sites is increased ~10 fold (Table I). This serves to slow the off rate of Ca^{2+} from both classes of sites. We have predicted that the off rate of Ca^{2+} from the $Ca^{2+}-Mg^{2+}$ and Ca^{2+}-specific sites of whole Tn will occur with half-lives of 3.6 sec (a rate of 0.2 sec^{-1}) and ~34 msec (a rate of ~20 sec^{-1}), respectively (Johnson *et al.,* 1979). Thus, in whole Tn, while Ca^{2+} exchange with the Ca^{2+}-specific sites is still sufficiently rapid to regulate the contraction–relaxation cycle of muscle contraction, Ca^{2+} exchange with the $Ca^{2+}-Mg^{2+}$ sites is far too slow. Recently we have verified this prediction using fluorescently labeled whole Tn, Tn_{IA}, labeled at a cysteine residue of troponin I. Using stopped-flow fluorimetry to follow the EGTA-induced reversal of the fluorescence changes which occur with Ca^{2+} binding to the $Ca^{2+}-Mg^{2+}$ and Ca^{2+}-specific sites of whole Tn, we calculated off rates of Ca^{2+} of 0.3–0.6 sec^{-1} and ~23 sec^{-1} for each class of site, respectively (Johnson *et al.,* 1980b, 1981a). Thus the rate of Ca^{2+} removal from each class of site is slower in Tn than in TnC and, as predicted, Ca^{2+} exchange with the Ca^{2+}-specific sites is still quite rapid, while Ca^{2+} exchange with the $Ca^{2+}-Mg^{2+}$ sites is too slow to be directly involved in the contraction–relaxation cycle.

The rapid structural changes we report in TnI, which occur with Ca^{2+} exchange with the Ca^{2+}-specific regulatory sites of TnC, may represent some part of the chain of structural changes which begin on the Ca^{2+}-specific side of TnC and propagate through the thin filament to regulate actomyosin interaction and contraction. A large part of our work is directed toward mapping the spatial and temporal propagation of this Ca^{2+} signal throughout thin-filament proteins. Thus far, only the structural changes which occur upon Ca^{2+} exchange with the Ca^{2+}-specific sites have the speed required to regulate contraction.

VIII. CONCLUSIONS

Troponin plays a key role in the thin-filament regulation of both skeletal and cardiac muscle contraction. Depending on its state of bound Ca^{2+}, it serves to position Tm along the F-actin filament to either allow or prevent the interaction of myosin with actin and contraction. Studies on the Ca^{2+} dependence of the interactions of Tn and its subunits with other thin-filament proteins suggest a model for their interactions and the Ca^{2+}-dependent changes in their interactions which allow contraction. Ca^{2+} binding to TnC (in particular the Ca^{2+}-specific regulatory sites) produces structural changes in TnI, Tm, and F-actin which play a key role in regulating the actomyosin interaction. STnC and STn have two classes of Ca^{2+}-binding sites: two $Ca^{2+}-Mg^{2+}$ structural sites and two Ca^{2+}-specific regulatory sites. Only the latter sites are sufficiently responsive to rapid Ca^{2+} transient to be directly involved in the regulation of skeletal muscle contraction. CTnC and Tn are very similar to their skeletal counterparts in terms of their Ca^{2+}-binding properties and interactions with the other thin-filament proteins. We have now shown that these cardiac proteins have a single Ca^{2+}-specific regulatory site which is similar in its Ca^{2+} affinity, Ca^{2+} exchange rates, and Ca^{2+}-dependent regulation of HMM ATPase to the two Ca^{2+}-specific regulatory sites of STn. This suggests that regulation of skeletal and cardiac muscle occurs by very similar mechanisms. We now have a reasonably clear picture of how both skeletal and cardiac thin-filament proteins respond to a Ca^{2+} transient, in terms of their Ca^{2+}-binding properties, Ca^{2+} exchange kinetics, and Ca^{2+}-induced structural changes in protein–protein interactions, to regulate muscle contraction. Future directions will concentrate on the finer details of these processes.

ACKNOWLEDGMENTS

We would like to thank Janet Duritsch for the excellent typing of the manuscript and Gwen Kraft for the excellent artwork. This work was supported by grants from the NIH (HL22619-3A,E) the American Heart Association (78-1167,79-1001), and the Muscular Dystrophy Association.

REFERENCES

Bremel, R. D., and Weber, A. (1972). Cooperation within actin filament in vertebrate skeletal muscle. *Nature (London) New Biol.* **238**, 97–101.

Collins, J. H., Potter, J. D., Horn, M. J., Wilshire, G., and Jackson, N. (1973). The amino acid sequence of rabbit skeletal muscle troponin C: Gene replication and homology with calcium-binding proteins from carp and hake muscle. *FEBS Lett.* **36**, 268–272.

Collins, J. H., Greaser, M. L., Potter, J. D., and Horn, M. J. (1977). Determination of the amino acid sequence of troponin C from rabbit skeletal muscle. *J. Biol. Chem.* **252**, 6356–6362.

Drabikowski, W., and Barylko, B. (1971). Calcium binding by troponin. *Acta Biochim. Pol.* **18**, 698–711.

Ebashi, S. (1963). Third component participating in the superprecipitation of natural actomyosin. *Nature (London)* **200**, 1010.

Ebashi, S., and Ebashi, F. (1964). A new protein component participating in the superprecipitation of myosin B. *J. Biochem. (Tokyo)* **55**, 604–613.

Ebashi, S., Kodama, A., and Ebashi, F. (1968). Troponin I: Preparation and physiological function. *J. Biochem. (Tokyo)* **64**, 465–477.

Ebashi, S., Endo, M., and Ohtsuki, I. (1969). Control of muscle contraction. *Q. Rev. Biophys.* **2**, 351–384.

Eigen, M. (1963). Fast elementary steps in chemical reaction mechanisms. *Pure Appl. Chem.* **6**, 97–115.

Fuchs, F. (1972). A rapid ultrafiltration procedure for the estimation of calcium-protein binding constants. *Int. J. Pept. Protein Res.* **4**, 147–149.

Fuchs, F. (1977a). The binding of calcium to glycerinated muscle fibers in rigor: The effect of filament overlap. *Biochim. Biophys. Acta* **491**, 523–531.

Fuchs, F. (1977b). Cooperative interactions between calcium-binding sites on glycerinated muscle fibers: The influence of cross-bridge attachment. *Biochim. Biophys. Acta* **492**, 314–322.

Fuchs, F. (1978). On the relation between filament overlap and the number of calcium-binding sites on glycerinated muscle fibers. *Biophys. J.* **21**, 273–277.

Fuchs, F., and Bayuk, M. (1976). Cooperative binding of calcium to glycerinated skeletal muscle fibers. *Biochim. Biophys. Acta* **440**, 448–455.

Fuchs, F., and Briggs, F. N. (1968). The site of calcium binding in relation to the activation of myofibrillar contraction. *J. Gen. Physiol.* **51**, 665–676.

Greaser, M. L., and Gergely, J. (1971). Reconstitution of troponin activity from three protein components. *J. Biol. Chem.* **246**, 4226–4233.

Greaser, M. L., and Gergely, J. (1973). Purification and properties of the components from troponin. *J. Biol. Chem.* **248**, 2125–2133.

Hartshorne, D. J., and Pyun, H. Y. (1971). Calcium binding by the troponin complex and the purification and properties of troponin A. *Biochim. Biophys. Acta* **229**, 698–711.

Hartshorne, D. J., Theiner, M., and Mueller, H. (1968). Studies on troponin. *Biochim. Biophys. Acta* **175**, 320–330.

Hincke, M. T., McCubbin, W. D., and Kay, C. M. (1978). Calcium-binding properties of cardiac and skeletal troponin C as determined by circular dichroism and ultraviolet difference spectroscopy. *Can. J. Biochem.* **56**, 384–395.

Hincke, M. T., McCubbin, W. D., and Kay, C. M. (1979). The interaction between beef cardiac troponin T and troponin I as demonstrated by ultraviolet absorption difference spectroscopy, circular dichroism and gel filtration. *Can. J. Biochem.* **57**, 768–775.

Hitchcock, S. E. (1975a). Cross-linking of troponin with dimethylimido esters. *Biochemistry* **14**, 5162–5167.

Hitchcock, S. E. (1975b). Regulation of muscle contraction: Binding of troponin and its components to actin and tropomyosin. *Eur. J. Biochem.* **52**, 255–263.

Hitchcock, S. E. (1981). Study of the structure of troponin-C by measuring the relative reactivities of lysines with acetic anhydride. *J. Mol. Biol.* **147**, 153–173.

Hitchcock, S. E., Zimmerman, C. J., and Smalley, C. (1981). Study of the structure of troponin-T by measuring the relative reactivities of lysines with acetic anhydride. *J. Mol. Biol.* **147**, 125–151.

Hitchcock, S. E., Huxley, H. E., and Szent-Györgyi, A. G. (1973). Calcium sensitive binding of troponin to actin-tropomyosin: A two-site model for troponin action. *J. Mol. Biol.* **80**, 825–836.

Holroyde, M. J., Robertson, S. P., Johnson, J. D., Solaro, R. J., and Potter, J. D. (1980). The calcium and magnesium binding sites on cardiac troponin and their role in the regulation of myofibrillar adenosine triphosphatase. *J. Biol. Chem.* **255**, 1168–1193.

Horwitz, J., Bullard, B., and Mercola, D. (1979). Interaction of troponin subunits. *J. Biol. Chem.* **254**, 350–355.

Iio, T., and Kondo, H. (1980). Comparison of the kinetic properties of troponin-C and dansylaziridine-labeled troponin-C. *J. Biochem. (Tokyo)* **58**, 547–556.

Iio, T., Mihashi, K., and Kondo, H. (1976). Kinetics of conformational change of troponin C induced by calcium. *J. Biochem. (Tokyo)* **79**, 689–691.

Johnson, J. D., and Potter, J. D. (1978). Detection of two classes of Ca^{2+}-binding sites in troponin C with circular dichroism and tyrosine fluorescence. *J. Biol. Chem.* **253**, 3775–3777.

Johnson, J. D., Collins, J. H., and Potter, J. D. (1978a). Dansylaziridine-labeled troponin C. *J. Biol. Chem.* **253**, 6451–6458.

Johnson, J. D., Collins, J. H., Robertson, S. P., and Potter, J. D. (1978b). A fluorescent probe study of Ca^{2+} exchange with the Ca^{2+}-specific site of cardiac troponin. *Circulation* **58**, II, 71.

Johnson, J. D., Charlton, S. C., and Potter, J. D. (1979). A fluorescence stopped-flow analysis of Ca^{2+} exchange with troponin C. *J. Biol. Chem.* **254**, 3497–3502.

Johnson, J. D., Collins, J. H., Robertson, S. P., and Potter, J. D. (1980a). A fluorescent probe study of Ca^{2+}-binding to the Ca^{2+}-specific sites of cardiac troponin and troponin C. *J. Biol. Chem.* **255**, 9635–9640.

Johnson, J. D., Robertson, S. P., and Potter, J. D. (1980b). Ca^{2+}-exchange rates of the Ca^{2+}–Mg^{2+} and the Ca^{2+}-specific sites of whole troponin. *Fed. Proc., Fed. Am. Soc. Exp. Biol.* **39**, 1621.

Johnson, J. D., Robinson, D. E., Robertson, S. P., Schwartz, A. and Potter, J. D. (1981a). Ca^{2+} exchange with troponin and the regulation of muscle contraction. *In* "The Regulation of Muscle Contraction: Excitation-Contraction Coupling" (A. Grinnel, ed.), pp. 241–259. Academic Press, New York.

Johnson, J. D. *et al.* (1981b). In preparation.

Katayama, E. (1979). Interaction of troponin-I with troponin-T and its fragment. *J. Biochem. (Tokyo)* **85**, 1379–1381.

Kawasaki, Y., and Van eerd, J. P. (1972). The effect of Mg^{++} on the conformation of the Ca^{++}-binding component of troponin. *Biochem. Biophys. Res. Commun.* **49**, 898–905.

Kretsinger, R. H., and Barry, C. D. (1975). The predicted structure of the calcium-binding component of troponin. *Biochim. Biophys. Acta* **405**, 40–52.

Kretsinger, R. H., and Nockolds, C. E. (1973). Carp muscle calcium-binding protein. II. Structure determination and general description. *J. Biol. Chem.* **248**, 3313–3326.

Leavis, P. C., and Kraft, E. L. (1978). Calcium binding to cardiac troponin C. *Arch. Biochem. Biophys.* **186**, 411–415.

Leavis, P. C., Rosenfeld, S. S., Gergely, J., Grabarek, Z., and Drabikowski, W. (1978). Proteolytic fragments of troponin C. *J. Biol. Chem.* **253**, 5452–5459.

Levine, B. A., Mercola, D., and Thornton, J. M. (1976). Proton magnetic resonance study of troponin C. *FEBS Lett.* **61**, 218–222.

Levine, B. A., Mercola, D., Coffman, D., and Thornton, J. M. (1977). Calcium binding by troponin-C: A proton magnetic resonance study. *J. Mol. Biol.* **115**, 743–760.

McCubbin, W. D., and Kay, C. M. (1973). Physicochemical and biological studies on the metal-induced conformational change in troponin A: Implication of carboxyl groups in the binding of calcium ion. *Biochemistry* **12**, 4228–4232.

McCubbin, W. D., Mani, R. S., and Kay, C. M. (1974). Physicochemical studies on the interaction of the calcium-binding protein (troponin C) with the inhibitory protein (troponin I) and calcium ions. *Biochemistry* **13**, 2689–2694.

Margossian, S. S., and Cohen, C. (1973). Troponin subunit interactions. *J. Mol. Biol.* **81**, 409–413.

Miki, M. (1979). Conformation change in reconstituted thin filament studied with fluorescence energy transfer between ϵ-ADP bound to F-actin and NBD-Cl bound to troponin C. *Biochim. Biophys. Acta* **578**, 96–106.

Murray, A. C., and Kay, C. M. (1972). Hydrodynamic and optical properties of troponin A: Demonstration of a conformational change upon binding calcium ion. *Biochemistry* **11**, 2622–2627.

Ohnishi, S., Maruyama, K., and Ebashi, S. (1975). Calcium induced conformational changes and mutual interactions of troponin components as studied by spin labeling. *J. Biochem. (Tokyo)* **78**, 73–81.

Ohtsuki, I. (1975). Distribution of troponin components in the thin filament studied by immunoelectron microscopy. *J. Biochem. (Tokyo)* **77**, 633–639.

Ohtsuki, I. (1979). Molecular arrangement of troponin-T in the thin filament. *J. Biochem. (Tokyo)* **86**, 491–497.

Ohyashiki, T., Kanaoka, Y., and Sekine, T. (1976). Studies on calcium ion-induced conformation changes in the actin-tropomyosin-troponin system by fluorimetry. *Biochim. Biophys. Acta* **420**, 27–36.

Pearlstone, J. R., Carpenter, M. R., Johnson, P., and Smillie, L. B. (1976). Amino-acid sequence of tropomyosin binding component of rabbit skeletal muscle troponin. *Proc. Natl. Acad. Sci. U.S.A.* **73**, 1902–1906.

Perry, S. V., Cole, H. A., Head, J. F., and Wilson, J. F. (1972). Localization and mode of action of the inhibitory protein component of the troponin complex. *Cold Spring Harbor Symp. Quant. Biol.* **37**, 251–262.

Potter, J. D. (1974). The content of troponin, tropomyosin, actin and myosin in rabbit skeletal muscle myofibrils. *Arch. Biochem. Biophys.* **162**, 436–441.

Potter, J. D., and Gergely, J. (1974). T:oponin, tropomyosin, and actin interactions in the Ca^{2+} regulation of muscle contraction. *Biochemistry* **13**, 2697–2703.

Potter, J. D., and Gergely, J. (1975). The calcium and magnesium binding sites on troponin and their role in the regulation of myofibrillar ATPase. *J. Biol. Chem.* **250**, 4628–4683.

Potter, J. D., Seidel, J. C., Leavis, P., Lehrer, S. S., and Gergely, J. (1976). Effect of Ca^{2+}-binding on troponin C. *J. Biol. Chem.* **251**, 7551–7556.

Potter, J. D., Johnson, J. D., Dedman, J. R., Schreiber, W. E., Mandel, F., Jackson, R. L., and Means, A. R. (1977a). Calcium-binding proteins: Relationship of binding, structure, conformation and biological function. *In* "Calcium-Binding Proteins and Calcium Function" (R. H. Wasserman, R. A. Corradino, E. Carafoli, R. H. Kretsinger, D. H. MacLennan, and F. L. Siegel, eds.), pp. 239–250. Elsevier/North-Holland Publ., Amsterdam.

Potter, J. D., Hsu, F. J., and Pownall, H. J. (1977b). Thermodynamics of Ca^{2+}-binding to troponin-C. *J. Biol. Chem.* **252,** 2452–2454.

Potter, J. D., Robertson, S. P., Collins, J. H., and Johnson, J. D. (1980). The role of the Ca^{2+} and Mg^{2+} binding sites on troponin and other myofibrillar proteins in the regulation of muscle contraction. *In* "Calcium-Binding Proteins: Structure and Function" (F. L. Siegel *et al.*, eds.), pp. 279–288. Elsevier/North-Holland Publ., Amsterdam.

Potter, J. D., Holroyde, M. J., Robertson, S. P., Solaro, R. J., Kranias, E. G., and Johnson, J. D. (1981). The regulation of cardiac muscle contraction. *In* "Current Topics in Muscle and Non-Muscle Motility" (J. W. Shay and R. W. Dowben, eds.). Plenum, New York pp. 245–256.

Prendergast, F. G., and Potter, J. D. (1979). Solution conformation and hydrodynamic properties of rabbit skeletal TnT. *Biophys. J.* **25,** 250a.

Robertson, S. P., Johnson, J. D., and Potter, J. D. (1981). The time course of Ca^{2+} exchange with calmodulin, troponin, parvalbumin and myosin to transient increases in Ca^{2+}. *Biophys. J.* **34,** 559–569.

Robertson, S. P., Johnson, J. D., Holroyde, M. J., Kranias, E. G., Potter, J. D., and Solaro, J. R. (1982). The effect of troponin I phosphorylation on the Ca^{2+}-binding properties of the Ca^{2+}-regulatory site of bovine cardiac troponin. *J. Biol. Chem.* (in press).

Schaub, M. C., and Perry, S. V. (1969). The relaxing protein system of striated muscle. *Biochem. J.* **115,** 993–1004.

Seamon, K. B., Hartshorne, D. J., and Bothner-by, A. A. (1977). Ca^{2+} and Mg^{2+} dependent conformations of troponin C as determined by ^1H and ^{19}F nuclear magnetic resonance. *Biochemistry* **16,** 4039–4046.

Solaro, R. J., Robertson, S. P., Johnson, J. D., Holroyde, M. J., and Potter, J. D. (1981). Troponin I phosphorylation: A unique regulation of the amounts of calcium required to activate cardiac myofibrils. *Cold Spring Harbor Symp. Quant. Biol.* Vol. 8: Protein Phosphorylation, pp. 901–911.

Strasburg, G. M., Greaser, M. L., and Sundaralingam, M. (1980). X-ray diffraction studies of troponin-C crystals from rabbit and chicken skeletal muscles. *J. Biol. Chem.* **255,** 3806–3808.

Sutoh, K., and Matsuzaki, F. (1980). Millisecond photo-cross-linking of protein components in vertebrate striated muscle thin filaments. *Biochemistry* **19,** 3878–3882.

Syska, H., Perry, S. V., and Trayer, I. P. (1974). A new method of preparation of troponin I (inhibitory protein) using affinity chromatography: Evidence for three different forms of troponin I in striated muscle. *FEBS Lett.* **40,** 253–257.

Syska, H., Wilkinson, J. M., Grand, R. J. A., and Perry, S. V. (1976). The relationship between biological and primary structure of troponin I from white skeletal muscle of the rabbit. *Biochem. J.* **153,** 375–387.

Taylor, K. A., and Amos, L. A. (1981). A new model for the geometry of the binding of myosin cross-bridges to muscle thin filaments. *Biophys. J.* **33,** 84a.

Tonomura, Y., Watanabe, S., and Morales, M. (1969). Conformational changes in the molecular control of muscle contraction. *Biochemistry* **8,** 2171–2176.

Van eerd, J. P., and Kawasaki, Y. (1972). Ca^{++} induced conformational changes in the Ca^{++}-binding component of troponin. *Biochem. Biophys. Res. Commun.* **47,** 859–865.

Van eerd, J. P., and Takahshi, K. (1976). Determination of the complete amino acid sequence of bovine cardiac troponin C. *Biochemistry* **15,** 1171–1180.

Weber, A., and Herz, R. (1963). The binding of calcium to actomyosin systems in relation to their biological activity. *J. Biol. Chem.* **238,** 599–605.

Weeks, R. A., and Perry, S. V. (1977). A region of the troponin C molecule involved in interaction with troponin I. *Biochem. Soc. Trans.* **5,** 1391–1392.

Wilkinson, J. M., and Grand, R. J. A. (1975). The amino acid sequence of troponin I from rabbit skeletal muscle. *Biochem. J.* **149,** 493–496.

Wilkinson, J. M., and Grand, R. J. A. (1978). Comparison of amino acid sequence of troponin I from different striated muscles. *Nature (London)* **271,** 31–35.

Yamada, K. (1978). The enthalpy titration of troponin C with calcium. *Biochim. Biophys. Acta* **535,** 342–347.

Chapter 6

Vitamin D-Induced
Calcium-Binding Protein

R. H. WASSERMAN
C. S. FULLMER

I.	Introduction	175
II.	Species and Tissue Distribution	176
III.	Properties of Calcium-Binding Proteins	181
IV.	Cellular Localization of Calcium-Binding Proteins	187
V.	Physiological Factors Affecting CaBP	188
	A. Effect of Low Dietary Calcium or Phosphorus Intake	189
	B. Effect of Diet Containing a High Concentration of Strontium Salts	190
	C. Aging and CaBP	190
	D. Cortisol and CaBP	194
	E. CaBP in Laying Birds	195
	F. Gonadal Hormones and CaBP	197
	G. Insulin, CaBP, and Diabetes	197
VI.	*In Vitro* Synthesis of CaBP	198
VII.	Embryonic Development	201
VIII.	CaBP and Calcium Reabsorption in the Kidney	202
IX.	Temporal Responses of CaBP to Acute Doses of $1,25(OH)_2D_3$	202
X.	Discussion	205
	References	207

I. INTRODUCTION

A vitamin D-dependent calcium-binding (CaBP) substance was first detected in the intestine of the chick about 18 years ago, and the fact that it was a protein was reported in 1966 by Wasserman and Taylor. At about

175

CALCIUM AND CELL FUNCTION, VOL. II

the same time, the dependency of vitamin D action on a protein synthetic event was also reported (Zull *et al.*, 1965; Norman, 1965). Since then an increasing number of research investigators have become involved in the vitamin D story, as well as in the science (or art) of "CaBPology." As will be documented in this chapter, a considerable amount of information on CaBP has become available: information that runs the usual gamut from nutritional aspects, to physiological relevance, to biochemical and molecular characterization. Unfortunately, only part of the available data on CaBP is included herein, and other reviews on CaBP can be consulted for additional information (Wasserman and Corradino, 1973; Wasserman *et al.*, 1978; Wasserman, 1980; Taylor, 1980b). Further studies on the metabolism, function, and therapeutic efficacy of vitamin D and its various metabolites and analogues have increased almost exponentially over the past 10 years or so and, because of the relation of vitamin D to CaBP, recent reviews and monographs on this subject are referenced here (De-Luca, 1979; Norman, 1979, 1980b; Lawson, 1978).

II. SPECIES AND TISSUE DISTRIBUTION

While vitamin D-dependent CaBP was first detected in, and isolated from, chick intestinal mucosal tissue (Wasserman and Taylor, 1966), analogous proteins have subsequently been shown to be present in a variety of species and tissues. It may be stated, in general, that two main subclasses of vitamin D-dependent CaBPs have been documented—those resembling chick intestinal CaBP (MW ~28,000) and those more characteristic of mammalian intestinal proteins (MW ~9000).

Chick intestinal CaBP has repeatedly been demonstrated to be dependent on vitamin D or active metabolites of vitamin D for its formation. It has been shown to be present in the greatest amount in the duodenum and to decrease in tissue concentration distally along the intestinal tract (Taylor and Wasserman, 1967), and this corresponds to the efficiency of calcium absorption in these segments. Apparently identical proteins have also been detected in the chick kidney (Taylor and Wasserman, 1967) and brain (Taylor and Brindak, 1974) and in the egg shell gland (uterus) of the laying hen (Corradino *et al.*, 1968). The chick brain protein, occurring predominately in the cerebellum, has recently been localized specifically in Purkinje cells (Jande *et al.*, 1980) and has been shown to be identical to the intestinal protein in terms of immunological reactivity, molecular size, and electrophoretic properties (Taylor and Brindak, 1974). This protein has also been demonstrated to be responsive to vitamin D (Taylor, 1974).

Hen uterine CaBP has also been shown to be identical to intestinal CaBP with respect to electrophoretic and immunological properties, molecular size, and amino acid composition (Fullmer et al., 1976). The uterine protein is also responsive to vitamin D (Corradino et al., 1968).

More recently, with the advent of sensitive radioimmunoassay procedures, immunoreactive CaBP has also been detected in chick bone tissue, pancreas, hypothalamus, blood, parathyroid glands, and possibly several other tissues (Christakos and Norman, 1978, 1980a; Christakos et al., 1979). These studies and others have also indicated that CaBPs from these various tissues are all responsive to vitamin D. All chick CaBPs studied to date possess a molecular weight of 28,000 with the possible exception of bone CaBP which was reported to be about 34,000 (Christakos and Norman, 1978).

A vitamin D-dependent CaBP was first detected in rat intestine by Kallfelz et al. (1967), and its existence subsequently confirmed by others (Harmeyer and DeLuca, 1969; Schachter, 1970; Ooizumi et al., 1970; Freund and Bronner, 1975). Drescher and DeLuca (1971) isolated this protein and reported a molecular weight of 8000–9000 by sedimentation studies and 13,000 by gel filtration.

Hermsdorf and Bronner (1975) succeeded in partially purifying a rat kidney CaBP. This CaBP was present in the cortex, but not the medulla, possessed a molecular weight of 28,000, and was reported to be vitamin D-dependent, occurring in normal but not vitamin D-deficient animals.

Bruns et al. (1977) purified rat intestinal CaBP and later (Bruns et al., 1978) developed an antiserum against this protein. In the screening of a variety of rat tissues with this antiserum, immunoreactive CaBP was only observed in the intestine and placenta at the level of detection. No immunoreactivity was observed in rat kidney.

Moriuchi et al. (1975) compared mucosal preparations from the duodenum, jejunum, and ileum and documented the existence of two vitamin D-responsive CaBPs in rat intestine. One of these proteins, presumably identical to that studied by others, predominated in the proximal intestine and exhibited a molecular weight of about 12,000. The other, occurring primarily in the distal intestine and similar to a protein reported earlier by Ooizumi et al. (1970), was shown to have a molecular weight of 27,000. Failure of a number of other studies to detect this latter protein may stem from the insensitivity of the standard calcium-binding assay procedure (Chelex-100), or from the fact that efforts to isolate CaBPs usually concentrate on the duodenum or proximal intestine.

A somewhat similar pattern was noted for bovine intestine (Wasserman et al., 1978) when two peaks of calcium-binding activity were observed along the length of the intestine, one in the proximal portion and the other in the distal portion. Only the binding activity in the proximal intestine coincided with immunoreactive CaBP (MW 9000). Binding activity occurring in the distal intestine was not further examined.

Hitchman and Harrison (1972) examined, in a comparative manner, duodenal mucosal preparations from the chick, rat, human, and pig. Gel filtration chromatography convincingly demonstrated calcium-binding activities corresponding to proteins with a molecular weight of about 10,000 for the rat, human, and pig, and greater than 25,000 for the chick.

In a similar fashion, Fullmer and Wasserman (1975) purified and partially characterized intestinal CaBPs from the cow, pig, horse, guinea pig, and chick. Amino acid compositional analyses and gel filtration chromatography established the similarity of the mammalian proteins (MW ~9000) which all possess a single tyrosine residue and are devoid of cysteine, methionine, histidine, tryptophan, and with the exception of porcine CaBP, arginine. Porcine CaBP contained a single arginine residue. Gel filtration estimates of molecular weight were about 11,000 for the mammalian CaBPs, possibly indicating a nonspherical shape and explaining the slightly higher values generally reported by this procedure. The molecular weight of chick CaBP was 27,000–28,000 by either method.

Using radioimmunoassay procedures with antiserum prepared in response to porcine intestinal CaBP, Arnold et al. (1975) and Murray et al. (1975) detected immunoreactive protein (MW ~12,000) in porcine kidney, liver, thyroid, pancreas, and blood. Intestinal CaBP was found in greater amounts in the proximal than in the distal small intestine. Immunoreactive CaBP was not detected in other tissues, including the parathyroid glands, bone, skeletal muscle, or brain. Nonimmunoreactive calcium-binding activity corresponding to larger proteins was found in the kidney (MW ~25,000) and the parathyroid gland (MW ~17,000). This latter activity was probably associated with the non-vitamin D-responsive parathyroid CaBP detected by Oldham et al. (1974).

Piazolo and co-workers (1971) reported CaBPs in human duodenal and kidney tissue with a molecular weight of greater than 21,500. Two intestinal CaBPs of similar size (MW ~20,000–25,000) were detected in human intestine by Alpers et al. (1972).

Morrissey and Rath (1974) isolated a human kidney CaBP (MW ~28,000) and later prepared antiserum to this protein (Morrissey et al., 1975). Using the indirect peroxidase-labeled antibody procedure, these authors reported the detection of immunoreactive CaBP in human jejunum, but not human duodenum or any intestinal segments from several species of animals. Immunoreactive CaBP was observed in kidney and pancreatic tissue from the dog, cat, rat, mouse, and chick, as well as kidney tissue from human and monkey.

Baimbridge et al. (1980) have demonstrated immunoreactivity between antiserum prepared in response to chick intestinal CaBP and human kidney and cerebellular tissue. The immunoreactive CaBP, in both cases, was shown to be similar in size to chick intestinal CaBP.

The mammary gland is another organ known to transfer significant amounts of calcium. Bauman *et al.* (1972) have reported the presence of a CaBP in both rat and bovine mammary tissue, the concentration of which was dependent on lactational status, but no information was provided regarding vitamin D dependency.

Recently, Hosoya *et al.* (1980) have reported the isolation of two CaBPs from bovine milk (MW ~15,000). There was no indication of vitamin D dependency, and antisera elicited in response to these proteins did not react with intestinal mucosal preparations.

Pavlovitch *et al.* (1980) have recently isolated a CaBP (MW ~9000) from rat skin which was shown to be responsive to vitamin D administration, low dietary calcium, and ultraviolet (UV) irradiation. There were no immunological similarities noted between this protein and rat intestinal CaBP.

It is clear from these observations that two main subclasses of vitamin D-dependent CaBPs, distinguishable primarily by molecular size and immunoreactivity, do exist. Despite incomplete and conflicting data in some cases, several general statements may be made regarding this subclassification. All available data are summarized in Table I. All CaBPs in avian tissue (with the possible exception of chick bone) studied to date exhibit a molecular weight of 28,000 and are immunoreactive against chick intestinal CaBP antisera. This group constitutes one major CaBP subclass. Numerous studies on mammalian intestinal CaBPs place their molecular weights variably at 8000–13,000, values actually close to 9000 on the basis of chemical studies on the bovine and porcine proteins. The disparity of reported molecular weights in this subclass is undoubtedly the result of anomalous behavior of these small proteins on gel filtration media. Clear examples of this second subclass are found in bovine and porcine intestine and kidney, porcine liver, thyroid, and pancreas, and rat intestine and placenta. No species immunological cross-reactivity is observed among these mammalian CaBPs.

Complicating any general species subclassification scheme based simply on avian-type (MW ~28,000) versus mammalian-type (MW ~9000) CaBPs are the observations that mammalian tissues from several species (human and rat kidney; bovine, human, and rat brain; rat pancreas; and human and rat intestine) contain CaBP of the 28,000-molecular-weight variety.

The observations of Moriuchi *et al.* (1975) on the rat intestinal proteins and Arnold *et al.* (1975) on porcine kidney further indicate that certain tissues from some species may contain both subclasses of CaBPs.

In the case of human CaBP, the issue remains confused. The predominant CaBP in human kidney and brain is apparently similar to the avian protein, as evidenced by size (MW 28,000) and immunological cross-reactivity with chick intestinal CaBP antisera. Human intestinal CaBP has

TABLE I

Properties of Calcium-Binding Proteins from Various Species and Tissues [a]

Species, tissue	Means of detection[b]	Estimated molecular weight[c]	Crossreactivity with antisera prepared in response to	
			Chick intestinal CaBP[c]	Other CaBPs[c]
Chick (hen)				
Intestine	CBA, IA, RIA, LOC	28,000	+	—
Kidney	IA, RIA, LOC	28,000	+	Human kidney
Uterus	CBA, IA, LOC	28,000	+	—
Brain	CBA, IA, RIA, LOC	28,000	+	—
Pancreas	RIA, LOC	28,000	+	Human kidney
Bone	RIA	34,000	+	—
Blood	RIA	ND	+	—
Hypothalamus	RIA	ND	+	—
Parathyroid	RIA	ND	+	—
Rat				
Intestine	CBA, IA	8000–13,000	ND	ND
Intestine	CBA	28,000	ND	Rat intestine
Kidney	CBA, LOC	27,000	+	Human kidney
Brain	IA	ND	+	—
Pancreas	LOC	ND	ND	Human kidney
Placenta	CBA, IA	10,000	ND	Rat intestine
Pig				
Intestine	CBA, RIA, LOC	9000–13,000	ND	Porcine intestine
Kidney	CBA	25,000	ND	—
Kidney	RIA	12,000	ND	Porcine intestine
Thyroid	RIA	12,000	ND	Porcine intestine
Pancreas	RIA	12,000	ND	Porcine intestine
Blood	RIA	ND	ND	Porcine intestine
Liver	RIA	12,000	ND	Porcine intestine
Cow				
Intestine	CBA, IA	9000–11,000	—	Bovine intestine
Kidney	CBA, IA	9000–11,000	—	Bovine intestine
Brain	IA	ND	+	—
Human				
Intestine	CBA, LOC	12,000–25,000	+	Human kidney
Kidney	CBA, IA, LOC	28,000	+	Human kidney
Brain	IA	28,000	+	—

[a] See text for references.

[b] CBA, calcium-binding assay; IA, immunoassay; RIA, radioimmunoassay; LOC, localization.

[c] ND, Not determined.

been reported in both the 9000- and 28,000-molecular-weight ranges, indicating the possible presence of both subclasses in this tissue. In support of these observations are the data of Hitchman and Harrison (1972), who detected the smaller CaBP in human duodena, and of Morrissey et al. (1975), who detected the larger CaBP in human jejunum but not in duodenum. These observations may bear on the findings of Moriuchi et al. (1975) on rat intestine. Indeed, it is possible that, through the use of specific immunological assays and specific portions of the small intestine, only a single subclass of CaBP may be detected in tissues where both in actuality exist. Clearly, more information is required in this area.

In general, indications are that the characteristic avian CaBP (MW ~28,000) may have been modified during evolution to yield a protein about one-third the size in specific tissues of some species. In the case of avian tissues, as well as mammalian brain and rat and human kidney, this apparently did not occur, whereas evolution to the smaller protein was complete in rat, bovine, and porcine duodena. Quite possibly the transition was incomplete in the rat jejunum and porcine kidney, resulting in the existence of both CaBP types.

III. PROPERTIES OF CALCIUM-BINDING PROTEINS

Chick intestinal CaBP (MW 28,000) binds 4 Ca^{2+} atoms with a high affinity ($K_a = 2 \times 10^6 \ M^{-1}$) and about 32 Ca^{2+} atoms with a lower affinity ($K_a \sim 10^2 \ M^{-1}$) (Bredderman and Wasserman, 1974). In addition, a number of other cations are bound, including the alkaline earths in the selectivity series, $Ca^{2+} > Sr^{2+} > Ba^{2+} > Mg^{2+}$ (Ingersoll and Wasserman, 1971), as well as La^{3+} and Nd^{3+} (Wasserman et al., 1974). Ca^{2+} binding is inhibited by both urea (Ingersoll and Wasserman, 1971) and lysolecithin (Wasserman, 1970), situations reversible by the removal of urea in the first case and sequestration of lysolecithin by taurocholate in the second. Sulfhydryl groups are apparently not involved in Ca^{2+} binding, since N-ethylmaleimide, iodoacetate, and β-mercaptoethanol are not inhibitory (Ingersoll and Wasserman, 1971). Heat treatment of the protein to 80°C does not result in significant alteration of immunological, electrophoretic, or Ca^{2+}-binding properties, indicating high thermal stability (Bredderman and Wasserman, 1974). There is a biphasic pH optimum for Ca^{2+}-binding activity (pH 6.3 and 9.2), and the protein exhibits an experimentally determined isoelectric point at pH 4.2–4.3 (Ingersoll and Wasserman, 1971), a value in agreement with that calculated from the amino acid composition (Bredderman and Wasserman, 1974). Circular dichroism

(CD) measurements of chick CaBP indicate 30–40% α-helicity which is not significantly altered by the presence or absence of Ca^{2+}.

Bovine intestinal CaBP binds two Ca^{2+} atoms per molecule with a high affinity ($K_a = 4.3 \times 10^6 \ M^{-1}$) and two to three additional atoms with a lesser affinity ($K_a \sim 10^4 \ M^{-1}$) (Fullmer and Wasserman, 1980). This protein also binds other divalent cations, including Sr^{2+}, Ba^{2+}, Cd^{2+}, and Mn^{2+}, as well as a number of the lanthanide series elements (Fullmer and Wasserman, 1977). Binding affinity was clearly demonstrated to be a function of cation ionic radius (charge density).

Birdsall *et al.* (1980), applying proton nuclear magnetic resonance (NMR) spectroscopy to the bovine intestinal CaBP, observed two mutually exclusive sets of spectral effects during titration of the protein with calcium. These were interpreted to represent Ca^{2+} binding to two nonidentical high-affinity sites ($K_a \sim 10^5 \ M^{-1}$) differing in K_a by a factor of 10–20. A slow conformational rearrangement of the protein occurred for both binding steps involving reorganization of the tertiary fold, with little alteration in secondary structure. The structural transition from the Ca_1^{2+}-bound form to the Ca_2^{2+}-bound form resulted in a compact conformation stable to denaturation at high temperature and pH. Further confirmation of the nonidentity of these sites came with the observation that Mg^{2+} was bound only to the higher-affinity Ca^{2+}-binding site. The lone tyrosine residue was reported to exhibit an anomalously high apparent pK_a, and its local environment was influenced only during the second binding step. The homologous porcine CaBP was shown to behave similarly in solution.

Bruns *et al.* (1977) isolated rat intestinal CaBP and determined a molecular weight of about 10,000 by gel filtration. This protein bound 2 moles Ca^{2+} mole^{-1} protein with an estimated K_d for Ca^{2+} of 0.3 μM ($K_a \sim 3.3 \times 10^6 \ M^{-1}$). Similarly, the rat placental protein (MW $\sim$ 10,000) was reported to bind about 2 moles Ca^{2+} mole^{-1} protein with an estimated K_d of 0.12 μM ($K_a \sim 8.3 \times 10^6 \ M^{-1}$) and to exhibit electrophoretic and immunological properties identical to those of intestinal CaBP (Bruns *et al.*, 1978).

Hitchman and Harrison (1972) reported only one high-affinity calcium-binding site ($K_a = 3.5$–$5.5 \times 10^6 \ M^{-1}$) for porcine intestinal CaBP. Circular dichroism and ultraviolet (UV) absorption studies (Dorrington *et al.*, 1974, 1978) also reported the existence of a single site which also showed affinity for several other divalent and trivalent cations. Circular dichroism measurements indicated $\sim$30% α-helicity, a value not significantly affected by Ca^{2+} binding, while changes in both the CD and UV spectra in the aromatic region, suggested perturbance of the sole tyrosine and one or more phenylalanine residues. These authors have concluded that no gross conformational changes occur on binding Ca^{2+} and that the observed changes arise from local effects on tyrosine (primarily) at or near the Ca^{2+}-binding site. These results are not consistent with those obtained for bovine and rat CaBPs in terms of the stoichiometry of Ca^{2+} binding or the effect of Ca^{2+} binding on solution conformation. It appears that the porcine CaBP may not have been completely free of bound Ca^{2+} at the onset of these studies, a possibility alluded to by the authors in a subsequent publication (Hofmann et al., 1979). Therefore, the meaning of the observed spectral changes, with respect to Ca^{2+} binding by porcine CaBP, is not clear. Based on sequence homology, it is unlikely that any significant differences in Ca^{2+}-binding properties exist between bovine and porcine CaBPs.

Fullmer and Wasserman (1973) have reported the isolation of three distinct CaBPs from bovine intestine, which are easily distinguishable on the basis of electrophoretic mobility, differ slightly in molecular size and amino acid composition, but are identical in terms of immunological and calcium-binding properties. Two of these proteins (originally termed the minor A and minor B components) were shown to be alteration products arising from the native molecule (originally termed the major component) during purification and storage (Fullmer and Wasserman, 1973). It was subsequently demonstrated (Fullmer et al., 1975) that selective conversion to the minor A and B components could be effected experimentally by exposure of the calcium-loaded native protein to trypsin, with no loss of calcium binding or immunological activity. In the absence of bound calcium, the native protein (as well as the minor A and B components) was rapidly degraded to the constituent tryptic peptides with complete loss of these activities (Fullmer et al., 1975; Fullmer and Wasserman, 1980). Other di- and trivalent cations shown to displace bound Ca^{2+} also provide varying degrees of protection against tryptic digestion (Fullmer and Wasserman, 1977).

Based on these results, the term "components" now appears to be inappropriate, as is the term "degradation products." Perhaps the most appropriate descriptive term for the minor A and B CaBPs is "alteration products," which shall henceforth be employed here.

Hitchman and Harrison (1972) were also successful in isolating two CaBPs from porcine intestine which were similar in both size (MW ~12,000) and calcium-binding properties but differed in electrophoretic mobility. The observed differences between these two proteins have recently been defined chemically (Hofmann et al., 1979).

Alteration products have been found in other species as well, which may be analogous to those defined in the cow and pig. Ueng and Bronner (1979) reported that, in growing rats, intestinal CaBP occurred consistently as two protein bands by conventional gel electrophoresis in material

Porcine

$$\overset{\text{B}}{\underset{5}{\downarrow}}$$

Ac-Ser-Ala- Gln -Lys- Ser -Pro- Ala-Glu- Leu-Lys- Ser -Ile- Phe -Glu- Lys -Tyr- Ala -Ala- Lys -Glu-
1   10   15   20

Bovine

$$\overset{\text{A}}{\downarrow}\ \overset{\text{B}}{\downarrow}$$

BG[a]-Ala-Lys-Lys- Ser ——— Glu ——— Gly ——————

Porcine

Gly-Asp- Pro -Asn-Gln-Leu- Ser-Lys- Glu -Glu- Leu-Lys- Gln -Leu- Ile -Gln- Ala -Glu- Phe -Pro-
30   35   40

Bovine

——————— Leu ——— Leu —— Thr ——————

Porcine

Ser-Leu- Leu-Lys- Gly -Pro- Arg -Thr- Leu-Asp- Asp-Leu- Phe -Gln- Glu- Leu-Asp-Lys- Asn-Gly-
45   50   55   60

Bovine

——————— Ser ——— Glu —— Glu ——————

Porcine

Asn-Gly- Glu - Val- Ser- Phe- Glu-Glu- Phe -Gln- Val -Leu- Val -Lys- Lys -Ile- Ser -Gln- OH
65   70   75   78

Bovine

Asp ———————————————— OH

Fig. 1. Amino acid sequences of porcine and bovine CaBPs. The N-terminal blocking group is designated BG, and A and B denote the beginning of the minor A and minor B forms, respectively.

from mucosal scrapings but as only one band in preparations from isolated intestinal cells. These authors concluded that the single protein species from isolated cells represented the native, cellular form of CaBP, part of which is transformed enzymatically to the second form in the presence of luminal material. As evidence of this, it was noted that formation of the second (altered) form was enhanced, at the expense of the first, by incubation at 37°C for 2 hr of the partially purified mucosal preparation, a conversion inhibited by phenylmethylsulfonyl fluoride.

Bruns *et al.* (1978) reported the appearance of minor electrophoretic protein bands following storage of rat placental CaBP, which were minimally altered from the native protein and remained immunoreactive.

C. S. Fullmer and R. H. Wasserman (unpublished) have observed alteration products, similar to those arising from bovine CaBP, in all species studied thus far, including the chick, pig, horse, and guinea pig.

Complete amino acid sequences of native porcine (Hofmann *et al.*, 1979) and bovine minor A (Fullmer and Wasserman, 1980) CaBPs have recently been reported and are compared in Fig. 1. Native porcine CaBP is a 78-residue protein (computed MW 8799) with an acetyl blocking group at the N-terminus. Bovine minor A CaBP, comprised of 75 residues (MW 8501) is not blocked. As expected on the basis of compositional analyses (Fullmer and Wasserman, 1975), these proteins show considerable sequence homology. All 9 residue differences between the bovine minor A form and the corresponding region of porcine native CaBP result from single nucleotide mutations and all are consistent with compositional differences.

Partial sequence analyses of the three bovine CaBPs (Fullmer and Wasserman, 1977) provided a chemical basis for the differences observed among them. The native protein was shown to be blocked at the N-terminus, while the minor A and B forms provided extended sequences which differed from each other only in the absence of the N-terminal lysine residue in the minor B form. Amino acid compositional analyses of the N-terminal *N*-bromosuccinimide peptides from the three forms confirmed this difference and further indicated that the native protein possessed one additional residue each of lysine and alanine in comparison to the minor A protein. Subsequent comparative peptide mapping studies (Fullmer and Wasserman, 1980) have confirmed that the only differences are at the N-termini of these three forms.

From these data and the known specificity of trypsin, the following sequence was deduced for bovine native CaBP: BG-Ala-Lys-Minor A CaBP. The blocking group (BG), as yet not definitely established, is most likely identical to that determined for native porcine CaBP (acetate).

Hofmann *et al.* (1979) also determined the sequence of the second form of CaBP isolated from porcine intestine by Hitchman and Harrison (1972). This form was shown to be unblocked, to begin at Ser-5 of the native protein, but to be otherwise identical for the 35 cycles investigated. It is clear from these results that this protein is analogous to the minor B form of bovine CaBP and probably arises in a similar fashion.

These results are interpreted to indicate that the minor A and B forms of bovine CaBP and the analogous alternate forms from the rat, pig, and other species arise by specific limited cleavage by trypsin or trypsin-like enzymes. In the case of bovine CaBP, at least, protection against total tryptic digestion with consequent loss of activity is conferred by Ca^{2+}, presumably via alteration of the three-dimensional structure that induces a tightly folded compact solution conformation in which 9 of the 11 lysine residues are masked from attack. The only substantial points for tryptic cleavage remaining are the two peptide bonds immediately adjacent to the lysine residue at position-1 of the bovine minor A form. Two Ca^{2+} atoms bound per molecule are apparently required for manifestation of this protective effect (Fullmer and Wasserman, 1977). These results are supported by NMR data (Levine *et al.*, 1977; Birdsall *et al.*, 1980) which indicate that binding of the second Ca^{2+} atom to the protein induces a tightly folded compact structure, rendering the molecule resistant to heat denaturation and tryptic digestion.

In the case of porcine (and probably rat) CaBP, the presence of only a single alteration product, corresponding to the bovine minor B form, may be explained by the absence of the Lys-Lys configuration presumed to be present near the N-terminus of bovine native CaBP. In Ca^{2+}-loaded porcine CaBP, tryptic cleavage is limited to Lys-4 to yield a single, unblocked alteration product.

There are presently no indications that these alteration products are of any physiological significance, since they most surely result from limited enzymatic activity arising from (as may be expected) the original mucosal tissue or luminal fluid. They must, however, be reckoned with, since they represent a possible source of confusion during purification and subsequent characterization. Certainly, for bovine and porcine CaBPs, their existence greatly facilitated, or aided in, confirmation of sequence determination.

Hofmann *et al.* (1979) also reported the isolation of two peptides from the N-terminus of native porcine CaBP, one beginning with acetylserine and another, which was one residue shorter, beginning with acetylalanine. This latter may indicate the existence of a naturally occurring minor variant form of porcine CaBP, the significance of which is not clear.

IV. CELLULAR LOCALIZATION OF CALCIUM-BINDING PROTEINS

Any assessment of the true role played by intestinal CaBP in the vitamin D-mediated metabolism of calcium has been severely hampered by disparate reports on the cellular localization of the protein. Chick intestinal CaBP was first localized by indirect fluorescent antibody techniques and shown to be present in the goblet cells and at the absorptive cell surface (Taylor and Wasserman, 1970). At this time, it was suggested that the goblet cells represented a site of synthesis of the protein which, along with the mucin granules, was extruded to become associated with the absorptive surface (mucous coat, microvillar surface) of the enterocytes.

Subsequent studies have either confirmed these results or reported other localization sites, including: in goblet cells only (Noda *et al.*, 1978); in goblet cells and at the surface of absorptive cells (Lippiello and Wasserman, 1975; Taylor and McIntosh, 1977); at the basal and apical membranes of cells (Helmke *et al.*, 1974); at the basal and apical membranes of absorptive cells and in goblet cells (Piazolo *et al.*, 1975); in the cytoplasm of absorptive cells and at the absorptive cell surface (Arnold *et al.*, 1976); in the intercellular spaces of absorptive cells and at the apical and basal membranes (Morrissey *et al.*, 1975); in the cytoplasm and nuclei of absorptive cells (Morrissey *et al.*, 1978a,b).

There emerges from these reports two general extreme viewpoints on the cellular localization of CaBP: (1) CaBP is present primarily in the cytoplasm of enterocytes, and (2) CaBP is present in the goblet cells and at the absorptive surface of enterocytes. Several recent reports seem to have eliminated confusion resulting from these extremes of localization and appear to have established that CaBP is a cytoplasmic protein of enterocytes.

Taylor (1980a, 1981) employed three different sectioning procedures in conjunction with the indirect fluorescent antibody technique to study the localization of chick intestinal CaBP. Intestinal sections prepared by the freeze-thaw procedure produced CaBP-specific fluorescence associated with the goblet cells and the absorptive surface, whereas sections prepared by both the freeze-dry and freeze-substitution methods showed specific fluoroscence to be within the cytoplasm of the enterocytes. It was concluded that, during the brief thaw period in the freeze-thaw procedure, CaBP migrated in the aqueous environment to become associated preferentially with goblet cell mucin and probably mucin along the absorptive surface. To test this hypothesis, sections prepared by the freeze-dry procedure were rehydrated prior to fixation. In this case, CaBP-specific fluorescence was no longer associated with the cell cytoplasm but became situated over goblet cells and along the absorptive surface, indicating these localization sites to be the artifactual consequence of the extreme water solubility of CaBP as well as the apparent affinity of the protein for mucin.

Jande *et al.* (1980), using the immunoperoxidase method, agreed with the localization of chick CaBP in the cytoplasm of the absorptive cell. The terminal web region was reported to give a stronger reaction, whereas the absorptive surface, basement membranes, and goblet cells' were completely negative. Marche *et al.* (1979) arrived at a similar conclusion for rat duodenum, detecting CaBP primarily in the terminal web region of the enterocyte cytoplasm but also near the basal laminae.

In the chick cerebellum, Jande *et al.* (1980) showed CaBP to be present only in Purkinje cells. Entire dendritic trees contained reaction product, and no other neurons in the molecular or granular layer were stained. In the deep cerebellar nuclei, all neurons were negative but were outlined by deeply staining Purkinje cell axons and their synaptic endings.

On the basis of these recent studies, it now seems certain that vitamin D-dependent CaBPs may be considered intracellular proteins and, therefore, more akin to other members of the troponin C superfamily. Additional studies, perhaps at the ultrastructural level, may provide additional information on any discrete intracellular localization.

V. PHYSIOLOGICAL FACTORS AFFECTING CaBP

One of the major physiological effects of vitamin D is stimulation of calcium absorption by the intestine. The response of the animal to the administration of vitamin D is preceded by a variable lag period, dependent upon animal species and other factors. Detailed studies (cf. reviews by DeLuca, 1979; Norman, 1979) disclose that, during the lag phase and prior to onset of the stimulation of calcium absorption, vitamin D is transformed from the parent to the monohydroxy form in the liver, followed by formation of the dihydroxy derivative in the kidney. This hormonal form, $1,25(OH)_2D_3$, localizes in the target tissue and, most prominently in the nucleus and in association with chromatin. The next step is the synthesis of specific proteins that are undoubtedly involved in stimulating the transfer of calcium and/or phosphate across the epithelial membrane.

A role of CaBP in vitamin D-dependent calcium absorption has been supported over the years by the correlation between the concentration of CaBP in intestinal mucosa and the rate of calcium absorption. It has been shown that, under many different circumstances, an increase in CaBP levels is accompanied by an increase in calcium absorption, and the reverse also holds true (Wasserman and Corradino, 1973; Wasserman *et al.*, 1978; Wasserman, 1980; Taylor, 1980b). Examples of these observations follow. It has also been shown that, under certain circumstances, CaBP concentration does not correlate with calcium absorption. An example of this noncorrelation will also be described.

A. Effect of Low Dietary Calcium or Phosphorus Intake

Chicks and other animal species have the ability to adapt to a low-calcium or a low-phosphorus diet, in the sense that the efficiency of calcium absorption is greater than when a normal calcium-normal phosphorus diet is ingested. In the chick, this increase in the efficiency of calcium absorption is accompanied by an increase in the concentration of intestinal CaBP (Morrissey and Wasserman, 1971; Swaminathan *et al.*, 1977; Friedlander *et al.*, 1977). The stimulation of intestinal CaBP concentrations in the rat by a low calcium intake was also demonstrated by Edelstein *et al.* (1978), Buckley and Bronner (1980), Thomasset *et al.* (1977), and Freund and Bronner (1975).

The pig, another species in which the presence of an intestinal vitamin D-dependent CaBP was demonstrated, also absorbs calcium more efficiently after adaptation to a low calcium intake, with a significant increase in jejunal and ileal CaBP but not duodenal CaBP (Fox *et al.*, 1977; Thomasset *et al.*, 1979b).

Phosphate deficiency also stimulates the synthesis of CaBP and calcium transport in the chick in the presence of an adequate intake of vitamin D (Morrissey and Wasserman, 1971; Bar and Wasserman, 1973; Montecuccoli *et al.*, 1977; Friedlander *et al.*, 1977), a phenomenon shown also to occur in the rat (Thomasset *et al.*, 1976, 1977) and pig (Fox *et al.*, 1978). Edelstein *et al.* (1978), however, were unable to observe an effect of phosphate deficiency on CaBP synthesis in the rat, a finding in contrast to that of Thomasset *et al.* (1976, 1977) in the same species.

Considerable evidence has indicated that the phenomenon of adaptation to a low calcium intake involves the increased synthesis of $1,25(OH)_2D_3$ which subsequently affects the intestinal calcium absorptive mechanism (cf. DeLuca, 1979; Norman, 1979). Other evidence further suggests that a key modulator of kidney 25-hydroxycholecalciferol-1α-hydroxylase is parathyroid hormone, secreted in response to the hypocalcemic state (Ribovich and DeLuca, 1976; Fraser and Kodicek, 1973; Rasmussen *et al.*, 1972). The increased absorption of calcium that occurs in response to a low phosphorus intake is only partially dependent upon the increased formation of $1,25(OH)_2D_3$, since adaptation under these circumstances occurs when either bona fide $1,25(OH)_2D_3$ or pseudo vitamin D hormone (dihydrotachysterol) is supplied as the only source of vitamin D (Bar and Wasserman, 1973; Ribovich and DeLuca, 1975; Friedlander et al., 1977). This was borne out by a direct measurement of the 1α-hydroxylase activity of chick kidney in vitro; it was noted that a low phosphorus intake did not significantly affect the hydroxylase enzyme, in contrast to the effect of a low calcium intake (Montecuccoli et al., 1977). However, the intestinal concentration of CaBP and calcium absorption were elevated by a factor of 2–3 after 10 days on a low phosphate diet (Montecuccoli et al., 1977). Thus, the concept that the mechanism of adaptation to a low calcium diet differs from that occurring in response to a low phosphate intake, initially proposed by Bar and Wasserman (1973), has been verified in subsequent studies.

B. Effect of a Diet Containing a High Concentration of Strontium Salts

A diet containing a high concentration of $SrCl_2$ (2.3%) depresses calcium absorption in the chick and, in the same time frame, there is an inhibition of CaBP synthesis by gut tissue (Corradino et al., 1971). A similar effect of strontium-containing diets on depressing calcium transport and CaBP levels was recently shown to occur in rats (Armbrecht et al., 1979a). The reversal of the inhibitory effect of the strontium diet by the administration of $1,25(OH)_2D_3$ was consistent with the proposal that a high strontium intake inhibits the formation of $1,25(OH)_2D_3$ by the kidney 25-hydroxycholecalciferol–1α-hydroxylase system (Omdahl and DeLuca, 1972). This high-strontium regime in essence produces a vitamin D-deficient state even in the presence of adequate vitamin D and has provided a useful tool in studies designed to determine the nature of the toxic principle in the calcinogenic plants Solanum malacoxylon and Cestrum diurnum (Wasserman and Nobel, 1980). The use of the strontium model and the detection of CaBP synthesis in response to the administration of plant extract provided the first evidence that the factor was $1,25(OH)_2D_3$-like in its action (Fig. 2). Subsequent biochemical analyses, including mass spectrometry, established the factor in both plants to be a glycoside of $1,25(OH)_2D_3$ (Wasserman et al., 1976; Hughes et al., 1977).

C. Aging and CaBP

The inverse relation between growth rate, age, and calcium absorption has been well-documented (cf. Nicolaysen et al., 1953). With advancing age, the requirement for calcium declines and a homeostatically mediated

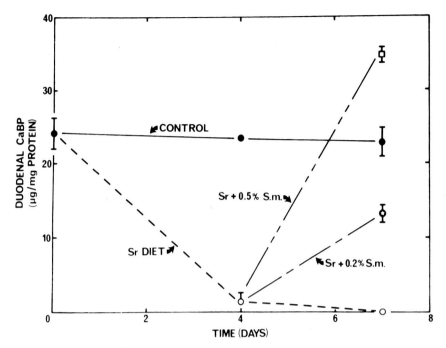

Fig. 2. Reversal of the inhibitory effect of dietary strontium on calcium absorption and net synthesis of vitamin D-dependent CaBP by *S. malacoxylon* (S.m.). Chicks at 3½ weeks of age were fed a normal diet for 7 days or a diet containing strontium (2.5%) for 4 days. On day 4, some of the strontium-fed chicks were continued on this diet or given the same diet without vitamin D_3, but with the addition of *S. malacoxylon* powder (either 0.2 or 0.5%), for 3 days more. Duodenal CaBP was determined at the times indicated. Since the strontium diet blocks $1,25(OH)_2D_3$ synthesis (see text), the presence of CaBP in chicks fed the strontium- and S.m.-containing diets provided evidence for a $1,25(OH)_2D_3$-like factor in S.m. Each point represents the mean of six chicks with ±SEM given. From Wasserman (1974).

decrease in calcium absorption occurs. Early studies clearly showed that, in the chick, the duodenal concentration of CaBP also decreased with age (Wasserman and Taylor, 1968).

Bruns *et al.* (1977) observed that intestinal CaBP, when expressed as units per gram of body weight, decreased nearly exponentially with the age of rats. A direct relationship between growth rate and CaBP levels was also observed.

More recently, detailed studies on aging and calcium metabolism in the rat were reported by Armbrecht *et al.* (1980a). In this investigation, CaBP was determined by radioimmunoassay and calcium transport by the everted gut procedure. As shown in Fig. 3, it can be seen that there was a

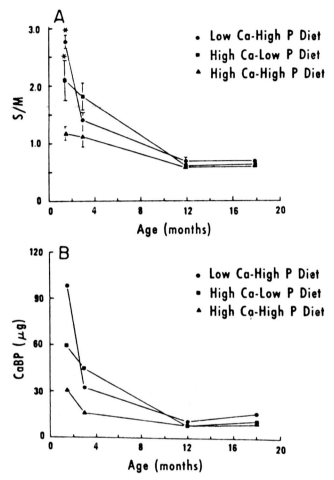

Fig. 3. (A) Active transport of calcium by proximal duodenum from rats of different ages on different diets. Active transport is expressed as the serosal/mucosal ratio (SPM) of ^{45}Ca activity as determined by the everted sac technique. Points are means ± SE for 508 rats. The standard error is smaller than the symbol where there is no error bar. An asterisk indicates that a point is significantly different ($p <$ .05) from that for the high-calcium–high-phosphorus diet for same age group. (B) Immunoreactive CaBP content of the first 5 cm of proximal duodenum from rats of different ages on different diets. CaBP was quantified in pooled intestinal supernates from the same segments used to measure active transport of calcium (A). The calcium and phosphorus contents of the diets were as follows: high Ca–high P, 0.6% Ca, 0.6% P; low Ca–high P, 0.02% Ca, 0.6%; high Ca–low P, 0.6% Ca, 0.1% P. Reproduced by permission from Armbrecht *et al.* (1980a).

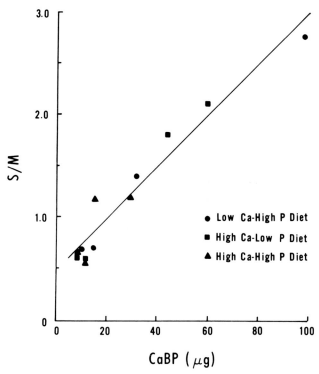

Fig. 4. Correlation plot of intestinal calcium transport (S/M) versus intestinal CaBP content for animals from 1.5 to 18 months. Data are the same as those shown in Fig. 3A and B. The linear least-squares regression line is $y - 0.025\,x + 0.46$ ($r = 0.978$). Reproduced with permission from Armbrecht *et al.* (1980a).

consistent decrease with age in the calcium transport capacity of the proximal duodenum and the amount of CaBP in the duodenal tissue. In younger rats, the expected adaptative response to either a low calcium or a low phosphorus intake occurred but, in older rats, the response to these dietary deficiencies was minimal. The parallelism between calcium transport and CaBP levels is apparent, and the correlation display given in Fig. 4 accentuates this point. Armbrecht *et al.* (1980b) further demonstrated that there was an extensive decrease in $1,25(OH)_2D_3$ synthesis by the kidney of adult rats when compared to young rats. However, adult rats do respond to exogenous $1,25(OH)_2D_3$, resulting in an increase in the formation of CaBP (Armbrecht *et al.*, 1979b).

These recent investigations suggest that the kidney 1-hydroxylase enzyme system in the adult rat is less active than that of younger animals

TABLE II

Effect of Hydrocortisone, Vitamin D_3, and $1,25(OH)_2D_3$ on Calcium Absorption, Intestinal CaBP, and Alkaline Phosphatase[a,b]

Group	Vitamin D activity	Hydrocortisone	^{47}Ca absorption (%)	CaBP ($\mu g/mg$)
1	D_3	−	77 ± 10	32 ± 3
2	D_3	+	50 ± 5	23 ± 3
3	$1,25(OH)_2D_3$	−	69 ± 4	9 ± 2
4	$1,25(OH)_2D_3$	+	36 ± 3	5 ± 1

[a] From Feher and Wassermann (1979a).

[b] Values are mean ± SEM for six chicks per group. Vitamin D_3 (500 IU) was given 72 hr, and $1,25(OH)_2D_3$ (1 μg) 24 hr, before the experiment. Hydrocortisone acetate (2 mg) given intramuscularly 48 and 24 hr before the experiment.

and, perhaps more importantly, less responsive to mineral deficiencies as compared to that of the younger animal. This provides support for the therapeutic potential of exogenous 1α-hydroxylated vitamin D_3 metabolites in senile osteoporosis.

D. Cortisol and CaBP

The inhibitory effect of glucocorticoids on the intestinal absorption of calcium has been well-documented (e.g., Kimberg et al., 1971; Carré et al., 1974; Edelstein et al., 1977). The relationship of this effect on the CaBP content of the intestine has been uncertain. In a series of experiments the effect of varying doses of cortisol was investigated in rachitic chicks given either vitamin D_3 (given 72 hr before the experiment) or $1,25(OH)_2D_3$ (given 24 hr before the experiment) (Feher and Wasserman, 1979a,b). As shown in Table II, cortisol depressed both calcium absorption and duodenal CaBP levels whether $1,25(OH)_2D_3$ or vitamin D_3 was administered. When the data from the $1,25(OH)_2D_3$ series or the vitamin D_3 series were plotted, two distinct regression lines were evident, with a high correlation (0.93 or better) between CaBP and calcium absorption within each set (Fig. 5).

Additional studies showed that vitamin D_3 at lower dosage levels fell on the $1,25(OH)_2D_3$ line but, as the dosage was increased, the values swerved toward the line generated by the 72 hr–500 IU vitamin D_3 time–dosage line (Fig. 6). Further, at 24 hr after administration, the 500-IU vitamin D_3 dose yielded values that also fell on the $1,25(OH)_2D_3$ regression line. These

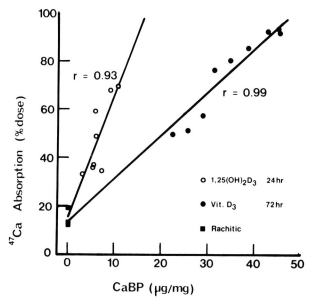

Fig. 5. Correlation between CaBP concentration and calcium absorption in rachitic chicks (■), chicks given 500 IU vitamin D_3 72 hr before the experiment (●) or 1,25(OH)$_2$D$_3$ 24 hr before the experiment (○). Various degrees of calcium absorption were produced by the administration of cortisol acetate intramuscularly in various dosages. Values given are means from at least six chicks. Reproduced from Feher and Wasserman (1979a).

observations suggested that the CaBP synthesized after 1,25(OH)$_2$D$_3$, after low doses of vitamin D_3, or by vitamin D_3 at short times after its administration, might be more directly involved in the calcium absorptive process. CaBP synthesized beyond this so-called essential level might have a lesser effect or have a subsidiary role in calcium transport. Another alternative is also quite feasible; i.e., another vitamin D-responsive or induced factor is limiting in the overall transport process under the above conditions.

E. CaBP in Laying Birds

During the maturation of the hen and female quail, considerable hormonally directed changes in calcium metabolism occur that precede or coincide with initiation of the physiological process of egg laying. This is illustrated with the quail. As shown in Table III, there is an increase in plasma calcium levels, intestinal concentration of 1,25(OH)$_2$D$_3$, and intestinal CaBP in the mature nonlaying quail as compared to the immature

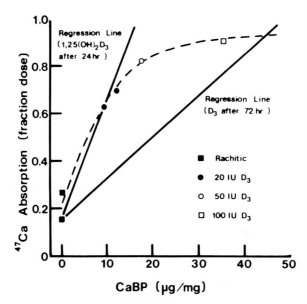

Fig. 6. Effect of vitamin D_3 dosage on the correlation between calcium absorption and soluble CaBP. Various doses of vitamin D_3 were given as indicated 72 hr before experiment. Results from two separate experiments are given ($n = 6$ chicks per group). Regression lines correspond to those in Fig. 5. Note that, as the amount of vitamin D_3 was increased, there was tendency for the plotted values to approach the regression line generated by the 500-IU vitamin D_3 dose given 72 hr before the experiment. Reproduced from Feher and Wasserman (1979b).

TABLE III

Alterations in Intestinal CaBP Levels and Other Parameters Associated with Maturation of the Female Quail[a]

	Physiological status		
Parameters	Immature	Mature nonlaying	Laying
Intestinal CaBP (mg/g)	0.5^a	1.1^b	2.0^c
Intestinal 1,25(OH)$_2$D$_3$ (pmoles/g)	0.10	0.30	1.0
Plasma calcium (mg/100 ml)	10.2^a	23.2^b	24.3^b
Kidney 1-hydroxylase (pmoles/g^{-1} 15 min^{-1})	1.10^a	1.36^a	1.70^b

[a] Modified from Bar *et al.* (1978a). Values designated by different superscript letters are significantly different at $p < .01$.

bird (Bar *et al.*, 1978a). At the onset of the laying period, there is a substantial increase in the activity of the kidney $25(OH)D_3$–1-hydroxylase system, a further increase in intestinal $1,25(OH)_2D_3$ levels, and a doubling of the concentration of CaBP in the intestine. Similar data are available for the laying hen (Bar *et al.*, 1978b). In addition, it has been shown that CaBP in the shell gland (uterus) increases by a factor of about 9 in the laying hen as compared to the immature bird.

F. Gonadal Hormones

The injection of gonadal hormones into birds was earlier shown to stimulate the formation of $1,25(OH)_2D_3$ (Tanaka *et al.*, 1976; Baksi and Kenny, 1977). Navickis *et al.* (1979) observed that estrogen increased the levels of calcium in the duodenum of vitamin D_3-replete female chicks; in the absence of vitamin D_3, no effect of estrogen occurred. Testosterone was also without effect but tended to inhibit the stimulatory effect of estrogen in vitamin D_3-replete chicks.

Were these direct hormonal effects on CaBP synthesis and $1,25(OH)_2D_3$ synthesis or indirect effects via changes in calcium metabolism known to accompany the action of gonadal hormones in birds? As pointed out by Bar and Hurwitz (1979a), the formation of medullary bone in male chicks is stimulated by the administration of estrogen and testosterone and represents a considerable "sink" for available calcium. Rackovsky (1978) noted that these hormones lowered ionic blood calcium levels, despite a considerable increase in total blood calcium levels, due to an increase in a serum CaBP (vitellogenin) of liver origin (Deeley *et al.*, 1975). The increased medullary bone formation and, possibly as a consequence, the decrease in ionic blood calcium levels, could provide the stimulus for $1,25(OH)_2D_3$ synthesis and subsequently enhanced CaBP formation. Bar and Hurwitz (1979a) further reported that the gonadal hormone response was considerably blunted when the chicks were fed a diet with an abnormally high concentration of calcium. If the estrogen–testosterone combination had a direct effect on $1,25(OH)_2D_3$ synthesis and/or CaBP synthesis, the dietary level of calcium would not be expected to interfere with their action. It therefore appears that the response of the chick to gonadal hormones is an indirect effect, operating by way of known effects of the hormones on calcium metabolism.

G. Insulin, CaBP, and Diabetes

Experimental diabetes can be induced in rats by injection of the drugs streptozotocin and alloxan. Schneider and Schedl (1972) provided evidence that, in the diabetic rat, the absorption of calcium was depressed, despite increased growth of the intestinal mucosa. Later it was reported that CaBP concentrations were also depressed and that $1,25(OH)_2D_3$ corrected the defect, providing evidence of a blockage in the synthesis of this hormone in these animals (Schneider *et al.*, 1977). Of considerable importance was the observation that insulin also increased the capacity of the intestine to absorb calcium, indicating that exogenous insulin restored the $1,25(OH)_2D_3$-synthesizing capability of the diabetic kidney.

The pancreas also contains vitamin D-induced CaBP (see above). The role of CaBP in insulin secretion is unknown, although it was recently reported that the secretion of insulin was depressed in vitamin D-deficient rats; no effect of vitamin D deficiency on glucagon synthesis was noted (Norman *et al.*, 1980).

VI. *IN VITRO* SYNTHESIS OF CaBP

The induction of CaBP synthesis *in vitro* in response to vitamin D was first observed in an organ culture preparation of embryonic chick intestine (Corradino and Wasserman, 1971). The duodenum of the chick embryo contains little or no CaBP but, when cultured in the presence of vitamin D_3, the formation of CaBP was readily shown to occur. Subsequently this chick embryonic system proved advantageous in determining the relative potencies of a wide variety of vitamin D metabolites and analogues (Corradino, 1980). The most potent forms capable of inducing CaBP synthesis were those analogues containing the hydroxyl group in the 1α-position of the steroid, and an equivalency in potency was observed between $1\alpha,25(OH)_2D_3$; $1\alpha,24R,25(OH)_3D_3$; and $1\alpha(OH)D_3$. Compounds devoid of the 1α-hydroxyl group had lower potencies by a factor of 70 or more.

In addition to assessing the CaBP-inducing potency of vitamin D_3 and vitamin D_3-related molecules, cultured chick duodenum provided a useful system for study of the molecular mechanism of CaBP formation, the control of its synthesis, and the relationship between calcium transport and CaBP concentration. Consonant with the concept that CaBP synthesis is a consequence of the formation of a unique mRNA in response to $1\alpha,25(OH)_2D_3$ treatment, it was observed that actinomycin D (an inhibitor of DNA-directed mRNA synthesis) and α-amanitin (an inhibitor of RNA polymerase II activity) inhibited both CaBP synthesis and ^{45}Ca uptake by cultured intestine (Corradino, 1973). Furthermore, this system has proved useful in determination of the effect of a wide variety of peptide and steroid hormones and other substances on CaBP synthesis, calcium transport, phosphate transport, adenylate cyclase activity, and alkaline phosphatase activity (Corradino, 1976, 1979; Corradino *et al.*, 1976).

Cultured embryonic chick intestine has also been employed by others in the study of factors affecting CaBP synthesis (Parkes and Reynolds, 1977; Parkes and DeLuca, 1979).

Another *in vitro* approach was the use of epithelial cells isolated from rat duodenum (Freund and Bronner, 1975). When the cells were derived from vitamin D-deficient rats and subsequently incubated in the presence of $1,25(OH)_2D_3$, both the uptake of ^{45}Ca by these cells and the synthesis of CaBP were increased. The same *in vitro* methodology proved of value in attempting to uncover an effect of $1,25(OH)_2D_3$ on translational or post-translational events in CaBP synthesis. In previous studies, it was observed that the initiation of synthesis of intestinal CaBP by rats fed a high-calcium, vitamin D-replete diet occurred more rapidly than in that by rats fed a vitamin D-deficient diet after $1,25(OH)_2D_3$ treatment (Buckley and Bronner, 1980). One hypothesis put forth was that the translation of CaBP-specific mRNA, in addition to the transcriptional steps in the formation of messenger, was dependent on $1,25(OH)_2D_3$. This was tested with isolated intestinal cells in which it was noted that actinomycin D did not inhibit CaBP synthesis by added $1,25(OH)_2D_3$, whereas cycloheximide was an effective inhibitor (Singh and Bronner, 1980). These results therefore supported the above hypothesis, although other explanations of the results are possible, as noted by the authors.

The *in vitro* synthesis of chick CaBP by isolated polyribosomes from chick intestine was first reported by Emtage *et al.* (1973, 1974a). In the presence of labeled leucine, ATP and an ATP-generating system, GTP, Mg^{2+}, a complement of nonlabeled amino acids, and rat liver cell sap, a product precipitable by specific anti-CaBP antiserum was detected, but only from polyribosomes derived from vitamin D-replete chicks. The authenticity of the synthesized product as CaBP was established by additional criteria. These studies support the hypothesis that vitamin D is required for the formation of an mRNA that codes for CaBP. The synthesis of no other soluble protein with vitamin D dependency was observed in this system. A high correlation between the initiation of CaBP synthesis by vitamin D and stimulation of calcium absorption in the chick was also reported (Emtage *et al.*, 1974b).

A subsequent study by this group (Spencer *et al.*, 1978a) was concerned with events occurring after the administration of $1,25(OH)_2D_3$ to rachitic chicks. As will be recalled, $1,25(OH)_2D_3$ yields a faster response and has a greater potency than vitamin D_3. The polyribosomal synthesis of CaBP *in vitro* and its relation to calcium transport by intestine were determined. The stimulation of calcium transport preceded by 2.5 hr any detectable increase in CaBP synthesis by isolated polyribosomes, although thereafter the increase in the rate of synthesis of CaBP paralleled that of calcium transport. The presumption from these studies is that CaBP synthesis follows, but does not precede, a change in the calcium permeability properties of the intestinal membrane due to $1,25(OH)_2D_3$ treatment. The Spencer study suggests that other events, in addition to CaBP synthesis, are required for the expression of $1,25(OH)_2D_3$ action on the intestinal calcium transport system.

Poly A-containing mRNA from chick intestine, when translated in a wheat germ cell-free system, yielded several proteins, only one of which was vitamin D-dependent and precipitable with antibody against CaBP (Spencer *et al.*, 1976). The size of this mRNA was considerably greater than required to code for a 28,000-molecular-weight protein, suggesting that a precursor protein might be the translated product. However, further experiments did not yield any evidence for the synthesis of such a precursor protein in the *in vitro* system and provided evidence that CaBP was not an "exported" protein because of the lack of a characteristic "signal" polypeptide. The discrepancy between the size of CaBP mRNA and the size of CaBP has not been explained.

Polyribosomes from chick kidney or duodenum were also shown to synthesize CaBP in a reticulocyte assay system (Christakos and Norman, 1980b). The CaBP product from polysomes from both tissues appeared identical to CaBP previously isolated from intestine. However, a higher fraction of the active polysomes from duodenum were involved in CaBP synthesis than those from kidney. Further, the CaBP synthesized *in vitro* by polysomes from duodenal tissue taken from chicks adapted to a low-calcium diet exceeded that from chicks fed a high-calcium diet. This latter finding is consistent with the known enhancement of intestinal CaBP levels in chicks adapted to a low-calcium diet. Charles *et al.* (1981), using a wheat germ lysate system, also translated polysomes from chick duodenum and also demonstrated the vitamin D dependency of the synthesis of CaBP by isolated polysomes.

In each of the above studies with polyribosomes derived from chick intestine, all evidence indicated the CaBP product to be identical with CaBP isolated from chick mucosa. No evidence of a precursor form of chick CaBP was therefore obtained.

Chick intestinal nuclear mRNA was translated in a wheat germ system by Spencer *et al.* (1978b), and the specific synthesis of CaBP determined by immunoprecipitation. After a pulse of $1,25(OH)_2D_3$, the first evidence for the transcription of CaBP-coded nuclear mRNA occurred at 2 hr, the same time period during which CaBP synthesis was first observed when polysomal mRNA was translatable *in vitro*. The absence of a lag between the appearance of CaBP-coded nuclear mRNA and polysomal mRNA indicated a rapid processing of nuclear mRNA to functional ribosomal mRNA. However, accumulation of the $1,25(OH)_2D_3$ hormone within the nucleus occurs rapidly and maximizes at 2 hr, suggesting that other $1,25(OH)_2D_3$-dependent cytoplasmic or nuclear events might control the formation and/or maturation of CaBP-coded nuclear mRNA.

A further step in the molecular biology of CaBP was taken when mRNAs were isolated from pig intestine and translated in a wheat germ lysate preparation in the presence of L-[^{35}S]methionine or L-[4,5-^{3}H]leucine (Tomlinson and Mellersh, 1980; Mellersh et al., 1980). Radiolabeled products were immunoprecipitated by specific anti-pig CaBP antiserum. The products that were precipitated expectedly contained the leucine label and, unexpectedly, the methionine label. The latter was not expected, since the amino acid compositional analysis of authentic pig CaBP does not disclose the presence of methionine. The methionine-containing product differed from authentic CaBP by a different elution volume from a gel filtration column and in its electrophoretic mobility on sodium dodecyl sulfate (SDS)–polyacrylamide gels. The apparent molecular weight of the methionine-containing polypeptide was 11,500, as compared to about 9000 for authentic CaBP. The larger product might represent a pre-CaBP that is modified by limited posttranslational proteolysis in intact epithelial cells.

VII. EMBRYONIC DEVELOPMENT

In the chick, the embryonic intestine during development contains little or no CaBP but, at the time of hatching (21 days), CaBP appears (Corradino et al., 1969). This occurs despite the known production of $1,25(OH)_2D_3$ by the embryonic kidney at about 12–13 days (Bishop and Norman, 1975; T. Oku and R. H. Wasserman, unpublished data) and even the presence of vitamin D-induced CaBP in kidney tissue as early as 10 days of embryonic life (Taylor and Wasserman, 1972).

The CaBP-synthesizing system in intestine is apparently intact, since exogenous $1,25(OH)_2D_3$ can induce CaBP synthesis, at least by 18 days of age (Corradino and Wasserman, 1974), and the $1,25(OH)_2D_3$ receptor is present in this tissue at least by day 17 (Oku et al., 1976). The refractoriness of the intestinal CaBP-synthesizing system to endogenous $1,25(OH)_2D_3$ in ovo is not known but might be merely a consequence of insufficient levels of the hormone available to the receptor.

In the rat, CaBP was undetectable in the prenatal animal but detectable in the postnatal stage at age 11 days (Ueng and Bronner, 1979), using a relatively insensitive binding assay. However, with a considerably more sensitive radioimmunoassay, DeLorme et al. (1979) detected small amounts of CaBP in fetal intestine at a gestation age of 17.5 days. Just prior to birth (20.5–21.5 days gestation), the CaBP level rose by a factor of 2, where it remained during the first 18 days of postnatal life at which time it increased substantially. This corresponded to the period just prior to weaning and, after weaning, CaBP decreased in concentration. The distribution of CaBP along the intestine of the neonate was similar to that in the adult rat (duodenum > jejunum > ileum).

In the pig, CaBP determined by radial immunoassay was detectable in all intestinal segments (duodenum, jejunum, ileum) at 18 days before birth (R. H. Wasserman, W. G. Pond, M. E. Brindak, J. F. Zimmer, and S. Raddi, unpublished data). Between that time period and birth, duodenal and jejunal CaBP increased slightly, whereas ileal CaBP declined. At 4 days after birth duodenal and jejunal CaBP greatly increased (a 5- to 10-fold enhancement), and ileal CaBP further decreased in concentration.

Thus, in the three species mentioned, there were considerable differences in maturation of the CaBP-synthesizing system. In the chick, intestinal CaBP was not detectable until the time of hatching, whereas CaBP was present in the intestine of the prenatal pig and rat. In both the chick and pig, there was a massive increase in CaBP at the time of birth or hatching. The rat, known to be less developed at birth than the other two species, did not show an increase in intestinal CaBP levels until just prior to weaning.

VIII. CaBP AND CALCIUM REABSORPTION IN THE KIDNEY

As previously mentioned, the presence of CaBP in the kidney of several animal species has been documented. Immunohistochemical methods demonstrated that renal CaBP was localized in the cells of the distal tubule region of the chick kidney (Lippiello, 1974). Costanzo et al. (1974) reported that the kidney site at which vitamin D enhances calcium reabsorption in the rat was in the same region. If extrapolation across species can be made, it is apparent that there is a correspondence between the location of CaBP and the site that physiologically responds to vitamin D.

IX. TEMPORAL RESPONSES OF CaBP TO ACUTE DOSES OF $1,25(OH)_2D_3$

Of the vitamin D analogs uncovered thus far, this metabolite is the most active in the intact animal. As previously noted, it acts faster and is considerably more potent than the parent steroid.

It was of interest to assess, in the rachitic animal, the temporal relationship between the stimulation of calcium absorption by $1,25(OH)_2D_3$ and the appearance of CaBP. The thesis is that, if CaBP is *the* macromolecule involved in calcium absorption, CaBP should appear before or at least at the same time that a change in calcium absorption occurs.

Three studies on this question have been reported. Spencer *et al.* (1978a) injected $1,25(OH)_2D_3$ intracardially into vitamin D-deficient chicks and determined the transport of ^{45}Ca by the *in vitro* everted gut sac procedure or by detecting changes in plasma ^{45}Ca after an oral ^{45}Ca dose at various times following the $1,25(OH)_2D_3$. At the same time points, the CaBP content of the intestinal mucosa and the CaBP-synthesizing capacity of intestinal polyribosomes were determined. In their hands, an increase in calcium absorption (or transport) preceded the synthesis of CaBP by isolated polysomes and the appearance of CaBP per se. This study indicated that the formation of CaBP by chick intestine followed the increase in ^{45}Ca absorption and thus suggested that another, early $1,25(OH)_2D_3$-mediated change in the calcium transport mechanisms was responsible for the increased absorption of calcium. Also, it was noted that intestinal CaBP persisted in the mucosa long after the $1,25(OH)_2D_3$-stimulated absorption of calcium had returned to the rachitic baseline.

We had performed similar experiments in the rachitic chick (Fig. 7) and observed that, when the first significant increase in calcium absorption occurred, CaBP was detectable in the intestinal mucosa (Wasserman *et al.*, 1977). However, as in the Spencer study, CaBP was present in the mucosa after calcium absorption returned to baseline (not shown). These results again indicate that the absorption of calcium is by a multicomponent process, only part of which is related to the total amount of CaBP present in the intestinal mucosa.

Results similar to those of Spencer *et al.* (1978a) were reported by Thomasset *et al.* (1979a) using the rat; i.e., calcium transport, as measured by the *in vitro* everted gut sac method, preceded the appearance of CaBP as determined by a radioimmunoassay. Specifically, an increase in CaBP (unlike the situation in the chick, residual CaBP was present in their vitamin D-deficient rats) was noted at 6 hr, and an increase in transport capacity was observed in the preceding experimental period, i.e., 3 hr.

Each of these studies points to a noncorrelation between CaBP concentrations in the intestine and the capacity of this organ to transport calcium. It should be understood that all these studies were performed in the nonsteady state after an acute dose of $1,25(OH)_2D_3$, when several different molecular changes might be taking place, one of which was stimulation of the synthesis of CaBP.

When a physiological steady state has been achieved, e.g., after an animal has adapted to a dietary calcium deficiency, the correlation between CaBP and calcium absorption is usually of the order of 0.9 or greater. This suggests that, under these circumstances, CaBP becomes

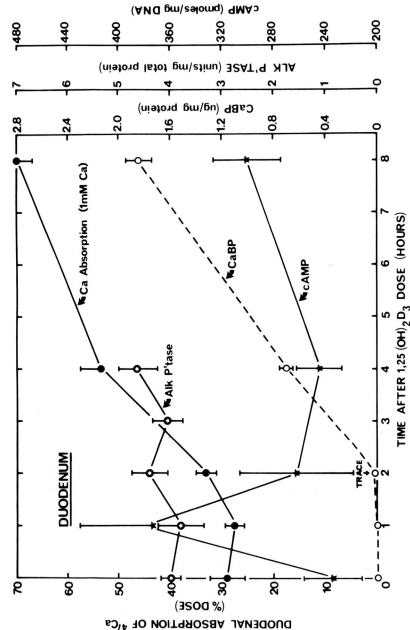

Fig. 7. Early response of rachitic chicks to intravenous 1,25(OH)$_2$D$_3$ (1 μg). Shown are time patterns of the duodenal absorption of ^{47}Ca (ligated loop technique), duodenal cyclic adenosine-3′,5′-phosphate (cAMP), calcium-binding protein (CaBP), and alkaline phosphatase (Alk P'tase) after 1,25(OH)$_2$D$_3$ administration. Each value is given as the mean ± SEM for six chicks. From Wasserman *et al.* (1977).

limiting in the transport process, whereas some other vitamin D-responsive event can assume the role of the limiting step under certain conditions.

X. DISCUSSION

The foregoing documents various physical and chemical properties of vitamin D-induced CaBP, its localization in different tissues of various species, the physiological relationship between CaBP and calcium transport, and factors that affect CaBP and calcium transport.

Overriding questions are What is the function of CaBP? and How does it relate to the calcium transport process? Obviously the answers to these questions are not easy or now readily available and, with new information on the tissue localization of CaBP and the uncertainty of the exact localization of CaBP within the cell itself, the problem becomes even more difficult. Initially, CaBP appeared to be associated only with epithelial-type tissues involved in the transmural transport of calcium. Since then, the studies from our group, that of Christakos and Norman (1980b), and that of Arnold et al. (1975) have shown that CaBP is located in such nonepithelial tissues as the pancreas, parathyroid glands, brain, and bone. Although each of these tissues is not seemingly involved in transport across the whole organ (as in the intestine), calcium is certainly a significant factor in their function. The pancreatic secretion of insulin is certainly dependent upon exogenous (and endogenous) Ca^{2+}, the secretion of hormone by the parathyroid gland is dependent on plasma calcium levels, and calcium is a major component of bone mineral. In fact, calcium is a significant effector of many physiological and biochemical reactions and has assumed the role of an important intracellular second messenger (Carafoli and Crompton, 1978). Calmodulin, reviewed in the previous volume of this series, appears to be a key molecular species that confers calcium sensitivity on many of these reactions. CaBP, in contrast, is apparently not as ubiquitous as calmodulin and is distinguished from calmodulin and other CaBPs by its dependency on vitamin D, although there is sequence homology among many CaBPs (Kretsinger, 1980).

The NMR studies of Birdsall et al. (1980) have shown that bovine CaBP undergoes structural changes in the Ca-binding state, as does troponin C and calmodulin. The structural rearrangements of troponin C and calmodulin are undoubtedly related to their interaction with recipient macromolecules, and it is suspected that CaBP might function in a similar way, i.e., confers Ca^{2+} sensitivity on another molecule. This has some support from the study of Freund and Borzemsky (1977) in which CaBP conferred calcium sensitivity on intestinal alkaline phosphatase. If this mechanism has validity in the intact cell, it will be important to know whether the modulation of alkaline phosphatase is indeed the physiologically significant event or whether CaBP interacts endogenously with other substances involved in the calcium transport reaction.

It has been suggested that CaBP is a bona fide transport protein, *directly* enhancing the permeability of biomembranes to calcium. This is unlikely, because of its solubility properties (hydrophilic) and its intracellular localization. However, part of the intestinal CaBP can be solubilized only with detergent (Triton X-100), possibly suggesting an intimate association with membrane components (Feher and Wasserman, 1978). Furthermore, CaBP interacts with lysolecithin, the result being a decrease in its calcium-binding capacity and electrophoretic mobility (Wasserman, 1970). Recent data also reveal that CaBP is present in isolated, highly purified intestinal microvilli, as visualized by electron microscopy and by the double antibody–horseradish peroxidase technique (Roth *et al.*, 1980). The above information suggests that CaBP could interact with membrane components, particularly if lysolecithin were present, and perhaps also associate with structures within the microvillus core. The relationship of these observations to calcium transport is highly speculative at best.

Another mechanism often proposed is that CaBP serves as an intracellular buffer, protecting the cell from the potentially deleterious effects of high intracellular concentrations of Ca^{2+} that might occur during calcium absorption. An argument in favor of this proposal is that CaBP in the intestine of the vitamin D-replete chick represents about 1–3% of the soluble protein and, by estimation, at a concentration on the order of 10^{-4} M, assuming equal distribution of CaBP throughout the intestinal cell. This concentration of CaBP could sequester much of the Ca^{2+} during the course of its translocation through the intracellular milieu. However, an argument against the ''buffer'' proposal relates to the change in CaBP as a result of adaptation of the chick to a calcium-deficient diet. After adaptation to a low-calcium diet, the CaBP concentration in the intestinal mucosa increases considerably, perhaps by a factor of 10 (Bar and Hurwitz, 1979b). This is in the wrong direction to be in accordance with the buffer hypothesis. In the calcium-deficient situation, considerably less calcium is actually transferred across the intestine than under high or normal dietary calcium circumstances. Thus, the CaBP concentration in the intestine is actually related more to the *efficiency* of the calcium transport mechanism (percentage of dietary calcium absorbed) than to the amount of calcium absorbed (milligrams of calcium absorbed per day).

Still another mechanism is that proposed by Hamilton and Holdsworth (1975). *In vitro* experiments revealed that chick CaBP was able to increase the rate of release of Ca^{2+} from isolated intestinal mitochondria. This effect was beyond the capacity of CaBP merely to bind the extruded Ca^{2+} and suggested an interaction of CaBP with the mitochondrial membrane(s), resulting in increased permeability of mitochondrial membranes to Ca^{2+}. These observations were thought to bear on vitamin D-mediated calcium transport as follows: Mitochondria have the capacity to accumulate Ca^{2+} against a considerable concentration gradient, and this sequestered calcium would be less available for transport through the intestinal cell as compared to cytoplasmic calcium. CaBP, by increasing the release of mitochondrial calcium, could have positive consequences on the overall transport of Ca^{2+}. Whether or not this proposal bears on the *in situ* transfer of calcium, these findings indicate that CaBP can affect the calcium permeability of a membrane. Extrapolation of these observations to the plasma membrane might not be warranted because of the known molecular differences between mitochondrial and plasma membranes.

Still another proposal is that CaBP might serve to shuttle calcium from the apical aspect of the intestinal cell to the basal-lateral membrane (Schachter, 1970). Ca^{2+} entering the cell is immediately sequestered by the high-affinity binding sites of CaBP. The release mechanism of CaBP-bound Ca^{2+} at the basal-lateral membrane could involve a calcium pump, a sodium–calcium exchange system, or one associated with a Ca-ATPase.

The mechanism of action of CaBP has proved thus far to be an enigma for researchers investigating its properties and behavior. Certainly it has a significant role in calcium metabolism and calcium transport but, again, what is this mysterious role? Despite this, CaBP represents one of the most studied molecular events occurring as a consequence of vitamin D action and has proved useful, as noted above, in the study of the molecular responses of the animal to vitamin D.

REFERENCES

Alpers, D. H., Lee, S. W., and Avioli, L. V. (1972). Identification of two calcium-binding proteins in human small intestine. *Gastroenterology* **62,** 559–564.

Armbrecht, H. J., Wasserman, R. H., and Bruns, M. E. H. (1979a). Effect of 1,25-dihydroxyvitamin D_3 on intestinal calcium absorption in strontium-fed rats. *Arch. Biochem. Biophys.* **192,** 466–473.

Armbrecht, H. J., Zenser, T. V., Bruns, M. E. H., and Davis, B. B. (1979b). Effect of age on intestinal calcium absorption and adaptation to dietary calcium. *Am. J. Physiol.* **236,** E769–E774.

Armbrecht, H. J., Zenser, T. B., Gross, C. J., and Davis, B. B. (1980a). Adaptation to dietary calcium and phosphorus restriction: Changes with age in the rat. *Am. J. Physiol.* **239,** E322–E327.

Armbrecht, H. J., Zenser, T. V., and Davis, B. B. (1980b). Effect of age on the conversion of

25-hydroxyvitamin D_3 to 1,25-dihydroxyvitamin D_3 by kidney of rat. *J. Clin. Invest.* **66**, 1118–1123.

Arnold, B. M., Kuttner, M., Willis, D. M., Hitchman, A. J. W., Harrison, J. E., and Murray, T. M. (1975). Radioimmunoassay studies of intestinal calcium-binding protein in the pig. II. The distribution of intestinal CaBP in pig tissue. *Can. J. Physiol. Pharmacol.* **53**, 1135–1140.

Arnold, B. M., Kovacs, K., and Murray, T. M. (1976). Cellular localization of intestinal calcium-binding protein in pig duodenum. *Digestion* **14**, 77–84.

Baimbridge, K. G., Selke, P. A., Ferguson, N., and Parkes, C. O. (1980). Human calcium-binding protein. *In* "Calcium-Binding Proteins: Structure and Function" (F. L. Siegel *et al.*, eds.), pp. 401–404. Elsevier/North Holland Publ., Amsterdam.

Baksi, S. N., and Kenny, A. D. (1977). Vitamin D_3 metabolism in immature Japanese quail: Effects of ovarian hormones. *Endocrinology* **101**, 1216–1227.

Bar, A., and Hurwitz, S. (1979a). The interaction between dietary calcium and gonadal hormone in their effect on plasma calcium, bone, 25-hydroxycholecalciferol-1-hydroxylase, and duodenal calcium-binding protein, measured by a radioimmunoassay in chicks. *Endocrinology* **104**, 1455–1460.

Bar, A., and Hurwitz, S. (1979b). Relationship of intestinal calcium absorption to intestinal and plasma calcium-binding protein measured by radioimmunoassay. *In* "Vitamin D: Basic Research and its Clinical Application" (A. W. Norman, K. Schafer, J. W. Coburn, H. F. DeLuca, D. Fraser, H. G. Grigoleit, and D. von Herrath, eds.), pp. 679–682. de Gruyter, Berlin.

Bar, A., and Wasserman, R. H. (1973). Control of calcium absorption and intestinal calcium-binding protein synthesis. *Biochem. Biophys. Res. Commun.* **54**, 191–196.

Bar, A., Cohen, A., Edelstein, S., Shemesh, M., Montecuccoli, G., and Hurwitz, J. (1978a). Involvement of cholecalciferol metabolism in birds in the adaptation of calcium absorption to the needs during reproduction. *Comp. Biochem. Physiol. B* **59B**, 245–249.

Bar, A., Cohen, A., Eisner, U., Risenfeld, G., and Hurwitz, S. (1978b). Differential response of calcium transport systems in laying hens to exogenous and endogenous changes in vitamin D status. *J. Nutr.* **108**, 1322–1328.

Bar, A., Dubrov, D., Eisner, U., and Hurwitz, S. (1978c). Calcium-binding protein and kidney 25-hydroxycholecalciferol-1-hydroxylase activity in turkey poults. *J. Nutr.* **108**, 1501–1507.

Bauman, V. K., Valinience, M. Y., and Pastuhob, M. V. (1972). Calcium-binding protein in bovine mammary gland. *Latv. PSR Zinat. Akad. Vestis* **294**, 133–134.

Birdsall, W. J., Dalgarno, D. C., Levine, B. A., Williams, R. J. P., Fullmer, C. S., and Wasserman, R. H. (1980). Structural study in solution of the vitamin D-induced calcium-binding protein. *In* "Calcium-Binding Proteins: Structure and Function" (F. L. Siegel *et al.*, eds.), pp. 405–406. Elsevier/North Holland Publ., Amsterdam and New York.

Bishop, J. E., and Norman, A. W. (1975). Metabolism of 25-hydroxy-vitamin D_3 by the chick embryo. *Arch. Biochem. Biophys.* **167**, 769–773.

Bredderman, P. J., and Wasserman, R. H. (1974). Chemical composition, affinity for calcium, and some related properties of the vitamin D dependent calcium-binding protein. *Biochemistry* **13**, 1687–1694.

Bruns, M. E. H., Fliecher, E. B., and Avioli, L. V. (1977). Control of vitamin D-dependent calcium-binding protein in rat intestine by growth and fasting. *J. Biol. Chem.* **252**, 4145–4150.

Bruns, M. E. H., Fausto, A., and Avioli, L. V. (1978). Placental calcium-binding protein in rats: Apparent identity with vitamin D-dependent calcium-binding protein from rat intestine. *J. Biol. Chem.* **253**, 3186–3190.

Buckley, M., and Bronner, F. (1980). Calcium-binding protein biosynthesis in the rat: Regulation by calcium and 1,25-dihydroxyvitamin D_3. *Arch. Biochem. Biophys.* **202**, 235–241.

Carafoli, E., and Crompton, M. (1978). Regulation of intracellular calcium. *Curr. Top. Membr. Transp.* **10**, 151–216.

Carré, M., Ayigbedé, O., Miravet, L., and Rasmussen, H. (1974). The effect of prednisolone upon the metabolism and action of 25-hydroxy- and 1,25-dihydroxyvitamin D_3. *Proc. Natl. Acad. Sci. U.S.A.* **71**, 2996–3000.

Charles, M. A., Martial, J., Zolock, D., Morrissey, R., and Baxter, J. D. (1981). Regulation of calcium-binding protein messenger RNA by 1,25-dihydroxycholecalciferol. *Calcif. Tissue Res.* **33**, 15–18.

Christakos, S., and Norman, A. W. (1978). Vitamin D_3-induced calcium-binding protein in bone tissue. *Science* **202**, 70–71.

Christakos, S., and Norman, A. W. (1980a). Vitamin D-dependent calcium-binding protein and its relation to 1,25-dihydroxyvitamin D receptor. *In* "Calcium-Binding Proteins: Structure and Function" (F. L. Siegel *et al.,* eds.), pp. 371–378. Elsevier/ North Holland Publ., Amsterdam and New York.

Christakos, S., and Norman, A. W. (1980b). Vitamin D-dependent calcium-binding protein synthesis by chick kidney and duodenal polysomes. *Arch. Biochem. Biophys.* **203**, 809–815.

Christakos, S., Friedlander, J., Frandsen, B. R., and Norman, A. W. (1979). Studies on the mode of action of calciferol. XIII. Development of a radioimmunoassay for vitamin D-dependent chick intestinal calcium-binding protein and tissue distribution. *Endocrinology* **104**, 1495–1503.

Corradino, R. A. (1973). 1,25-Dihydroxycholecalciferol: Inhibition of action in organ-cultured intestine by actinomycin D and α-amanitin. *Nature (London)* **243**, 41–43.

Corradino, R. A. (1976). Embryonic chick intestine in organ culture: Hydrocortisone, vitamin D_3 and phosphate transport. *In Vitro* **12**, 299.

Corradino, R. A. (1979). Embryonic chick intestine in organ culture: Hydrocortisone and vitamin D-mediated processes. *Arch. Biochem. Biophys.* **192**, 302–310.

Corradino, R. A. (1980). Structure activity relationships of vitamin D metabolites and analogues in the induction of calcium-binding protein in organ-cultured embryonic chick duodenum. *In* "Calcium-Binding Proteins: Structure and Function" (F. L. Siegel *et al.,* eds.), pp. 385–392. Elsevier/North-Holland Publ., Amsterdam and New York.

Corradino, R. A., and Wasserman, R. H. (1971). Vitamin D_3: Induction of calcium-binding protein in embryonic chick intestine *in vitro*. *Science* **172**, 731–733.

Corradino, R. A., and Wasserman, R. H. (1974). *Solanum malacoxylon* extract: 1,25-Dihydroxycholecalciferol-like activity on calcium transport in organ-cultured embryonic chick duodenum. *Nature (London)* **252**, 716–718.

Corradino, R. A., Wasserman, R. H., Pubols, M. H., and Chang, S. I. (1968). Vitamin D_3 induction of a calcium-binding protein in the uterus of the laying hen. *Arch. Biochem. Biophys.* **125**, 378–380.

Corradino, R. A., Taylor, A. N., and Wasserman, R. H. (1969). Appearance of vitamin D_3-induced calcium-binding protein (CaBP) in chick intestine during development. *Fed. Proc., Fed. Am. Soc. Exp. Biol.* **28**, 760.

Corradino, R. A., Ebel, J. G., Craig, P. H., Taylor, A. N., and Wasserman, R. H. (1971).

Calcium absorption and the vitamin D_3-dependent calcium-binding protein. 1. Inhibition by dietary strontium. *Calcif. Tissue Res.* **7**, 81–92.

Corradino, R. A., Fullmer, C. S., and Wasserman, R. H. (1976). Embryonic chick intestine in organ culture: Stimulation of calcium transport by exogenous vitamin D_3-induced calcium-binding proteins. *Arch. Biochem. Biophys.* **174**, 738–743.

Costanzo, L. S., Sheeke, P. R., and Weiner, I. M. (1974). Renal actions of vitamin D in D-deficient rats. *Am. J. Physiol.* **226**, 1490–1495.

Deeley, R. G., Mullinex, K. P., Wetekam, W., Kronenberg, H. M., Meyers, M., Eldridge, J. D., and Goldberger, R. F. (1975). Vitellogenin synthesis in the avian liver: Vitellogenin is the precursor of the egg yolk phosphoproteins. *J. Biol. Chem.* **250**, 9060–9066.

Delorme, A. C., Marche, P., and Garcel, J. M. (1979). Vitamin D-dependent calcium-binding protein changes during gestation, prenatal and postnatal development in rats. *J. Dev. Physiol.* **1**, 181–194.

DeLuca, H. F. (1979). "Vitamin D: Metabolism and Function." Springer-Verlag, Berlin and New York.

Dorrington, K. J., Hui, A. H., and Hofmann, T. (1974). Porcine intestinal calcium-binding protein: Molecular properties and the effect of binding calcium ions. *J. Biol. Chem.* **219**, 199–204.

Dorrington, K. J., Kells, D. I. C., Hitchman, A. J. W., Harrison, J. E., and Hofmann, T. (1978). Spectroscopic studies on the binding of divalent cations to porcine intestinal calcium-binding protein. *Can. J. Biochem.* **56**, 492–499.

Drescher, D., and DeLuca, H. F. (1971). Vitamin D-stimulated calcium-binding protein from rat intestinal mucosa: Purification and some properties. *Biochemistry* **10**, 2302–2307.

Edelstein, S., Noff, D., Matitiahu, A., Sapir, R., and Harell, A. (1977). Functional metabolism of vitamin D in rats treated with cortisol. *FEBS Lett.* **82**, 115–117.

Edelstein, S., Noff, D., Sinai, L., Harell, A., Puschett, J. B., Golub, E. E., and Bronner, F. (1978). Vitamin D metabolism and expression in rats fed on low-calcium and low-phosphorus diets. *Biochem. J.* **170**, 227–233.

Emtage, J. S., Lawson, D. E. M., and Kodicek, E. (1973). Vitamin D-induced synthesis of mRNA for calcium-binding protein. *Nature (London)* **246**, 100–101.

Emtage, J. S., Lawson, D. E. M., and Kodicek, E. (1974a). The response of the small intestine to vitamin D: Isolation and properties of chick intestinal polyribosomes. *Biochem. J.* **140**, 239–247.

Emtage, J. S., Lawson, D. E. M., and Kodicek, E. (1974b). The response of the small intestine to vitamin D: Correlation between calcium-binding protein production and increased calcium absorption. *Biochem. J.* **144**, 339–346.

Feher, J. J., and Wasserman, R. H. (1978). Evidence for a membrane-bound form of chick intestinal calcium-binding protein. *Biochim. Biophys. Acta* **540**, 134–144.

Feher, J. J., and Wasserman, R. H. (1979a). Intestinal calcium-binding protein and calcium absorption in cortisol-treated chicks: Effects of vitamin D_3 and 1,25-dihydroxyvitamin D_3. *Endocrinology* **104**, 547–551.

Feher, J. J., and Wasserman, R. H. (1979b). Calcium absorption and intestinal calcium-binding protein: Quantitative relationship. *Am. J. Physiol.* **235**, E556–E561.

Fox, J., Swaminathan, R., Murray, T. M., and Care, A. D. (1977). Role of the parathyroid glands in the enhancement of intestinal calcium absorption in response to a low calcium diet. *J. Endocrinol.* **74**, 345–354.

Fox, J., Pickard, D. W., Care, A. W., and Murray, T. M. (1978). Effect of low phosphorus diets on intestinal calcium absorption and the concentration of calcium-binding protein in intact and parathyroidectomized pigs. *J. Endocrinol.* **78**, 379–387.

Fraser, D. R., and Kodicek, E. (1973). Regulation of 25-hydroxycholecalciferol-1α-hydroxylase activity in kidney by parathyroid hormone. *Nature (London), New Biol.* **241**, 163–166.

Freund, T. S., and Borzemsky, G. (1977). Vitamin D-dependent intestinal calcium-binding protein: A regulatory protein. *In* "Calcium-Binding Proteins and Calcium Function" (R. H. Wasserman, R. A. Corradino, E. Carafoli, R. H. Kretsinger, D. H. MacLennan, and F. L. Siegel, eds.), pp. 353–356. Elsevier/North Holland Publ., Amsterdam and New York.

Freund, T., and Bronner, F. (1975). Regulation of intestinal calcium-binding protein by calcium intake in the rat. *Am. J. Physiol.* **228**, 861–869.

Friedlander, E. J., Henry, H. L., and Norman, A. W. (1977). Effects of dietary calcium and phosphorus on the relationship between the 25-hydroxyvitamin D_3-1α-hydroxylase and production of chick intestinal calcium-binding protein. *J. Biol. Chem.* **252**, 8677–8683.

Fullmer, C. S., and Wasserman, R. H. (1973). Bovine intestinal calcium-binding proteins: Purification and some properties. *Biochim. Biophys. Acta* **317**, 172–186.

Fullmer, C. S., and Wasserman, R. H. (1975). Isolation and partial characterization of intestinal calcium-binding proteins from the cow, pig, horse, guinea pig and chick. *Biochim. Biophys. Acta* **393**, 134–142.

Fullmer, C. S., and Wasserman, R. H. (1977). Bovine intestinal calcium-binding protein: Cation-binding properties, chemistry and trypsin resistance. *In* "Calcium-Binding Proteins and Calcium Function" (R. H. Wasserman, R. A. Corradino, E. Carafoli, R. H. Kretsinger, D. H. MacLennan, and F. L. Siegel, eds.), pp. 303–312. Elsevier/North Holland Publ., Amsterdam and New York.

Fullmer, C. S., and Wasserman, R. H. (1980). The amino acid sequence of bovine intestinal calcium-binding protein. *In* "Calcium-Binding Proteins: Structure and Function" (F. L. Siegel *et al.*, eds.), pp. 363–370. Elsevier/North Holland Publ., Amsterdam and New York.

Fullmer, C. S., Wasserman, R. H., Hamilton, J. W., Huang, W. Y., and Cohn, D. V. (1975). The effect of calcium on the tryptic digestion of bovine intestinal calcium-binding protein. *Biochim. Biophys. Acta* **412**, 256–261.

Fullmer, C. S., Brindak, M. E., Bar, A., and Wasserman, R. H. (1976). The purification of calcium-binding protein from the uterus of the laying hen. *Proc. Soc. Exp. Biol. Med.* **152**, 237–241.

Hamilton, J. W., and Holdsworth, E. S. (1975). Role of calcium-binding protein in the mechanism of action of cholecalciferol (vitamin D). *Aust. J. Exp. Biol.* **53**, 469–478.

Harmeyer, J., and DeLuca, H. F. (1969). Calcium-binding protein and calcium absorption after vitamin D administration. *Arch. Biochem. Biophys.* **133**, 247–254.

Helmke, K., Federlin, K., Piazolo, P., Stroder, J., Jeschke, R., and Franz, H. E. (1974). Localization of calcium-binding protein in intestinal tissue by immunofluorescence in normal, vitamin D-deficient and uraemic subjects. *Gut* **15**, 875–879.

Hermsdorf, C. L., and Bronner, F. (1975). Vitamin D-dependent calcium-binding protein from rat kidney. *Biochim. Biophys. Acta* **379**, 553–561.

Hitchman, A. J. W., and Harrison, J. E. (1972). Calcium-binding proteins in the duodenal mucosa of the chick, rat, pig and human. *Can. J. Biochem.* **50**, 758–765.

Hofmann, T., Kawakami, M., Hitchman, A. J. W., Harrison, J. E., and Dorrington, K. J.

(1979). The amino acid sequence of porcine intestinal calcium-binding protein. *Can. J. Biochem.* **57**, 737–748.

Hosoya, N., Tamura, M., and Oku, T. (1980). Purification and characterization of calcium-binding protein from bovine milk. *In* "Calcium-Binding Proteins: Structure and Function" (F. L. Siegel *et al.*, eds.), pp. 489–490. Elsevier/North Holland Publ., Amsterdam and New York.

Hughes, M. R., McCain, T. A., Chang, S. Y., Haussler, M. R., Villareale, M., and Wasserman, R. H. (1977). Presence of 1,25-dihydroxyvitamin D_3-glycoside in the calcinogenic plant, *Cestrum diurnum. Nature (London)* **268**, 347–349.

Ingersoll, R. J., and Wasserman, R. H. (1971). Vitamin D_3-induced calcium-binding protein: Binding characteristics, conformation effects, and other properties. *J. Biol. Chem.* **246**, 2808–2814.

Jande, S. S., Tolnai, S., and Lawson, D. E. (1980). Cellular localization of vitamin D-dependent CaBP in intestine and cerebellum of chicks. *In* "Calcium-Binding Proteins: Structure and Function" (F. L. Siegel *et al.*, eds.), pp. 409–411. Elsevier/North-Holland Publ., Amsterdam and New York.

Kallfelz, F. A., Taylor, A. N., and Wasserman, R. H. (1967). Vitamin D-induced calcium-binding factor in rat intestinal mucosa. *Proc. Soc. Exp. Biol. Med.* **125**, 54–58.

Kimberg, D. V., Baerg, R. D., Gershon, E., and Graudusius, R. T. (1971). Effect of cortisone treatment on the active transport of calcium by the small intestine. *J. Clin. Invest.* **50**, 1309–1321.

Kretsinger, R. H. (1980). Structure and function of calcium-modulated proteins. *CRC Crit. Rev. Biochem.* **8**, 119–174.

Lawson, D. E. M., ed. (1978). "Vitamin D." Academic Press, New York.

Levine, B. A., Williams, R. J. P., Fullmer, C. S., and Wasserman, R. H. (1977). NMR studies of various calcium-binding proteins. *In* "Calcium-Binding Proteins and Calcium Function" (R. H. Wasserman, R. A. Corradino, E. Carafoli, R. H. Kretsinger, D. H. MacLennan, and F. L. Siegel, eds.), pp. 29–37. Elsevier/North-Holland Publ., Amsterdam and New York.

Lippiello, L. (1974). Vitamin D-induced calcium-binding protein: Fluorescent antibody localization in the shell gland (uterus) and kidney and related studies. Ph.D. Thesis, Cornell University, Ithaca, New York.

Lippiello, L., and Wasserman, R. H. (1975). Fluorescent localization of the vitamin D-dependent calcium-binding protein in the oviduct of the laying hen. *J. Histochem. Cytochem.* **23**, 111–116.

Marche, P., LeGuern, C., and Cassier, P. (1979). Immunocytochemical localization of a calcium-binding protein in the rat duodenum. *Cell Tissue Res.* **197**, 69–77.

Mellersh, H., Tomlinson, S., and Pollack, A. (1980). Messenger ribonucleic acids from pig intestinal mucosa: Direct synthesis of calcium-binding protein in a cell-free translation system. *Biochem. J.* **185**, 601–607.

Montecuccoli, G., Bar, A., Risenfeld, G., and Hurwitz, S. (1977). The response response of 25-hydroxycholecalciferol-1-hydroxylase, intestinal calcium absorption and calcium-binding protein to phosphate deficiency. *Comp. Biochem. Biophys. A* **57A**, 331–334.

Moriuchi, S., Yamanouchi, T., and Hosoya, N. (1975). Demonstration of two different vitamin D-dependent calcium-binding proteins in rat intestinal mucosa. *J. Nutr. Sci. Vitaminol.* **21**, 251–259.

Morrissey, R. L., and Rath, D. F. (1974). Purification of human renal calcium binding protein from necropsy specimens. *Proc. Soc. Exp. Biol. Med.* **145**, 699–703.

Morrissey, R. L., and Wasserman, R. H. (1971). Calcium absorption and calcium-binding protein in chicks on differing calcium and phosphorus intake. *Am. J. Physiol.* **220,** 1509–1515.

Morrissey, R. L., Bucci, T. J., Empson, R. N., and Lufkin, E. G. (1975). Calcium-binding protein: Its cellular localization in jejunum, kidney and pancreas. *Proc. Soc. Exp. Biol. Med.* **149,** 56–60.

Morrissey, R. L., Empson, R. N., Zolock, D. T., Bikle, D. D., and Bucci, T. J. (1978a). Intestinal response to 1α,25-dihydroxycholecalciferol II. A timed study of the intracellular localization of calcium binding protein. *Biochim. Biophys. Acta* **538,** 34–41.

Morrissey, R. L., Zolock, D. T., Bucci, T. J., and Bikle, D. D. (1978b). Immunoperoxidase localization of vitamin D dependent calcium-binding protein. *J. Histochem. Cytochem.* **26,** 628–634.

Murray, T. M., Arnold, B. M., Kuttner, M., Kovacs, K., Hitchman, A. J. W., and Harrison, J. E. (1975). Radioimmunoassay studies of porcine intestinal calcium-binding protein (CaBP). *In* "Calcium-Regulating Hormones" (R. V. Talmage, M. Owen, and J. A. Parson, eds.), pp. 371–375. Excerpta Medica, Amsterdam.

Navickis, R. J., Dial, O. K., Katzenellenbogen, B. S., and Nalbandov, A. V. (1979). Effects of gonadal hormones on calcium-binding protein in chick duodenum. *Am. J. Physiol.* **237,** E409–E417.

Nicolaysen, R., Eeg-Larsen, N., and Malm, O. J. (1953). Physiology of calcium metabolism, *Physiol. Rev.* **33,** 424–444.

Noda, S., Kubota, K., Yoshizawa, S., Moriuchi, S., and Hosoya, N. (1978). Visualization of vitamin D-dependent calcium-binding protein in chick intestinal tissue by immuno-scanning electron microscopy. *J. Nutr. Sci. Vitaminol.* **24,** 331–334.

Norman, A. W. (1965). Actinomycin D and the response to vitamin D. *Science* **149,** 184–186.

Norman, A. W. (1979). "Vitamin D: The Calcium Homeostatic Steroid Hormone." Academic Press, New York.

Norman, A. W., ed. (1980). "Vitamin D: Molecular Biology and Clinical Nutrition." Dekker, New York.

Norman, A. W., Frankel, B. J., Heldt, A. M., and Grodsky, G. M. (1980). Vitamin D deficiency inhibits pancreatic secretion of insulin. *Science* **209,** 823–825.

Oku, T., Shimura, F., Moriuchi, S., and Hosoya, N. (1976). Development of 1,25-Dihydroxycholecalciferol receptor in the duodenal cytosol of chick embryo. *Endocrinol. Jpn.* **23,** 375–381.

Oldham, S. B., Fischer, J. A., Shen, L. H., and Arnaud, C. D. (1974). Isolation and properties of a calcium-binding protein from porcine parathyroid glands. *Biochemistry* **13,** 4790–4796.

Omdahl, J. L., and DeLuca, H. F. (1972). Rachitogenic activity of dietary strontium. I. Inhibition of intestinal calcium absorption and 1,25-dihydroxycholecalciferol synthesis. *J. Biol. Chem.* **247,** 5520–5526.

Ooizumi, K., Moriuchi, S., and Hosoya, N. (1970). Vitamin D dependent calcium-binding protein in rat intestinal mucosa. *J. Vitaminol.* **16,** 228–234.

Parkes, C. O., and DeLuca, H. F. (1979). The influence of substitution at C_{24} on the CaBP stimulating activity of vitamin D metabolites. *Arch. Biochem. Biophys.* **194,** 271–274.

Parkes, C. O., and Reynolds, J. J. (1977). An *in vitro* bioassay for $1,25(OH)_2D_3$ and other anti-rachitic agents. *Mol. Cell. Endocrinol.* **7,** 25–31.

Pavlovitch, J. H., Lawuari, D., Diderjean, W., Saurat, J. H., and Balsan, S. (1980). Vitamin D-dependent calcium-binding protein in rat epidermis. *In* "Calcium-Binding Proteins: Structure and Function'' (F. L. Siegel *et al.*, eds.), pp. 417–419. Elsevier/ North Holland Publ., Amsterdam and New York.

Piazolo, P., Hotz, J., Helmke, K., Franz, H. E., and Schleyer, M. (1975). Calcium-binding protein in the duodenal mucosa of uremic patients and normal subjects. *Kidney Int.* **8**, 110–118.

Rackovsky, N. S. (1978). Aspects of the effect of estrogen and testosterone on calcium metabolism and calcium-binding protein synthesis in the chick: *In vivo* and *in vitro* studies. M.S. Thesis, Cornell University, Ithaca, New York.

Rasmussen, H., Wong, D., Bikle, D., and Goodman, D. B. P. (1972). Hormonal control of the renal conversion of 25-hydroxycholecalciferol to 1,25-dihydroxycholecalci-ferol. *J. Clin. Invest.* **51**, 2502–2504.

Ribovich, M. L., and DeLuca, H. F. (1975). The influence of dietary calcium and phosphorus on intestinal calcium transport in rats given vitamin D metabolites. *Arch. Biochem. Biophys.* **170**, 529–535.

Ribovich, M. L., and DeLuca, H. F. (1976). Intestinal calcium transport: Parathyroid hormone and adaptation to dietary calcium. *Arch. Biochem. Biophys.* **174**, 256–261.

Roth, S. I., Futrell, J. M., Wasserman, R. H., Brindak, M. E., Fullmer, C. S., Su, S. P. C., and Brakhop, N. H. (1980). Ultrastructural (EM) immunocytochemical localization of calcium-binding proteins (CaBP) in the chick intestine. *Fed. Proc., Fed. Am. Soc. Exp. Biol.* **39**, 559.

Schachter, D. (1970). Calcium transport, vitamin D and the molecular basis of active transport. *In* ''Fat Soluble Vitamins'' (H. F. DeLuca and J. W. Suttie, eds.), pp. 55–65. Univ. of Wisconsin Press, Madison.

Schneider, L. E., and Schedl, H. P. (1972). Diabetes and intestinal calcium absorption in the rat. *Am. J. Physiol.* **223**, 1319–1323.

Schneider, L. E., Nowosielski, L. M., and Schedl, H. P. (1977). Insulin treatment of diabetic rats: Effects on duodenal calcium absorption. *Endocrinology* **100**, 67–73.

Singh, R. P., and Bronner, F. (1980). Vitamin D acts post-transcriptionally: *In vitro* studies with the vitamin D-dependent calcium-binding protein of rat duodenum. *In* ''Calcium-Binding Proteins: Structure and Function'' (F. L. Siegel *et al.*, eds.), pp. 379–383. Elsevier/North-Holland Publ., Amsterdam and New York.

Spencer, R., Charman, M., Emtage, J. S., and Lawson, D. E. M. (1976). Production and properties of vitamin D-induced mRNA for chick calcium-binding protein. *Eur. J. Biochem.* **71**, 399–409.

Spencer, R., Charman, M., Wilson, P. N., and Lawson, D. E. M. (1978a). Relationship between vitamin D-stimulated calcium transport and intestinal calcium-binding protein in the chicken. *Biochem. J.* **170**, 93–101.

Spencer, R., Charman, M., and Lawson, D. E. M. (1978b). Stimulation of intestinal calcium-binding protein mRNA synthesis in the nucleus of vitamin D-deficient chicks by 1,25-dihydroxycholecalciferol. *Biochem. J.* **175**, 1089–1094.

Swaminathan, R., Sommerville, B. A., and Care, A. D. (1977). The effect of dietary calcium on the activity of 25-hydroxycholecalciferol-1-hydroxylase and Ca absorption in vitamin D-replete chicks. *Br. J. Nutr.* **38**, 47–54.

Tanaka, Y., Castillo, L., and DeLuca, H. F. (1976). Control of renal vitamin D hydroxylases in birds by sex hormones. *Proc. Natl. Acad. Sci. U.S.A.* **73**, 2701–2705.

Taylor, A. N. (1974). Chick brain calcium-binding protein: Response to vitamin D and anticonvulsant drugs. *Fed. Proc., Fed. Am. Soc. Exp. Biol.* **33**, 1551.

Taylor, A. N. (1980a). Immunocytochemical localization of the vitamin D-induced calcium-binding protein: Resolution of two divergent distribution patterns. *In* ''Calcium-

Binding Proteins: Structure and Function'' (F. L. Siegel *et al.,* eds.) pp. 393–400. Elsevier/North-Holland Publ., Amsterdam and New York.

Taylor, A. N. (1980b). Vitamin D-dependent calcium-binding proteins. *In* "Vitamin D: Molecular Biology and Clinical Nutrition" (A. W. Norman, ed.), Chapter 8. Dekker, New York.

Taylor, A. N. (1981). Immunocytochemical localization of the vitamin D-induced calcium-binding protein: Relocation of antigen during frozen section processing. *J. Histochem. Cytochem.* **29,** 65–73.

Taylor, A. N., and Brindak, M. E. (1974). Chick brain calcium-binding protein: Comparison with intestinal vitamin D-induced calcium-binding protein. *Arch. Biochem. Biophys.* **161,** 100–108.

Taylor, A. N., and McIntosh, J. E. (1977). Light and electron microscopic immunoperoxidase localization of chick intestinal vitamin D-induced calcium-binding protein. *In* "Vitamin D: Biochemical, Chemical and Clinical Aspects Related to Calcium Metabolism" (A. W. Norman, K. Schafer, J. W. Coburn, H. F. DeLuca, D. Fraser, H. G. Grigoleit, and D. von Herrath, eds.), pp. 303–312. de Gruyter, Berlin.

Taylor, A. N., and Wasserman, R. H. (1967). Vitamin D_3-induced calcium-binding protein: Partial purification, electrophoretic visualization, and tissue distribution. *Arch. Biochem. Biophys.* **119,** 536–540.

Taylor, A. N., and Wasserman, R. H. (1970). Immunofluorescent localization of vitamin D-dependent calcium-binding proteins. *J. Histochem. Cytochem.* **18,** 107–115.

Taylor, A. N., and Wasserman, R. H. (1972). Vitamin D-induced calcium-binding protein: Comparative aspects in kidney and intestine. *Am. J. Physiol.* **223,** 110–114.

Thomasset, M., Cuisinier-Gleizes, P., and Mathieu, H. (1976). Duodenal calcium-binding protein (CaBP) and phosphorus deprivation in growing rats. *Biomedicine* **25,** 333–337.

Thomasset, M., Cuisinier-Gleizes, P., and Mathieu, H.(1977). Difference in duodenal calcium-binding protein (CaBP) in response to low calcium or a low phosphorus intake. *Calcif. Tissue Res.* **22,** 45–50.

Thomasset, M., Cuisinier-Gleizes, P., and Mathieu, H. (1979a). 1,25-Dihydroxycholecalciferol: Dynamics of the stimulation of duodenal calcium-binding protein, calcium transport and bone calcium mobilization in vitamin D and calcium-deficient rats. *FEBS Lett.* **107,** 91–94.

Thomasset, M., Pointillart, A., Cuisinier-Gleizes, P., and Gueguen, L. (1979b). Effect of vitamin D or calcium deficiency on duodenal, jejunal and ileal calcium-binding protein and on plasma calcium and 25-hydroxycholecalciferol levels in the growing pig. *Ann. Biol. Anim., Biochim., Biophys.* **19,** 769–773.

Tomlinson, S., and Mellersh, H. (1980). Cell-free synthesis of pig intestinal calcium-binding protein. *In* "Calcium-Binding Proteins: Structure and Function" (F. L. Siegel *et al.,* eds.), pp. 503–505. Elsevier/North-Holland Publ., Amsterdam and New York.

Ueng, T. H., and Bronner, F. (1979). Cellular and luminal forms of rat intestinal calcium-binding protein as studied by counter ion electrophoresis. *Arch. Biochem. Biophys.* **197,** 205–217.

Wasserman, R. H. (1970). Interaction of vitamin D-dependent calcium binding protein with lysolecithin: Possible relevance to calcium transport. *Biochim. Biophys. Acta* **203,** 176–179.

Wasserman, R. H. (1974). Calcium absorption and calcium-binding protein synthesis: *Solanum malacoxylon* reverses strontium inhibition. *Science* **183,** 1092–1094.

Wasserman, R. H. (1980). Molecular aspects of the intestinal absorption of calcium and phosphorus. *In* "Pediatric Diseases Related to Calcium" (H. F. DeLuca and C. S. Anast, eds.), pp. 107–132. Elsevier/North-Holland Publ., Amsterdam and New York.

Wasserman, R. H., and Corradino, R. A. (1973). Vitamin D, calcium and protein synthesis. *Vitam. Horm. (N.Y.)* **31**, 43–103.

Wasserman, R. H., and Nobel, T. (1980). Vitamin D-related compounds as the toxic principle in calcinogenic plants. *In* "Vitamin D: Molecular Biology and Clinical Nutrition" (A. W. Norman, ed.), pp. 455–487. Dekker, New York.

Wasserman, R. H., and Taylor, A. N. (1966). Vitamin D_3-induced calcium-binding protein in chick intestinal mucosa. *Science* **152**, 791–793.

Wasserman, R. H., and Taylor, A. N. (1968). Vitamin D-dependent calcium-binding protein: Response to some physiological variables. *J. Biol. Chem.* **243**, 3987–3993.

Wasserman, R. H., Corradino, R. A., Fullmer, C. S., and Taylor, A. N. (1974). Vitamin D-dependent Ca^{2+} binding protein and its function. *Vitam. Horm. (N.Y.)* **32**, 299–324.

Wasserman, R. H., Henion, J. D., Haussler, M. R., and McCain, T. A. (1976). Evidence that a calcinogenic factor in *Solanum malacoxylon* is 1,25-dihydroxyvitamin D_3-glycoside. *Science* **194**, 853–855.

Wasserman, R. H., Corradino, R. A., Feher, J., and Armbrecht, H. J. (1977). Temporal patterns of response of the intestinal calcium absorptive system and related parameters to 1,25-dihydroxycholecalciferol. *In* "Vitamin D: Biochemical, Chemical and Clinical Aspects Related to Calcium Metabolism" (A. W. Norman, K. Schafer, J. W. Coburn, H. F. DeLuca, D. Fraser, H. G. Grigoleit, and D. von Herrath, eds.), pp. 331–340. de Gruyter, Berlin.

Wasserman, R. H., Fullmer, C. S., and Taylor, A. N. (1978). The vitamin D-dependent calcium-binding proteins. *In* "Vitamin D" (D. E. M. Lawson, ed.), pp. 133–166. Academic Press, New York.

Zull, J. E., Czarnowska-Misztal, E., and DeLuca, H. F. (1965). Actinomycin D inhibition of vitamin D action. *Science* **149**, 182–184.

Chapter 7

γ-Carboxyglutamic Acid-Containing Ca²⁺-Binding Proteins

BARBARA C. FURIE
MARIANNE BOROWSKI
BRUCE KEYT
BRUCE FURIE

I. Introduction 217
   A. History 218
   B. γ-Carboxyglutamic Acid as a Metal Ligand 219
   C. γ-Carboxyglutamic Acid-Containing Peptides as Models
      for Metal-Binding Studies 220
II. γ-Carboxyglutamic Acid-Containing Proteins
   of Blood Plasma 221
   A. Prothrombin 221
   B. Factor X 229
   C. Factor IX 232
   D. Factor VII 234
   E. Protein C 235
III. γ-Carboxyglutamic Acid-Containing Proteins of Bone 236
IV. Other γ-Carboxyglutamic Acid-Containing Proteins 237
V. Summary 238
   References 238

I. INTRODUCTION

γ-Carboxyglutamic acid has recently been identified in a unique class of calcium-binding proteins (Stenflo *et al.*, 1974; Nelsestuen *et al.*, 1974; Magnusson *et al.*, 1974). Since its discovery the presence of this amino

217

CALCIUM AND CELL FUNCTION, VOL. II
Copyright © 1982 by Academic Press, Inc.
All rights of reproduction in any form reserved.
ISBN 0-12-171402-0

acid has been definitively established in fluids or tissues from many verte-brates including such evolutionarily primitive species as the lamprey eel. In mammals synthesis of proteins containing γ-carboxyglutamic acid re-quires vitamin K, hence these proteins have been termed the vitamin K-dependent proteins (for review, see Stenflo and Suttie, 1977; Suttie, 1980; Gallop *et al.*, 1980; Burnier *et al.*, 1981). γ-Carboxyglutamic acid residues having a dicarboxylic acid side chain may confer different calcium-binding properties on γ-carboxyglutamic acid-containing proteins than those found in traditional calcium-binding proteins. Calcium ion-binding sites have in the past been described as containing polypeptide backbone carbonyls and carboxyl groups of aspartic and glutamic acids (Kretsinger, 1976). This chapter will describe the participation of γ-carboxyglutamic acid residues in metal binding.

A. History

The identification of γ-carboxyglutamic acid as a constituent of proteins had its initiation in the study of bleeding abnormalities in several animal species (for review, see Burnier *et al.*, 1981). In the first quarter of this century it was noted that chickens fed a vitamin K-deficient diet and cows ingesting spoiled sweet clover developed a bleeding diathesis. The hay contained a vitamin K antagonist, dicoumarol. The disorders were even-tually traced to low levels of several coagulant activities, those of factor VII, factor IX, factor X, and prothrombin. It was subsequently demon-strated that, although the biological activity of prothrombin was low in the plasma of humans treated with sodium warfarin, also a vitamin K an-tagonist, the level of prothrombin measured immunologically was normal (Ganrot and Nilehn, 1968). A biologically inactive protein, termed abnor-mal prothrombin, antigenically cross-reactive with prothrombin, was iso-lated. The identification of γ-carboxyglutamic acid in prothrombin and its absence in abnormal prothrombin defined the structural difference be-tween the two molecules. In contradistinction to prothrombin, abnormal prothrombin does not bind calcium ions or phospholipid in the presence of these ions (Stenflo and Ganrot, 1972, 1973; Nelsestuen and Suttie, 1972a; Esmon *et al.*, 1975). This difference taken in conjunction with the lack of activity of abnormal prothrombin indicated an important biological role for the newly identified amino acid.

γ-Carboxyglutamic acid was thus implicated as a calcium ion ligand in the vitamin K-dependent proteins. Definition of this role insofar as it is known and its importance in biology are the subjects of this chapter. As the vitamin K-dependent proteins of blood coagulation are the best stud-ied examples of this class of proteins, emphasis will be on these proteins.

Reviews dealing with other aspects of these proteins have recently appeared (Davie and Hanahan, 1977; Jackson and Nemerson, 1980; Nemerson and Furie, 1980; Suttie and Jackson, 1977).

B. γ-Carboxyglutamic Acid as a Metal Ligand

In order to appreciate the unique role of γ-carboxyglutamic acid in proteins it is necessary to understand its chemical properties. As indicated above, indirect evidence that γ-carboxyglutamic acid binds metal ions came from observations of the properties of normal and abnormal prothrombin. Normal prothrombin contains γ-carboxyglutamic acid and binds metal ions (Bajaj *et al.*, 1975; Benson *et al.*, 1973; Benarous *et al.*, 1976; Stenflo and Ganrot, 1973; Nelsestuen and Suttie, 1972b; Henriksen and Jackson, 1975), while abnormal prothrombin, lacking γ-carboxyglutamic acid, does not bind metal ions. In addition, the chemical structure of γ-carboxyglutamic acid suggested that it might be a metal ion chelator.

Direct observation of metal ion binding to γ-carboxyglutamic acid was first reported by Marki *et al.* (1977). A dissociation constant K_d of 50 μM was determined for the Ca²⁺–γ-carboxyglutamic acid complex by potentiometric titration. In order to explore further the nature of the interaction between free γ-carboxyglutamic acid and metal ions, Sperling *et al.* (1978) used lanthanide ions as substitutes for calcium ions. Lanthanide ions have been shown to substitute for Ca²⁺ in the Ca²⁺-dependent proteolytic activation of prothrombin to thrombin by another γ-carboxyglutamic acid-containing blood coagulation enzyme, activated factor X (Furie *et al.*, 1976). With the use of the spectral and magnetic properties of these metal ions the stoichometry and geometry of the metal ion–γ-carboxyglutamate complex was studied. Data from fluorescence and ultraviolet (UV) spectroscopic studies suggest that, when γ-carboxyglutamic acid is in excess over metal ions, a 2:1 complex of γ-carboxyglutamate to metal ions is formed with a K_d of approximately 55 μM. However, when metal ions are in excess, a 1:1 complex of metal to γ-carboxyglutamate (lower limit) is formed with a K_d of 54 μM.

The solution geometry of the metal–γ-carboxyglutamate complex was determined by observing the effect of Gd³⁺ upon the relaxation rate of carbon nuclei in γ-carboxyglutamic acid. The results of these studies indicate that metal ions are bound symmetrically between the two γ-carboxyl carbons at a metal-to-carbon interatomic distance of 3.2 Å (Fig. 1). As the metal-to-oxygen distance cannot be determined by the methods employed, there is no way to ascertain whether coordination is unidentate or bidentate with regard to each carboxyl group. Crystallo-

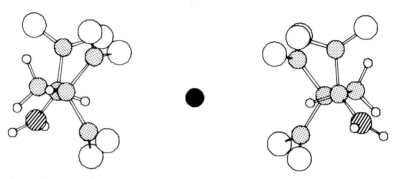

Fig. 1. Ternary complex of γ-carboxyglutamate–Gd^{3+}. Model based upon NMR-derived distances for the γ-carbon and two γ-carboxyl carbons in the binary γ-carboxyglutamate–Gd^{3+} complex. Atom code: Gd^{3+} (solid circle), oxygen (open circles), carbon (dotted circles), nitrogen (hatched circles). The smallest atoms are hydrogen. The conformation of the backbone of the amino acid has not been determined, and the structure shown is for descriptive purposes only. From Sperling *et al.*, 1978.

graphic analysis of Ca^{2+}–γ-carboxyglutamic acid complexes has yielded a Ca^{2+}-to-oxygen distance of 2.47 Å, suggesting unidentate interaction in free amino acid complexes (Satyshur, 1978). Of interest in terms of the role that γ-carboxyglutamic acid may play as a metal ligand in proteins is the fact that the amino acid occupies one half of the coordination sphere of a metal ion. The other half remains free to interact with other metal ligands. Thus metal ions could serve to link two γ-carboxyglutamic acid residues in the same polypeptide chain, a form of liganding which may be observed in γ-carboxyglutamic acid-containing proteins (vide infra). Alternatively, metal ions could serve to bridge γ-carboxyglutamic acid residues in different polypeptide chains or to link a polypeptide to some other molecule which contains a metal ligand.

C. γ-Carboxyglutamic Acid-Containing Peptides as Models for Metal-Binding Studies

Small peptides containing γ-carboxyglutamic acid have been used as model systems to explore further the interaction of γ-carboxyglutamic acid with metal ions. Since γ-carboxyglutamic acid residues often appear in pairs in polypeptide chains, di-γ-carboxyglutamic acid peptides have been of particular interest. When metal ion nuclear magnetic resonance (NMR) spectroscopy and ^{43}Ca^{2+} were employed, a K_d of 0.6 mM was determined for N-CBZ-D-γ-carboxyglutamate-D-γ-carboxyglutamate-OMe–Ca^{2+} (Robertson *et al.*, 1978). This is in good agreement with the value determined potentiometrically for L-γ-carboxyglutamate-L-γ-carboxyglutamate by Marki *et al.* (1977). These investigators suggest that di-γ-carboxyglutamic acids bind metal ions more tightly than free γ-carboxyglutamic acid. Although such peptides appeared to be promising models for γ-carboxyglutamic acid-containing proteins, the extension of peptide studies by Sarasua *et al.* (1980) suggest the contrary. Using luminescence decay, these investigators studied the interaction of Eu^{3+} with several blocked peptides containing γ-carboxyglutamic acid up to four amino acid residues in length. Neither the number of γ-carboxyglutamic acid residues in the peptide nor the sequence of these residues correlates in an obvious manner with the number or affinity of metal-binding sites. Small polypeptides may not therefore be useful predictors of the metal-binding behavior of γ-carboxyglutamic acid-containing proteins.

II. γ-CARBOXYGLUTAMIC ACID-CONTAINING PROTEINS OF BLOOD PLASMA

The importance of Ca^{2+} as a cofactor in the conversion of vitamin K-dependent blood coagulation zymogens to their enzymatic forms has long been appreciated. Calcium ions are required for assembly of the macromolecular complexes in which these proteins are activated. For example, prothrombin is proteolytically cleaved *in vitro* to form thrombin by the action of the serine protease, activated factor X. A protein cofactor, activated factor V, phospholipid, and calcium ions are required to form a complex in which optimal rates of activation are achieved (Davie and Hanahan, 1977; Nemerson and Furie, 1980). The metal-binding properties of prothrombin, factor IX, factor X, factor VII, and protein C, all plasma glycoproteins that contain γ-carboxyglutamic acid, have been studied. As prothrombin is the most abundant vitamin K-dependent protein in plasma, its metal-binding characteristics have been studied most extensively. The mode and role of metal liganding to prothrombin will be described first.

A. Prothrombin

Data from a large number of laboratories indicate that prothrombin has from 6 to 15 Ca^{2+}-binding sites and that its affinity for Ca^{2+} is weak, with a K_d of about 1 mM (Stenflo and Ganrot, 1973; Benson and Hanahan, 1975; Bajaj *et al.*, 1975, 1976; Nelsestuen *et al.*, 1975; Henrikson and Jackson, 1975; Brittain *et al.*, 1976). More precise definition of the numbers and classes of Ca^{2+}-binding sites has been complicated by the low affinity of the interactions involved. Several laboratories have reported heterogeneous, noncooperative binding of Ca^{2+} to prothrombin (Fig. 2B) (Nelsestuen and Suttie, 1972b; Benson *et al.*, 1973). However, most laboratories have reported heterogeneous, positively cooperative binding (Fig. 2A) (Stenflo and Ganrot, 1973; Bajaj *et al.*, 1975; Nelsestuen *et al.*, 1975; Benarous *et al.*, 1976). The data from these laboratories are characterized by the type of Scatchard plot shown in Fig. 2A. The binding of Ca^{2+} to fragment 1, the N-terminal 156 amino acids of prothrombin, was observed by monitoring quenching of protein intrinsic fluorescence (Prendergast and Mann, 1977). Since fragment 1 contains all the γ-carboxyglutamic acid residues of prothrombin and is of significantly lower molecular weight, it has often been used as a model for prothrombin–metal ion binding studies. Although prothrombin undergoes quenching of intrinsic fluorescence upon metal binding, the phenomenon is more pronounced with fragment 1 (6% quenching versus 43% quenching at maximum). However, binding curves similar to those obtained for fragment-1–Ca^{2+} binding are obtained for prothrombin–Ca^{2+} interaction using a variety of experimental approaches. The Scatchard analysis suggests that fragment 1 (and prothrombin) contain Ca^{2+}-binding sites of different affinities and that positive cooperativity is exhibited by these sites. Some investigators have suggested that at the concentrations of prothrombin or fragment 1 and Ca^{2+} required to make these determinations self-association of fragment 1 or prothrombin may occur, resulting in the curvilinear plots observed in the Scatchard analysis. Others have suggested that cooperative interaction between metal-binding sites is the result of a conformational change in the protein, which occurs upon initial binding of Ca^{2+}. In this view, protein association, when it occurs, is subsequent to the change in secondary structure. An understanding of the number, nature, and role of metal-binding sites on pro-

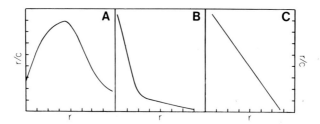

Fig. 2. Representative plots of Scatchard analyses of ligand protein binding, where r is moles of ligand bound per mole of protein and C is the free ligand concentration. (A) A curvilinear plot of this form results from positive cooperativity among binding sites. (B) A curvilinear plot of this form may indicate either two classes of noninteracting sites of different affinity for the ligand or negative cooperativity among binding sites. (C) This plot indicates a single class of binding sites with equal affinity for the ligand.

thrombin has been achieved by studying the interaction of metal ions other than Ca^{2+} with both prothrombin and fragment 1.

With either Mn^{2+} or Gd^{3+} two high-affinity metal-binding sites were observed on prothrombin. Electron paramagnetic resonance spectroscopy was used to observe the binding of Mn^{2+} to prothrombin, fragment 1, and prethrombin 1 (amino acid residues 157–582 of prothrombin) (Bajaj *et al.*, 1976). Prothrombin contains two high-affinity sites (K_d, $1.2 \times 10^{-5} M$) and two or three lower-affinity sites (K_d, $1.3 \times 10^{-4} M$). At least one of the high-affinity sites exhibits cooperativity of binding (Fig. 2A). Similarly, fragment 1 contains two high-affinity sites (K_d, $2.2 \times 10^{-5} M$) and two or three lower-affinity sites (K_d, $2.5 \times 10^{-4} M$). The prethrombin-1 portion of prothrombin contains one Mn^{2+}-binding site with a K_d of $3.2 \times 10^{-4} M$ (Fig. 2C). In competition experiments Ca^{2+} can replace all the Mn^{2+} bound to prothrombin, suggesting that Mn^{2+} and Ca^{2+} bind to the same sites on the protein.

A similar distribution of metal ion-binding sites is observed in dialysis experiments employing Gd^{3+} as the metal ligand (Furie *et al.*, 1976). Scatchard analysis of the data yielded a plot of the form shown in Fig. 2B. Prothrombin has two high-affinity Gd^{3+}-binding sites (K_d, $7.5 \times 10^{-7} M$). The number and affinity of sites which bind Gd^{3+} less tightly was not ascertained. At concentrations of protein and metal ions required to make such measurements Scatchard analysis of the binding data yielded plots similar to that in Fig. 2A. Fragment 1 has two high-affinity binding sites for Gd^{3+} (K_d, $1.6 \times 10^{-7} M$), while prethrombin 1 has one Gd^{3+}-binding site (K_d, $1 \times 10^{-7} M$). It is clear from both the Mn^{2+}- and Gd^{3+}-binding data that, with regard to affinity, prothrombin has at least two classes of metal ion-binding sites. In addition, although metal ion binding to the fragment-1 region of the prothrombin molecule containing all 10 γ-carboxyglutamic acid residues is vital for biological function, the prethrombin-1 portion of the prothrombin molecule, containing no γ-carboxyglutamic acid residues, contains a metal-binding site. The nature of this site must be similar to that found in other Ca^{2+}-binding proteins in which carboxyl oxygens of aspartic and glutamic acid residues and backbone carbonyl oxygens are the primary ligands for Ca^{2+}.

Under optimal conditions Mn^{2+} or Gd^{3+} can act as the obligatory metal cofactor for the generation of thrombin from prothrombin in the presence of activated factor X, activated factor V, and phospholipid (Bajaj *et al.*, 1976; Furie *et al.*, 1976). However, the rate of thrombin generation in the presence of these metal ions is significantly diminished from that attained with Ca^{2+} present at optimal concentration. With optimal metal ion concentrations the maximum rate of thrombin generation obtained with Mn^{2+} is 5% and that with Gd^{3+} 25% of the rate in the presence of Ca^{2+}. Furthermore, all lanthanide ions are not equally effective in supporting thrombin generation. Finally Mn^{2+} can inhibit thrombin generation by activated factor X in the presence of Ca^{2+}, activated factor V, and phospholipid. It is unlikely that metal ions participate directly in the hydrolase activity of activated factor X. More likely these metal ions support formation of enzyme substrate or enzyme cofactor complexes. While some metal ions other than Ca^{2+} can fill metal-binding sites on prothrombin (vide infra) required for formation of the activation complex, there may be subtle changes in their liganding which alter the reactivity of the complex.

Several spectroscopic and physical techniques have been used to evaluate metal ion binding to prothrombin and to explore the nature of cooperativity in Ca^{2+} binding. Nelsestuen (1976) and Bloom and Mann (1978) have observed that the intrinsic fluorescence of bovine and human prothrombin is decreased in the presence of Ca^{2+}. As mentioned above, the maximal fluorescence decrease is 6%, a change too small to be useful in a detailed study of the phenomenon. However, the maximal intrinsic fluorescence quenching of 40% observed for human or bovine fragment 1 in the presence of Ca^{2+} is an alteration of sufficient magnitude to be useful as a means of studying the process of Ca^{2+} binding to these molecules. The marked difference in fluorescence quenching for prothrombin versus fragment 1 suggests that the process inducing the intrinsic fluorescence quenching upon Ca^{2+} binding might be confined to the fragment-1 region of prothrombin. Fragment 1 contains 2 of 11 tryptophan residues in prothrombin. Immunological studies to be discussed later have confirmed this interpretation of the fluorescence data (Furie et al., 1978; Tai et al., 1980). Fragment 1 has been used as a model for prothrombin in the studies described below.

Human and bovine fragment 1 both undergo rapid quenching of intrinsic fluorescence upon binding Ca^{2+}. Maximal fluorescence quenching in human fragment 1 appears to occur in a single rapid process, while for bovine fragment 1 there appears to be a rapid and then a slower subsequent fluorescence quenching. Nelsestuen (1976) and Marsh et al. (1979a) have speculated that for bovine fragment 1 in the absence of Ca^{2+} two isomeric populations exist in solution. Marsh et al. (1979a) have further speculated that the isomers may result from isomerization about Pro-22. If only one conformer of fragment 1 binds metal ions, the pattern of intrinsic fluorescence quenching can be explained. Human prothrombin having threonine at residue 22 does not exhibit the same behavior.

Several metal ions can substitute for Ca^{2+} in facilitating quenching of the intrinsic fluorescence of prothrombin. The concentrations of metal ions required for half-maximal fluorescence quenching have been reported by several laboratories and are in general agreement. The ability of the metal ions tested to facilitate the change in the protein resulting in fluorescence quenching reflects the affinity of the metal ions for prothrombin. Concentrations of metal ions required for half-maximal quenching of fluorescence for prothrombin (and/or fragment 1) are approximately 0.2 mM for Ca^{2+}, 0.01–0.02 mM for Mn^{2+}, 0.4 mM for Mg^{2+}, and 5 μM for Gd^{3+} (Prendergast and Mann, 1977; Bloom and Mann, 1978; Nelsestuen et al., 1981). Dissociation constants for the higher-affinity sites of prothrombin and fragment 1, respectively, are: Ca^{2+}, 0.3 and 0.6 mM; Mn^{2+}, 12 and 22 μM; and Gd^{3+}, 0.75 and 0.16 μM (Bajaj et al., 1975, 1976; Furie et al., 1976).

Calcium ion binding to prothrombin fragment 1, in addition to causing quenching of intrinsic fluorescence (Prendergast and Mann, 1977; Bloom and Mann, 1978; Marsh et al., 1979b; Nelsestuen et al., 1981), also results in alteration in the circular dichroism (CD) spectrum (Bloom and Mann, 1978; Marsh et al., 1979b) and facilitates fragment-1 self-association and fragment-1 binding to phospholipid (Prendergast and Mann, 1977; Jackson et al., 1979). Most recent data suggest that the concentrations of Ca^{2+} required to reach half-maximal fluorescence quenching, half-maximal CD perturbation, or half-maximal dimer formation are all approximately 0.3–0.4 mM (Nelsestuen et al., 1981). However, for Mn^{2+}, the concentration of metal ions required to achieve half-maximal fluorescence quenching or CD spectral alteration is well removed from that required to induce fragment-1 dimerization, 0.02 versus 0.1 mM. The concentration of Mn^{2+} required to produce half-maximal spectral changes is approximately the same as the dissociation constant K_d determined for the two high-affinity metal binding sites on fragment 1. Magnesium ions, while promoting the same spectral alterations upon binding to fragment 1 as Ca^{2+} and Mn^{2+}, do not promote protein self-association. Neither Mn^{2+} nor Mg^{2+} support the binding of protein to phospholipid (Nelsestuen et al., 1976). Finally, Gd^{3+} facilitates the process reflected spectrally when fragment 1 binds metal ions, induces self-association, and promotes metal-dependent binding to phospholipid as well.

From these data several conclusions have been drawn. One view is that two different types of metal-binding sites exist on fragment 1. All the divalent and trivalent metal ions discussed above can bind to one class of higher-affinity sites and induce a conformational change. A second class of sites, of lower affinity, responsible for promoting self-association of fragment 1 and phospholipid binding, exhibits greater ion selectivity. These sites may not be filled by all the cations that satisfactorily bind to the first class of sites (Nelsestuen, 1976; Prendergast and Mann, 1977; Bloom and Mann, 1978). Recently evidence has been presented that there are equal numbers of sites for all the cations, Ca^{2+}, Mg^{2+}, Mn^{2+}, Cd^{2+}, and La^{3+}, tested on fragment 1. A model has been suggested in which all the cations can be accommodated at all the sites. The number of sites is unaltered by protein self-association, and different cations fill the different classes of sites in varying order. In general, it is filling of the higher-affinity sites which results in protein conformational transition, while filling of lower-affinity sites results in protein association (or phospholipid binding). Perhaps because of the overlapping affinities of the two classes of sites, segregation of these two processes cannot be seen for all metal ions (Nelsestuen *et al.*, 1981). Finally, while it appears that all the cation-binding sites may be filled in fragment 1 by all the cations listed above, only Gd^{3+} and to a lesser extent Mn^{2+} are substitutes for Ca^{2+} in producing biological activity when bound to prothrombin. Several possibilities for the distinction among metal ions exist. On the one hand, subtle differences in liganding among the metal ions may result in protein–metal ion complexes with varying behavior. On the other hand, although certain of the cations studied cause fragment-1 dimerization (Jackson *et al.*, 1979; Nelsestuen *et al.*, 1981), most recent data indicate that prothrombin does not dimerize in the presence of these cations (Nelsestuen *et al.*, 1981). This may reflect differences in the liganding of metal ions to the lower-affinity sites on fragment 1 and prothrombin. Differences could result from subtle changes in the mode of metal ion binding, binding sites on prothrombin which exhibit greater metal selectivity than those of fragment 1, steric differences, or perhaps even additional sites on prothrombin. Such differences could be reflected in the ability of metal ions to promote prothrombin–phospholipid binding. If such differences in metal-binding sites between prothrombin and fragment 1 exist, evidence suggests that the differences are *not* in the metal-binding sites responsible for alterations in the protein accompanied by fluorescence quenching or CD transition.

Nuclear magnetic resonance studies on fragments 12–44 from prothrombin in the presence of the paramagnetic lanthanide ion Gd^{3+} have resulted in the identification of a potential high-affinity binding site on prothrombin. Fragment 12–44 contains 8 of the 10 γ-carboxyglutamic acid residues of prothrombin and an internal disulfide loop (Fig. 3). Fragment 12–44 has one high-affinity (K_d, 0.55 μM) and four to six lower-affinity (K_d, ~6 μM) lanthanide ion-binding sites. Reduced and carboxymethylated fragment 12–44 has no high-affinity binding sites and five or six low-affinity sites (K_d, 8 μM) (Furie *et al.*, 1979). The loss of high-affinity metal-binding sites in fragment 1 upon reduction of disulfide bridges has also been reported (Henrikson and Jackson, 1975).

The effect of titration of fragment 12–44 and S-carboxylated fragment 12–44 with paramagnetic Gd^{3+} has been observed by natural abundance

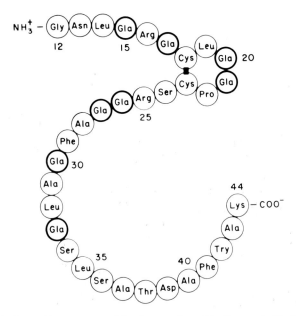

Fig. 3. Amino acid sequence of bovine prothrombin fragment 12–44. γ-Carboxy-glutamic acid residues are indicated as Gla; Cys-18 and Cys-23 are bridged by a disulfide bond. From Furie *et al.*, 1979.

^{13}CNMR spectroscopy. Results indicate that in native fragment 12–44 the two arginine residues are in close proximity to bound metal ions, while in the reduced peptide arginine residues are not in close proximity to bound metal. On the basis of these data a model for the high-affinity binding site in fragment 12–44 has been proposed. The high-affinity site is composed of two γ-carboxyglutamic acid residues, possibly residues 15 and 26, which form an intramolecular metal-dependent bridge between two regions of the polypeptide chain (Fig. 4). It is further suggested that the state of occupancy of such a site could influence protein conformation. Un-liganded γ-carboxyglutamic acid residues would repel one another, while liganded residues would be drawn together by the metal bridge (Furie *et al.*, 1979). The model shown suggests that each γ-carboxyglutamic acid residue donates four oxygen ligands to Gd^{3+}. While speculative, this may be supported by the results of luminescence decay studies on fragment-1–Eu^{3+} binding. These studies suggested that, for the high-affinity sites on fragment 1, binding of Eu^{3+} results in displacement of eight of the nine water molecules in the Eu^{3+}–aquo complex (Scott *et al.*, 1980).

   Several laboratories have observed that, for antisera raised in rabbits against prothrombin, greater precipitation of antigen by antibody occurs

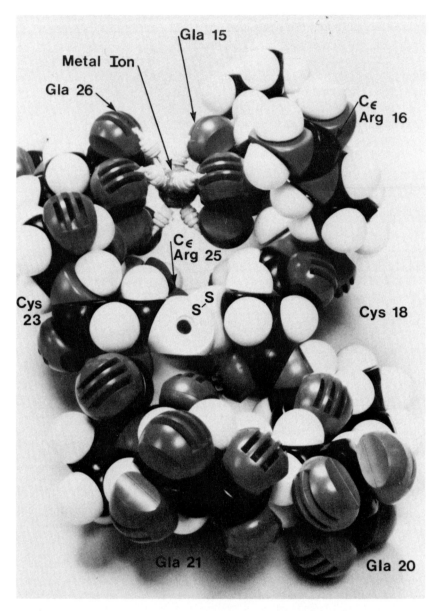

Fig. 4. CPK model of prothrombin fragment 12–44. In this model the high-affinity metal-binding site is formed by Gla-16 and Gla-26 and is occupied by a single metal ion. The disulfide bond is indicated by S—S. From Furie *et al.*, 1979.

in the presence of Ca^{2+} than in their absence (Stenflo and Ganrot, 1972; Furie *et al.*, 1978; Tai *et al.*, 1980). From some of these antisera a subpopulation of antibodies has been isolated which bind prothrombin only in the presence of Ca^{2+}. It has been suggested that this phenomenon reflects a Ca^{2+}-induced conformational change in prothrombin upon metal binding. The antibody subpopulation termed anti-prothrombin Ca^{2+} recognizes only the conformation assumed by prothrombin in the presence of Ca^{2+}. Fragment 1 in the presence of Ca^{2+} can completely displace prothrombin from anti-prothrombin Ca^{2+}. Other studies employed anti-fragment 12–44 an anti-prothrombin antibody subpopulation isolated from anti-prothrombin and specific for the fragment 12–44 domain of prothrombin (Furie *et al.*, 1978; Furie and Furie, 1979). The antibody population bound to prothrombin better in the presence of metal ions than in their absence. The concentrations of various metal ions required to achieve half-maximal binding of anti-fragment 12–44 and prothrombin correspond quite closely to the metal ion concentrations required for the achievement of half-maximal fluorescence quenching or CD transition in fragment 1 or prothrombin. Thus it appears that the conformational alteration reported by the selected antibody populations when prothrombin or prothrombin fragments bind metal ions is closely related to that observed spectroscopically. Conformation-specific antibodies can thus be used to identify portions of the polypeptide chain involved in a conformational alteration. It appears from the studies described here that the metal-dependent conformational transition occurs in the fragment-1 region of the molecule. The region 12–44 is significantly involved in this process, a result that might be predicted from the metal-binding model described above.

B. Factor X

Factor X plays a central role in blood coagulation, acting at the confluence of two pathways of zymogen activation which result in clot formation. A glycoprotein of molecular weight 56,000, factor X is made up of two polypeptide chains of molecular weight 17,000 and 38,000 linked by a disulfide bond. The light chain of factor X contains 12 γ-carboxyglutamic acid residues. Conversion of factor X from its zymogen form to its enzymatically active form, activated factor X, can be accomplished physiologically by either activated factor VII in the presence of tissue factor and Ca^{2+} or by activated factor IX in the presence of activated factor VIII, phospholipid, and Ca^{2+}. In addition, a protease from the venom of *Vipera russelli,* the coagulant protein of Russell's viper venom (RVV), has been a useful *in vitro* activator of factor X. Activated factor X serves as enzyme in the Ca^{2+}-dependent activation of prothrombin in the presence of activated factor V and phospholipid (Davie and Hanahan, 1977; Nemerson and Furie, 1980).

The binding of Ca^{2+} to factor X has been studied in a number of laboratories. Several laboratories have obtained data which suggest positive cooperativity of Ca^{2+} binding similar to that observed for prothrombin (Henrikson and Jackson, 1975; Lindhout and Hemker, 1978). Under restricted conditions of protein and Ca^{2+} concentration, two to three high-affinity Ca^{2+}-binding sites and a large number of low-affinity Ca^{2+}-binding sites have been identified on factor X (Furie and Furie, 1975; Yue and Gertler, 1978a). Binding of Ca^{2+} to factor X results in an alteration of the UV spectrum in the aromatic region (Furie and Furie, 1976). This has been interpreted to be the result of a conformational change in the protein upon metal ion binding.

The binding of Ca^{2+} to the isolated heavy and light chain of factor X has been studied. Reduced and carboxymethylated light chain binds Ca^{2+} weakly and in a noncooperative manner, while the reduced and carboxymethylated heavy chain does not appear to bind Ca^{2+}. Loss of cooperativity of metal binding in the light chain of factor X upon reduction and carboxymethylation parallels results obtained with prothrombin fragment 1, a polypeptide with which the light chain shares a high degree of sequence homology (Henrikson and Jackson, 1975). Decarboxy-factor X (des-γ-carboxy-factor X), again in parallel with des-γ-carboxy prothrombin, also binds metal ions with weak affinity and in a noncooperative manner (Lindhout and Hemker, 1978). These results suggest that both γ-carboxyglutamic acid residues and a tertiary protein structure stabilized by disulfide linkages are required to preserve the higher-affinity metal-binding sites in factor X.

In attempts to further explore the nature and function of metal ion binding to factor X the binding of several other metal ions has been studied. Factor X has two high-affinity metal ion-binding sites which bind lanthanide ions with a dissociation constant K_d of $4 \times 10^{-7} M$ and four to six lower-affinity sites (K_d, $1.5 \times 10^{-5} M$). Scatchard analysis of the data yields a plot of the form shown in Fig. 2B. Lanthanide ions cannot substitute for Ca^{2+} in the calcium-dependent activation of factor X by the coagulant protein of RVV but are competitive inhibitors of this reaction (K_i, $1–4 \times 10^{-6} M$) (Furie and Furie, 1975). In contrast, Scatchard analysis of Mn^{2+} binding to factor X suggests positive cooperativity among the binding sites (Fig. 2A). Manganese ions are capable of substituting for Ca^{2+} in the activation of factor X by coagulant protein of RVV (Bajaj *et al.*, 1977; Yue and Gertler, 1978b).

Activation of decarboxy-factor X by the coagulant protein of RVV has been compared to activation of factor X by the same enzyme (Lindhout *et al.*, 1978a). Decarboxy-factor X can be converted to an hydrolase, activated decarboxy-factor X, by the coagulant protein of RVV, but its conversion is 60-fold slower than conversion of native factor X. The activation of decarboxy-factor X is a calcium-dependent reaction, although the calcium dependence observed is altered from that of the native protein. It has been reported that the coagulant protein of RVV does not bind metal ions (Furie and Furie, 1975), and decarboxy-factor X does so only weakly (Lindhout *et al.*, 1978a). Calcium ions increase the V_{max} for the activation of native factor X by the coagulant protein. Calcium ions exhibit a cooperative effect with approximately three ions participating (Kosow, 1976). It thus appears that the substrate for the coagulant protein is a factor X–Ca²⁺ complex. A similar complex may be formed with decarboxy-factor X far less efficiently utilizing metal ligands other than oxygens of γ-carboxyglutamic acid residues. The dependence of the rate of activation of decarboxy-factor X on Ca²⁺ is not cooperative (Lindhout *et al.*, 1978a). Alternatively the metal-binding site required for substrate formation may be on either the light or heavy chain of factor X and may not involve γ-carboxyglutamic acid residues. Occupation of binding sites involving these residues in native factor X may further stabilize a desirable substrate conformation. The proportion of time decarboxy-factor X can achieve this conformation may be reduced.

Metal ion binding to factor X, as to prothrombin, results in the quenching of intrinsic protein fluorescence (Nelsestuen *et al.*, 1976). As with prothrombin, the selectivity of the metal ion-binding sites which induce conformational change is not great, and Mg²⁺, Ca²⁺, Sr²⁺, or Mn²⁺ will cause quenching of intrinsic protein fluorescence at half-maximal concentrations of 2.0, 1.0, 14.0, or <0.2 mM, respectively. However, the effectiveness of these metal ions in facilitating protein–phospholipid complex formation is quite different, half-maximal metal ion concentrations being >10.0, 0.22, 0.50, and >4 mM for Mg²⁺, Ca²⁺, Sr²⁺, and Mn²⁺ respectively. These results suggest that metal ions are required for two different roles in formation of the complexes in which factor X is activated. First, metal ions bring about a conformational change in factor X. Subsequently, this factor X–metal ion complex binds *in vitro* to phospholipid micelles with the participation of additional metal ions (Nelsestuen and Lim, 1977). These two processes may be distinguished by the diversity of their metal requirements. It is interesting to note that Mg²⁺ or Mn²⁺, which facilitate the protein conformational transition alone, may substitute completely for Ca²⁺ in the activation of factor X by the coagulant protein of RVV. This suggests that it is this initial protein–metal ion complex which is the substrate of the coagulant protein (Bajaj *et al.*, 1977; Yue and Gertler, 1978b).

In contrast to the conversion of prothrombin to thrombin, activation of factor X to activated factor X results in the production of an enzyme containing γ-carboxyglutamic acid residues. The presence of γ-carboxyglutamic residues in the enzyme is clearly required for biological activity, probably to facilitate macromolecular complex formation. The role, if any, of γ-carboxyglutamate-mediated Ca^{2+} binding in catalytic function is unclear. Several laboratories have explored this question by observing the effect of Ca^{2+} on the rate of hydrolysis of benzoyl-Ile-Glu-Gly-Arg-p-nitroanilide. At pH 7.5 and 30°C the rate of hydrolysis of benzoyl-Ile-Glu-Gly-Arg-p-nitroanilide by human activated factor Xa has been reported to be enhanced by Ca^{2+}, Mg^{2+}, or Mn^{2+} (Orthner and Kosow, 1978). At optimal concentrations these metal ions increased the rate of hydrolysis two- to threefold. The metal ions decrease the K_m for the peptide substrate. On the other hand, the initial velocity of bovine factor Xa hydrolysis of the same peptide substrate, benzoyl-Ile-Glu-Gly-Arg-p-nitroanilide, at pH 8.2 and 37°C appears to decrease in the presence of Ca^{2+} (Lindhout et al., 1978b). This has been ascribed to a metal-dependent association of the enzyme. In addition, bovine des-γ-carboxy-activated factor X and bovine activated factor X demonstrate essentially the same kinetic parameters for hydrolysis of benzyl-Ile-Glu-Gly-Arg-p-nitroanilide and for p-nitrophenyl-p-guanidinobenzoate. Molecular models of activated factor X (heavy chain) place the light chain and its γ-carboxyglutamic acid residues in a position distant from the active site (Furie et al., 1982). These results suggest that occupation of Ca^{2+}-binding sites formed by γ-carboxyglutamic acid residues in bovine activated factor X does not affect the catalytic process.

C. Factor IX

Factor IX is a single polypeptide chain glycoprotein of about 55,000 molecular weight. The N-terminal region of factor IX contains 12 γ-carboxyglutamic acid residues and shares sequence homology with the N-terminal region of prothrombin and the light chain of factor X. In parallel with factor X activation, the γ-carboxyglutamic acid-containing portion of factor IX is retained upon activation of the zymogen to the enzyme, activated factor IX. The physiological activator of factor IX appears to be activated factor XI, a serine protease (Davie and Hanahan, 1977; Nemerson and Furie, 1980).

Activation of factor IX by activated factor XI in vitro is the result of two proteolytic cleavages. Cleavage of an Arg-Ala peptide bond in factor IX results in the formation of factor IXα, a molecule with the same molecular weight as factor IX but composed of two disulfide-linked polypeptide chains. Cleavage of this bond in factor IX by activated factor XI is facilitated by metal ions but does not have an absolute requirement for them. Subsequent cleavage of an Arg-Val bond in one of the chains yields activated factor IX, factor IXaβ, an enzymatically active form of the protein, with a molecular weight of about 45,000. Cleavage of the Arg-Val bond by activated factor XI requires metal ions. Alternatively, *in vitro* activation of factor IX may be accomplished by the coagulant protein of RVV. Activation of factor IX by this protease occurs with a single peptide bond cleavage at the Arg-Val site. The rate of this reaction is enhanced by metal ions. The resultant enzyme factor IXaα has a molecular weight of about 55,000 and is composed of two disulfide-linked polypeptide chains (Davie and Hanahan, 1977; Nemerson and Furie, 1980).

Factor IX, the activation intermediate factor IXα, factor IXaβ, and factor IXaα all bind Ca²⁺. All the forms of factor IX listed appear to have two classes of calcium-binding sites, two high-affinity sites with a dissociation constant of 0.1–0.2 mM, and 10 or 11 lower-affinity binding sites with a dissociation constant of approximately 1.9 mM (Fig. 2B). This is in marked distinction from the positive cooperativity of the binding of Ca²⁺ to prothrombin or factor X. It is interesting to note that, also in contradistinction to the activation of prothrombin or factor X, activation of factor IX *in vitro* is not facilitated by the presence of phospholipid (Amphlett *et al.*, 1978).

Several other metal ions bind to factor IX. The binding of Mn²⁺ to factor IX appears qualitatively similar to that of Ca²⁺ occurring at 2 sites with a dissociation constant of approximately 13 μM and at 5–6 sites with a dissociation constant of approximately 0.16 mM. However, in contrast to the situation with prothrombin or factor X, Mn²⁺ cannot completely displace Ca²⁺ from fully liganded factor IX. Lanthanide ions do appear to displace Ca²⁺ completely from the fully liganded protein. Finally, Sr²⁺ binds to factor IX, but in a manner qualitatively different from that of the other metal ions. There appears to be only one class of Sr²⁺-binding sites on factor IX (Fig. 2C), 16 sites with a dissociation constant of approximately 3 mM. Evidence suggests that Sr²⁺ binds at the lower-affinity Ca²⁺-binding sites. Strontium ions facilitate the activation of factor IX by activated factor XI, albeit less effectively than Ca²⁺, whereas Mn²⁺, La³⁺, and Co²⁺ are ineffective in supporting this reaction (Amphlett *et al.*, 1978).

The conversion of factor IX to factor IXaα by the coagulant protein of RVV is relatively nonselective with regard to metal ions. In this reaction Co²⁺, Sr²⁺, and Mn²⁺ all effectively substitute for Ca²⁺. Even Mg²⁺, Ba²⁺, or Tb³⁺ enhance the rate of this reaction. In contrast, the rates of overall conversion of factor IX by activated factor XI are more selective with regard to metal ions. Calcium ions produce maximal enhancement, while

Sr^{2+} facilitates conversion but at a rate diminished from that in the presence of Ca^{2+}. However, in the presence of suboptimal concentrations of Ca^{2+}, Sr^{2+}, Mn^{2+}, Co^{2+} or Mg^{2+} will stimulate activation of factor IX by activated factor XI. Conversion of factor $IX\alpha$ to factor $IXa\beta$ by activated factor XI similarly has stringent metal ion specificity. The same conversion when catalyzed by the coagulant protein is supported by Ca^{2+}, Sr^{2+}, Co^{2+}, or Mn^{2+}. The nature of the complementarity of metal ions in facilitating this reaction has not been defined. It also remains unclear why activation by the coagulant protein of RVV exhibits less stringent metal specificity than activation by activated factor XI (Byrne *et al.*, 1980).

Although cooperativity of metal ion binding to factor IX has not been observed, immunological studies suggest that factor IX may undergo a conformational change upon binding of Ca^{2+} (Lewis *et al.*, 1980). If such a conformational alteration occurs, then understanding its metal requirements may help to understand the complex metal dependence of the activation reaction.

D. Factor VII

Factor VII is a single-chain protein which participates in the extrinsic phase of blood coagulation. The protein contains γ-carboxyglutamic acid and exhibits N-terminal sequence homology with the other vitamin K-dependent proteins of plasma. Although the single-chain molecule appears to have coagulant activity, it is greatly enhanced by proteolytic cleavage of the molecule to activated factor VII. Several blood coagulation proteases—activated factor X, activated factor XII, and thrombin—can convert factor VII to its active form. The activation of factor VII by activated factor X in the presence of phospholipid requires Ca^{2+} (Davie and Hanahan, 1977; Nemerson and Furie, 1980).

Metal-binding data have been interpreted by Scatchard analysis (Fig. 2B) to indicate that factor VII has two high-affinity Ca^{2+}-binding sites (K_d, 0.1 mM) and about five weaker-affinity sites (K_d, 1.7 mM). The intrinsic protein fluorescence of factor VII is quenched by about 11% upon Ca^{2+} binding. This quenching is suggested to be the result of a conformational change upon Ca^{2+} binding, which results in alteration of the environment of one or more tryptophan residues in the protein. In contradistinction to factor X and prothrombin, where positive cooperativity of Ca^{2+} binding was suggested to be the result of a conformational change, factor VII does not exhibit positive cooperativity of Ca^{2+} binding. Indeed, an alternate explanation for the observed form of the Scatchard plot may be negative cooperativity of metal binding. However, it may be appropriate to suggest for factor VII, as has been suggested for factor X and prothrombin, that a conformational change resulting from initial binding of Ca^{2+} facilitates interaction of the protein with phospholipid (Strickland and Castellino, 1980).

E. Protein C

Protein C, a recently discovered γ-carboxyglutamic acid-containing protein of blood, is unique in inhibiting rather than promoting blood coagulation (Stenflo, 1976; Kisiel, 1979). Protein C functions as an anticoagulant by proteolytically inactivating two protein cofactors of blood coagulation factor V and factor VIII (Kiesiel et al., 1977; Walker et al., 1979; Vehar and Davie, 1980). Protein C contains two polypeptide chains with molecular weights of 41,000 and 21,000 (Kisiel et al., 1977). Like many of the other plasma proteins containing γ-carboxyglutamic acid, protein C circulates in a zymogen form and is activated by limited proteolysis of the heavy chain. The 11 γ-carboxyglutamic acid residues are in the N-terminal portion of the light chain (Fernlund et al., 1978).

A Scatchard plot constructed from the data obtained in binding experiments between Ca^{2+} and protein C takes the form shown in Fig. 2C. Protein C appears to contain about 16 Ca^{2+} ion-binding sites all of equal affinity (K_d, ~0.87 mM). Activated protein C binds about 9 Ca^{2+} (K_d, ~0.43 mM). The loss of sites upon activation is probably not entirely a reflection of the loss of liganding amino acids upon activation (loss of tetradecapeptide) but may be the result of conformational rearrangement. Calcium ions have the opposite effect on activation of protein C by either thrombin or the coagulant protein of RVV. The activation by thrombin is inhibited by metal ions, while that by coagulant protein is enhanced (Amphlett et al., 1981).

It is of interest to note at this juncture that, although prothrombin, factor IX, and the light chains of factor X and protein C have 60–71% sequence homology in their N-terminal regions (residues 1–44) with maintenance of position of the γ-carboxyglutamic acid residues and an internal disulfide loop (Katayama et al., 1979), they exhibit marked differences in their modes of Ca^{2+} binding. It is clear that other factors—participation of other Ca^{2+}-liganding amino acids and conformational differences in remote portions of the molecule—play a role in determining the way in which each of these proteins binds metal ions. Studies on free γ-carboxyglutamic acid or small peptides containing this amino acid are not entirely predictive of its behavior in proteins. To understand fully the varying metal-binding patterns in these proteins will require further study.

III. γ-CARBOXYGLUTAMIC ACID-CONTAINING PROTEIN OF BONE

The affinity of γ-carboxyglutamic acid-containing proteins for Ca^{2+} ions suggested that such proteins might be present in calcium-rich tissues such as bone. Indeed low-molecular-weight (5000–6000) γ-carboxyglutamic acid-containing proteins have been isolated from the noncollagenous protein matrix of chicken, bovine, swordfish, and human bone (Hauschka *et al.*, 1975; Price *et al.*, 1976a; Poser *et al.*, 1980). These proteins have been designated osteocalcin or the γ-carboxyglutamic acid-containing protein of bone. Current evidence indicates that the γ-carboxyglutamic acid-containing protein of bone is synthesized in the bone, most likely in osteoblasts (Nishimoto and Price, 1979, 1980). Its synthesis in bone is dependent on vitamin K (Hauschka and Reid, 1978; Nishimoto and Price, 1980). There appears to be a higher-molecular-weight precursor of osteocalcin synthesized and processed in these cells (Nishimoto and Price, 1980).

The swordfish, bovine, and human γ-carboxyglutamic acid-containing bone proteins have been sequenced and show a high degree of sequence homology (Price *et al.*, 1976b, 1980; Poser *et al.*, 1980). Calf and swordfish osteocalcin each contain three residues of γ-carboxyglutamic acid at positions 17, 21, and 24. While human osteocalcin has γ-carboxyglutamic acid at residues 20 and 24, the glutamic acid at residue 17 is only partially carboxylated. As little is known about the mechanism for initiating and terminating carboxylation in general, no firm conclusions can be drawn about this difference. As all the glutamic acids at the N-terminal of the plasma γ-carboxyglutamic acid-containing proteins are carboxylated, it has been suggested that the partial carboxylation of position 17 in human osteocalcin is the result of loss of a carboxyl group. The γ-carboxyglutamic acid-containing protein of bone is probably part of the extracellular matrix of bone and, like collagen, probably has a half-life of several years. While the osteocalcin analyzed from the swordfish and the dog came from young animals, the osteocalcin used for sequence analysis of the human protein came from an older individual. The hypothesis is that decarboxylation may be a natural process of aging. Chicken bone osteocalcin has not been sequenced, but amino acid analysis indicates four residues of γ-carboxyglutamic acid per molecule.

The γ-carboxyglutamic acid-containing proteins of bone have been demonstrated to bind metal ions. Chicken osteocalcin has two Ca^{2+}-binding sites (K_d, ~0.8 mM) and two or three lower-affinity sites (K_d, ~3 mM) (Hauschka and Gallop, 1977). The γ-carboxyglutamic acid-containing protein of bovine bone binds three calcium ions with an average dissociation constant of approximately 4 mM (Price *et al.*, 1977). Differences in Ca^{2+} binding between species may be related to the differences in γ-carboxyglutamic acid content of the bone proteins.

Bovine γ-carboxyglutamic acid-containing bone protein binds to hydroxyapatite but not to amorphous calcium phosphate *in vitro* (Price *et al.*, 1976a). While thermally decarboxylated or reduced and carboxymethylated bone protein also binds to hydroxyapatite (as do many anionic proteins), it is felt that there is direct participation of the γ-carboxyglutamic acid residues in the binding of the native protein to hydroxyapatite (Poser and Price, 1979). Hydroxyapatite when bound to the native protein protects against thermal decarboxylation; however, reduced and S-carboxymethylated bone protein is not fully protected. It has been suggested that the bone protein interacts with Ca^{2+} on the hydroxyapatite surface. Reduction of the disulfide bond in the bone protein may alter the spatial array of γ-carboxyglutamic acid residues on the protein surface and perhaps their ability to bind the Ca^{2+} in hydroxyapatite. Native γ-carboxyglutamic acid-containing bone protein inhibits hydroxyapatite precipitation from solutions containing K_2HPO_4 and $CaCl_2$; decarboxylated or reduced bone protein are unable to inhibit hydroxyapatite formation.

The role of the γ-carboxyglutamic acid-containing protein of bone has not yet been clearly defined. Possible roles include participation in mineralization or maturation of mineral, skeletal homeostasis, or participation in matrix structure. While current experimental evidence may sway the argument one way or another, no conclusive statement on the role of osteocalcin can yet be made.

IV. OTHER γ-CARBOXYGLUTAMIC ACID-CONTAINING PROTEINS

γ-Carboxyglutamic acid has been identified in a number of other calcified tissues beside bone. Tissues of mammalian origin include atherosclerotic plaque, ectopic calcification, and renal calculi. Little is known of the metal-liganding properties of the γ-carboxyglutamic acid-containing proteins of these tissues. Hydroxyapatite known to bind the osteocalcin of bone is found in the calcium-containing exudate of patients with scleroderma and ectopic calcifications of patients with dermatomyositis. Only renal stones that contain Ca^{2+} consist of proteins with γ-carboxyglutamic acid (for review, see Gallop *et al.*, 1980).

Calcium-containing mineralized tissue from invertebrates has been found to be devoid of γ-carboxyglutamic acid. However, γ-carboxyglutamic acid has been found even in very early vertebrates (e.g., lamprey eel, shark cartilage, fossils of mammoth bone). This amino acid is found associated with mineral phases other than hydroxyapatite and with noncollagen-containing matrixes (for review, see Burnier *et al.*, 1981).

V. SUMMARY

Vitamin K-dependent carboxylation systems have been identified in a wide variety of tissues other than bone and liver. Tissues in which such systems have been identified include lung, kidney, spleen, pancreas, placenta, cultured fibroblasts, and chick chorioallantoic membrane (Burnier *et al.*, 1981). Little is known about the γ-carboxyglutamic acid-containing proteins synthesized in these systems. However, it is clear that γ-carboxyglutamic residues can confer unique metal-liganding properties on proteins. The importance of the interaction between Ca^{2+} and the γ-carboxyglutamic acid-containing proteins of blood coagulation in the stabilization of protein structure, macromolecular complex assembly, and control of physiological events has already been demonstrated. Postulated roles for other γ-carboxyglutamic acid-containing proteins include participation in calcium or magnesium homeostasis, mineralization, metal ion transport, and other physiological processes regulated by metal ions.

REFERENCES

Amphlett, G. W., Byrne, R., and Castellino, F. J. (1978). The binding of metal ions to bovine factor IX. *J. Biol. Chem.* **253**, 6774–6779.

Amphlett, G. W., Kisiel, W., and Castellino, F. J. (1981). Interaction of calcium with bovine protein C. *Biochemistry* **20**, 2156–2161.

Bajaj, S. P., Butkowski, R. J., and Mann, K. G. (1975). Prothrombin fragments: Calcium binding and activation kinetics. *J. Biol. Chem.* **250**, 2150–2156.

Bajaj, S. P., Nowak, T., and Castellino, F. J. (1976). Interaction of manganese with bovine prothrombin and its thrombin-mediated cleavage products. *J. Biol. Chem.* **251**, 6294–6299.

Bajaj, S. P., Byrne, R., Nowak, T., and Castellino, F. J. (1977). Interaction of manganese with bovine factor X. *J. Biol. Chem.* **252**, 4758–4761.

Benarous, R., Elion, J., and Labie, D. (1976). Calcium binding properties of human prothrombin. *Biochimie* **58**, 391–394.

Benson, B. J., and Hanahan, D. J. (1975). Structural studies on bovine prothrombin: Isolation and partial characterization of the Ca^{2+} binding and carbohydrate containing peptides of the N-terminus. *Biochemistry* **14**, 3265–3277.

Benson, B. J., Kisiel, W., and Hanahan, D. J. (1973). Calcium binding and other characteristics of bovine factor II and its activation intermediates. *Biochim. Biophys. Acta* **329**, 81–87.

Bloom, J. W., and Mann, K. G. (1978). Metal ion induced conformational transitions of prothrombin and prothrombin fragment 1. *Biochemistry* **17**, 4430–4438.

Brittain, H. G., Richardson, F. S., and Martin, R. B. (1976). Terbium(III) emission as a probe of Ca(II) binding sites in proteins. *J. Am. Chem. Soc.* **98**, 8255–8260.

Burnier, J., Borowski, M., Furie, B. C., and Furie, B. (1981). Gamma-carboxyglutamic acid. *Mol. Cell. Biochem.* (in press).

Byrne, R., Amphlett, G. W., and Castellino, F. J. (1980). Metal ion specificity of the conversion of bovine factors IX, IXα and IXaα to bovine factor IXaβ. *J. Biol. Chem.* **255**, 1430–1435.

Davie, E. W., and Hanahan, D. J. (1977). Blood coagulation proteins. In "The Plasma Proteins" (F. W. Putnam, ed.), Vol. 3, pp. 421–544. Academic Press, New York.

Esmon, C. T., Suttie, J. W., and Jackson, C. M. (1975). The functional significance of vitamin K action. *J. Biol. Chem.* **250**, 4095–4099.

Fernlund, T., Stenflo, J., and Tufuesson, A. (1978). Bovine protein C: Amino acid sequence of the light chain. *Proc. Natl. Acad. Sci. U.S.A.* **75**, 5889–5992.

Furie, B., and Furie, B. C. (1976). Spectral changes in bovine factor X associated with activation by the venom coagulant protein of *Vipera russelli*. *J. Biol. Chem.* **251**, 6807–6814.

Furie, B., and Furie, B. C. (1979). Conformation-specific antibodies as probes for the γ-carboxyglutamic acid-rich region of bovine prothrombin: Studies of metal-induced structural changes. *J. Biol. Chem.* **254**, 9766–9771.

Furie, B., Provost, K. L., Blanchard, R. A., and Furie, B. C. (1978). Antibodies directed against a γ-carboxyglutamic acid-rich region of bovine prothrombin: Preparation, isolation, and characterization. *J. Biol. Chem.* **253**, 8980–8987.

Furie, B., Bing, D. H., Feldmann, R. J., Robison, D. J., Burnier, J. P., and Furie, B. C. (1981). Computer-generated models of blood coagulation factor Xa, factor IXa, and thrombin based upon structural homology with other serine proteases. *J. Biol. Chem.* (in press).

Furie, B. C., and Furie, B. (1975). Interaction of lanthanide ions with bovine factor X and their use in the affinity chromatography of the venom coagulant protein of *Vipera russelli*, *J. Biol. Chem.* **250**, 601–608.

Furie, B. C., Mann, K. G., and Furie, B. (1976). Substitution of lanthanide ions for calcium ions in the activation of bovine prothrombin by activated factor X: High affinity metal binding sites of prothrombin and the derivatives of prothrombin activation. *J. Biol. Chem.* **251**, 3235–3241.

Furie, B. C., Blumenstein, M., and Furie, B. (1979). Metal binding sites of a γ-carboxyglutamic acid-rich fragment of bovine prothrombin. *J. Biol. Chem.* **254**, 12521–12530.

Gallop, P. M., Lian, J. B., and Hauschka, P. V. (1980). Carboxylated calcium-binding proteins and vitamin K. *N. Engl. J. Med.* **302**, 1460–1466.

Ganrot, P. O., and Nilehn, J. E. (1968). Plasma prothrombin during treatment with dicumarol: Demonstration of an abnormal prothrombin fraction. *Scand. J. Clin. Lab. Invest.* **22**, 23–28.

Hauschka, P. V., and Gallop, P. M. (1977). Purification and calcium binding properties of osteocalcin, the γ-carboxyglutamate-containing protein of bone. In "Calcium-Binding Proteins and Calcium Function" (R. H. Wasserman, R. A. Corradino, E. Carafoli, R. H. Kretsinger, D. H. MacLennan, and F. L. Siegel, eds.), pp. 338–347. North-Holland Publ., Amsterdam.

Hauschka, P. V., and Reid, M. L. (1978). Vitamin K dependence of a calcium-binding protein containing γ-carboxyglutamic acid in chicken bone. *J. Biol. Chem.* **253**, 9063–9068.

Hauschka, P. V., Lian, J. B., and Gallop, P. M. (1975). Direct identification of the calcium-binding amino acid, γ-carboxyglutamate, in mineralized tissue. *Proc. Natl. Acad. Sci. U.S.A.* **72**, 3925–3929.

Henrikson, R. A., and Jackson, C. M. (1975). Cooperative calcium binding by the phospholipid binding region of bovine prothrombin: A requirement for intact disulfide bridges. *Arch. Biochem. Biophys.* **170**, 149–159.

Jackson, C. M., and Nemerson, Y. (1980). Blood coagulation. *Annu. Rev. Biochem.* **49**, 765–811.

Jackson, C. M., Peng, C., Brenckle, G. M., Jonas, A., and Stenflo, J. (1979). Multiple modes of association in bovine prothrombin and its proteolysis products. *J. Biol. Chem.* **254**, 5020–5026.

Katayama, K., Ericcson, L. H., Enfield, D. L., Walsh, K. A., Neurath, H., Davie, E. W., and Titani, K. (1979). Comparison of amino acid sequence of bovine coagulation factor IX (Christmas factor) with that of other vitamin K-dependent plasma proteins. *Proc. Natl. Acad. Sci. U.S.A.* **76**, 4990–4994.

Kisiel, W. (1979). Human plasma protein C: Isolation, characterization, and mechanism of activation by α-thrombin. *J. Clin. Invest.* **64**, 761–769.

Kisiel, W., Canfield, W. M., Ericsson, L. H., and Davie, E. W. (1977). Anticoagulant properties of bovine plasma protein C following activation by thrombin. *Biochemistry* **16**, 5824–5831.

Kosow, D. P. (1976). Purification and activation of human factor X: Cooperative effect of Ca^{2+} on the activation reaction. *Thromb. Res.* **9**, 565–573.

Kretsinger, R. H. (1976). Calcium binding proteins. *Annu. Rev. Biochem.* **45**, 239–266.

Lewis, R. M., Reisner, H. M., Chung, K., and Roberts, H. R. (1980). Detection of factor IX antibodies by radioimmunoassay: Effect of calcium on antibody-factor IX interaction. *Blood* **56**, 608–614.

Lindhout, M. J., and Hemker, H. C. (1978). The role of γ-carboxyglutamyl residues in the positive cooperative binding of Ca^{2+} to blood coagulation factor X. *Biochim. Biophys. Acta* **533**, 318–326.

Lindhout, M. J., Kop-Klassen, B. H. M., and Hemker, H. C. (1978a). Activation of decarboxy factor X by a protein from Russell's viper venom: Purification and partial characterization of decarboxy factor X. *Biochim. Biophys. Acta* **533**, 327–341.

Lindhout, M. J., Kop-Klassen, B. H. M., and Hemker, H. C. (1978b). The effect of γ-carboxyglutamate residues on the enzymatic properties of the activated blood clotting factor X. *Biochim. Biophys. Acta* **533**, 342–354.

Magnusson, S., Sottrup-Jensen, L., Petersen, T. E., Morris, H. R., and Dell, A. (1974). Primary structure of the vitamin K-dependent part of prothrombin. *FEBS Lett.* **44**, 189–193.

Marki, W., Oppliger, M., and Schwyzer, R. (1977). Chemical synthesis, proton NMR parameters, hydrogen and calcium-ion complexation of L-γ-carboxyglutamyl-L-γ-carboxyglutamic acid and D-γ-carboxyglutamyl-L-leucine. *Helv. Chim. Acta* **60**, 807–815.

Marsh, H. C., Scott, M. E., Hiskey, R. G., and Koehler, K. A. (1979a). The nature of the slow metal ion-dependent conformational transition in bovine prothrombin. *Biochem. J.* **183**, 513–517.

Marsh, H. C., Robertson, P., Jr., Scott, M. E., Koehler, K. A., and Hiskey, R. G. (1979b). Magnesium and calcium ion binding to prothrombin fragment 1: A circular dichroism, fluorescence and $^{43}Ca^{2+}$ and $^{25}Mg^{2+}$ nuclear magnetic resonance study. *J. Biol. Chem.* **254**, 10268–10275.

Nelsestuen, G. L. (1976). Role of γ-carboxyglutamic acid: An unusual protein transition required for the calcium-dependent binding of prothrombin to phospholipid. *J. Biol. Chem.* **251**, 5648–5656.

Nelsestuen, G. L., and Lim, T. K. (1977). Equilibria involved in prothrombin- and blood-clotting factor X-membrane binding. *Biochemistry* **16**, 4164–4171.

Nelsestuen, G. L., and Suttie, J. W. (1972a). The purification and properties of an abnormal prothrombin protein produced by dicumarol-treated cows. *J. Biol. Chem.* **247**, 8176–8182.

Nelsestuen, G. L., and Suttie, J. W. (1972b). Mode of action of vitamin K: Calcium binding properties of bovine prothrombin. *Biochemistry* **11**, 4961–4964.

Nelsestuen, G. L., Zytkovicz, T. H., and Howard, J. B. (1974). The mode of action of vitamin K: Identification of γ-carboxyglutamic acid as a component of prothrombin. *J. Biol. Chem.* **249**, 6347–6350.

Nelsestuen, G. L., Broderius, M., Zytkovicz, T. H., and Howard, J. B. (1975). On the role of γ-carboxyglutamic acid in calcium and phospholipid binding. *Biochem. Biophys. Res. Commun.* **65**, 233–240.

Nelsestuen, G. L., Broderius, M., and Martin, G. (1976). Role of γ-carboxyglutamic acid: Cation specificity of prothrombin and factor X-phospholipid binding. *J. Biol. Chem.* **251**, 6886–6893.

Nelsestuen, G. L., Resnick, R. M., Wei, G. J., Pletcher, C. H., and Bloomfield, V. A. (1981). Metal ion interactions with bovine prothrombin and prothrombin fragment 1: Stoichiometry of binding, protein self-association and conformational change induced by a variety of metal ions. *Biochemistry* **20**, 351–358.

Nemerson, Y., and Furie, B. (1980). Zymogens and cofactors of blood coagulation. *CRC Crit. Rev. Biochem.* **9**, 45–85.

Nishimoto, S. K., and Price, P. A. (1979). Proof that the γ-carboxyglutamic acid-containing bone protein is synthesized in calf bone. *J. Biol. Chem.* **254**, 437–441.

Nishimoto, S. K., and Price, P. A. (1980). Secretion of the vitamin K-dependent protein of bone by rat osteosarcoma cells. *J. Biol. Chem.* **255**, 6579–6583.

Orthner, C. L., and Kosow, D. P. (1978). The effect of metal ions on the amidolytic activity of human factor Xa (activated Stuart-Prower factor). *Arch. Biochem. Biophys.* **185**, 400–406.

Poser, J. W., and Price, P. A. (1979). A method for decarboxylation of γ-carboxyglutamic acid in proteins: Properties of the decarboxylated γ-carboxyglutamic acid protein from calf bone. *J. Biol. Chem.* **254**, 431–436.

Poser, J. W., Esch, F. S., Ling, N. C., and Price, P. A. (1980). Isolation and sequence of the vitamin K-dependent protein from human bone. *J. Biol. Chem.* **255**, 8685–8691.

Prendergast, F. G., and Mann, K. G. (1977). Differentiation of metal ion-induced transitions of prothrombin fragment 1. *J. Biol. Chem.* **252**, 840–850.

Price, P. A., Otsuka, A. S., Poser, J. W., Kristaponis, J., and Raman, N. (1976a). Characterization of a γ-carboxyglutamic acid-containing protein from bone. *Proc. Natl. Acad. Sci. U.S.A.* **73**, 1447–1451.

Price, P. A., Poser, J. W., and Raman, N. (1976b). Primary structure of the γ-carboxyglutamic acid-containing protein from bovine bone. *Proc. Natl. Acad. Sci. U.S.A.* **73**, 3374–3375.

Price, P. A., Otsuka, A. S., and Poser, J. W. (1977). Comparison of γ-carboxyglutamic acid-containing proteins from bovine and swordfish bone: Primary structure and Ca²⁺-binding. *In* "Calcium-Binding Proteins and Calcium Function" (R. H. Wasserman, R. A. Corradino, E. Carafoli, R. H. Kretsinger, D. H. MacLennan, and F. L. Siegel, eds.), pp. 333–337. North-Holland Publ., Amsterdam.

Price, P. A., Epstein, D. J., Lothringer, J. W., Nishimoto, S. K., Poser, J. W., and Williamson, M. K. (1980). Structure and function of the vitamin K-dependent protein of bone. *In* "Vitamin K Metabolism and Vitamin K-Dependent Proteins" (J. W. Suttie, ed.), pp. 219–225. Univ. Park Press, Baltimore, Maryland.

Robertson, P., Jr., Hiskey, R. G., and Koehler, K. A. (1978). Calcium and magnesium binding of γ-carboxyglutamic acid-containing peptides via metal ion nuclear magnetic resonance. *J. Biol. Chem.* **253,** 5880–5883.

Sarasua, M. M., Scott, M. E., Helpern, J. A., Ten Kortenaar, P. B. W., Boggs, N. T., III, Pedersen, L. G., Koehler, K. A., and Hiskey, R. G. (1980). Europium ion coordination with γ-carboxyglutamic acid containing ligand systems. *J. Am. Chem. Soc.* **102,** 3404–3412.

Satyshur, K. A. (1978). Structural studies of modified amino acids by x-ray crystallography: γ-Carboxyglutamic acid and methylated amino acids. Ph.D. Thesis, University of Wisconsin, Madison.

Scott, M. E., Sarasua, M. M., Marsh, H. C., Harris, D. L., Hiskey, R. G., and Koehler, K. A. (1980). Interaction of lanthanide (III) ions with bovine prothrombin fragment 1: A luminescence and nuclear magnetic resonance study. *J. Am. Chem. Soc.* **102,** 3413–3419.

Sperling, R., Furie, B. C., Blumenstein, M., Keyt, B., and Furie, B. (1978). Metal binding properties of γ-carboxyglutamic acid: Implications for the vitamin K-dependent blood coagulation proteins. *J. Biol. Chem.* **253,** 3898–3906.

Stenflo, J. (1976). A new vitamin K-dependent protein: Purification from bovine plasma and preliminary characterization. *J. Biol. Chem.* **251,** 355–363.

Stenflo, J., and Ganrot, P. O. (1972). Vitamin K and the biosynthesis of prothrombin. *J. Biol. Chem.* **247,** 8160–8166.

Stenflo, J., and Ganrot, P. O. (1973). Binding of Ca^{2+} to normal and dicumarol-induced prothrombin. *Biochem. Biophys. Res. Commun.* **50,** 98–104.

Stenflo, J., and Suttie, J. W. (1977). Vitamin K-dependent formation of γ-carboxyglutamic acid. *Annu. Rev. Biochem.* **46,** 157–172.

Stenflo, J., Fernlund, P., Egan, W., and Roepstorff, P. (1974). Vitamin K-dependent modifications of glutamic acid residues in prothrombin. *Proc. Natl. Acad. Sci. U.S.A.* **71,** 2730–2733.

Strickland, D. R., and Castellino, F. J. (1980). The binding of calcium to bovine factor VII. *Arch. Biochem. Biophys.* **199,** 61–66.

Suttie, J. W. (1980). Mechanism of action of vitamin K: Synthesis of γ-carboxyglutamic acid. *CRC Crit. Rev. Biochem.* **8,** 191–223.

Suttie, J. W., and Jackson, C. M. (1977). Prothrombin structure, activation, and biosynthesis. *Physiol. Rev.* **57,** 1–70.

Tai, M. M., Furie, B. C., and Furie, B. (1980). Conformation-specific antibodies directed against the bovine prothrombin : calcium complex. *J. Biol. Chem.* **255,** 2790–2795.

Vehar, G. A., and Davie, E. W. (1980). Preparation and properties of bovine factor VIII (antihemophilic factor). *Biochemistry* **19,** 401–410.

Walker, F. J., Sexton, P. W., and Esmon, C. J. (1979). The inhibition of blood coagulation by activated protein C through the selective inactivation of activated factor V. *Biochim. Biophys. Acta* **571,** 333–342.

Yue, R. H., and Gertler, M. M. (1978a). The binding of calcium to bovine factor X by rate dialysis. *Thromb. Haemostasis* **40,** 350–357.

Yue, R. H., and Gertler, M. M. (1978b). Activation of bovine factor X in the presence of calcium, magnesium, barium or manganese ion. *Thromb. Haemostasis* **40,** 358–367.

Chapter 8

Parvalbumins and Other Soluble High-Affinity Calcium-Binding Proteins from Muscle

WLODZIMIERZ WNUK
JOS A. COX
ERIC A. STEIN

I. Introduction	243
II. Historical Review	244
III. Distribution of Sarcoplasmic Calcium-Binding Proteins in the Animal Kingdom	245
A. Evolutionary Aspects	245
B. Functional Aspects	250
IV. Parvalbumins	250
A. Characterization	250
B. Tissue and Intracellular Distribution	251
C. Structure	254
D. Calcium and Magnesium Binding	259
V. Sarcoplasmic Calcium-Binding Proteins from Invertebrates	261
A. Characterization	261
B. Muscular and Intracellular Distribution	264
C. Structure	265
D. Calcium and Magnesium Binding	266
VI. Physiological Implications	270
References	273

I. INTRODUCTION

The calcium ion is now generally recognized as being involved in regulation of the major step of muscle contraction, namely, the interaction of the myosin head with actin (for recent reviews, see Perry, 1979; Taylor,

CALCIUM AND CELL FUNCTION, VOL. II
Copyright © 1982 by Academic Press, Inc.
All rights of reproduction in any form reserved.
ISBN 0-12-171402-0

1979; Adelstein and Eisenberg, 1980). The onset of contraction is triggered by the binding of Ca^{2+} to one of the following proteins: (1) troponin C, in the actin-linked regulation found in vertebrate skeletal and cardiac muscle; (2) the regulatory light chain of myosin in the myosin-linked regulation found in mollusks; (3) calmodulin, which in turn activates myosin light chain kinase in the myosin-linked regulation via phosphorylation found in vertebrate smooth muscle and in nonmuscle cells.

Beside the above three proteins, there is another important calcium-binding system in both vertebrate (Pechère *et al.*, 1973) and invertebrate (Cox *et al.*, 1976a) muscle, namely, soluble sarcoplasmic calcium-binding proteins. Those found in vertebrates are called parvalbumins and are believed to belong to the family of intracellular calcium-binding proteins, such as troponin C, myosin light chains, and calmodulin, that have evolved from a common ancestor (Goodman *et al.*, 1979; Kretsinger, 1980) to perform specific functions. Although parvalbumin has been studied extensively, including determination of its structure to atomic resolution (Moews and Kretsinger, 1975), its physiological role has not yet been satisfactorily elucidated. The sarcoplasmic Ca^{2+}-binding proteins found in invertebrates display properties distinct from those of parvalbumins and will be designated here as SCPs; their involvement in physiological events also remains to be clarified.

II. HISTORICAL REVIEW

Proteins of remarkably low molecular weight, detected by ultracentrifugation of low-ionic-strength extracts of frog muscle, were first described by Deuticke (1934). Henrotte (1955) later reported the presence of similar proteins in carp myogen; they were acidic, and their ultraviolet (UV) spectrum showed a predominant contribution of phenylalanine. Because of their small size and high solubility, they are now referred to as parvalbumins. In lower vertebrates, two to five electrophoretically distinct components were usually observed in the same animal (Focant and Pechère, 1965; Pechère *et al.*, 1971a). Their physicochemical properties (Konosu *et al.*, 1965) and peptide maps (Pechère and Capony, 1969) suggested that these proteins were homologous. This was later confirmed by the presence of many isologies in their amino acid sequence (Capony *et al.*, 1973; Coffee and Bradshaw, 1973; Pechère *et al.*, 1973). More important, Pechère *et al.* (1971b) showed that these proteins had a high affinity for Ca^{2+}. Parvalbumins were long thought to occur exclusively in the white skeletal muscle of fish and amphibians. However, in 1974, Lehky *et al.* discovered parvalbumin in the skeletal muscle of higher vertebrates, including the turtle, chicken, rabbit and human. At the same time the presence of parvalbumin in the rabbit was described by Pechère (1974). The unsuccessful search for parvalbumins among invertebrates led to the discovery of a new group of SCPs present in crustaceans (Benzonana *et al.*, 1974), mollusks (Lehman and Szent-Györgyi, 1975), annelids, and cephalochordates (Cox *et al.*, 1976a). In many respects the SCPs found in invertebrates are different from parvalbumins and also from the other Ca^{2+}-binding proteins known so far.

III. DISTRIBUTION OF SARCOPLASMIC CALCIUM-BINDING PROTEINS IN THE ANIMAL KINGDOM

The ignorance of precise biological activities constitutes an obstacle to the elaboration of specific and sensitive methods for the detection of parvalbumins and SCPs. They can be perceived because of their high affinity for calcium (Cox *et al.*, 1976b; Le Peuch *et al.*, 1978). Immunological methods of detection and estimation, which are much more sensitive, are not applicable in the absence of specific antisera. Furthermore, cross-reactivities of various parvalbumins (Baron *et al.*, 1975) and SCPs (Kohler *et al.*, 1978) with an antiserum against a particular parvalbumin or SCP are usually weak. Thus the presence of a calcium-binding protein in sarcoplasm can be established unambiguously only by its isolation and characterization. Since calmodulin is widely distributed in eukaryotes, if not ubiquitous, the possibility of mistaking this protein for a SCP exists unless one of the enzymes known to be specifically stimulated by calmodulin is tested.

The distribution pattern of sarcoplasmic calcium-binding proteins (SCPs) is simple (Tables I–III, Fig. 1). Parvalbumins and SCPs have not yet been found to coexist in the same animal: The former are restricted to vertebrates (Table I), whereas the latter are present only in various invertebrate phyla (Table II). In some phyla or classes there are instances where the detection of parvalbumin or SCP has been unsuccessful (Table III). Distribution of parvalbumins is essentially confined to striated skeletal muscle, whereas SCPs are present in a wide range of muscle types: cross-striated (crustaceans, cephalochordates, mollusks), obliquely striated (annelids), and smooth (mollusks), irrespective of the type of calcium regulation of muscular contraction—actin or myosin-linked (Lehman and Szent-Györgyi, 1975).

A. Evolutionary Aspects

Calcium-binding proteins are thought to have evolved from a four-domain ancestor; the latter arose from two successive tandem duplications of the precursor gene coding for a single-domain polypeptide consist-

TABLE I

Vertebrate Classes in Which Parvalbumins Have Been Found

Class	Species	Muscle[a]	Reference
Mammals	Human	Pectoral and iliac	Lehky *et al.*, 1974
		Brachial	Le Peuch *et al.*, 1978
	Rabbit	Leg	Lehky *et al.*, 1974; Pechère, 1974
Birds	Chicken	Leg	Lehky *et al.*, 1974
Reptiles	Turtle	Leg and neck	Lehky *et al.*, 1974
Amphibians	Frog	Leg	Pechère and Capony, 1969
Bony fishes	Carp[b]	Dorsal	Konosu *et al.*, 1965
	Coelacanth[b]	Dorsal	Hamoir *et al.*, 1973
Cartilage fishes	Thornback ray	Dorsal	Pechère and Capony, 1969
	Dogfish	Dorsal	Heizmann *et al.*, 1974

[a] All parvalbumins from the muscles listed were purified to homogeneity except that from human brachial muscle, which was detected by radioelectrophoresis.

[b] Parvalbumins from many other bony fishes were purified and characterized.

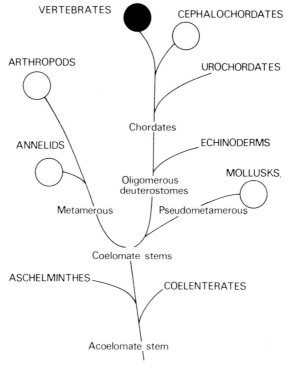

Fig. 1. Distribution of sarcoplasmic calcium-binding proteins. Simplified evolutionary tree modified from Valentine (1980). Only the animal phyla examined are shown. Solid circle, presence of parvalbumin; open circle, presence of SCP; no circle, absence of either one.

TABLE II

Invertebrate Phyla in Which Sarcoplasmic Calcium-Binding Proteins Have Been Found

Phylum or subphylum	Class	Species	Common name	Muscle	Reference[a]
Cephalochordata		Branchiostoma lanceolatum	Amphioxus	Body wall[a]	Kohler et al., 1978
Arthropoda	Crustacea	Astacus leptodactylus	Crayfish	Tail,[a] claw	Benzonana et al., 1974
		Homarus grammarus	Lobster	Tail, claw	Benzonana et al., 1974
		Hippolyte zostericola	Shrimp	Tail[b]	
		Callinectes sapidus	Blue crab	Claw[b]	
		Balanus balanoides	Barnacle	Depressor[c]	
Mollusca	Pelecypoda	Pecten maximus	Scallop	Striated adductor[a]	Lehman and Szent-Györgyi, 1975
		Venus verrucose	Clam	Striated and smooth[a]	Cox et al., 1976a
		Venerupis decussata	Clam	Striated adductor[b]	
		Mytillus edulis	Mussel	Adductor[b]	
		Ostrea edulis	Oyster	Adductor[b]	
Annelida	Polychaeta	Nereis virens	Sandworm	Body wall[a]	Cox et al., 1976a
	Oligochaeta	Lumbricus terrestris	Earthworm	Body wall[a]	Cox et al., 1976a

[a] SCP purified to homogeneity.

[b] After centrifugation of the muscle homogenate, the concentrated supernatant was chromatographed, first on Sephadex G-100 and then on DE-52 cellulose. All emerging fractions were analyzed for calcium. The detection limit was 5 μmoles of calcium bound per kilogram of muscle.

[c] SCP detected by immuno cross-reactivity with anti-crayfish SCP antibodies.

[a] Unless otherwise stated, unpublished results from this laboratory.

247

TABLE III

Organisms in Which Neither Parvalbumin Nor Sarcoplasmic Calcium-Binding Proteins Could be Detected

Phylum[a]	Class	Species	Common name	Muscle[b]
Chordata (Vertebrata)	Agnatha	*Petromyzon fluviatilis*	Lamprey	Dorsal
		Myxine flutinosa	Hagfish	Dorsal
Chordata (Urochordata)		*Ciona intestinales*	Sea squirt	Body wall
Echinodermata	Holothuroidea	*Cucumaria frondosa*	Sea cucumber	Lantern retractor
	Echinoidea	*Centrostephanus longispinus*	Sea urchin	Body wall
	Asteroida	*Echinaster sepositus*	Starfish	Arm
Arthropoda	Insecta	*Schistocerca gregaria*	Locust	Leg and flight
		Locusta viridis	Grasshopper	Leg
Mollusca	Gastropoda	*Helix pomatia*	Wineyard snail	Foot
	Cephalopoda	*Loligo vulgaris*	Squid	Branch and head
		Sepia officinalis	Cuttlefish	Branch and head
		Octopus vulgaris	Octopus	Branch and head
Aschelminthes	Nematoda	*Ascaris suum*	Intestinal roundworm	Longitudinal
Coelenterata	Anthozoa	*Matridium marginatum*	Sea anemone	Whole animal
	Scyphozoa	*Aurelia aurita*	Jellyfish	Whole animal

[a] Subphylum is shown in parentheses.
[b] See Table II, footnotes *b* and *d*.

ing of a calcium-binding region flanked by two helices (Goodman *et al.*, 1979). A major divergence separated the lineages of calmodulin, troponin C, and the alkali light chain of myosin in one branch from the lineages of parvalbumin and the regulatory light chain of myosin in the other branch. The parvalbumin gene became shorter as a result of incomplete copying of the precursor and led to the production of a smaller protein lacking domain I and deprived of the calcium-binding properties of domain II. Since parvalbumins are found in all classes of vertebrates except agnaths (jawless fishes), the most ancient group of living vertebrates (Tables I and III), they probably appeared later than the first vertebrates, but earlier than the cartilage fish–bony fish divergence (425 million years ago). Parvalbumins are coded by two major genetic lines designated α and β (Goodman *et al.*, 1979).

The sarcoplasm of the cephalochordate amphioxus, which is considered the invertebrate closest to the early ancestor of vertebrates, is rich in SCP but does not contain parvalbumin (Cox *et al.*, 1976a). Sarcoplasmic calcium-binding proteins were also found in invertebrate phyla which appeared earlier in evolution. Wnuk and Cox (1978) compared the amino acid composition of SCPs (Table VII) and those of parvalbumins, troponin C, calmodulin, and myosin light chains, using an index for assessing the extent of composition divergence between two proteins (Cornish-Bowden, 1977). The statistical analysis indicated potential isologies in the sequences for the SCPs from the following animals: clam–lobster, clam–amphioxus, clam–sandworm, lobster–crayfish, lobster–amphioxus. The Cornish-Bowden index suggested also a sequence homology between the scallop EDTA* light chain of myosin and the SCPs from sandworm, clam, and amphioxus, as well as between amphioxus SCP and three other myosin light chains: scallop SH, rabbit DTNB, and rabbit A2. Thus invertebrate SCPs seem to constitute a group of homologous proteins that are somehow related to the family of intracellular calcium-binding proteins. The widespread presence of SCPs in various coelomates, animals that possess an internal body cavity (Fig. 1), suggests that these proteins appeared in protocoelomates at least about 700 million years ago. Their absence in vertebrates (this laboratory, unpublished observations) and the appearance of parvalbumins suggest that the latter proteins have evolved from an invertebrate SCP stem, but more precise genealogical links between SCPs and the other calcium-binding proteins cannot be established unless amino acid sequence data for SCPs become available. In any case,

---

* Key to abbreviations: EDTA light chain of myosin, the chain extractable with ethylenediamine tetraacetic acid; SH light chain, the thiol-containing chain; DTNB light chain, the chain extractable with 5,5'-dithiobis(2-nitrobenzoic acid); A2 light chain, the shorter of the two light chains extractable at alkaline pH.

the retention of parvalbumin throughout vertebrate evolution and the presence of SCPs in several invertebrate phyla suggest that these proteins possess a definite physiological function (or functions), probably related to Ca^{2+}-activated processes.

B. Functional Aspects

Parvalbumins cannot substitute for the other Ca^{2+}-binding proteins tested so far: (1) troponin C in an *in vitro* ATPase assay system and in troponin C- and troponin T-binding studies, (2) EDTA light chain of myosin in regulating scallop myosin (Hitchcock and Kendrick-Jones, 1975), and (3) calmodulin in activating phosphorylase kinase (Cohen, 1980) and cyclic-nucleotide phosphodiesterase (Le Donne and Coffee, 1979). Similarly, SCPs tested in this laboratory replace neither troponin C nor calmodulin (unpublished results).

Sarcoplasmic calcium-binding proteins are not found in animals that are mostly sedentary in adulthood, as exemplified by urochordates and echinoderms. Therefore it is tempting to postulate that the presence of these proteins is somehow related to the development of an efficient system of locomotion. In this respect, one wonders why they have not been found yet in jawless fish, insects, molluscan gastropods, and cephalopods.

Sarcoplasmic calcium-binding proteins and parvalbumins do not belong to the contractile apparatus itself, since they are not present in all types of muscle. Extensive studies have restricted the range of functional possibilities to Ca^{2+}- and Mg^{2+}-binding properties. Pechère *et al.* (1975) have proposed that parvalbumins increase the speed of relaxation in muscle (soluble relaxing factor), since these proteins are found preferentially in fast skeletal muscle. In contrast, SCPs are also present in muscles (e.g., in annelids and mollusks) with a rate of action which is slow compared to that of vertebrate skeletal muscles. Moreover, the molecular and calcium-binding properties of parvalbumins and SCPs are distinct (Cox *et al.*, 1976a; Kohler *et al.*, 1978; Wnuk *et al.*, 1979), and their intracellular localizations are not identical (Benzonana *et al.*, 1977). Thus these proteins may have functions that are—at least in part—different.

IV. PARVALBUMINS

A. Characterization

One of the typical properties of parvalbumins is their polymorphism. For instance, five parvalbumin isotypes have been found in carp myogen (Pechère *et al.*, 1971a; Gosselin-Rey *et al.*, 1978). However, in chicken and rabbit, only a single isotype could be isolated (Lehky *et al.*, 1974). Parvalbumins can be purified by a three-step method which involves ammonium sulfate fractionation, gel filtration, and anion exchange (Pechère *et al.*, 1971a), or by heat treatment followed by anion exchange, a method which takes advantage of the remarkable heat stability of parvalbumins in the presence of micromolar Ca^{2+} (Haiech *et al.*, 1979a). Both methods allow separation of the various parvalbumin isotypes. They are usually designated by their isoelectric point (pI).

Parvalbumins are highly acidic proteins with pI values ranging from 3.95 (carp) to 5.5 (rabbit). Their molecular weight based upon the amino acid sequence ranges from 9980 (chub, isotype V, 106 amino acids; Gerday *et al.*, 1978) to 12,150 (cod 4.75, 113 amino acids; Elsayed and Bennich, 1975). Sedimentation equilibrium and gel electrophoresis in the presence of sodium dodecyl sulfate (SDS) have also given molecular weight values in the range 10,000–12,000. Thus parvalbumins exist as monomers. Their small size is reflected in a low sedimentation rate constant and a high diffusion constant ($D_{20,w} = 1.2 \times 10^{-6}$ cm^2 sec^{-1}; Konosu *et al.*, 1965). Various physical parameters of parvalbumins from the different vertebrate classes are presented in Table IV.

The amino acid composition of parvalbumins is characterized by a high number of acidic (23–27) and phenylalanine (8–11) residues, usually a single arginine residue (frog 4.5 isotype with three arginines seems to be an exception), and little or no tyrosine, histidine, proline, cysteine, methionine, or tryptophan (Pechère *et al.*, 1973; Kretsinger, 1980). The high ratio of phenylalanine to tyrosine and tryptophan yields a characteristic UV absorption spectrum in which the bands of the phenylalanine residues are readily visible. Parvalbumins containing no aromatic amino acids other than phenylalanine have an $\epsilon_{259\,nm}^{1\%}$ of 1.5–1.7. The circular dichroic (CD) spectrum in the far UV is characteristic of a protein with a high content of α helix. Binding of Ca^{2+} to parvalbumin causes an increase in the ellipticity of the peptide backbone and induces changes in both the fluorescence and the absorption spectra of the aromatic chromophores (Section IV,D).

B. Tissue and Intracellular Distribution

Comparison of parvalbumin distribution in various muscles and other tissues (Table V) gives some insight into the possible role of this protein. Parvalbumins are essentially muscle proteins. However, as they are not usually found in cardiac and smooth muscles (Section III,b), they are not indispensable components of the contractile mechanism. High parvalbumin contents are frequent in white skeletal muscle, whereas low levels prevail in red muscles. Since parvalbumin is essentially absent in breast

TABLE IV

Physicochemical Properties of Parvalbumins

Property	Thornback ray	Carp	Frog	Turtle	Chicken	Rabbit
pI	4.45[a]	4.25[d]	4.50[i]	4.4[l]	4.9[l]	4.9[i]; 5.5[n]
Molecular weight						
Sequence	11,820[b]	11,580[e]	11,620[j]	—	—	12,140[o]
Sedimentation equilibrium	11,470[a]	9,830[d]	—	—	12,000[l]	12,000[l]
SDS–gel electrophoresis	—	—	—	11,000[l]	12,000[l]	12,000[l]
$s_{20,w}$ (S)	1.7[a]	1.6[d]	—	1.45[l]	1.45[l]	1.45[l]
$\epsilon_{259}^{1\%}$ nm	1.92[a]	1.76[d]	2.06[i]	2.04[m]	1.88[m]	1.51[n]
$\epsilon_{280}/\epsilon_{259}$	0.60[a]	0.05[d]	0.59[i]	0.52[m]	0.57[m]	0.06[n]
θ_{222} (deg cm² dmole⁻¹ 10⁻³)	-9.5[c]	-12.5[f], -15.5[g]	—	—	—	—
Calcium bound (mole mole⁻¹)	1.96[a]	2[h]	2.32[k]	1.9[l]	2.1[l]	2.0[p]

[a] Gerday and Teuwis (1972).
[b] Thatcher and Pechère (1977).
[c] Parello and Pechère (1971).
[d] Konosu et al. (1965).
[e] Coffee and Bradshaw (1973).
[f] Donato and Martin (1974).
[g] Cox et al. (1976b).
[h] Kretsinger and Nockolds (1973).
[i] Pechère et al. (1973).
[j] Capony et al. (1975).
[k] Benzonana et al. (1972).
[l] Blum et al. (1977).
[m] Calculated from amino acid composition (from Ref. l).
[n] Capony et al. (1976).
[o] Enfield et al. (1975).
[p] Lehky et al. (1974).

TABLE V

Parvalbumin Content in Various Tissues

Species	Muscle or other tissue	Micromoles per kilogram wet weight
Human	Brachioradialis, quadriceps femoris[a]	~17
	Pectoralis major, latissimus dorsi[a]	None
Rabbit	Psoas (white)[b]	25
	Adductor magnus (white)[c]	66
	Soleus (red)[c]	1.3
	Diaphragm[d]	3.1
	Myocardium[d]	<0.01
	Uterus, bladder, small intestine, lung, liver, adipose tissue, erythrocytes, serum[d]	<0.01
	Brain[d]	0.15
	Spleen, kidney, ovary[d]	0.02
Shrew	Hind leg[a]	30
	Myocardium[a]	19
Chicken	Adductor magnus (white)[c]	33
	Breast (white)[e]	0.26
Turtle	Neck (white)[c]	916
Frog	Skeletal (white)[f]	~240
Carp	Skeletal (white)[g]	~500
	Skeletal (red)[h]	~100
	Myocardium[g]	Traces
	Brain[i]	<0.15
Hake	Skeletal (white)[j]	480
Coelacanth	Skeletal (white)[k]	940

[a] Determined by radioelectrophoresis (Le Peuch et al., 1978).

[b] Purification yield (Lehky et al., 1974).

[c] Quantified by densitometry of Coomassie blue-stained polyacrylamide gels of muscle extract (Blum et al., 1977).

[d] Estimated by passive hemagglutination using antibodies against rabbit skeletal parvalbumin (Baron et al., 1975).

[e] Determined by radioimmunoassay (Le Peuch et al., 1979).

[f] Estimated by immunodiffusion using antibodies against frog skeletal parvalbumin (Gosselin-Rey and Gerday, 1977).

[g] Estimated by passive hemagglutination using antibodies against carp skeletal parvalbumin (Gosselin-Rey, 1974).

[h] Purification yield (Gosselin-Rey et al., 1978).

[i] Estimated by immunodiffusion using antibodies against pike III parvalbumin isotype (Gosselin-Rey et al., 1978).

[j] Estimated by passive hemagglutination using antibodies against hake 4.36 parvalbumin (Baron et al., 1975).

[k] Purification yield (Hamoir et al., 1973).

white muscle from chicken (Le Peuch *et al.*, 1979) and present in significant amounts in the red muscles of electric eel (Childers and Siegel, 1976)
and of carp (Gosselin-Rey *et al.*, 1978), there seems to be little correlation
between parvalbumin content and aerobic versus anaerobic metabolism in
muscle. The fact that parvalbumins are usually more abundant in fast
muscles and that significant amounts are found in the fast-beating
myocardium of the shrew suggests that the fastest muscles have the
greatest need for parvalbumins (Pechère *et al.*, 1975). However, this hypothesis is not fully supported by the distribution shown in Table V.
Among the nonmuscular tissues examined, brain contains detectable
amounts of a protein cross-reacting with an antiserum against a muscular
parvalbumin (Baron *et al.*, 1975; Gosselin-Rey *et al.*, 1978). Although parvalbumins are absent from liver, it is interesting to note that a calcium-
binding parvalbumin-like protein was purified from Morris hepatoma
(MacManus, 1980).

There is an array of observations indicating that parvalbumins are
genuine sarcoplasmic proteins; they are not tightly bound to an insoluble
matrix and seem to constitute an autonomous regulatory system. Baron *et
al.* (1975) obtained comparable parvalbumin yields from frog muscle irrespective of the extraction procedure. By indirect immunofluorescence,
using antibodies specific for carp 4.25 parvalbumin, Benzonana *et al.*
(1977) found a uniform distribution of the marker in cryosections of the
carp muscle. Figure 2 shows the same observation for perch muscle.
Parvalbumin could not be localized within isolated myofibrils from perch
(Benzonana and Gabbiani, 1978) or from chicken (Heizmann and Strehler,
1979) skeletal muscle. Gillis *et al.* (1979) reported that parvalbumin could
diffuse entirely out of frog skinned muscle fiber. In primary myogenic cell
cultures from chicken embryos, synthesis of parvalbumin does not start at
the same time as that of proteins forming the myofibrillar structure; parvalbumin was not even detected in myotubes in which myofibrils and
sarcoplasmic reticulum were already functioning (Heizmann and Strehler,
1979). Active synthesis of parvalbumin begins at about the time of hatching (Le Peuch *et al.*, 1979).

C. Structure

The ease with which parvalbumins can be purified and their small size
makes them well suited for amino acid sequencing; at this time, 14 parvalbumin sequences have already been determined. Figure 3 shows the sequence of rabbit parvalbumin (Enfield *et al.*, 1975) and of carp 4.25 isotype
(Coffee and Bradshaw, 1972), as representatives of the α and β lineages,
respectively. The secondary structure of the carp protein is also given, as

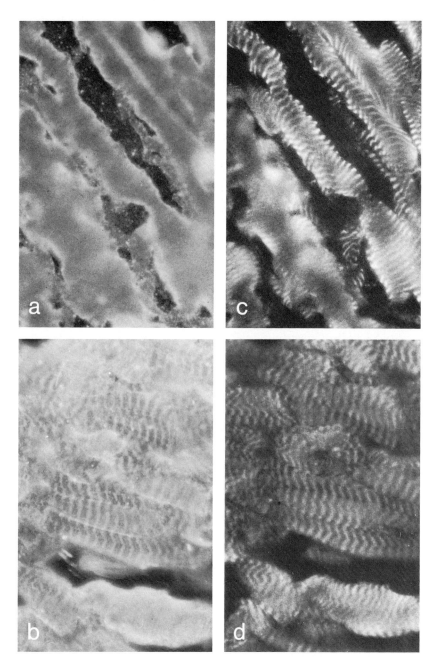

Fig. 2. Immunofluorescence localization of perch parvalbumin (a) and crayfish SCP (b) using antibodies specific for perch parvalbumin and for crayfish SCP, respectively. Actin staining with antiactin antibodies, performed on the same muscle sections as in (a) and (b), is shown in (c) and (d), respectively. ×700. (a) and (c) from Benzonana and Gabbiani (1978); (b) and (d) from Benzonana *et al.* (1977); courtesy of G. Benzonana.

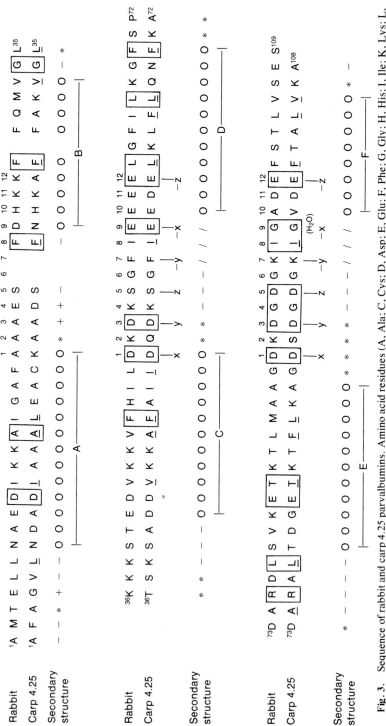

Fig. 3. Sequence of rabbit and carp 4.25 parvalbumins. Amino acid residues (A, Ala; C, Cys; D, Asp; E, Glu; F, Phe; G, Gly; H, His; I, Ile; K, Lys; L, Leu; M, Met; N, Asn; P, Pro; Q, Gln; R, Arg; S, Ser; T, Thr; V, Val) are presented in three rows of which corresponds to the alignment of the three homologous domains of the proteins (Kretsinger, 1972). Residues judged to be in an α helix are shown as ○, in a β sheet as /, in a right-handed turn as *, in a left-handed turn as +, and in an undefined secondary structure as − (Levitt and Greer, 1977). The helices are labeled from A to F, the calcium-binding loop sequences are numbered 1–12 from the N-terminal, the calcium-coordinating positions are labeled from x to $-z$, and residues contributing to the hydrophobic core are underlined (Kretsinger and Nockolds, 1973). Residues identical for the 14 known parvalbumin sequences (Kretsinger, 1980) are framed.

computed by Levitt and Greer (1977) on the basis of the three-dimensional structure, refined at 1.9-Å resolution by Moews and Kretsinger (1975). Sixty-three of the 108 residues form six α-helical regions (A through F). There is a nonhelical N-terminal region and five loops between the six α helices. The loops between helices C and D, and between E and F, each bind a calcium ion. The AB domain does not bind calcium, although it has a structure similar to the CD and EF domains.

Carp 4.25 parvalbumin is a globular molecule without pits or crevasses characteristic of the enzyme structures determined to date. This molecule possesses a well-defined hydrophobic core, one-seventh of its total volume, composed mainly of side chains of phenylalanine, isoleucine, leucine, and valine. The compact hydrocarbon core has a remarkable stabilizing effect on the structure of calcium-saturated parvalbumin. This results, for instance, in a high temperature of heat-induced structural transition ($70°$–$80°C$; Burstein *et al.*, 1975; Cavé *et al.*, 1979) and in a considerable resistance to proteolysis (Cox *et al.*, 1979). All the polar side chains are at the surface of the molecule except those associated with calcium binding and the residues Arg-75 and Glu-81 forming an intramolecular salt bridge; the latter is shielded from contact with water by the AB loop. The crystal structure of the two Ca^{2+}-binding domains of carp 4.25 parvalbumin is shown in Fig. 4. The side chains in positions 1, 3, 5, 9, and 12 (as numbered in Fig. 3), as well as the peptide carbonyl oxygen of residue 7, are the dentates forming a near-octahedral arrangement around the cation. Residues 1 and 9 occupy the $\pm x$ vertices, residues 3 and 7, the $\pm y$ vertices, and residues 5 and 12, the $\pm z$ vertices. The CD loop chelates the calcium ion through four oxygen from four acidic side chains ($+x$, Asp-51; $+y$, Asp-53; $-x$, Glu-59; $-z$, Glu-62), one peptide carbonyl oxygen ($-y$, Phe-57) and one hydroxyl oxygen from a Ser-55 side chain ($+z$). In the EF loop the cation dentate coordination sphere consists of six oxygens from four acidic side chains ($+x$, Asp-90; $+y$, Asp-92; $+z$, Asp-94; $-z$, Glu-101), one peptide carbonyl oxygen ($-y$, Lys-96), and one water molecule ($-x$), as residue 98 is glycine. The calcium in the EF domain is thus formally eight-coordinated, because both oxygen atoms of the carboxylate groups of Asp-92 and Glu-101 chelate the cation. The CD and EF domains are coupled by an antiparallel β-sheet conformation containing two hydrogen bonds between Ile-58 and Ile-97. As illustrated in Fig. 4, the CD domain is related to the EF domains by the twofold axis. Thus there is structural isology of the two calcium-binding domains.

The alignment of all the known parvalbumin sequences reveals many isologies in primary structure (Kretsinger, 1980). Their N-terminal is usually blocked by an acetyl group. The sole exception is coelacanth 5.44 isotype, which has a free threonine as an N-terminal (Jauregui-Adell and

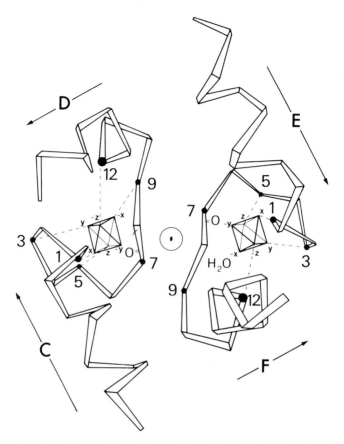

Fig. 4. Drawing of the peptide backbone of the CD and the EF domains of carp 4.25 parvalbumin as seen by x-ray diffraction (Kretsinger and Nockolds, 1973). Solid circles represent α carbons (numbered as in Fig. 3) bearing the side chains involved in calcium coordination; their direction is indicated by dashed lines. The point shown in the central circle represents an approximate twofold axis.

Pechère, 1978). Twenty-four of 109 positions are invariant (Fig. 3); 17 positions can be considered functionally conserved. The greater variability of the N-terminal region and of the AB domain, which has apparently suffered a three-amino acid deletion, is obvious. In contrast, the calcium-binding CD and EF loops are remarkably conserved. Among 12 calcium-coordinating positions, 9 are invariant and 2 conservative; the only variable position is residue 7 in the EF loop, but this residue chelates the calcium cation via its peptide carbonyl oxygen and its side chain is oriented outward. The invariable glycyl side chain at position 9 in the EF loop permits a water molecule to coordinate the calcium ion. The hydrophobic core also seems to be well preserved. Among 28 core residues, 10 positions are invariable and 8 conservative; the variation at the remaining 10 positions is confined to hydrophobic amino acids. The buried Arg-75–Glu-81 pair of carp 4.25 isotype is also invariant in other 13 parvalbumin sequences. Thus there is little doubt that all parvalbumins have an architecture very similar to that of the carp 4.25 protein.

D. Calcium and Magnesium Binding

Besides Ca^{2+} (Benzonana *et al.*, 1972), parvalbumin also binds Mg^{2+} (Cox *et al.*, 1977; Potter *et al.*, 1977). At physiological levels of Mg^{2+} (1 mM) and K^+ (80 mM), and at levels of Ca^{2+} corresponding to those of resting muscle ($\sim10^{-8}\ M$), parvalbumin binds 2 moles Mg^{2+} mole^{-1} and none of Ca^{2+}; a rise in free Ca^{2+} levels causes calcium uptake, which is accompanied by a release of Mg^{2+}, as shown in Fig. 5a. Thus two high-affinity Ca^{2+}-binding sites accommodate Mg^{2+} competitively (Ca^{2+}–Mg^{2+} sites). This implies that, in the muscle cell, parvalbumin may have either

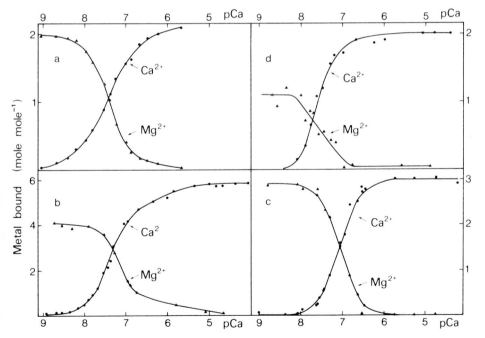

Fig. 5.   Calcium-binding (●) and magnesium release (▲) in the presence of 1 mM Mg^{2+}. (a) Carp parvalbumin isotype 4.25 (Moeschler *et al.*, 1980). (b) Crayfish SCP isotype α_2 (Wnuk *et al.*, 1979). (c) Sandworm SCP isotype α (Cox *et al.*, 1977). (d) Amphioxus SCP (Kohler *et al.*, 1978).

Mg^{2+} or Ca^{2+} bound at the CD and EF loops and cannot remain metal-depleted under physiological conditions. The $Ca^{2+}–Mg^{2+}$ competition explains why much higher equilibrium constants for the binding of Ca^{2+} have been obtained in the absence of Mg^{2+} than in its presence. The apparent constants found by Benzonana et al. (1972) with 2 mM Mg^{2+} ranged from 2.5×10^6 to $1 \times 10^7 M^{-1}$, whereas Haiech et al. (1979b) and Moeschler et al. (1980) reported values ranging from 1.3×10^8 to $2.7 \times 10^9 M^{-1}$, respectively, for Mg^{2+}-free systems. Similarly, values scattered from 1.1×10^4 (Potter et al., 1977) to $1.2 \times 10^5 M^{-1}$ (Lehky and Stein, 1979) were found for the binding of Mg^{2+} in the absence of Ca^{2+}. The variations in the values obtained for each of the three constants may result from differences in Ca^{2+}-binding properties among parvalbumin isotypes (Haiech et al., 1979b) and from the diversity of the physicochemical conditions and techniques of measurements. It should be pointed out that different EGTA affinity constants for Ca^{2+} are available in the literature and that this may lead to discrepancies in equilibrium constants for Ca^{2+} (calculated from free Ca^{2+} concentrations) that can differ by a factor of 4 (Harafuji and Ogawa, 1980). On the other hand, if both EGTA and parvalbumin are used in the concentration range $0.1-1.0$ mM, their interaction ($K_a \sim 30 M^{-1}$; Haiech et al., 1979b) should not affect the determination of the affinity for Ca^{2+}. Indeed, similar values for the binding constants of Ca^{2+} to parvalbumin were obtained using polyacrylamide-immobilized parvalbumin without EGTA (Lehky et al., 1977) and equilibrium dialysis in combination with an EGTA buffer system (Moeschler et al., 1980).

Direct measurements of Ca^{2+} or Mg^{2+} binding to various parvalbumins show no evidence for cooperativity between the two sites; both sites can be considered equivalent with regard to affinity for Ca^{2+} and Mg^{2+} (Cox et al., 1977; Potter et al., 1977; Haiech et al., 1979b; Moeschler et al., 1980). No difference between sites was found when Ca^{2+}- or Mg^{2+}-induced changes in conformation were followed by differential UV spectroscopy (Haiech et al., 1979b), fluorescence, and microcalorimetry (Moeschler et al., 1980). A similar conclusion was reached from studies on exchange kinetics using ^{43}Ca-NMR (Parello et al., 1978). On the other hand, circular dichroism (Donato and Martin, 1974), ^{13}C-NMR (Nelson et al., 1976), ^1H-NMR (Birdsall et al., 1979), and tryptophan fluorescence (Permyakow et al., 1980) studies have shown that certain structural changes are not proportional to the Ca^{2+} content of the parvalbumin molecule. To interpret the latter results, it was assumed that the calcium complex at the EF site was less stable than that at the CD loop, since the EF site contains a water molecule as one of its dentates (Kretsinger and Nockholds, 1973). However, the use of very high concentrations of EGTA and of parvalbumin in the NMR experiments may result in misleading interactions. As for lan-

thanides, the binding is different for the two sites. For instance, Sowadski
et al. (1978) found that, in crystals of parvalbumin, Tb^{3+} replaced the
calcium ion first at the EF site, and ^{113}Cd-NMR studies concluded a
nonequivalence of CD and EF sites when Cd^{2+} was replaced by Gd^{3+}
(Drakenberg *et al.*, 1978).

As parvalbumin is never metal-depleted *in vivo,* the only conformational
changes of physiological significance are those that may occur upon
Ca^{2+}–Mg^{2+} exchange. Circular dichroism, trypsin susceptibility, thiol titration, phenylalanine fluorescence (Cox *et al.,* 1979), and differential
spectra (Haiech *et al.,* 1979b) revealed little or no difference in the overall
conformation of parvalbumin whether in the Ca^{2+} or Mg^{2+} form. Nevertheless, microcalorimetric studies on Mg^{2+}–Ca^{2+} exchange (Moeschler *et
al.,* 1980) indicate that the two metal complexes of parvalbumin have a
different conformational entropy: that of Ca^{2+}–parvalbumin is significantly
higher than that of Mg^{2+}–parvalbumin. The higher structural order of the
Mg^{2+} form seems to result from a tightening of the loops when the smaller
Mg^{2+} ion ($r = 0.78$ Å) rather than the Ca^{2+} ion ($r = 1.06$ Å) occupies the
metal-binding site. The lower stability of the magnesium complex as compared to that of the calcium complex is probably due to the acid residues
being closer to the cation, thus increasing dentate–dentate repulsion and
steric interaction in the ligand. Birdsall *et al.* (1979) also have found differences in conformation between Ca^{2+}- and Mg^{2+}-parvalbumin, which are
visible in the ^1H-NMR signals corresponding to carboxylate groups, the
residues that coordinate Ca^{2+} or Mg^{2+}. Thus replacement of Mg^{2+} by Ca^{2+}
seems to induce structural changes that are essentially restricted to the
two metal-binding loops.

V. SARCOPLASMIC CALCIUM-BINDING PROTEINS FROM INVERTEBRATES[*]

A. Characterization

Similar to parvalbumins, SCPs can be purified by ammonium sulfate
fractionation, gel filtration, and anion exchange (Cox *et al.,* 1976b). High
amounts of mucopolysaccharides and pigments found in certain invertebrates occasionally constitute an obstacle to the purification of these proteins. Since SCPs display a high resistance to heat denaturation in the
presence of Ca^{2+}, heat treatment as an initial step facilitates their isola-

---

[*] Data and observations presented without references are results from this laboratory that
have not been published yet.

tion. As judged from Ca^{2+}-binding properties, crayfish SCPs purified after heat treatment are indistinguishable from those prepared at $+4°C$.

The last step of purification of SCPs usually reveals the existence of polymorphic polypeptide chains. The mollusks that have been examined, clam and oyster, each contain three SCP isotypes. Two SCP isotypes have been found in annelids such as the sandworm and earthworm. A single isotype could be detected in amphioxus myogen (Kohler *et al.*, 1978). As for crayfish, polymorphism is multiplied by the existence of dimeric forms of SCPs, in contrast to the monomeric SCPs mentioned above. Upon DE-52 cellulose chromatography, the SCP from crayfish emerges in three peaks in the proportion 14:1.5:1. Gel electrophoresis and isoelectrofocusing experiments, in the absence and in the presence of urea, have shown that the three SCP isotypes have a subunit composition of α_2, $\alpha\beta$, and β_2 (Wnuk, 1978). Tryptic peptide maps of these isotypes have confirmed such a polypeptide chain composition. This type of polymorphism seems to be common for crustacean SCPs, as similar results have been obtained in the instance of lobster and shrimp.

Like other intracellular calcium-binding proteins, SCPs are acidic, with pI values ranging from 4.75 (crayfish isotype β_2) to 5.23 (clam isotype α).

Crustacean SCPs have a molecular weight of 44,000, as determined by sedimentation equilibrium and Sephadex chromatography. They dissociate in the presence of SDS or urea into two subunits of 22,000 (Cox *et al.*, 1976b). For the SCPs from other invertebrate phyla (annelids, mollusks, and cephalochordates), both gel filtration under nondenaturating conditions and gel electrophoresis in the presence of SDS gave molecular weights in the range 20,000–22,000. These values are in agreement with those calculated from the amino acid compositions and indicate that the latter SCPs exist as monomers. A summary of physicochemical properties of SCPs from the different invertebrate phyla are presented in Table VI.

The amino acid composition of several SCPs is reported in Table VII; 25–30% of the residues are acidic, whereas the basic amino acids amount to 11–13%. Sarcoplasmic calcium-binding proteins contain little or no histidine or cystine, but they do not show the unusually high lysine/arginine ratio characteristic of parvalbumins. Proline exists in very variable amounts in SCPs, whereas little or none is found in parvalbumin, troponin C, and calmodulin. As compared to these three proteins, SCPs contain much more tyrosine and tryptophan, but a similar amount of phenylalanine residues. Consequently, SCPs have a higher UV absorbance with a maximum at about 280 nm. The phenylalanine absorption bands characteristic of the spectrum of most intracellular Ca^{2+}-binding proteins are scarcely visible in sandworm SCP and undetectable in crayfish and amphioxus SCPs.

TABLE VI

Physicochemical Properties of Sarcoplasmic Calcium-Binding Proteins

Property	Crayfish $\alpha_2{}^a$	Sandworm α^b	Clam β^c	Amphioxus α^d
pI	5.06	4.8	5.00	4.85
Molecular weight				
gel filtration				
(Sephadex G-100)	44,000	20,000	22,000	22,000
SDS–gel electrophoresis	$21,000^{e,f}$	$17,000^e$–$19,700^f$	21,000	22,000
$\epsilon_{280}^{1\%}$	10.6	9.7	7.1	20
$\epsilon_{280}/\epsilon_{260}$	2.3	1.5	1.3	2.5
θ_{222} (deg cm^2 dmole^{-1} 10^{-3})	15.4^e–13.0^f	11.8^e–9.8^f	14.0^e–11.5^f	$13.8^{e,f}$
Ca^{2+}-specific sites per molecule	2	0	1	1
Ca^{2+}–Mg^{2+} sites per molecule	4	3	0	1

[a] Cox *et al.* (1976b); Wnuk *et al.* (1979); Wnuk *et al.* (1981).
[b] Cox *et al.* (1976a); Cox *et al.* (1977); Cox and Stein, 1981.
[c] Cox *et al.* (1976a).
[d] Kohler *et al.* (1978).
[e] In the presence of 1 m*M* CaCl$_2$.
[f] In the presence of 1 m*M* EDTA.

As indicated by circular dichroism in the far UV, SCPs possess a high content of α helix. Interaction of SCPs with Ca^{2+} usually induces a significant increase in ellipticity (attributed to both peptide backbone and aromatic chromophores), as well as a change in tryptophan fluorescence. Amphioxus SCP seems to be an exception, as no change in the circular dichroic spectrum could be detected whether the protein was in the metal-free or Ca^{2+} form (Kohler *et al.*, 1978). In contrast, sandworm SCP shows a calcium-dependent change in conformation even in the presence of SDS, as indicated by its electrophoretic mobility (Table VI). This behavior is reminiscent of calmodulin (Klee *et al.*, 1979). Indeed, the backbone circular dichroism reveals that in the presence of SDS sandworm SCP has retained a significant helicity which is sensitive to calcium (Cox and Stein, 1981).

Sarcoplasmic calcium-binding proteins form a less homogeneous group of calcium-binding proteins than parvalbumins; besides amino acid composition, which differs significantly from one invertebrate phylum to another, and the unique dimer formation of crustacean SCPs, the number of Ca^{2+}-binding sites varies from one to three per polypeptide chain. Furthermore, while some of the sites accommodate Mg^{2+} as well as Ca^{2+} (Ca^{2+}–Mg^{2+} sites), there are also sites that display a striking selectivity for Ca^{2+}, i.e., do not bind Mg^{2+} in the millimolar range.

TABLE VII

Amino Acid Composition and pI Values of Sarcoplasmic Calcium-Binding Proteins[a]

	Crayfish		Lobster		Sandworm	Earthworm	Clam (*Venus*)			Amphioxus
	α[b]	β	α	β	α	β	α	β	γ	α[c]
Asx	34	35	37	36	28	19	27	31	31	30
Thr	5	5	7	7	10	9	11	10	11	8
Ser	8	9	8	8	12	14	18	18	18	9
Glx	20	21	21	22	16	18	26	24	24	24
Pro	3	3	—[d]	—[d]	5	12	0	0	0	6
Gly	11	11	11	13	13	17	18	14	15	12
Ala	23	20	16	15	13	10	17	18	18	13
Cys	3	4	—[d]	—[d]	0	0	0	0	0	5
Val	13	13	17	16	11	10	12	12	14	10
Met	2	2	2	2	9	2	6	9	5	6
Ile	11	10	9	9	8	8	8	7	8	7
Leu	13	15	13	12	12	15	11	10	11	15
Tyr	9	10	10	10	2	3	4	3	3	7
Phe	12	12	12	12	15	15	8	9	10	10
Trp	2	2	2	2	3	1	—[d]	2	—[d]	6
Lys	16	13	14	14	12	17	20	20	19	16
His	0	0	0	0	1	0	4	3	3	1
Arg	7	8	8	8	5	4	5	4	4	6
	192	193	187	186	175	174	195	194	194	191
pI	5.06	4.75	4.94	4.75	4.80	–	5.23	5.00	4.90	4.85

[a] Unless otherwise stated, unpublished results from this laboratory.
[b] Cox *et al.* (1976b).
[c] Kohler *et al.* (1978).
[d] Not determined.

B. Muscular and Intracellular Distribution

Contents of SCPs in various muscles presented in Table VIII show that the sarcoplasmic Ca^{2+}-binding capacity provided by SCPs for crayfish tail and amphioxus longitudinal body muscle, for instance, is comparable to that of vertebrate muscles rich in parvalbumins (Table V). Comparison of concentrations within different crayfish muscles shows a ratio of approximately 1 : 10 : 40 for heart, claw, and tail muscle SCP, respectively. With respect to tension development, contraction speed, and mitochondria content, the crustacean tail muscle is analogous to vertebrate white fast muscle; in the crayfish claw, the larger crusher muscle has the characteristics of a red slow muscle, and the minor cutter those of a white fast muscle (Atwood, 1972). Thus SCP distribution among various muscles is somewhat akin to that of parvalbumin in certain vertebrates, such as carp, and

TABLE VIII

Sarcoplasmic Calcium-Binding Protein Content in Various Muscles

Species	Muscle	SCP (μmoles kg^{-1} wet weight)	Ca^{2+}-binding capacity (μmoles kg^{-1} wet weight)[a]
Crayfish[b]	Tail	62.0	372
	Claw	16.3	98
	Heart	1.6	10
Amphioxus[c]	Body wall	~114	228
Sandworm[d]	Body wall	~25	75
Clam (*Venus*)[d]	Striated adductor	~40	40

[a] See Table VI.
[b] Determined by quantitative immunoelectrophoresis (Cox *et al.*, 1976b).
[c] Purification yield (Kohler *et al.*, 1978).
[d] Purification yield.

a similar role in muscular activity for crayfish SCP and parvalbumin cannot be ruled out.

Sarcoplasmic calcium-binding proteins are easily extracted from muscle homogenates by centrifugation in a physiological salt solution or in water. In the case of crayfish, the first extraction yields about 90–95% total SCP (Cox *et al.*, 1976b), indicating that this protein is essentially sarcoplasmic. However, in contrast to parvalbumin, all of which is evenly distributed in muscle cells, there is a portion of SCP that appears to be localized at the site of the isotropic bands in crayfish muscle (Fig. 2). This indicates that crayfish SCP binds more tightly to the particulate structure of muscle than parvalbumin.

C. Structure

Four domains were found in the amino acid sequence of rabbit skeletal troponin C (Collins *et al.*, 1973) and myosin alkali light chains (Weeds and McLachlan, 1974). These domains, similar in sequence, are related to those of carp parvalbumin isotype 4.25. Kretsinger (1975) postulated that the structure of the domains found in parvalbumin exists in all the intracellular calcium-binding proteins that are homologous to parvalbumin. He also proposed the term "EF hand" for the arrangement consisting of two helical regions connected by a loop of 12 amino acids with suitably located Ca^{2+}-coordinating residues. As of now, the tertiary structure of carp 4.25 parvalbumin provides the only model for interpreting the sequence data of the other calcium-binding proteins (Kretsinger, 1980). It is tempting to propose that SCP monomers with a molecular weight of 20,000–22,000 are four-domain polypeptide chains; some of their domains would have lost their calcium-binding capacity during evolution, as happened in myosin light chains, parvalbumin, etc. Kretsinger *et al.* (1980) have grown crystals of crayfish SCP, and it will be interesting to see whether this hypothesis is verified when the structure is resolved. In any case, the α-helix content calculated from circular dichroic spectrum is 51% for the calcium-saturated crayfish SCP. This corresponds to about 98 amino acids per subunit, which yields about 8 helices per monomer, the expected number for 4 domains, assuming that the helical segments contain 9–12 residues as in parvalbumin (Fig. 3) and troponin C (Nagy and Gergely, 1979).

D. Calcium and Magnesium Binding

1. Crayfish Sarcoplasmic Calcium Binding Protein

The SCP isotype α_2 binds 6 moles Ca^{2+} mole^{-1}. This is the first intracellular protein for which pronounced cooperativity in calcium binding has been demonstrated (Wnuk *et al.*, 1975). The cooperativity is positive for four sites, as indicated (Table IX) by increasing logarithmic intrinsic binding constants (8.08–8.70), and negative for the other two sites (decreasing logarithmic constants from 8.32 to 6.89). In the presence of 1 mM Mg^{2+}, the binding constants are reduced by about one order of magnitude, but cooperativity is conserved. Mg^{2+} competes with Ca^{2+} for four out of six sites on SCP (Fig. 5b). In the absence of Ca^{2+}, magnesium binds to SCP with positive cooperativity for two sites and with negative cooperativity for two others, the four logarithmic intrinsic binding constants for Mg^{2+} ranging from 4.60 to 5.00. All three SCP isotypes from crayfish have the same Ca^{2+}- and Mg^{2+}-binding properties (Wnuk, 1978). With its two Ca^{2+}-specific sites and two Ca^{2+}–Mg^{2+} sites that interact with positive cooperativity, and two Ca^{2+}–Mg^{2+} sites that interact with negative cooperativity, crayfish SCP displays metal-binding properties that are markedly different from those of parvalbumin. In this respect, crayfish SCP resembles more troponin C from striated muscle of vertebrates, which also has both Ca^{2+}-specific and Ca^{2+}–Mg^{2+} sites (Potter and Gergely, 1975; Wnuk and Stein, 1978; Holroyde *et al.*, 1980).

The conformations of the SCP isotype α_2 in the calcium- (SCP · Ca_6), magnesium- (SCP · Mg_4) and metal-free states, as well as those of transition species, have been monitored by circular dichroism in the far- and near-UV regions, tryptophan fluorescence, sulfhydryl reactivity, and tryptic susceptibility. An analysis of the Ca^{2+}- and Mg^{2+}-induced changes

TABLE IX

Binding of Ca^{2+} to Sarcoplasmic Calcium-Binding Proteins in the Presence (1 mM) and Absence of Mg^{2+}

Calcium-binding constants[a]	Crayfish SCP α_2[b]		Sandworm SCP α[c]		Amphioxus SCP[d]	
	$-Mg^{2+}$	$+Mg^{2+}$	$-Mg^{2+}$	$+Mg^{2+}$	$-Mg^{2+}$	$+Mg^{2+}$
Log K_1	8.86	7.34	8.70	6.93	8.09	6.65
Log K_2	8.75	7.36	8.23	7.11	7.49	8.30
Log K_3	8.50	7.36	7.74	7.18	—	—
Log K_4	8.58	7.32	—	—	—	—
Log K_5	7.92	6.00	—	—	—	—
Log K_6	6.11	4.85	—	—	—	—
Log K_1'	8.08	6.56	8.23	6.46	7.79	6.35
Log K_2'	8.17	6.96	8.23	7.11	7.79	8.60
Log K_3'	8.38	7.24	8.23	7.65	—	—
Log K_4'	8.70	7.45	—	—	—	—
Log K_5'	8.32	6.40	—	—	—	—
Log K_6'	6.89	5.62	—	—	—	—

[a] The calcium-binding measurements were obtained by equilibrium dialysis in combination with an EDTA or EGTA buffer system. The values of the stoichiometric binding constants K_i (M^{-1}) were computed using the nonlinear least squares curve-fitting procedure. The values of the intrinsic binding constants K_i' values were obtained from the K_i values by correction for the statistical factors.

[b] Wnuk et al., 1979.

[c] Cox et al., 1977; Cox and Stein, 1981.

[d] Kohler et al., 1978.

in the backbone circular dichroism indicates that the SCP subunit possesses three short α-helical segments (each of about 10 amino acid residues) which require calcium to become helical. One of them is unaffected by Mg^{2+}. It seems therefore that only one of the two α helices that supposedly flank each Ca^{2+}-binding loop, responds to divalent metals. This is reminiscent of the changes in α-helicity induced by Ca^{2+} in the domains containing Ca^{2+}–Mg^{2+} sites in troponin C (Nagy et al., 1978; Nagy and Gergely, 1979). When crayfish SCP is saturated by Ca^{2+}, the three calcium-sensitive helices are stable even in the presence of urea. The remaining five helical segments are not sensitive to Ca^{2+} or Mg^{2+} in the native state and are not protected by divalent cations against denaturing agents such as urea (Wnuk et al., 1981).

The extent of ellipticity attributed to the aromatic residues suggests that 8 to 9 phenylalanine residues out of 12 (per subunit) and a considerable proportion of tyrosine residues are immobilized in α-helical segments. The Ca^{2+}- and Mg^{2+}-induced changes in the circular dichroic bands of phenylalanine and tyrosine chromophores show that two phenylalanine side chains per subunit are included in the helices affected by Ca^{2+} in the domain containing the Ca^{2+}-specific site; the tyrosyl side chains are present in the Ca^{2+}- or Mg^{2+}-sensitive helices that belong to the domains containing the Ca^{2+}-specific and $Ca^{2+}-Mg^{2+}$ sites. The marked decrease in thiol reactivity with 5,5'-dithiobis (2-nitrobenzoic acid) observed when SCP binds Ca^{2+} (or Mg^{2+}) is consistent with the presence of two cysteinyl side chains out of three (per subunit) in the vicinity of the helices affected by divalent metals (Wnuk *et al.*, 1981).

The major structural changes induced by Ca^{2+} in the presence of physiological levels of Mg^{2+} ($SCP \cdot Mg_4 \rightarrow SCP \cdot Ca_6$) contrast sharply with the situation prevailing in parvalbumin where the $Ca^{2+}-Mg^{2+}$ exchange results in very limited conformational changes. As computed by an iterative procedure, at a millimolar concentration of Mg^{2+}, Ca^{2+} first binds to $SCP \cdot Mg_4$ at the two Ca^{2+}-specific sites ($SCP \cdot Mg_4Ca_2$) and then replaces Mg^{2+} at the four $Ca^{2+}-Mg^{2+}$ sites ($SCP \cdot Ca_6$). This is consistent with the shape of the calcium-binding curve as compared with that of the simultaneous release of Mg^{2+} (Fig. 5b). A comparison of the structural changes accompanying Ca^{2+} binding to SCP in the presence of 1 mM $MgCl_2$ with the fractional occupancy curves depicted in Fig. 6 indicates that the overall changes in conformation can be attributed to the binding of Ca^{2+} to the Ca^{2+}-specific sites rather than to the replacement of Mg^{2+} by Ca^{2+} at the $Ca^{2+}-Mg^{2+}$ sites. These changes are consistent with a coil–helix transition involving domains containing Ca^{2+}-specific sites and resulting in a more compact structure for the entire molecule (Wnuk *et al.*, 1981).

2. Sandworm

The SCP isotype α binds 3 moles Ca^{2+} mole^{-1}. In the absence of Mg^{2+}, the binding of Ca^{2+} does not display cooperativity and can be characterized by a single logarithmic intrinsic binding constant of 8.23 (Table IX). In the absence of Ca^{2+}, 3 moles Mg^{2+} mole^{-1} bind to the protein with positive cooperativity, as indicated by increasing logarithmic intrinsic constants from 3.60 to 4.82. Consequently, as seen in Fig. 5c, at 1 mM Mg^{2+} and at low concentration of Ca^{2+} ($\sim 10^{-8} M$), the protein is essentially saturated with Mg^{2+}. Calcium uptake causes a release of Mg^{2+}, so that SCP saturated with 3 moles Ca^{2+} mole^{-1} is free of Mg^{2+}. The protein exhibits cooperativity in Ca^{2+} binding, as seen by increasing logarithmic intrinsic constants (6.46–7.65). Thus sandworm SCP has three $Ca^{2+}-Mg^{2+}$ mixed sites, and cooperativity in the binding of Ca^{2+} is induced by Mg^{2+}. This would constitute a remarkable example of cooperativity in a monomeric protein. However, the molecular weight of the magnesium form of SCP from the sandworm is not known, and therefore monomer–

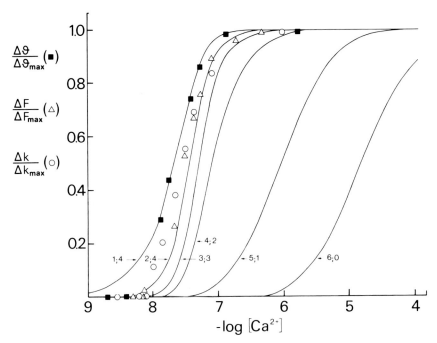

Fig. 6. Fractional occupancy curves for the crayfish SCP complexes with divalent cations compared with the calcium dependence of relative conformational changes. ■, ellipticity at 220 nm; △, tryptophan fluorescence intensity at 333 nm; ○, thiol reactivity expressed as a pseudo-first-order rate constant. The crayfish SCP isotype α_2 was used in the presence of 1 mM MgCl$_2$. The divalent cation–SCP complexes are: SCP · Ca$_1$Mg$_4$ (1;4), SCP · Ca$_2$Mg$_4$ (2;4), SCP · Ca$_3$Mg$_3$ (3;3), SCP · Ca$_4$Mg$_2$ (4;2), SCP · Ca$_5$Mg$_1$ (5;1) and SCP · Ca$_6$ (6;0).

polymer equilibria cannot be ruled out as the reason for such cooperativity. Alternatively, at least two affinities for Mg^{2+} (but not for Ca^{2+}) are observed in this protein, and binding of Ca^{2+} could displace the equilibrium between the two conformations that determine these affinities (Cox and Stein, 1981).

The α-helical contents calculated from circular dichroism are 44.6 and 41.5% for the Mg^{2+}- and the Ca^{2+}-saturated proteins, respectively, and 34% for the metal-free SCP. Taking into account that 44.6% corresponds to about 78 amino acids and assuming that the helical segments contain about 10 residues, one again obtains about eight helices per molecule. Circular dichroism measurements suggest that 13 and 18 amino acids become helical upon saturation of the protein by Ca^{2+} and Mg^{2+}, respectively (i.e., four and six residues per site, assuming equal contributions for the three sites). Near-uv circular dichroism and tryptophan fluorescence also indicate that the Mg^{2+} form of sandworm SCP is significantly more structured than the Ca^{2+} form. This contrasts with the changes in conformation of parvalbumin (which also has only Ca^{2+}–Mg^{2+} sites) that are minor and essentially unaffected by Ca^{2+}–Mg^{2+} exchange.

3. Amphioxus

The SCP binds 2 moles Ca^{2+} mole^{-1}. In the absence of Mg^{2+}, the binding of Ca^{2+} is not cooperative and is described by a single logarithmic intrinsic binding constant of 7.79 (Table IX). In the presence of 1 mM Mg^{2+}, a strong positive cooperativity appears, as indicated by increasing logarithmic intrinsic constants from 6.35 to 8.60. Mg^{2+} competes with Ca^{2+} for one of two sites (Fig. 5d). In the absence of Ca^{2+}, magnesium binds to SCP at the Ca^{2+}–Mg^{2+} site with an affinity characterized by a logarithmic constant of 4.87. Hence amphioxus SCP displays metal-binding properties intermediate between those of crayfish SCP (Ca^{2+}-specific site) and of sandworm SCP (Mg^{2+}-induced cooperativity in Ca^{2+} binding). The secondary structure of amphioxus SCP does not change significantly upon Ca^{2+} binding. Such unusual behavior among intracellular Ca^{2+}-binding proteins has been reported for brain S-100 protein (Callissano et al., 1969) and for certain myosin light chains (Stafford and Szent-Györgyi, 1978). The magnesium form of amphioxus SCP possesses a slightly higher α-helical content than the calcium form (Kohler et al., 1978).

4. Mollusks

The SCP isotypes from the Venus clam bind 1 mole Ca^{2+} mole^{-1} in the presence of 24 μM free Ca^{2+}. No magnesium binds to the calcium-free or to the calcium-containing proteins even at 1.5 mM free Mg^{2+}. The circular dichroic spectrum confirms that the clam SCP is not affected by 1 mM Mg^{2+}, whereas the presence of calcium causes a significant increase in ellipticity (Table VI). Lehman and Szent-Györgyi (1975) have reported that SCP from scallop muscle also binds 1 mole Ca^{2+} mole^{-1}.

VI. PHYSIOLOGICAL IMPLICATIONS

Since the exact role of SCPs is not known, we shall conclude this chapter with an evaluation of the regulatory potential resulting from the ability of these proteins to bind Ca^{2+} and Mg^{2+}. Two physiological functions may be exercised by parvalbumin and SCPs: (1) regulation of the Ca^{2+} and Mg^{2+} fluxes in muscle, and (2) participation in Ca^{2+}-activated processes through calcium-dependent interaction with enzymes or proteins. In the sarcoplasm, the concentration of free Ca^{2+} changes from about 0.01 μM at rest to at least 1 μM during contraction (Hasselbach,

1976); as for free Mg^{2+}, its concentration is believed to be in the range of 0.5–5 mM (Brinley et al., 1977; Gupta and Moore, 1980). Consequently, assuming that the results depicted in Fig. 5 are also valid in vivo, it is likely that at rest the Ca^{2+}-specific sites are free of divalent metals, whereas the Ca^{2+}–Mg^{2+} sites are occupied by Mg^{2+}. This view is supported by experiments where Ca^{2+} was removed from rabbit parvalbumin and crayfish SCP by sarcoplasmic reticulum and replaced by Mg^{2+} at the Ca^{2+}–Mg^{2+} sites (Wnuk et al., 1979). When the free calcium concentration rises in the sarcoplasm, the Ca^{2+}-specific sites become saturated, and the Ca^{2+}–Mg^{2+} sites exchange their Mg^{2+} for Ca^{2+}, provided that the rate of Mg^{2+} dissociation is fast enough to allow this exchange to take place during the contraction–relaxation cycle.

Important physiological consequences resulting from the binding of Ca^{2+} to Ca^{2+}-specific sites have been demonstrated in at least two cases: (1) the triggering of muscle contraction by troponin C (Potter and Gergely, 1975), and (2) the control of several enzymes by calmodulin (Klee et al., 1980; Cox et al., 1981). In both instances the biological activity results from the calcium-induced conformational changes which are transmitted to the proteins of the contractile apparatus, or to various enzymes, as the case may be. In crayfish and clam SCPs also, important structural changes occur upon Ca^{2+} binding to the Ca^{2+}-specific sites at physiological levels of Mg^{2+}. Indeed, a fraction of crayfish SCP is localized at the site of the isotropic bands in muscle and might well interact in a calcium-dependent fashion with other proteins. Alternatively, the Ca^{2+}-specific sites could modulate the rapid Ca^{2+} fluxes in muscle. Indeed, it may reasonably be assumed that the association rate of Ca^{2+} is essentially controlled by diffusion (on rate constant $= 3 \times 10^8\,M^{-1}\,sec^{-1}$; Eigen and Hammes, 1963). The off rate constant of the Ca^{2+}-specific site of crayfish, calculated using the intrinsic equilibrium constant of $6 \times 10^6\,M^{-1}$ (Table IX), equals 50 sec^{-1}, which corresponds to a $t_{1/2}$ of about 14 msec. These rates are comparable to the speed of contraction and relaxation. As a matter of fact, the Ca^{2+} transient during relaxation decays with a $t_{1/2}$ of 50 msec (Miledi et al., 1977), and the duration of contraction and relaxation in fast skeletal muscle is as short as 27 and 70 msec, respectively (Buller et al., 1960). Hence SCPs with Ca^{2+}-specific sites may influence the free Ca^{2+} concentration in muscle and increase the speed of relaxation by a fast uptake of Ca^{2+}; thereafter the sarcoplasmic reticulum would pump away calcium from SCP. In this respect, the longitudinal body muscle of amphioxus, which contains neither a T system nor sarcoplasmic reticulum (Peachey, 1961) and displays a rapid rate of contraction, constitutes a good model for the study of SCP as a soluble relaxing factor.

The physiological function of the Ca^{2+}–Mg^{2+} mixed sites seems to reside in metal exchange itself. The effect of these sites on the Ca^{2+} and Mg^{2+} balance in muscle is obvious, since the level of proteins with such sites (e.g., parvalbumins) may be as high as 1 mmole kg^{-1}. The amount of Mg^{2+} released from these proteins can reach 2–3 mM in terms of intracellular water. Hence the mixed sites may be involved in the control of magnesium transients and subsequently of magnesium-dependent enzymes (Lehky and Stein, 1979). Interestingly enough, the existence of a cellular mechanism responsible for the regulation of intracellular free Mg^{2+} has just been reported (Gupta and Moore, 1980) for frog muscle, which is rich in parvalbumin. However, the dissociation of Mg^{2+} is slow. In the case of parvalbumin, it occurs with a $t_{1/2}$ of ~60–600 msec, for $K_{Mg} = 0.1–1.2 \times 10^5 \ M^{-1}$ and a diffusion-limited on rate constant of $1.3 \times 10^5 \ M^{-1} \ sec^{-1}$ (Eigen and Hammes, 1963). A $t_{1/2}$ value of 260 msec has been obtained from kinetic studies on conformational changes in parvalbumin (Potter et al., 1978). Because of the resulting delay in Ca^{2+} binding to parvalbumin, the threshold levels of Ca^{2+} needed for activation of troponin C are reached faster.

The slow rate of Mg^{2+} dissociation from the $Ca^{2+}–Mg^{2+}$ sites raises doubts as to whether parvalbumin may constitute a soluble relaxing factor, as suggested by Pechère et al. (1975) and Gillis and Gerday (1977), i.e., whether Ca^{2+} flows from troponin C to parvalbumin during the short time of muscular relaxation. An experiment performed by Pechère et al. (1977) showed that, upon addition of Ca^{2+}, the activation of myofibrilar ATPase was indeed followed by inactivation resulting from the uptake of Ca^{2+} by parvalbumin. By means of the data of the latter authors, we have calculated from the amount of ATP hydrolyzed during transient activation of the myofibrils and from their maximal activity in the presence of Ca^{2+} that at least 25 sec would be required to displace Ca^{2+} from myofibrils to parvalbumin. Moreover, in a simulation analysis of calcium transients in muscle, Robertson and Potter (1980) have calculated that the Mg^{2+} content of parvalbumin hardly diminishes during a few muscle twitches. The dissociation of Ca^{2+} from $Ca^{2+}–Mg^{2+}$ sites is also slow. Potter et al. (1978) obtained a $t_{1/2}$ value of 1.4 sec, in agreement with the $t_{1/2} = $ ~0.3–5.7 sec calculated from K_{Ca} (Section IV,D) and from the diffusion-controlled on rate constant for Ca^{2+}. Consequently repeated stimulations of muscle are required for the saturation with Ca^{2+} of the $Ca^{2+}–Mg^{2+}$ sites. This will occur progressively in proportion to the rate and intensity of muscle activity. The slow Mg^{2+} and Ca^{2+} dissociation rates have a significant advantage for the energy economy of muscle: If $Ca^{2+}–Mg^{2+}$ exchange does not follow each muscular twitch, then less ATP is consumed in removing calcium from the sarcoplasm upon relaxation (Lehky and Stein, 1979). Thus $Ca^{2+}–Mg^{2+}$ mixed sites may be instrumental in regulation of the

$Ca^{2+}-Mg^{2+}$ balance in muscle during prolonged contraction, e.g., smooth tetanus.

In summary, SCPs possessing Ca^{2+}-specific sites can exchange Ca^{2+} with a rate comparable to that of a single twitch and have thus the potential to provide short-range modulation. Proteins with $Ca^{2+}-Mg^{2+}$ sites respond much slower and may have a long-range modulatory role in prolonged contraction.

REFERENCES

Adelstein, R. S., and Eisenberg, E. (1980). Regulation and kinetics of the actin-myosin-ATP interaction. *Annu. Rev. Biochem.* **49**, 921–956.

Atwood, H. W. (1972). Crustacean muscle. *In* "The Structure and Function of Muscle" (G. H. Bourne, ed.), 2nd ed., Vol. 1, Part 1, pp. 421–489. Academic Press, New York.

Baron, G., Demaille, J., and Dutruge, E. (1975). The distribution of parvalbumins in muscle and in other tissues. *FEBS Lett* **56**, 156–160.

Benzonana, G., and Gabbiani, G. (1978). Immunofluorescence subcellular localization of some muscle proteins: A comparison between tissue sections and isolated myofibrils. *Histochemistry* **55**, 61–76.

Benzonana, G., Capony, J.-P., and Pechère, J.-F. (1972). The binding of calcium to muscular parvalbumins. *Biochim. Biophys. Acta* **278**, 110–116.

Benzonana, G., Cox, J. A., Kohler, L. G., and Stein, E. A. (1974). Caractérisation d'une nouvelle métalloprotéine calcique du myogène de certains crustacés. *C. R. Hebd. Seances Acad. Sci.* **279**, 1491–1493.

Benzonana, G., Wnuk, W., Cox, J. A., and Gabbiani, G. (1977). Cellular distribution of sarcoplasmic calcium-binding proteins by immunofluorescence. *Histochemistry* **51**, 335–341.

Birdsall, W. J., Levine, B. A., Williams, R. J. C., Demaille, J. G., Haiech, J., and Pechère, J.-F. (1979). Calcium and magnesium binding by parvalbumin: A proton magnetic resonance spectral study. *Biochimie* **61**, 741–750.

Blum, H. E., Lehky, P., Kohler, L., Stein, E. A., and Fischer, E. H. (1977). Comparative properties of vertebrate parvalbumins. *J. Biol. Chem.* **252**, 2834–2838.

Brinley, F. J., Scarpa, A., and Tiffert, T. (1977). The concentration of ionized magnesium in barnacle muscle fibers. *J. Physiol. (London)* **266**, 545–565.

Buller, A. J., Eccles, J. C., and Eccles, R. M. (1960). Differentiation of fast and slow muscles in the cat hind limb. *J. Physiol. (London)* **150**, 399–416.

Burstein, E. A., Permyakov, E. A., Emelyanenko, V. I., Bushueva, T. L., and Pechère, J.-F. (1975). Investigation of some physico-chemical properties of muscular parvalbumins by means of the luminescence of their phenylalanyl residues. *Biochim. Biophys. Acta* **400**, 1–16.

Callisano, P., Moore, B. W., and Friesen, A. (1969). Effect of calcium ion on S-100, a protein of the nervous system. *Biochemistry* **8**, 4318–4326.

Capony, J.-P., Rydén, L., Demaille, J., and Pechère, J.-F. (1973). The primary structure of the major parvalbumin from hake muscle: Overlapping peptides obtained with chemical and enzymatic methods—The complete amino acid sequence. *Eur. J. Biochem.* **32**, 97–108.

Capony, J.-P., Demaille, J., Pina, C., and Pechère, J.-F. (1975). The amino-acid sequence of the most acidic major parvalbumin from frog muscle. *Eur. J. Biochem.* **56,** 215–227.

Capony, J.-P., Pina, C., and Pechère, J.-F. (1976). Parvalbumin from rabbit muscle: Isolation and primary structure. *Eur. J. Biochem.* **70,** 123–135.

Cavé, A., Pages, M., and Morin, P., (1979). Conformational studies on muscular parvalbumins. Cooperative binding of calcium (II) to parvalbumins. *Biochimie* **61,** 607–613.

Childers, S. R., and Siegel, F. L. (1976). Calcium-binding protein in electroplax and skeletal muscle: Comparison of the parvalbumin and phosphodiesterase activator protein of *Electrophorus electricus. Biochim. Biophys. Acta* **439,** 316–325.

Coffee, C. J., and Bradshaw, R. A. (1973). Carp muscle calcium-binding protein B. I. Characterization of the tryptic peptides and the complete amino acid sequence. *J. Biol. Chem.* **248,** 3305–3312.

Cohen, P. (1980). The role of calcium ions, calmodulin and troponin in the regulation of phosphorylase kinase. *Eur. J. Biochem.* **111,** 563–574.

Collins, J. H., Potter, J. D., Horn, M. J., Wilshire, G., and Jockman, N. (1973). The amino acid sequence of rabbit skeletal muscle troponin C: Gene replication and homology with calcium-binding proteins from carp and hake muscle. *FEBS Lett.* **36,** 268–272.

Cornish-Bowden, A. (1977). Assessment of protein sequence identity from amino acid composition data. *J. Theor. Biol.* **65,** 735–742.

Cox, J. A., and Stein, E. A. (1981). Characterization of a new sarcoplasmic calcium-binding protein with magnesium-induced cooperativity in the binding of calcium. *Biochemistry* **20,** 5430–5436.

Cox, J. A., Winge, D. R., Wnuk, W., and Stein, E. A. (1976a). Soluble calcium-binding proteins from invertebrate muscle. *Fed. Proc., Fed. Am. Soc. Exp. Biol.* **35,** 1363.

Cox, J. A., Wnuk, W., and Stein, E. A. (1976b). Isolation and properties of a sarcoplasmic calcium-binding protein from crayfish. *Biochemistry* **15,** 2613–2618.

Cox, J. A., Wnuk, W., and Stein, E. A. (1977). Regulation of calcium-binding by magnesium. *In* "Calcium-Binding Proteins and Calcium Function" (R. H. Wasserman, R. A. Corradino, E. Carafoli, R. H. Kretsinger, D. H. MacLennan, and F. L. Siegel, eds.), pp. 266–269. Elsevier/North-Holland Publ., Amsterdam and New York.

Cox, J. A., Winge, D. R., and Stein, E. A. (1979). Calcium, magnesium and the conformation of parvalbumin during muscular activity. *Biochimie* **61,** 601–605.

Cox, J. A., Malnoë, A., and Stein, E. A. (1981). Regulation of brain cyclic nucleotide phosphodiesterase by calmodulin: A quantitative analysis. *J. Biol. Chem.* **256,** 3218–3222.

Deuticke, H. J. (1934). Uber die Sedimentationskonstante von Muskelproteine. *Hoppe-Seyler's Z. Physiol. Chem.* **224,** 216–228.

Donato, H., and Martin, R. B. (1974). Conformations of carp muscle calcium-binding parvalbumin. *Biochemistry* **13,** 4575–4579.

Drakenberg, T., Lindman, B., Cavé, A., and Parello, J. (1978). Non-equivalence of the CD and EF sites of muscular parvalbumins: A ^{113}Cd-NMR study. *FEBS Lett.* **92,** 346–350.

Eigen, M., and Hammes, G. G. (1963). Elementary steps in enzyme reactions. *Adv. Enzymol.* **25,** 1–38.

Elsayed, S., and Bennich, H. (1975). The primary structure of allergen M from cod. *Scand. J. Immunol.* **4,** 203–208.

Enfield, D. L., Ericsson, L. H., Blum, H. E., Fischer, E. H., and Neurath, H. (1975). Amino-acid sequence of parvalbumin from rabbit skeletal muscle. *Proc. Natl. Acad. Sci. U.S.A.* **72**, 1309–1313.

Focant, B., and Pechère, J.-F. (1965). Contribution à l'étude de protéines de faible poids moléculaire des myogènes de vertébrés inférieurs. *Arch. Int. Physiol. Biochim.* **73**, 334–354.

Gerday, C., and Teuwis, J.-C. (1972). Isolation and characterization of the main parvalbumins from *Raja clarata* and *Raja montagni* white muscles. *Biochim. Biophys. Acta* **271**, 320–331.

Gerday, C., Collins, S., and Piron, L. A. (1978). Phylogenetic relationships between Cyrinidae parvalbumins. II. The amino acid sequence of the parvalbumin V of chub (*Leviseus cephalis*). *Comp. Biochem. Physiol. B* **61B**, 459–461.

Gillis, J. M., and Gerday, C. (1977). Calcium movement between myofibrils, parvalbumins and sarcoplasmic reticulum in muscle. *In* "Calcium-Binding Proteins and Calcium Function" (R. H. Wasserman, R. A. Corradino, E. Carafoli, R. J. Kretsinger, D. H. MacLennan, and F. L. Siegel, eds.), pp. 193–196. Elsevier/North-Holland Publ., Amsterdam and New York.

Gillis, J. M., Piront, A., and Gosselin-Rey, C. (1979). Parvalbumins: Distribution and physical state inside the muscle cell. *Biochim. Biophys. Acta* **585**, 444–450.

Goodman, M., Pechère, J.-F., Haiech, J., and Demaille, J. G. (1979). Evolutionary diversification of structure and function in the family of intracellular calcium-binding proteins. *J. Mol. Evol.* **13**, 331–352.

Gosselin-Rey, C. (1974). Fish parvalbumins: Immunochemical reactivity and biological distribution. *In* "Calcium-Binding Proteins" (W. Drabikowski, H. Strzelecka-Golaszewka, and E. Carafoli, eds.), pp. 679–707. Elsevier/North-Holland Publ., Amsterdam and New York.

Gosselin-Rey, C., and Gerday, C. (1977). Parvalbumin from frog skeletal muscle (*Rana temporaria* L.): Isolation and characterization, structural modification associated with calcium-binding. *Biochim. Biophys. Acta* **492**, 53–63.

Gosselin-Rey, C., Piront, A., and Gerday, C. (1978). Polymorphism of parvalbumins and tissue distribution: Characterization of component I, isolated from red muscles of *Cyprinus carpio* L. *Biochim. Biophys. Acta* **532**, 294–304.

Gupta, R. K., and Moore, R. D. (1980). ^{31}P-NMR studies of intracellular free Mg^{2+} in intact frog skeletal muscle. *J. Biol. Chem.* **255**, 3987–3992.

Haiech, J., Derancourt, J., Pechère, J.-F., and Demaille, J. G. (1979a). A new large-scale purification procedure for muscular parvalbumins. *Biochimie* **61**, 583–587.

Haiech, J., Derancourt, J., Pechère, J.-F., and Demaille, J. G. (1979b). Magnesium and calcium binding to parvalbumins: Evidence for differences between parvalbumins and an explanation of their relaxing function. *Biochemistry* **13**, 2752–2758.

Hamoir, B., Piront, A., Gerday, C., and Dande, P. R. (1973). Muscle proteins of the coelacanth *Latimeria chalumnae*. *J. Mar. Biol. Assoc. U.K.* **53**, 763–784.

Harafuji, H., and Ogawa, Y. (1980). Re-examination of the apparent binding constant of ethylene glycol bis(β-aminoethyl ether)-N,N,N',N'-tetraacetic acid with calcium around neutral pH. *J. Biochem. (Tokyo)* **87**, 1305–1312.

Hasselbach, W. (1976). Release and uptake of calcium by the sarcoplasmic reticulum. *In* "Molecular Basis of Motility" (L. M. G. Heilmeyer, J. C. Rüegg, and T. Wieland, eds.), pp. 81–92. Springer-Verlag, Berlin and New York.

Heizmann, C. W., and Strehler, E. E. (1979). Chicken parvalbumin: Comparison with parvalbumin-like protein and three other components (M_r = 8,000 to 13,000). *J. Biol. Chem.* **254**, 4296–4303.

Heizmann, C. W., Malencik, D. A., and Fischer, E. H. (1974). Generation of parvalbumin-like proteins from troponin. *Biochem. Biophys. Res. Commun.* **57**, 162–168.

Henrotte, J. G. (1955). A crystalline component of carp myogen precipitating at high ionic strength. *Nature (London)* **176**, 1221.

Hitchcock, S. E., and Kendrick-Jones, J. (1975). Myosin light chains, carp calcium binding proteins and troponin components: Do they interact to form functional complexes? *In* "Calcium Transport in Contraction and Secretion" (E. Carafoli, F. Clementi, W. Drabikowski, and A. Margreth, eds.), pp. 447–458. North-Holland Publ., Amsterdam.

Holroyde, M. J., Robertson, S. P., Johnson, J. D., Solaro, R. J., and Potter, J. D. (1980). Characterization of the Ca^{2+} binding properties of bovine cardiac troponin and troponin C. *Fed. Proc., Fed. Am. Soc. Exp. Biol.* **39**, 1620.

Jauregui-Adell, J., and Pechère, J.-F. (1978). Parvalbumins from coelacanth muscle. I. General survey. *Biochim. Biophys. Acta* **536**, 263–268.

Klee, C. B., Crouch, T. H., and Krinks, M. H. (1979). Calcineurin: A calcium- and calmodulin-binding protein of the nervous system. *Proc. Natl. Acad. Sci. U.S.A.* **76**, 6270–6273.

Klee, C. B., Crouch, T. H., and Richman, P. G. (1980). Calmodulin. *Annu. Rev. Biochem.* **49**, 489–515.

Kohler, L., Cox, J. A., and Stein, E. A. (1978). Sarcoplasmic calcium-binding proteins in protochordate and cyclostome muscle. *Mol. Cell. Biochem.* **20**, 85–93.

Konosu, S., Hamoir, G., and Pechère, J.-F. (1965). Carp myogen of white and red muscles: Properties and amino acid composition of the main low molecular-weight components of white muscle. *Biochem. J.* **96**, 98–112.

Kretsinger, R. H. (1972). Gene triplication deduced from the tertiary structure of a muscle calcium binding protein. *Nature (London), New Biol.* **240**, 85–88.

Kretsinger, R. H. (1975). Hypothesis: Calcium modulated proteins contain EF hands. *In* "Calcium Transport in Contraction and Secretion" (E. Carafoli, F. Clementi, W. Drabikowski, and A. Margreth, eds.), pp. 469–478. North-Holland Publ., Amsterdam.

Kretsinger, R. H. (1980). Structure and evolution of calcium modulated proteins. *CRC Crit. Rev. Biochem.* **8**, 119–174.

Kretsinger, R. H., and Nockolds, C. E. (1973). Carp muscle calcium-binding protein. II. Structure determination and general description. *J. Biol. Chem.* **248**, 3313–3326.

Kretsinger, R. H., Rudnick, S. E., Smeden, D. A., and Schatz, V. B. (1980). Calmodulin, S-100 and crayfish sarcoplasmic calcium-binding protein crystals suitable for X-ray diffraction studies. *J. Biol. Chem.* **255**, 8154–8156.

Le Donne, N. C., and Coffee, C. J. (1979). Inability of parvalbumin to function as a calcium-dependent activator of cyclic nucleotide phosphodiesterase activity. *J. Biol. Chem.* **254**, 4317–4320.

Lehky, P., and Stein, E. A. (1979). Perch muscle parvalbumin: General characterization and magnesium-binding properties. *Comp. Biochem. Physiol. B* **63B**, 253–259.

Lehky, P., Blum, H. E., Stein, E. A., and Fischer, E. H. (1974). Isolation and characterization of parvalbumins from the skeletal muscle of higher vertebrates. *J. Biol. Chem.* **249**, 4332–4334.

Lehky, P., Comte, M., Fischer, E. H., and Stein, E. A. (1977). A new solid-phase chelator with high affinity and selectivity for calcium: Parvalbumin-polyacrylamide. *Anal. Biochem.* **82**, 158–169.

Lehman, W., and Szent-Györgyi, A. G. (1975). Regulation of muscular contraction: Distribution of actin control and myosin control in the animal kingdom. *J. Gen. Physiol.* **66**, 1–30.

Le Peuch, C. J., Demaille, J. G., and Pechère. J.-F. (1978). Radioelectrophoresis—A specific microassay for parvalbumins: Application to muscle biopsies from man and other vertebrates. *Biochim. Biophys. Acta* **537**, 152–159.

Le Peuch, C. J., Ferraz, C., Walsh, M. P., Demaille, J. G., and Fischer, E. H. (1979). Calcium and cyclic nucleotide dependent regulatory mechanisms during development of chick embryo skeletal muscle. *Biochemistry* **18**, 5267–5273.

Levitt, M., and Greer, J. (1977). Automatic identification of secondary structure in globular proteins. *J. Mol. Biol.* **114**, 181–239.

MacManus, J. P. (1980). The purification of a unique calcium-binding protein from Morris hepatoma 5123 tc. *Biochim. Biophys. Acta* **621**, 296–304.

Miledi, R., Parker, I., and Schalow, G. (1977). Measurement of calcium transients in frog muscle by the use of arsenazo(III). *Proc. R. Soc. London, Ser. B* **198**, 201–210.

Moeschler, H. J., Schaer, J.-J. and Cox, J. A. (1980). A thermodynamic analysis of the binding of calcium and magnesium ions to parvalbumin. *Eur. J. Biochem.* **111**, 73–78.

Moews, P. G., and Kretsinger, R. H. (1975). Refinement of the structure of carp muscle calcium binding parvalbumin by model building and difference Fourier analysis. *J. Mol. Biol.* **91**, 201–228.

Nagy, B., and Gergely, J. (1979). Extent and localization of conformational changes in troponin C caused by calcium binding. *J. Biol. Chem.* **254**, 12732–12737.

Nagy, B., Potter, J. D., and Gergely, J. (1978). Calcium-induced conformational changes in a cyanogen bromide fragment of troponin C that contains one of the binding sites. *J. Biol. Chem.* **253**, 5971–5974.

Nelson, D. J., Opella, S. J., and Jardetzky, O. (1976). ^{13}C Nuclear magnetic resonance study of molecular motions and conformational transitions in muscle calcium binding parvalbumins. *Biochemistry* **15**, 5552–5560.

Parello, J., and Pechère, J.-F. (1971). Conformational studies on muscular parvalbumins. I. Optical rotatory dispersion and circular dichroism analysis. *Biochimie* **53**, 1079–1083.

Parello, J., Lilja, H., Cavé, A., and Lindman, B. (1978). A ^{43}Ca-NMR study of the binding of calcium to parvalbumins. *FEBS Lett.* **87**, 191–195.

Peachey, L. D. (1961). Structure of the longitudinal body muscles of amphioxus. *J. Biophys. Biochem. Cytol.* **10**, Suppl., 159–176.

Pechère, J.-F. (1974). Isolement d'une parvalbumine du muscle de lapin. *C. R. Hebd. Seances Acad. Sci.* **278**, 2577–2579.

Pechère, J.-F., and Capony, J.-P. (1969). A comparison at the peptide level of muscular parvalbumins from several lower vertebrates. *Comp. Biochem. Physiol.* **28**, 1089–1102.

Pechère, J.-F., Demaille, J., and Capony, J.-P. (1971a). Muscular parvalbumins: Preparative and analytical methods of general applicability. *Biochim. Biophys. Acta* **236**, 391–408.

Pechère, J.-F., Capony, J.-P., and Rydén, L. (1971b). The primary structure of the major parvalbumin from hake muscle: Isolation and general properties of the protein. *Eur. J. Biochem.* **23**, 421–428.

Pechère, J.-F., Capony, J.-P., and Demaille, J. G. (1973). Evolutionary aspects of the structure of muscular parvalbumins. *Syst. Zool.* **22**, 533–548.

Pechère, J.-F., Demaille, J., Capony, J.-P., Dutruge, E., Baron, F., and Pina, C. (1975). Muscular parvalbumins: Some explorations into their possible biological significance. *In* "Calcium Transport in Contraction and Secretion" (E. Carafoli, F. Clementi, W. Drabikowski, and A. Margreth, eds.), pp. 459–468. North-Holland Publ., Amsterdam.

Pechère, J.-F., Derancourt, J., and Haiech, J. (1977). The participation of parvalbumins in the activation-relaxation cycle of vertebrate fast skeletal muscle. *FEBS Lett.* **75**, 111–114.

Permyakow, E. A., Yarmolenko, V. V., Emelyanenko, V. I., and Burstein, E. A. (1980). Fluorescence studies of the calcium binding to whiting (*Gadus merlangus*) parvalbumin. *Eur. J. Biochem.* **109**, 307–315.

Perry, S. V. (1979). The regulation of contractile activity in muscle. *Biochem. Soc. Trans.* **7**, 593–617.

Potter, J. D., and Gergely, J. (1975). The calcium and magnesium binding sites on troponin and their role in the regulation of myofibrillar adenosine triphosphatase. *J. Biol. Chem.* **250**, 4628–4633.

Potter, J. D., Johnson, J. D., Dedman, J. R., Schreiber, W. E., Mandel, F., Jackson, R. L., and Means, A. R. (1977). *In* "Calcium-Binding Proteins and Calcium Function" (R. H. Wasserman, R. A. Corradino, E. Carafoli, R. H. Kretsinger, D. H. MacLennan, and F. L. Siegel, eds.), pp. 266–269. Elsevier/North-Holland Publ., Amsterdam and New York.

Potter, J. D., Johnson, J. D., and Mandel, F. (1978). Fluorescence stopped flow measurements of Ca²⁺ and Mg²⁺ binding to parvalbumin. *Fed. Proc., Fed. Am. Soc. Exp. Biol.* **37**, 1608.

Robertson, S. P., and Potter, J. D. (1980). On the role of various myofibrillar metal binding sites in the Ca²⁺ regulation of vertebrate striated muscle. *Fed. Proc., Fed. Am. Soc. Exp. Biol.* **39**, 1621.

Sowadski, J., Cornick, G., and Kretsinger, R. H. (1978). Terbium replacement of calcium in parvalbumin. *J. Mol. Biol.* **124**, 123–132.

Stafford, W. F., and Szent-Györgyi, A. G. (1978). Physical characterization of myosin light chains. *Biochemistry* **17**, 607–614.

Taylor, E. W. (1979). Mechanism of actomyosin ATPase and the problem of muscle contraction. *CRC Crit. Rev. Biochem.* **6**, 103–164.

Thatcher, D. R., and Pechère, J.-F. (1977). The amino-acid sequence of the major parvalbumin from thornback-ray muscle. *Eur. J. Biochem.* **75**, 121–132.

Valentine, J. W. (1980). L'origine des grands groupes d'animaux. *Recherche* **11**, 666–674.

Weeds, A. G., and McLachlan, A. D. (1974). Structural homology of myosin alkali light chains, troponin C and carp calcium-binding protein. *Nature* (*London*) **252**, 646–649.

Wnuk, W. (1978). Polymorphism in sarcoplasmic Ca-binding proteins from crustaceans. *Experientia* **34**, 919.

Wnuk, W., and Cox, J. A. (1978). Homology between myosin light chains and invertebrate sarcoplasmic Ca-binding proteins. *Experientia* **34**, 920.

Wnuk, W., and Stein, E. A. (1978). Evolution of the Ca-binding properties of troponin C. *Experientia* **34**, 920.

Wnuk, W., Cox, J. A., Kohler, L. G., and Stein, E. A. (1979). Calcium- and magnesium-binding properties of a high affinity calcium-binding protein from crayfish sarcoplasm. *J. Biol. Chem.* **254**, 5284–5289.

Wnuk, W., Cox, J. A., and Stein, E. A. (1981). Structural changes induced by calcium and magnesium in a high affinity calcium-binding protein from crayfish sarcoplasm. *J. Biol. Chem.* **256** (In press).

Wnuk, W., Kohler, L. G., Cox, J. A., and Benzonana, G. (1975). Cooperative binding of calcium by a protein from crayfish muscle. *Experientia* **31**, 725.

Chapter 9

Myosin Light Chain Kinase in Skinned Fibers

W. GLENN L. KERRICK

I. Introduction 279
  A. *In Vitro* Evidence for a Light Chain Kinase–Phosphatase
    System 280
  B. *In Vivo* Evidence for a Light Chain Kinase–Phosphatase
    System 281
  C. Postulated Physiological Role 281
II. Usefulness of Skinned Fibers as a Model for Contraction 283
III. Evidence for a Light Chain Kinase–Phosphatase System
  in Skinned Fibers 285
  A. Correlations between Myosin Light Chain Phosphorylation
    and Contraction 285
  B. The Effect of Modulators of Light Chain Kinase–Phosphatase
    Activity on Contraction 286
IV. Summary 292
  References 293

I. INTRODUCTION

The work reviewed in this chapter is primarily concerned with the use of skinned muscle fibers to elucidate the physiological role of myosin light chain kinase in the regulation of muscle contraction. The skinned muscle fibers referred to are pieces or bundles of cells which have had their sarcolemma rendered nonfunctional. The ionic environment surrounding the contractile and regulatory proteins therefore can be controlled, and protein phosphorylation as well as the physiological variable, tension, can be measured. The use of skinned muscle fiber preparations in our labora-

CALCIUM AND CELL FUNCTION, VOL. II

tory has been an extremely useful tool in deciphering the role of Ca^{2+} and phosphorylation in the regulation of muscle contraction. Before discussing skinned fibers it is necessary to review the evidence for a light chain kinase–phosphatase system being involved in the regulation of muscle contraction.

A. *In Vitro* Evidence for a Light Chain Kinase–Phosphatase System

The *in vitro* evidence that a myosin light chain kinase–phosphatase system is involved in the regulation of muscle contraction has been well-documented in several recent reviews (Adelstein and Eisenberg, 1980; Bárány and Bárány, 1980; Stull *et al.*, 1980). Therefore, only a brief outline of this evidence will be presented in this chapter. Myosin is a molecule composed of two high-molecular-weight subunits called myosin heavy chains and four low-molecular-weight subunits called myosin light chains. Two of the myosin light chains from both skeletal and smooth muscle can be phosphorylated by a calcium-sensitive enzyme composed of two subunits—calmodulin and myosin light chain kinase.

Although myosin light chain kinase was originally isolated from skeletal muscle (Perrie *et al.*, 1972), indications of its physiological role in muscle contraction were discovered when phosphorylation of the myosin light chains from platelets was shown to facilitate the activity of platelet actomyosin ATPase (Adelstein and Conti, 1975). It was then discovered that in gizzard smooth muscle a calcium-sensitive light chain kinase could be isolated which phosphorylated myosin light chains, resulting in the activation of actomyosin ATPase (Sobieszek, 1977). Further work by many investigators substantiated this finding, and subsequent studies showed that a calcium–calmodulin–light chain kinase (Ca^{2+}–CaM–LCK) complex was required for activation of the enzyme (Dabrowska *et al.*, 1978). Furthermore, it was shown that irreversible phosphorylation (phosphorylation in the absence of a phosphatase) or thiophosphorylation of gizzard myosin light chains resulted in actomyosin activation even in the absence of Ca^{2+} (Sherry *et al.*, 1978). Similar studies show that a light chain kinase requiring Ca^{2+} and calmodulin for activation can be isolated from skeletal (Yagi *et al.*, 1978) or cardiac (Wolf and Hofmann, 1980) muscle and that these kinases will phosphorylate one myosin light chain of each muscle. However, phosphorylation of these myosin light chains of striated muscle, with one exception (Pemrick, 1980), has not been shown to affect the actomyosin kinetics (Stull *et al.*, 1980). Further studies using actomyosin threads and measuring the contractile properties of these threads under conditions of nonphosphorylated and phosphorylated myosin light chains show no differences in the forces these threads can develop or the Ca^{2+} requirements for activation (Stull *et al.*, 1980). Isolation and characterization of the phosphatases responsible for the dephosphorylation of myosin light chains has been difficult, but reports on such phosphatases for skeletal (Morgan *et al.*, 1976) and smooth muscle (Pato and Adelstein, 1980) exist and indicate that they are relatively specific for myosin light chains.

B. *In Vivo* Evidence for a Light Chain Kinase–Phosphatase System

In vivo evidence for a myosin light chain kinase–phosphatase system is excellent. Skeletal (Bárány and Bárány, 1977; Stull and High, 1977) and cardiac muscle (Frearson *et al.*, 1976; Kopp and Bárány, 1979) can be shown to exhibit phosphorylation of myosin light chains during activation. However, it is clear from later studies that relaxation of tetanized skeletal muscle occurs long before dephosphorylation of myosin light chains (Manning and Stull, 1979). In fact, phosphorylation of light chains continues to increase following the cessation of a tetanus. These studies by Stull and co-workers show a strong correlation between posttetanic potentiation in skeletal muscle and myosin light chain phosphorylation, suggesting that some unknown mechanism involving myosin light chain phosphorylation may be responsible for this potentiation.

Studies using arterial smooth muscle strips show a strong correlation between contraction and phosphorylation of myosin light chains (Barron *et al.*, 1980). In contrast to skeletal muscle, dephosphorylation of light chains occurs concomitantly with relaxation, and drugs known to inhibit the biological activity of calmodulin inhibit smooth muscle contraction and phosphorylation of these light chains.

C. Postulated Physiological Role

It is the evidence from the *in vitro* and *in vivo* studies that has led to the postulated physiological roles of the myosin light chain kinase–phosphatase system in muscle contraction. In the most clearly defined physiological role for this myosin light chain kinase–phosphatase system it functions as a Ca^{2+} switch for contractile protein interaction and contraction in smooth muscle. In addition, myosin light chain phosphorylation via some unknown mechanism may alter actomyosin kinetics in skeletal muscle in such a way as to potentiate the active force a muscle may develop during a twitch. As indicated earlier, data on this point are sketchy and ambiguous.

As has been shown for other enzyme systems, myosin light chain kinase can also serve as a substrate for cyclic AMP-dependent protein kinase (Adelstein *et al.*, 1978). Phosphorylation of myosin light chain kinase results in a decrease in the association constant for light chain kinase with the calcium–calmodulin (Ca^{2+}–CaM) complex (Conti and Adelstein, 1980). This inhibition of formation of the myosin light chain kinase Ca^{2+}–CaM complex by phosphorylation has been shown to occur in both smooth (Adelstein *et al.*, 1978) and cardiac muscle (Conti and Adelstein, 1980). This demonstrated *in vitro* mechanism for regulating myosin light chain kinase activity by cAMP-dependent protein kinase strongly suggests that modulation of muscle contraction by β-adrenergic stimulation could occur through the myosin light chain kinase–phosphatase system. Crude *in vitro* actomyosin preparations from aortic muscle show that phosphorylation of a high-molecular-weight protein believed to be myosin light chain kinase by cAMP-dependent protein kinase results in a decrease in the Ca^{2+} sensitivity of actomyosin ATPase and light chain phosphorylation (Silver and DiSalvo, 1979). Although phosphorylation of myosin light chain kinase in cardiac muscle (Kopp and Bárány, 1979) can be demonstrated, its physiological role is undefined.

A model for the regulation of smooth muscle contraction by a calcium–calmodulin–light chain kinase–phosphatase (Ca^{2+}–CaM–LCK–P) system based on *in vitro* and *in vivo* studies is depicted in Fig. 1. This model has many unique and predictable properties relative to direct Ca^{2+} control systems postulated for the activation of muscle contraction which are not regulated by calmodulin, light chain kinase, and phosphatase. Predictions for this proposed light chain kinase–phosphatase system are that (1) there should be a close correlation between phosphorylation and contraction, (2) irreversible phosphorylation or thiophosphorylation should result in contraction even in the absence of Ca^{2+}, (3) increasing either calmodulin or light chain kinase or both should result in less Ca^{2+} being required for a given level of myosin light chain phosphorylation and contraction, (4) inhibitors which would interfere with the biological activity of calmodulin, such as phenothiazines, troponin I, and an inhibitor of the Ca^{2+}–CaM complex and light chain kinase interaction such as the catalytic subunit of cyclic AMP-dependent protein kinase, should inhibit phosphorylation of myosin light chains and contraction of the muscle, and (5) the use of varying concentrations of ATP or the use of ATP analogues believed not to be used by the myosin light chain kinase as a substrate should affect light chain phosphorylation and tension in a predictable manner. In contrast, these predictions for a light chain kinase–phosphatase mechanism are not consistent with the proposed direct Ca^{2+}-binding mechanisms, with the possible exception of troponin I phosphorylation by the catalytic subunit

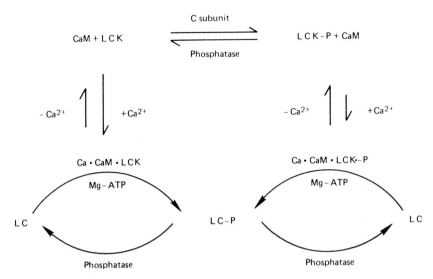

Fig. 1.   Model of the regulation of myosin light chain phosphorylation. CaM, Calmodulin; LCK, myosin light chain kinase; LCK-P, phosphorylated myosin light chain; LC, myosin light chain; LC-P, phosphorylated myosin light chain; C subunit, catalytic subunit of cAMP-dependent protein kinase.

of cAMP-dependent protein kinase reported to affect the Ca^{2+} sensitivity of the troponin–tropomyosin regulatory system for striated muscle contraction.

II. USEFULNESS OF SKINNED FIBERS AS A MODEL FOR CONTRACTION

Although considerable *in vitro* and *in vivo* evidence exists suggesting a light chain kinase–phosphatase system is responsible for the activation of smooth muscle, another control system has been shown *in vitro* to regulate smooth muscle actomyosin interactions (Mikawa *et al.*, 1978). This regulatory system has been postulated to regulate Ca^{2+}-activated contraction in smooth muscle. It consists of two subunits, leiotonin A and leiotonin C, and is believed to function in a manner similar to troponin in the Ca^{2+} regulation of contraction in striated muscle. This postulated direct Ca^{2+}-binding control system does not involve phosphorylation. The existence of two regulatory systems in the same muscle raises the question as to the relative importance of these control systems *in vivo*. Also, since a light chain kinase–phosphatase system has been shown to operate both *in vivo* and *in vitro* in mammalian striated muscles (Bárány and Bárány, 1980; Adelstein and Eisenberg, 1980; Stull *et al.*, 1980) believed to be regulated by the direct Ca^{2+} control system, troponin, it seems important to use a contractile model system which lies somewhere between purified contractile and regulatory protein systems and intact *in vivo* preparations in order to better understand the relative importance of the possible control systems in the regulation of muscle contraction. For this purpose we have made use of skinned muscle fiber preparations (no functional sarcolemma) which allow the ionic environment surrounding the contractile and regulatory proteins to be controlled while at the same time tension and phosphorylation can be measured.

Skinned muscle fibers for smooth muscle can be prepared in several ways such as using staphylococcal α-toxin (Cassidy *et al.*, 1979), non-ionic detergent Triton X-100 (Kerrick *et al.*, 1981a), saponin (Saida and Nonomura, 1978), and a combination of chemical and mechanical disruptions of the membranes by light homogenization (Kerrick and Krasner, 1975; Hoar *et al.*, 1979). Pieces of skinned single skeletal muscle cells (Kerrick and Krasner, 1975) or bundles of cardiac (Best *et al.*, 1977) or smooth muscle cells are inserted into a tension transducer and immersed in solutions of varying ionic, protein, drug, ATP, and ATP analogue concentrations, while at the same time tension and phosphorylation levels are monitored. Phosphorylation levels are determined by quickly immersing the fibers in boiling sodium dodecyl sulfate (SDS) and either identifying the degree of myosin light chain phosphorylation by a combination of SDS and isoelectric focusing polyacrylamide gel electrophoresis or determining the ^{32}P incorporation into the light chain from $[\gamma\text{-}^{32}P]$ATP chains by either liquid scintillation counting or autoradiography. The advantage of skinned fiber preparations in contrast to *in vivo* preparations is that they allow the ionic and protein environment surrounding the contractile and regulatory proteins to be controlled without interference of the sarcolemma or internal membrane structures such as sarcoplasmic reticulum and mitochondria. In contrast to purified protein systems, which discard all proteins except those of direct interest, skinned fiber preparations are relatively structurally intact, losing only soluble proteins that are able to diffuse out of the fiber and which can be added back to the fiber if desired. In addition, the physiological variable, tension, is measured in contrast to biochemical measures such as ATPase activity or superprecipitation. Such preparations readily contract and relax in the presence and absence of Ca^{2+} in solutions containing physiological levels of ATP and salt.

We have used skinned muscle fiber preparations to test the hypothesis that a light chain kinase–phosphatase system is involved in the regulation of muscle contraction in various muscle types. In carrying out these studies we have used muscles of different types [skeletal, cardiac, and smooth muscle (Cassidy *et al.*, 1979, 1980; Hoar *et al.*, 1979; Kerrick *et al.*, 1980, 1981a,b,c)], as well as muscles from different lines of evolution (Kerrick and Bolles, 1981). Our criteria in using a skinned muscle fiber preparation is that it must contract in the presence and relax in the absence of Ca^{2+}. Our underlying assumption is that the Ca^{2+} control protein responsible for the regulation of contraction in these fiber types are present and that predictions based on a light chain kinase–phosphatase system model regarding its physiological function (tension) and myosin light chain phosphorylation can be tested. By the use of agents known to affect a light chain kinase–phosphatase system *in vitro* we can test for the predicted results in skinned muscle fiber preparations.

III. EVIDENCE FOR A LIGHT CHAIN KINASE–PHOSPHATASE SYSTEM IN SKINNED FIBERS

A. Correlations between Myosin Light Chain Phosphorylation and Contraction

The first prediction from the model we tested was that there should be a close correlation between myosin light chain phosphorylation and activation of the fibers. Since most of the *in vitro* work had been carried out on chicken gizzard, we used bundles of skinned chicken gizzard muscle fibers and compared the relationship between Ca^{2+}-activated tension and phosphorylation of myosin light chains (Hoar *et al.*, 1979). This work showed that, in the presence of Ca^{2+} contraction and phosphorylation of myosin light chains occurred, and in the absence of Ca^{2+} the fibers relaxed concomitantly with dephosphorylation of myosin light chains. We also tested the relationship between varying degrees of Ca^{2+}-activated tension and phosphorylation levels of myosin light chains and again found a good correlation (Hoar *et al.*, 1979). In addition, we found a good correlation between the percentage of maximum Sr^{2+}-activated tension and the percentage of maximum Sr^{2+}-activated phosphorylation, although substantially higher concentrations of this cation than of Ca^{2+} were required to activate the fibers to the same degree. Similar studies were carried out on rabbit pulmonary artery (Kerrick *et al.*, 1980, 1981c) and shown to have the same close correlations between the percentage of maximum Ca^{2+}- and Sr^{2+}-activated tension and phosphorylation of myosin light chains.

B. The Effect of Modulators of Light Chain Kinase–Phosphatase Activity on Contraction

1. Irreversible Thiophosphorylation of Myosin Light Chains Using ATPγS

Our next goal was to phosphorylate myosin light chains of skinned muscle fiber preparations in the absence of a phosphatase and to show that the fibers remained activated when Ca^{2+} was removed from the solutions. Removal of the phosphatase activity in our preparations proved not to be practical, so we decided to make use of the ATP analogue ATPγS, which had been used initially by Gratecos and Fischer (1974) as a substrate for phosphorylation of phosphorylase, since the thiophosphorylated protein was resistant to the action of the phosphorylase phosphatase. Concomitant with our studies, experiments were being carried out in D. Hartshorne's laboratory showing that ATPγS could be used as a substrate by smooth muscle myosin light chain kinase to thiophosphorylate irreversibly gizzard myosin light chains (Sherry et al., 1978). Treatment of our skinned smooth muscle fiber preparations with this ATP analogue proved to be one of the most useful tests for a light chain kinase–phosphatase system. We were able to show that, in the presence of Ca^{2+} and ATPγS, irreversible thiophosphorylation of myosin light chains of chicken gizzard (Hoar et al., 1979), rabbit ileum (Cassidy et al., 1979), and rabbit pulmonary artery (Kerrick et al., 1980) occurred, but not in the absence of Ca^{2+}. Following removal of Ca^{2+} from the fibers after pretreatment with high Ca^{2+} and ATPγS, the tension in the fibers was irreversible activated and the only protein to remain thiophosphorylated in the fibers was the myosin light chains. This irreversible activated tension was unique to the use of ATPγS, since other thio reducing agents did not cause similar effects (Cassidy et al., 1979; Kerrick et al., 1980). The irreversible activated tension was not a rigor type, since the fibers could be made to relax, using the ATP analogue α, β-methylene ATP to dissociate actin and myosin, and then to contract and shorten again when immersed in ATP (Cassidy et al., 1979). In addition, redevelopment of tension and shortening of the muscle occurred after quickly releasing the muscle to a shorter length.

One of the most interesting results from these experiments was that the maximum force occurred when myosin light chains were only 10–20% phosphorylated (Kerrick et al., 1980). However, when these smooth muscle fibers were exposed to ATPγS and high Ca^{2+}, myosin light chains could be 80–90% thiophosphorylated (Cassidy et al., 1979, Kerrick et al., 1980). In spite of this difference in maximum phosphorylation and thiophosphorylation levels, the maximum Ca^{2+}-activated tensions in both cases are essentially the same. This suggested to us that only 10–20% phosphorylation of myosin light chains was required for maximum Ca^{2+}-activated tension. In order to test this hypothesis we exposed skinned smooth muscle fiber preparations to high Ca^{2+} and ATPγS for various time periods in order to vary the amount of thiophosphorylation of myosin light chains and to compare the resulting tension with the maximal Ca^{2+}-activated tension (Kerrick et al., 1980). We found results similar to those obtained with ATP: 10–20% thiophosphorylation of myosin light chains resulted in maximal activation of the fibers. We further manipulated the phosphorylation levels of the light chains for a given level of Ca^{2+} by adding exogenous calmodulin and again verified our previous results that low levels of phosphorylation resulted in maximal activation of the fibers (Kerrick et al., 1981b). The low level of phosphorylation of myosin light chains required for maximum activation of smooth muscle fiber preparations suggests a high degree of cooperativity between phosphorylation and activation of actomyosin interaction.

Experiments similar to those described above for vertebrate smooth muscle were carried out simultaneously in our laboratory on vertebrate striated muscle such as slow twitch, fast twitch, and cardiac (Kerrick et al., 1980, 1981a,b,c). In these muscles we found no correlation between phosphorylation of myosin light chains and the Ca^{2+}-activated tension. Pretreatment with high Ca^{2+} and ATPγS resulted in essentially no thiophosphorylation of myosin light chains, and did not affect in any way the Ca^{2+}-activated tension following this pretreatment. Although we saw a very small turnover of phosphate in myosin light chains of cardiac and slow-twitch skinned muscle fibers, it amounted to less than 0.5% of the total phosphate in the light chains and was not correlated with the Ca^{2+}-activated tension. Our conclusion from these experiments is that, although vertebrate striated muscle is regulated by Ca^{2+}, there is no evidence using skinned muscle fiber preparations that this Ca^{2+}-activated tension is switched on and off by a myosin light chain kinase–phosphatase system. In fact, our data suggest that an active light chain kinase–phosphatase system does not operate in skinned striated muscle fiber preparations. They do not rule out the possibility that such a system has an effect in the in vivo system where it is clearly shown that phosphorylation of myosin light chains occurs during sustained periods of activation. As mentioned earlier, these in vivo studies show that relaxation of these fibers is not correlated with dephosphorylation, in agreement with our own data which show that the Ca^{2+} switch for contraction in striated muscle is not a light chain kinase–phosphatase system. It appears that the physiological role of phosphorylation of myosin light chain in striated muscle has yet to be fully explained.

2. Calmodulin Regulation of the Myosin Light Chain Kinase

Although our data show a very strong correlation between phosphorylation and thiphosphorylation and activation of muscle contraction in smooth muscle, we have been concerned with testing other aspects of the model which would be important in its regulation. It was proposed from *in vitro* studies that the Ca^{2+}–CaM complex was responsible for the activation of myosin light chain kinase *in vivo*. We have checked this assumption in skinned smooth muscle fiber preparations using two approaches.

a. Inhibitors of Calmodulin Activity. Our first approach was to use phenothiazines (antipsychotic drugs) to inhibit selectively the biological activity of calmodulin in skinned smooth muscle fibers and to show that this would inhibit phosphorylation, thiophosphorylation, and tension in these fibers (Cassidy *et al.*, 1980, Kerrick *et al.*, 1980, 1981c). The rationale for these experiments came from previous studies by Weiss and Levin (1978) who showed that phosphodiesterase was inhibited by trifluoperazine binding to the calmodulin subunit of phosphodiesterase and inhibiting its biological activity. The same authors also showed that calmodulin had a much higher affinity for this drug than other Ca^{2+}-binding proteins such as troponin C. By using the phenothiazine trifluoperazine as well as other phenothiazines, we showed that we could selectively inhibit thiophosphorylation of myosin light chains and contractile force in these fibers at concentrations expected to inhibit the biological activity of calmodulin (Kerrick *et al.*, 1980). Once the fibers had been inhibited by the drug, the Ca^{2+} sensitivity of the fibers could be regained by the addition of exogenous calmodulin to the bathing solutions (Cassidy *et al.*, 1980; Kerrick *et al.*, 1980, 1981a,c). We also showed that the inhibitory effect of phenothiazines on skinned smooth muscle fibers followed the same selectivity sequence (trifluoperazine > chlorpromazine > promethazine) as their ability to inhibit phosphodiesterase, which is regulated by the Ca^{2+}–CaM complex (Kerrick *et al.*, 1980). Skinned fiber preparations which have been inhibited by trifluoperazine or chlorpromazine cannot be thiophosphorylated by ATPγS, showing that the light chain kinase system has been inhibited (Cassidy *et al.*, 1980; Kerrick *et al.*, 1980). When untreated fibers have been maximally activated by pretreatment with Ca^{2+} and ATPγS to thiophosphorylate irreversibly myosin light chains, trifluoperazine and chlorpromazine have no effect on the irreversibly activated tension, as expected if phosphorylation of myosin light chains alone were required for activation (Cassidy *et al.*, 1980; Kerrick *et al.*, 1980). Trifluoperazine and chlorpromazine have no effect on the Ca^{2+} sensitivity of vertebrate striated skinned fibers at the same concentration (50 μM)

which is effective on smooth muscle. Thus the phenothiazines, inhibitors of the biological activity of calmodulin, affect the Ca^{2+}-activated tension and phosphorylation of myosin light chains in smooth muscle fibers in a manner consistent with a light chain kinase–phosphatase system regulated by a Ca^{2+}–CaM complex.

b. Effects of Exogenous Calmodulin. The model also predicts that, if exogenous calmodulin were added to skinned muscle fiber preparations, the concentration of the Ca^{2+}–CaM complex would increase for a given level of Ca^{2+}, as would myosin light chain phosphorylation and submaximal tension. Thus one would expect that Ca^{2+}-activated myosin light chain phosphorylation and tension would occur at lower levels of Ca^{2+} than in skinned fibers not exposed to exogenous calmodulin. Our experimental data show that this prediction holds true and that the addition of 5 μM calmodulin to the fibers will shift the relationship between the percentage of maximum Ca^{2+}-activated tension and the Ca^{2+} concentration 1.2 log units toward lower Ca^{2+} concentrations (Kerrick *et al.*, 1981b,c). Also, calmodulin causes a dramatic increase in phosphorylation over control values, reaching a maximum of 0.6 mole phosphate mole^{-1} light chain as compared with 0.2 mole phosphate mole^{-1} of light chain in untreated fibers (Kerrick *et al.*, 1981b). In addition, the rate of activation of tension in the fibers when transferred to a high Ca^{2+} concentration is increased severalfold in the presence of calmodulin. When fibers are maximally activated in Ca^{2+}, no additional tension occurs if calmodulin is added, although there is a very large increase in the phosphorylation of myosin light chains (Kerrick *et al.*, 1981b). This agrees with our other data using untreated fibers that low levels of myosin light chain phosphorylation, 0.1–0.2 mole of phosphate per myosin light chain, are responsible for full activation of the fibers and that any additional phosphorylation of the light chains has no effect upon tension generation. Similar studies carried out using Sr^{2+} and calmodulin (P. E. Hoar and W. G. L. Kerrick, unpublished data) show the same results although, as mentioned earlier, smooth muscle fibers are less sensitive to Sr^{2+} than to Ca^{2+}.

3. Cyclic AMP-Dependent Protein Kinase

The evidence presented so far strongly supports the hypothesis that a Ca^{2+}–CaM-LCK complex is responsible for the activation of smooth muscle contraction. Another approach in testing this hypothesis is to use an inhibitor of Ca^{2+}–CaM–LCK complex formation and to show that the physiological effect is to inhibit both myosin light chain phosphorylation and tension in skinned fiber preparations. *In vitro* experiments by Conti and Adelstein (1980) show that the phosphorylation of myosin light chain kinase of smooth muscle by cAMP-dependent protein kinase results in inhibition of formation of the Ca^{2+}–CaM–LCK complex because of a decrease in the affinity of the Ca^{2+}–CaM complex for the light chain kinase. Addition of the exogenous catalytic subunit of cAMP-dependent protein kinase causes an inhibition of maximally Ca^{2+}-activated tension in gizzard (Kerrick *et al.*, 1981c) and pulmonary artery skinned smooth muscle fiber preparations (P. E. Hoar and W. G. L. Kerrick, unpublished data). This inhibition can be reversed by addition of the regulatory subunit of cAMP-dependent protein kinase to the fibers (Kerrick *et al.*, 1981c).

As predicted from the model, a decrease in the affinity of the Ca^{2+}–CaM complex for the light chain kinase should also result in an increase in the amount of calcium required to activate the fiber to a given level. Indeed, following partial inhibition of the fibers by cAMP-dependent protein kinase, it is clear that more calcium is required to activate skinned smooth muscle fibers to the same degree (Kerrick *et al.*, 1981c). The model also predicts that, if one were to increase the Ca^{2+}–CaM complex by the addition of exogenous calmodulin to the fibers, one should be able to override this inhibition. When exogenous calmodulin is added to fiber preparations that have been inhibited by the catalytic subunit of the cAMP-dependent protein kinase, tension is recovered in agreement with the predictions (Kerrick *et al.*, 1981a,c). Furthermore, when the muscle fibers are irreversibly activated by thiophosphorylation of myosin light chains, the catalytic subunit of cAMP-dependent protein kinase has no effect upon the tension, as would be expected if the effect of the catalytic subunit of cAMP-dependent protein kinase were only on the light chain kinase (Kerrick and Hoar, 1981). These data strongly support the hypothesis that it is the interaction between the Ca^{2+}–CaM complex and the light chain kinase which forms an active Ca^{2+}–CaM–LCK system responsible for the activation of smooth muscle contraction. In addition, these data show that, in the intact muscle fiber, the possibility of β-adrenergic stimulation causing relaxation via phosphorylation of myosin light chain kinase is distinct and real. In contrast, the catalytic subunit of cAMP-dependent protein kinase has no effect upon the submaximal or maximal Ca^{2+}-activated tension of striated muscles, although heavy incorporation of ^{32}P into the troponin I of cardiac muscle fibers from $[\gamma\text{-}^{32}P]ATP$ can be observed (Kerrick et al., 1981b).

The fact that exogenous calmodulin strongly affects Ca^{2+} sensitivity, light chain phosphorylation, and the rate of activation of skinned smooth muscle fiber preparations suggests that it is the activator of myosin light chain kinase in smooth muscle fiber preparations. However, for this to occur, the implications are that our fibers are somewhat calmodulin-depleted and that the addition of exogenous calmodulin would fully activate myosin light chain kinase, or that the Ca^{2+}–CaM–LCK complex is more effective in activating the light chain kinase than some other Ca^{2+}-binding protein. Therefore, these data do not totally exclude the possibility that another calmodulin-like protein is also an activator of myosin light chain kinase. The addition of exogenously added calmodulin to vertebrate striated muscle, in contrast to smooth muscle, has shown no effect upon the phosphorylation levels of myosin light chains or the Ca^{2+}-activated tension, again suggesting no participation of a Ca^{2+}–CaM-regulated light chain kinase in the Ca^{2+} activation of these skinned muscle fibers.

4. Inosine Triphosphate as a Test for Light Chain Kinase

Another prediction from the light chain kinase–phosphatase model is that the use of ATP analogues such as ITP, which cannot be employed as a substrate for myosin light chain kinase, will not result in activation of skinned smooth muscle fibers. In vitro studies have shown that ITP is not used as a substrate by myosin light chain kinase (Onishi and Watanabe, 1979) but can be used as a substrate for actomyosin ATPase. Therefore, one would predict that skinned smooth muscle fibers would not contract in the presence of Ca^{2+} and the ATP analogue ITP, since no phosphorylation of myosin light chains would occur. These results were borne out in experiments showing that in the presence of Ca^{2+} and ITP no contraction occurred (Kerrick et al., 1981b). In addition, we showed the same results for GTP. We also synthesized ITPγS and GTPγS and observed that no irreversible activation of tension occurred following pretreatment of fibers with ITPγS or GTPγS in the presence of Ca^{2+}. The use of either ITP or GTP in contracting and relaxing solutions of striated muscle demonstrates that they serve as a substrate for actomyosin interactions and muscle contraction in a system which is not regulated by a light chain kinase–phosphatase system (Kerrick et al., 1981b).

5. Changing Concentrations of ATP as a Test for a Light Chain Kinase–Phosphatase System

Another clear way of testing for a light chain kinase–phosphatase system model is to lower the Mg-ATP^{2-} substrate level below the K_m for myosin light chain kinase. This lowering of the substrate level should make the smooth muscle fibers appear much less sensitive to Ca^{2+} for a given level of the Ca^{2+}–CaM–LCK complex, since less phosphorylation and consequently less tension would occur. When the Mg-ATP^{2-} concentration in gizzard smooth muscle is lowered to 50 μM from 2 mM, the concentration of Ca^{2+} required to achieve half-maximal activation of the fibers is a whole order of magnitude higher and the maximum force the muscle will develop is considerably reduced. Both of these observed results are expected from a Ca^{2+}-sensitive light chain kinase–phosphatase control system and are consistent with the proposed *in vitro* model (Kerrick *et al.*, 1981b). In contrast, reducing the substrate level of Mg-ATP^{2-} to such low values in striated muscle results in a slight increase in Ca^{2+} sensitivity as well as an increase in the maximum force consistent with the known properties of contractile and regulatory proteins of vertebrate striated muscle (Best *et al.*, 1977).

IV. SUMMARY

The evidence discussed in this chapter for and against a myosin light chain kinase–phosphatase system playing a role in the regulation of muscle contraction primarily comes from data using skinned muscle fiber preparations in our laboratory. Our approach has been to correlate phosphorylation with activation of tension through the use of specific inhibitors of calmodulin activity (trifluoperazine and chlorpromazine) and myosin light chain kinase activity (the catalytic subunit of cAMP-dependent protein kinase), alteration of the calmodulin concentration and therefore light chain kinase–calmodulin complex activity, use of the ATP analogue ITP, which selects between myosin light chain kinase and myosin ATPases, and variation of the concentration of the substrate Mg-ATP^{2-}. All these data from vertebrate smooth muscle have been consistent with a Ca^{2+}-sensitive light chain kinase complex being responsible for the activation of smooth muscle. In contrast, our data for skinned vertebrate striated muscle are not consistent with myosin light chain kinase being responsible for Ca^{2+} activation. Therefore, it is our conclusion that vertebrate smooth muscle contraction is regulated by a Ca^{2+}-sensitive myosin light chain kinase–phosphatase system in contrast to the direct Ca^{2+}-binding control (leiotonin) proposed by Ebashi and colleagues, and that Ca^{2+} regulation in vertebrate striated muscle is by direct Ca^{2+} binding. However, we must point out that our system contains no sarcolemma, and soluble proteins such as calmodulin and the light chain kinase in the vertebrate striated muscles could presumably diffuse out from these preparations and render them insensitive to our tests. However, these striated skinned fiber preparations all contract and relax when the $[Ca^{2+}]$ is changed and certainly do not require the presence of a light chain kinase–phosphatase system for their activation. These data are also consistent with the *in vitro* data, as discussed earlier in this chapter, which suggest that, although phosphorylation of myosin light chains occurs *in vivo* in striated muscle, their functional significance is yet to be elucidated.

One report has been cited showing that a light chain kinase–phosphatase system exists in invertebrates (Sellers, 1981) and that such a system may occur elsewhere in the evolutionary tree as well. We have made preliminary investigations using the ATP analogue ATPγS to examine this possibility in two lines of evolutionary phyla, protostomes and deuterostomes. The ATPγS tension data show no evidence for a light chain kinase–phosphatase system being involved in regulation of the muscles so far tested in protostomes, which include arthropods, mollusks, and annelids, whereas in deuterostomes we have found evidence for a light chain kinase–phosphatase system in one of the earliest of these phyla to evolve (echinoderms), in the body wall muscle of the sea cucumber (*Parastichopus*) (Kerrick and Bolles, 1981).

Ca^{2+} activation in muscles is regulated by one or more of at least three types of Ca^{2+} control systems: thin filament control (troponin) and the myosin or thick filament regulation which involves direct Ca^{2+} binding to the myosin or a Ca^{2+}-sensitive light chain kinase–phosphatase system. As evidenced by the work reported in this chapter, skinned muscle fiber preparations provide a powerful means of elucidating the relative roles of the Ca^{2+} control systems described *in vitro* in muscle contraction.

REFERENCES

Adelstein, R. S., and Conti, M. A. (1975). Phosphorylation of platelet myosin increases actin-activated myosin ATPase activity. *Nature* (*London*) **256**, 597–598.

Adelstein, R. S., and Eisenberg, E. (1980). Regulation and kinetics of the actin-myosin-ATP interaction. *Annu. Rev. Biochem.* **49**, 921–956.

Adelstein, R. S., Conti, M. A., Hathaway, D. R., and Klee, C. B. (1978). Phosphorylation of smooth muscle myosin light chain kinase by the catalytic subunit of adenosine 3′:5′-monophosphate-dependent protein kinase. *J. Biol. Chem.* **253**, 8347–8350.

Bárány, K., and Bárány, M. (1977). Phosphorylation of the 18,000-dalton light chains of myosin during a single tetanus of frog muscle. *J. Biol. Chem.* **252**, 4752–4754.

Bárány, M., and Bárány, K. (1980). Phosphorylation of the myofibrillar proteins. *Annu. Rev. Physiol.* **42**, 275–292.

Barron, J. T., Bárány, M., Bárány, K., and Storti, R. V. (1980). Reversible phosphorylation and dephosphorylation of the 20,000-dalton light chain of myosin during the contraction-relaxation-contraction cycle of arterial smooth muscle. *J. Biol. Chem.* **255**, 6238–6244.

Best, P. M., Donaldson, S. K. B., and Kerrick, W. G. L. (1977). Tension in mechanically disrupted mammalian cardiac cells: Effects of magnesium adenosine triphosphate. *J. Physiol.* (*London*) **265**, 1–17.

Cassidy, P., Hoar, P. E., and Kerrick, W. G. L. (1979). Irreversible thiophosphorylation and activation of tension in functionally skinned rabbit ileum strips by [^{35}S]ATPγS. *J. Biol. Chem.* **254**, 11148–11153.

Cassidy, P., Hoar, P. E., and Kerrick, W. G. L. (1980). Inhibition of Ca^{2+}-activated tension and myosin light chain phosphorylation in skinned smooth muscle strips by the phenothiazines. *Pfluegers Arch.* **387**, 115–120.

Conti, M. A., and Adelstein, R. S. (1980). Phosphorylation by cyclic adenosine 3′:5′-monophosphate-dependent protein kinase regulates myosin light chain kinase. *Fed. Proc., Fed. Am. Soc. Exp. Biol.* **39**, 1569–1573.

Dabrowska, R., Sherry, J. M. F., Aromatorio, D. K., and Hartshorne, D. J. (1978). Modulator protein as a component of the myosin light chain kinase from chicken gizzard. *Biochemistry* **17**, 253–258.

Frearson, N., Solaro, R. J., and Perry, S. V. (1976). Changes in phosphorylation of P light chains of myosin in perfused rabbit heart. *Nature (London)* **264**, 801–802.

Gratecos, D., and Fischer, E. H. (1974). Adenosine 5′-O(3-thiotriphosphate) in the control of phosphorylase activity. *Biochem. Biophys. Res. Commun.* **58**, 960–967.

Hoar, P. E., Kerrick, W. G. L., and Cassidy, P. S. (1979). Chicken gizzard: Relation between calcium-activated phosphorylation and contraction. *Science* **204**, 503–506.

Kerrick, W. G. L., and Bolles, L. L. (1981). *Limulus* and *Parastichopus* skinned fibers: Is regulation of contraction via a Ca²⁺ sensitive light chain kinase/phosphatase system? *Biophys. J.* **33**, 85a.

Kerrick, W. G. L., and Hoar, P. E. (1981). Inhibition of smooth muscle tension by cyclic AMP-dependent protein kinase. *Nature (London)* **292**, 253–255.

Kerrick, W. G. L., and Krasner, B. (1975). Disruption of the sarcolemma of mammalian skeletal muscle fibers by homogenization. *J. Appl. Physiol.* **39**, 1052–1055.

Kerrick, W. G. L., Hoar, P. E., and Cassidy, P. S. (1980). Calcium-activated tension: The role of myosin light chain phosphorylation. *Fed. Proc., Fed. Am. Soc. Exp. Biol.* **39**, 1558–1563.

Kerrick, W. G. L., Hoar, P. E., Cassidy, P. S., Bolles, L., and Malencik, D. A. (1981a). Calcium-regulatory mechanisms: Functional classification using skinned fibers. *J. Gen. Physiol.* **77**, 177–190.

Kerrick, W. G. L., Hoar, P. E., Cassidy, P. S., and Bridenbaugh, R. L. (1981b). Skinned muscle fibers: The functional significance of phosphorylation and calcium-activated tension. *Cold Spring Harbor Conf. Cell Proliferation* **8**, 887–900.

Kerrick, W. G. L., Hoar, P. E., Cassidy, P. S., and Malencik, D. A. (1981c). Ca²⁺ regulation of contraction in skinned muscle fibers. *In* "Regulation of Muscle Contraction: Excitation-Contraction Coupling" (A. D. Grinnell and M. A. B. Brazier, eds.), pp. 227–240 Academic Press, New York .

Kopp, S. J., and Bárány, M. (1979). Phosphorylation of the 19,000-dalton light chain of myosin in perfused rat heart under the influence of negative and positive inotropic agents. *J. Biol. Chem.* **254**, 12007–12012.

Manning, D. R., and Stull, J. T. (1979). Myosin light chain phosphorylation and phosphatase A activity in rat extensor digitorum longus muscle. *Biochem. Biophys. Res. Commun.* **90**, 164–170.

Mikawa, T., Nonomura, Y., Hirata, M., Ebashi, S., and Kakiuchi, S. (1978). Involvement of an acidic protein in regulation of smooth muscle contraction by the tropomyosin-leiotonin system. *J. Biochem. (Tokyo)* **84**, 1633–1636.

Morgan, M., Perry, S. V., and Ottaway, J. (1976). Myosin light-chain phosphatase. *Biochem. J.* **157**, 687–697.

Onishi, H., and Watanabe, S. (1979). Calcium regulation in chicken gizzard muscle and inosine triphosphate-induced superprecipitation of skeletal acto-gizzard myosin. *J. Biochem. (Tokyo)* **86**, 569–573.

Pato, M. D., and Adelstein, R. S. (1980). Dephosphorylation of the 20,000-dalton light chain of myosin by two different phosphatases from smooth muscle. *J. Biol. Chem.* **255,** 6535–6538.

Pemrick, S. M. (1980). The phosphorylated L_2 light chain of skeletal myosin is a modifier of the actomyosin ATPase. *J. Biol. Chem.* **255,** 8836–8841.

Perrie, W. T., Smillie, L. B., and Perry, S. V. (1972). A phosphorylated light-chain component of myosin. *Biochem. J.* **128,** 105P–106P.

Saida, K., and Nonomura, Y. (1978). Characteristics of Ca^{2+}- and Mg^{2+}-induced tension development in chemically skinned smooth muscle fibers. *J. Gen. Physiol.* **72,** 1–14.

Sellers, J. R. (1981). Phosphorylation-dependent regulation of *Limulus* myosin. *J. Biol. Chem.* **256,** 9274–9278.

Sherry, J. M. F., Górecka, A., Aksoy, M. O., Dabrowska, R., and Hartshorne, D. J. (1978). Roles of calcium and phosphorylation in the regulation of activity of gizzard myosin. *Biochemistry* **17,** 4411–4418.

Silver, P. J., and DiSalvo, J. (1979). Adenosine 3′:5′-monophosphate-mediated inhibition of myosin light chain phosphorylation in bovine aortic actomyosin. *J. Biol. Chem.* **254,** 9951–9954.

Sobieszek, A. (1977). Ca-linked phosphorylation of a light chain of vertebrate smooth-muscle myosin. *Eur. J. Biochem.* **73,** 477–483.

Stull, J. T., and High, C. W. (1977). Phosphorylation of skeletal muscle contractile proteins *in vivo*. *Biochem. Biophys. Res. Commun.* **77,** 1078–1083.

Stull, J. T., Blumenthal, D. K., and Cooke, R. (1980). Regulation of contraction by myosin phosphorylation: A comparison between smooth and skeletal muscle. *Biochem. Pharmacol.* **29,** 2537–2543.

Weiss, B., and Levin, R. M. (1978). Mechanism for selectively inhibiting the activation of cyclic nucleutide phosphodiesterase and adenylate cyclase by antipsychotic agents. *Adv. Cyclic Nucleotide Res.* **9,** 285–303.

Wolf, H., and Hofmann, F. (1980). Purification of myosin light chain kinase from bovine cardiac muscle. *Proc. Natl. Acad. Sci. U.S.A.* **77,** 5852–5855.

Yagi, K., Yazawa, M., Kakiuchi, S., Ohshima, M., and Uenishi, K. (1978). Identification of an activator protein for myosin light chain kinase as the Ca^{2+}-dependent modulator protein. *J. Biol. Chem.* **253,** 1338–1340.

Chapter 10

Possible Roles of Calmodulin in a Ciliated Protozoan Tetrahymena

YOSHIO WATANABE
YOSHINORI NOZAWA

I. Introduction	297
II. Properties of *Tetrahymena* Calmodulin	299
A. Isolation of *Tetrahymena* Calmodulin	299
B. Similarities and Dissimilarities between Calmodulins from *Tetrahymena* and Other Sources	299
III. Activation of Membrane-Bound Guanylate Cyclase of *Tetrahymena*	303
A. Intracellular Distribution and Properties of Guanylate Cyclase	303
B. Activation of Particulate Guanylate Cyclase by *Tetrahymena* Calmodulin	303
C. Cell Cycle-Associated Changes in Guanylate Cyclase Activity	306
IV. Search for New Functions of Calmodulin in *Tetrahymena*	308
A. Intracellular Localization of *Tetrahymena* Calmodulin by Indirect Immunofluorescence	308
B. Localization of Calmodulin in Cilia as Revealed by Alkali Gel Electrophoresis and Immunoelectron Microscopy	308
C. Effects of Trifluoperazine on Cellular Functions	313
V. Concluding Remarks	319
References	319

I. INTRODUCTION

Among a great variety of protozoa, the ciliate *Tetrahymena pyriformis* is one of those known to resemble higher animal cells in many respects, so

CALCIUM AND CELL FUNCTION, VOL. II
Copyright © 1982 by Academic Press, Inc.
All rights of reproduction in any form reserved.
ISBN 0-12-171402-0

that it is considered one of the best experimental materials for analyzing the morphological, biochemical, biophysical, and genetic aspects of various biological phenomena. The existence of calmodulin in *Tetrahymena* raises new interest concerning the possible roles of the protein in this cell; four laboratories have succeeded in isolating calmodulin or calmodulin-like proteins from *Tetrahymena* during the past 2 years (Suzuki *et al.*, 1979; Jamieson *et al.*, 1979; Kumagai *et al.*, 1980; Kakiuchi *et al.*, 1981).

Calmodulins from various sources have been intensely studied with respect to regulatory role, multifunctionality, ubiquity, and structural conservativeness (for review, see Cheung, 1980). Although the lack of tissue and species specificity, and structural conservativeness, have been shown to be common characteristics of calmodulins, attention should be paid not only to the similarities among calmodulins from different sources but also to the specificities of the respective calmodulins in different cells. From this point of view, the similarities among calmodulins from *Tetrahymena* and higher organisms are discussed in this chapter. In particular, studies on the activation mechanism of *Tetrahymena* guanylate cyclase by calmodulin are discussed as a unique function of *Tetrahymena* calmodulin.

Calmodulins from some lower eukaryotes have been known to activate mammalian brain phosphodiesterase but not their own phosphodiesterase (Gomes *et al.*, 1979; Head *et al.*, 1979; Waisman *et al.*, 1978; Jones *et al.*, 1979), and the functions *in situ* remain to be established. As far as *Tetrahymena* is concerned, in addition to the regulation of guanylate cyclase, calmodulin may also be involved in endo- and exocytosis, and in ciliary movement and reversal (Jamieson *et al.*, 1979; Suzuki, 1980; Ohnishi *et al.*, 1981). This chapter deals with such roles of *Tetrahymena* calmodulin as well.

Ciliary reversal in protozoa such as *Paramecium* and *Tetrahymena* has been known to occur in a Ca^{2+}-dependent manner (Naitoh and Kaneko, 1972). Such a behavioral response seems to resemble in many respects the sensory responses of higher animals and in some respects bacterial chemotaxis. It has recently become apparent in the nervous system of higher animals that Ca^{2+}, calmodulin, cyclic nucleotides, and phosphorylated protein may play important roles in signal transduction (for reviews, see Greengard, 1976, 1978). In addition, Adler's group recently demonstrated that cyclic GMP and membrane protein methylation were involved in intracellular signaling in the chemotactic response of *Escherichia coli* (Springer *et al.*, 1979; Black *et al.*, 1980). The information obtained from such analogous phenomena would be useful for the study of behavioral responses in ciliates.

II. PROPERTIES OF *TETRAHYMENA* CALMODULIN

A. Isolation of *Tetrahymena* Calmodulin

Calmodulin has been isolated from the ciliated protozoan *T. pyriformis* by four different methods. Suzuki *et al.* (1979) purified a calmodulin-like protein from *Tetrahymena* as a Ca^{2+}-binding protein (TCBP) by preparative electrophoresis on alkali polyacrylamide gel following its Ca^{2+}-dependent mobility change as observed with troponin C (Head and Perry, 1974). Jamieson *et al.* (1979) and Kumagai *et al.* (1980) purified calmodulin from *Tetrahymena* by applying Ca^{2+}-dependent affinity chromatography on phenothiazine–Sepharose (Jamieson and Vanaman, 1979) and tubulin–Sepharose, respectively. Kakiuchi *et al.* (1981) also succeeded in the purification of *Tetrahymena* calmodulin by a trichloroacetic acid precipitation method (Yazawa *et al.*, 1980). These isolation methods are not specific for *Tetrahymena* calmodulin but have proved applicable in the isolation of calmodulins from various sources ranging from mammalian brain to lower eukaryotes (Jamieson and Vanaman, 1979; Yazawa *et al.*, 1980; Suzuki, 1980).

Tetrahymena calmodulins purified to homogeneity by four different methods appear to be identical, since they markedly stimulated mammalian brain cyclic-nucleotide phosphodiesterase in a Ca^{2+}-dependent manner (Suzuki *et al.*, 1981; Jamieson *et al.*, 1979; Kumagai *et al.*, 1980; Kakiuchi *et al.*, 1981).

B. Similarities and Dissimilarities between Calmodulins from *Tetrahymena* and Other Sources

Tetrahymena calmodulin bears a striking resemblance to calmodulins from higher organisms in many properties, such as a low molecular weight (Suzuki *et al.*, 1979; Jamieson *et al.*, 1979; Kumagai *et al.*, 1980), a low pI of 4.0 (Suzuki *et al.*, 1979), stability against heat, acetone (Suzuki *et al.*, 1979), and trichloroacetic acid (Kakiuchi *et al.*, 1981), a differential electrophoretic mobility in Ca^{2+} (Suzuki *et al.*, 1979), and binding to troponin I in the presence of Ca^{2+} (Suzuki *et al.*, 1979; Jamieson *et al.*, 1979). Moreover, the amino acid composition of *Tetrahymena* calmodulin (Jamieson *et al.*, 1979; Suzuki *et al.*, 1981) was very similar to that of brain (Watterson *et al.*, 1980), sea urchin eggs (Head *et al.*, 1979), earthworm (Waisman *et al.*, 1978), and sea pansy (Jones *et al.*, 1979), including the presence of an unusual amino acid, trimethyllysine.

Studies on Ca^{2+} binding of *Tetrahymena* calmodulin have been made with $^{45}Ca^{2+}$ using both alkali gel electrophoresis (Suzuki *et al.*, 1979) and equilibrium dialysis (Suzuki *et al.*, 1981). From the Scatchard plot obtained by equilibrium dialysis, it was found that the calmodulin had two high-affinity Ca^{2+}-binding sites with a dissociation constant (K_d) of $4.6 \times 10^{-6} M$. This value is comparable to that for calmodulin from bovine brain (Cheung *et al.*, 1978) and earthworm (Waisman *et al.*, 1978) and nearly coincides with the free Ca^{2+} level necessary for the activation of brain phosphodiesterase (Suzuki *et al.*, 1981; Teo and Wang, 1973).

Antipsychotic drugs such as chlorpromazine and trifluoperazine have been known as specific inhibitors of calmodulin (Levin and Weiss, 1976, 1977), and they also interact specifically with *Tetrahymena* calmodulin, as indicated by affinity chromatography (Jamieson *et al.*, 1979) and by the inhibition of calmodulin-mediated guanylate cyclase activity (Kakiuchi *et al.*, 1981). These indicate the similarities known to date between the calmodulins from *Tetrahymena* and from higher organisms.

In spite of these similarities, calmodulins from *Tetrahymena* and from higher animals display differences with respect to antigenicity and the capacity to activate guanylate cyclase of *Tetrahymena*.

Calmodulins from higher organisms have weak antigenicity, presumably because the proteins lack tissue and species specificity. In fact, only a few laboratories have so far produced antibodies against mammalian calmodulins. Dedman *et al.* (1978) obtained antiserum against rat testis calmodulin in a goat. Andersen *et al.* (1978) and Wallace and Cheung (1979) elicited antibodies in rabbits against bovine brain calmodulin which had been pretreated with alum and dinitrophenol, respectively. Among these antisera, those of Andersen *et al.* (1978) were the only ones that were precipitating. On the other hand, antibody against *Tetrahymena* calmodulin could easily be produced in rabbits (Suzuki, 1980).

An immunized serum against *Tetrahymena* calmodulin contained high titers of specific antibodies, since they reacted with a crude extract of *Tetrahymena* to give a single precipitin line in an Ouchterlony immunodiffusion test, and the line was completely confluent with that formed between the antibody and purified *Tetrahymena* calmodulin. Anti-*Tetrahymena* calmodulin cross-reacted with a partially purified *Paramecium* calmodulin but not with partially purified calmodulins from a cellular slime mold, sea anemone, sea urchin, porcine brain, or troponin C from rabbit skeletal muscle (Fig. 1). Based on the results, ciliate calmodulins appear to possess some common antigenic determinants which are absent in calmodulins from higher organisms. In contrast, the antibodies against mammalian calmodulins cross-reacted with calmodulins from var-

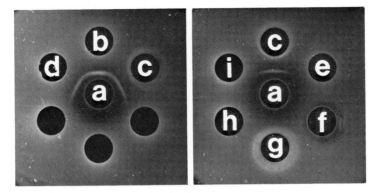

Fig. 1. Reactivity between anti-*Tetrahymena* calmodulin and calmodulins from several cells and tissues. Anti-*Tetrahymean* calmodulin was placed in the central wells (a) of Ouchterlony agar plates. *Tetrahymena* calmodulin (b) and crude calmodulin fractions from *Tetrahymena* (c), *Paramecium* (d), porcine brain (e), sea urchin testis (f), sea anemone (g), *Dictyostelium* (h), and troponin C from rabbit skeletal muscle (i) were placed in the circumferential wells. [*Paramecium* calmodulin has been identified by Kudo *et al.* (1981b) and Walter and Schultz (1981).] (From Suzuki, 1980, reproduced by permission.)

ious tissues and species, ranging from mammals to coelenterates (Chafouleas *et al.*, 1979), a finding highly suggestive of the presence of common antigenic determinants.

In addition to the dissimilarities in antigenicity and cross-reactivity, the capacity of these calmodulins to stimulate guanylate cyclase of *Tetrahymena* is another important difference. *Tetrahymena* calmodulin markedly activated its membrane-bound guanylate cyclase in a Ca^{2+}-dependent manner (Nagao *et al.*, 1979), whereas calmodulins from other sources were totally inactive (Kakiuchi *et al.*, 1981). More detailed results will be presented in the succeeding section. Thus it appears that *Tetrahymena* calmodulin has some unique properties that may not be shared by calmodulins from other sources.

Very recently, Yazawa *et al.* (1981) determined the complete amino acid sequence of *Tetrahymena* calmodulin and found that 12 amino acid residues diverged in comparison with the sequence of bovine brain calmodulin (Watterson *et al.*, 1980). The difference in amino acid sequence between calmodulins from *Tetrahymena* and bovine brain seems to be large, since only 3 replacements of amino acid residues were found between calmodulins from sea anemone (Takagi *et al.*, 1980) and bovine brain. The evidence is compatible well with the immunological and enzymatic data mentioned above.

TABLE I

Intracellular Distribution of Guanylate and Adenylate Cyclases in *Tetrahymena pyriformis*

Cell fractions	Protein (mg per 10^8 cells)	Guanylate cyclase		Adenylate cyclase	
		Total activity (pmoles min^{-1})[a]	Specific activity (pmoles min^{-1} mg^{-1} protein)	Total activity (pmoles min^{-1})[a]	Specific activity (pmoles min^{-1} mg^{-1} protein)
Cilia	5.6	176 (1)	31	0	0
Pellicles	20.3	13,688 (81)	673	3,335 (80)	163
Mitochondria	28.4	2,462 (15)	87	508 (12)	18
Microsomes	24.6	683 (4)	28	352 (8)	14
Postmicrosomal supernatant	70.2	0	0	0	0

[a] Values in parentheses express percentages of the total activity.

III. ACTIVATION OF MEMBRANE-BOUND GUANYLATE CYCLASE OF *TETRAHYMENA*

A. Intracellular Distribution and Properties of Guanylate Cyclase

It appears that all organisms have guanylate cyclase which catalyzes the formation of cyclic GMP from GTP. Guanylate cyclase is present in the membrane and the cytosol, and the relative amounts of the two forms vary from one tissue to another. Intracellular localization of guanylate cyclase was examined for *Tetrahymena* using the fractionation technique of Nozawa-Thompson (1971) with a slight modification (Nozawa, 1975). Table I shows the distribution of guanylate and adenylate cyclase activity in different membrane fractions. These results indicate that both cyclases are virtually associated with particulate fractions, in sharp contrast to earlier findings in sea urchin sperm (Garbers and Gray, 1974) and *Physarum polycephalum* (Lovely and Threlfall, 1979). The total and specific activities of guanylate cyclase are highest in the pellicular membrane which surrounds the cytoplasm and is equivalent to the plasma membrane in other cells.

Guanylate cyclase in microorganisms and mammalian tissues requires Mn^{2+} for maximum activity; Mg^{2+} is a very poor cofactor for the enzyme activity (Hardman and Sutherland, 1969; White and Aurbach, 1969). However, the cyclase activity in *Tetrahymena* was much higher with Mg^{2+} than with Mn^{2+} (Nakazawa *et al.*, 1979). The enzyme activity in the presence of 3 mM Mg^{2+} was nearly twofold that in the presence of 1 mM Mn^{2+}. Ca^{2+} or Co^{2+} was much less effective, and no activity was detected with Fe^{2+}, Cu^{2+}, or Zn^{2+}. The cyclase activity of human peripheral blood lymphocytes was comparable with Mg^{2+} or Mn^{2+} (Colfey *et al.*, 1978). This implies that Mg^{2+} whose intracellular concentration is more than 100 times higher than that of Mn^{2+}, is a physiological cofactor. The cyclase was not stimulated by Triton X-100 or Lubrol PX but was rather inhibited.

The guanylate cyclase activity exhibited a broad maximum between pH 10 and 11; the significance, if any, of such a high pH is not apparent at the present time. Double reciprocal plots of velocity versus GTP concentration were linear and gave an apparent K_m of 50 μM for GTP.

B. Activation of Particulate Guanylate Cyclase by *Tetrahymena* Calmodulin

A certain heat-stable protein factor present in the 105,000 g supernatant fraction of *Tetrahymena* caused great enhancement of the particulate guanylate cyclase in a Ca^{2+}-dependent manner (Nakazawa *et al.*, 1979).

TABLE II

Effect of *Tetrahymena* Calmodulin and Ca^{2+} on Guanylate and Adenylate Cyclase in *Tetrahymena pyriformis* [a]

Additions	Guanylate cyclase	Adenylate cyclase [b]
EGTA (0.1 mM)	2.3	62.5
Ca^{2+} (0.2 mM)	31.0	23.4
EGTA (0.1 mM) plus *Tetrahymena* calmodulin	2.7	53.2
Ca^{2+} (0.2 mM) plus *Tetrahymena* calmodulin	551.6	20.6
Ca^{2+} (0.2 mM) plus *Tetrahymena* calmodulin plus TFP [c] (25 μM)	363.0	20.4
Ca^{2+} (0.2 mM) plus *Tetrahymena* calmodulin plus TFP [c] (50 μM)	66.9	19.5

[a] Data reprinted by permission from Nagao *et al.* (1979).
[b] Values are in picomoles per minute per milligram of protein.
[c] Trifluoperazine.

Addition of EGTA prevented its stimulation, suggesting the possible involvement of a Ca^{2+}-dependent regulatory protein. This activating factor for guanylate cyclase (TCBP) was purified and examined in more detail (Suzuki *et al.*, 1979; Nago *et al.*, 1979) and was identified as calmodulin (Kakiuchi *et al.*, 1981; Suzuki *et al.*, 1981). The effect of Ca^{2+} and *Tetrahymena* calmodulin on the guanylate cyclase activity of *Tetrahymena* is demonstrated in Table II. Calmodulin did not affect the cyclase activity when Ca^{2+} was omitted. However, in the presence of Ca^{2+}, calmodulin exerted a marked increase in its effect on enzyme activity. Some enhancement by Ca^{2+} alone in the absence of the activator might be due to endogenous calmodulin which remained with the enzyme even after thorough washings with EGTA buffer. In the control containing no enzyme, calmodulin plus GTP produced no measurable amounts of cGMP.

A variety of Ca^{2+}-associated cellular events are controlled by a transient fluctuation in cytoplasmic Ca^{2+} in the range 10^{-7}–10^{-5} M (Kretsinger, 1977). The effect of various concentrations of Ca^{2+} on *Tetrahymena* calmodulin-mediated activation of guanylate cyclase was examined; a drastic increase in enzyme activity occurred over the pCa^{2+} range 6.0–4.5. To determine whether the activation process was readily reversible, the time course of the guanylate cyclase activity was determined after the addition of Ca^{2+} or EGTA (Fig. 2). When the particulate fraction was incubated with 0.2 mM EGTA and 6.25 μg/ml *Tetrahymena* calmodulin,

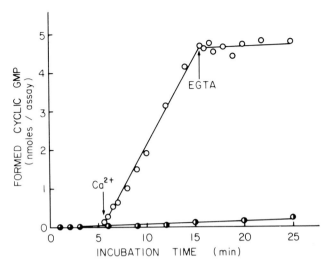

Fig. 2. Effect of Ca^{2+} on *Tetrahymena* guanylate cyclase. Guanylate cyclase activity was measured as described by Nakazawa *et al.* (1979) in the presence of *Tetrahymena* calmodulin (6.25 μg/ml). Aliquots were taken at different incubation times from the sample (○) containing the additives (Ca^{2+}, EGTA) and the untreated sample (◑). (From Nagao *et al.*, 1979, reproduced by permission.)

the enzyme activity was barely detectable. Upon adding Ca^{2+}, the rate of cyclic-GMP formation increased immediately and continued until excess EGTA was added to the reaction mixture, at which time guanylate cyclase activity was reduced to its prestimulated level. These results indicate that the activity of *Tetrahymena* guanylate cyclase responds to changes in free Ca^{2+} in an immediate and reversible manner.

Phenothiazine psychotropic agents such as trifluoperazine and chlorpromazine are potent inhibitors of calmodulins (Weiss and Levin, 1978). The two drugs inhibited in a dose-dependent fashion the calmodulin-activated fraction of guanylate cyclase (Table II) (S. Nagao and Y. Nozawa, unpublished observations). Of the two, trifluoperazine appears to be the more potent inhibitor. The Ca^{2+}-independent guanylate cyclase activity was not influenced by the agents. The trifluoperazine-induced repression of guanylate cyclase activation was overcome by increasing concentrations of calmodulin. For example, at a calmodulin concentration of 10 μg per sample, 50 μM trifluoperazine prevented activation by about 80%, while at 40 μg per sample the same concentration of trifluoperazine caused 35% inhibition. This indicates that calmodulin competitively antagonizes the trifluoperazine-induced inhibition of the cyclase. Adenylate cyclase, which is also membrane-bound, was not affected by calmodulin with or without Ca^{2+}; rather, it was inhibited by Ca^{2+} (0.2 mM)(Table II). In the concentration range examined (20–100 μM), trifluoperazine had little or no effect on the adenylate cyclase activity. Phosphodiesterase, which catalyzes the hydrolysis of cyclic AMP and cyclic GMP, is present in both soluble and particulate fractions of *Tetrahymena* cells (Kudo *et al.*, 1980) and is not affected by calmodulin at various concentrations of Ca^{2+}.

An important question now arises whether the differential response of *Tetrahymena* guanylate cyclase to its own calmodulin is due to the specificity of the activator or to the enzyme. To examine the former possibility, guanylate cyclase activities from various tissues were measured in the presence or absence of calmodulin; no enhancement of enzyme activities was observed. Calmodulins isolated from various sources were tested for their ability to stimulate *Tetrahymena* guanylate cyclase. They included bovine brain, sea anemone, scallop (Kakiuchi *et al.*, 1981) and bovine heart, and human placenta (S. Umeki and Y. Nozawa, unpublished observations), but none of these calmodulins mimicked *Tetrahymena* calmodulin in the activation of guanylate cyclase. Recently, Kudo *et al.* (1981b) have found that *Paramecium* calmodulin purified to homogeneity can activate both brain phosphodiesterase and *Tetrahymena* guanylate cyclase in a calcium-dependent manner as well as *Tetrahymena* calmodulin. These observations suggest that activation of *Tetrahymena* guanylate cyclase by calmodulins from the ciliated protozoa is highly specific, a finding consistent with the fact that the enzyme is exclusively associated with the particulate fraction and that Mg^{2+} is favored for enzyme activity.

C. Cell Cycle-Associated Changes in Guanylate Cyclase Activity

The role of cyclic nucleotides in cell growth and proliferation has been implied by a number of studies with synchronously growing cells (Pastan *et al.*, 1975; Friedman *et al.*, 1976). Gray *et al.* (1977) demonstrated that guanylate cyclase activity varied during hypoxia-induced cell cycle of *T. pyriformis* strain W. The enzyme activity in the homogenate of heat-synchronized cells showed a marked decline as cells entered the division phase (Fig. 3), followed by a gradual rise. When guanylate cyclase was assayed in EGTA (0.2 mM), the enzyme activity remained at the reduced level throughout the cell cycle. Since guanylate cyclase activity is markedly enhanced by *Tetrahymena* calmodulin, we determined whether the fluctuating activity of guanylate cyclase during the cell cycle was related to

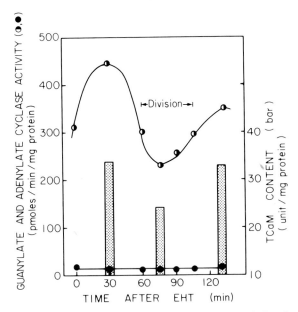

Fig. 3. Changes in guanylate and adenylate cyclase activities during the cell cycle of *T. pyriformis* strain GL. Cells were synchronized by heat shock treatment as described by Zeuthen (1964), and homogenates were prepared at different stages during the cell cycle. Each point expresses the mean of duplicate determinations of three to five different experiments. ◑, Guanylate cyclase; ●, adenylate cyclase. The content of calmodulin (TCaM) was measured, based on the extent of stimulation of guanylate cyclase (*Tetrahymena*) or phosphodiesterase (bovine heart), for the 105,000 *g* supernatant prepared at the indicated intervals. EHT, End of heat treatment. (Kudo *et al.*, 1981a, reproduced by permission.)

the cellular level of calmodulin. Since calmodulin was present in the supernatant fraction, whereas guanylate cyclase was exclusively particulate (Nakazawa *et al.*, 1979; Nagao *et al.*, 1979), the calmodulin content was determined in the soluble fraction isolated at various stages during the cell cycle. The level of calmodulin was determined by the degree of stimulation of guanylate cyclase in the particulate fraction from the control culture. One arbitrary unit was defined as the amount of protein required to double the cyclase activity in the presence of 0.1 mM Ca^{2+}. Changes in calmodulin content are depicted in Fig. 3, and they appear to be correlated with the variations in guanylate cyclase activity. In addition, it was recently observed with an asynchronous culture that a peak guanylate cyclase activity in mid-log phase coincided with the highest level of calmodulin (Kudo *et al.*, 1981a). The mechanism by which the level of calmodulin is regulated remains to be elucidated.

IV. SEARCH FOR NEW FUNCTIONS OF CALMODULIN IN *TETRAHYMENA*

A. Intracellular Localization of *Tetrahymena* Calmodulin by Indirect Immunofluorescence

As a possible lead in the identification of unknown functions of a protein, investigation of its intracellular localization by the use of specific antibody could be useful. For example, the finding that fluorescent antibody against calmodulin was localized in the mitotic spindle of the dividing cell (Welsh *et al.*, 1978) led to the finding that calmodulin enhanced Ca^{2+} sensitivity in microtubule disassembly (Marcum *et al.*, 1978). Accordingly, the localization of calmodulin in *Tetrahymena* cells was examined by indirect immunofluorescence. As shown in Fig. 4, a bandlike structure around the anterior end of the cell, the oral apparatus including deep fiber, ciliary basal bodies, and contractile vacuole pores fluoresced intensely. Cilia also fluoresced when the specimen was fixed with Formalin instead of ethanol (Suzuki, 1980). The pattern of localization may point to some possible functions of calmodulin: (1) Calmodulin in the oral apparatus might be involved in food vacuole formation, in other words, in nutrient uptake from the oral apparatus. (2) Calmodulin in contractile vacuole pores might function in the regulation of osmotic pressure, contraction of the contractile vacuole, or excretion of the vacuole contents. (3) Calmodulin in the cilia and ciliary basal bodies might play a role in ciliary movement and reversal. (4) Calmodulin in a head band structure at the anterior end of the cell might be associated with calcium channels on the somatic membrane, since the localization of calmodulin coincides with that of a mechanoreceptor relevant to the avoiding response in cell behavior (Naitoh and Eckert, 1969).

To reinforce the above notions, experiments were carried out to investigate the effects of trifluoperazine on certain cellular functions such as food vacuole formation, excretion of contractile vacuole contents, and ciliary reversal; the results of these studies will be presented in Section IV,C.

B. Localization of Calmodulin in Cilia as Revealed by Alkali Gel Electrophoresis and Immunoelectron Microscopy

The involvement of Ca^{2+} in ciliary movement has been well known from electrophysiological studies (for review, see Eckert, 1972). In ciliates such as *Paramecium* and *Tetrahymena*, ciliary reversal is seen when a transient calcium influx occurs via calcium channels on the ciliary membrane (Naitoh and Kaneko, 1972); and two or three types of nonreversal behav-

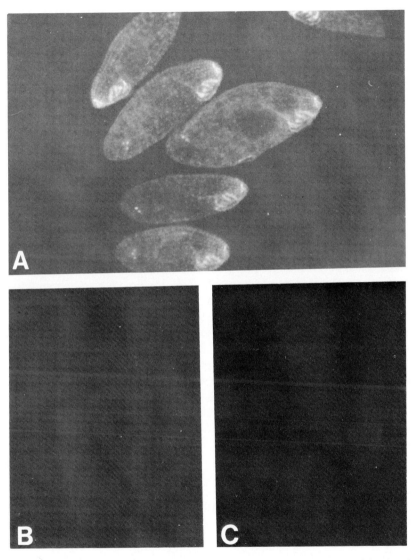

Fig. 4. Localization of calmodulin in *Tetrahymena* cells. Cells were fixed with ethanol–acetone at −20°C and stained by indirect fluorescent antibody using anti-*Tetrahymena* calmodulin purified on an antigen affinity column. The second antibody was fluorescein-labeled goat anti-rabbit IgG. Staining was carried out with affinity-purified IgG (A), IgG not adsorbed to the affinity column (B), and affinity-purified IgG preabsorbed with excess *Tetrahymena* calmodulin (C). Note that the controls showed no fluorescence (B and C). (From Suzuki, 1980, reproduced by permission.)

ioral mutants each having different mutation gene locus have been isolated from *Paramecium tetraurelia* (Chang and Kung, 1973), *P. caudatum* (Takahashi, 1979), and *Tetrahymena thermophila* (Takahashi *et al.*, 1980). These mutants have certain defects in their ciliary membrane calcium channels as revealed by electrophysiological studies and by studies on the reactivation of Triton-extracted models (Kung *et al.*, 1975; Takahashi, 1979; Takahashi and Naitoh, 1978; Takahashi *et al.*, 1980).

In wild-type cells, the free Ca^{2+} level necessary for ciliary reversal is about $10^{-6} M$ (Naitoh and Kaneko, 1973), which nearly coincides with the K_d of the Ca^{2+} binding of *Tetrahymena* calmodulin (Suzuki *et al.*, 1981). In addition, Ohnishi *et al.* (1981) succeeded for the first time in isolating calmodulins from the cilia and cell bodies of *Tetrahymena* and proved that these calmodulins were just the same with each other in properties. It is, therefore, probable that calmodulin plays a role in ciliary reversal as a Ca^{2+} sensor for intracellular Ca^{2+} transient. In this regard, localization of calmodulin and calmodulin-binding protein(s) within cilia was investigated by means of gel electrophoresis and immunoelectron microscopy. As shown in Fig. 5, a considerable amount of calmodulin was found to be included in cilia isolated from wild-type *T. thermophila*, as revealed by alkali glycerol gel electrophoresis in the presence of EGTA. However, in the presence of Ca^{2+}, the band corresponding to calmodulin completely disappeared, suggesting the existence of calmodulin-binding protein(s) in the cilia. This Ca^{2+}-dependent complex(es) was shown to be present on the top of the alkali gel when the gel was subjected to second-dimensional electrophoresis in the presence of EGTA (Ohnishi *et al.*, 1981).

The electrophoretic patterns of cilia from nonreversal mutants of *T. thermophila*, referred to as *tnrA* and *tnrB* (Takahashi *et al.*, 1980), were identical to those of wild-type cells (Fig. 6). The cilia of both mutants contained normal amounts of calmodulin which was capable of forming Ca^{2+}-dependent complex(es) with other proteins, in spite of the fact that their membrane calcium channels were defective (Takahashi *et al.*, 1980). These results suggest that calmodulin is not a protein encoded by the mutation gene.

Recently, Haga and Hiwatashi (1980) showed that a nonreversal mutant of *Paramecium caudatum*, *cnrC*, exhibited Ca^{2+}-dependent ciliary reversal similar to that of the wild type after it had been microinjected with a heat-labile, water-soluble, relatively small protein prepared from the wild type. The protein appeared to differ from calmodulin in its heat lability. Using a similar bioassay, K. Hiwatashi and Y. Watanabe (unpublished observations) noted that a small number of *cnrC* cells which had been injected with a *Tetrahymena* calmodulin showed Ca^{2+}-dependent reversal (Table III). The recovered cells were few in number, but the results were

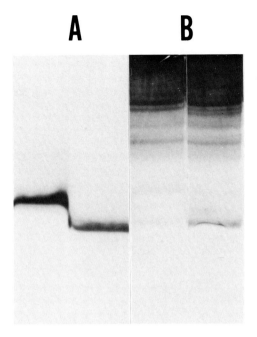

Fig. 5. Existence of calmodulin and Ca²⁺-dependent calmodulin-binding protein(s) in isolated cilia of *Tetrahymena*. Urea extract from isolated cilia (B) was electrophoresed on alkali glycerol gel electrophoresis (Suzuki *et al.*, 1979) with *Tetrahymena* calmodulin (A) as a reference. Each sample was electrophoresed in the presence of Ca²⁺ (2 mM CaCl₂, left lane) or absence of Ca²⁺ (2 mM EGTA, right lane). (Ohnishi *et al.*, 1981, reproduced by permission.)

very definite; injection of unrelated proteins always resulted in negative results. Although calmodulin does not seem to be dictated by the mutation gene, it is likely that excess calmodulin may disrupt the mutant membrane geometry and that a small fraction of it may play a role in Ca²⁺ influx. These results suggest that calmodulin may be a Ca²⁺ sensor in the membrane calcium channel and that a Ca²⁺-dependent calmodulin-binding protein (CaMBP) or a CaMBP-associated protein may be defective in a nonreversal mutant.

It has been generally accepted that inside the cilium there exists besides membrane calcium channels other Ca²⁺-sensitive apparatus responsible for ciliary reversal. Membrane-disrupted Triton models of all nonreversal mutants exhibit Ca²⁺-dependent reversal when exposed to 10⁻⁶ M Ca²⁺ (Kung *et al.*, 1975; Takahashi, 1979; Takahashi *et al.*, 1980) just like those of wild-type cells (Naitoh and Kaneko, 1973). Two types of mutants having normal membrane responsiveness but an abortive Ca²⁺-dependent re-

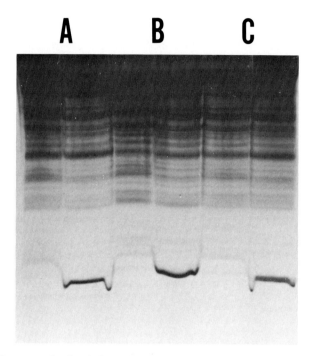

Fig. 6. Presence of calmodulin and Ca^{2+}-dependent calmodulin-binding protein(s) in isolated cilia of *Tetrahymena* nonreversal mutants. Urea extracts from the *T. thermophila* mutants *tnrA* and *tnrB* (Takahashi *et al.*, 1980) were electrophoresed together with that of wild-type cells (inbred strain B1868). A, Wild type; B, *tnrA*; C, *tnrB*. Each sample was electrophoresed in the presence (left lane) or absence (right lane) of Ca^{2+}. Any defect could not be detected in the mutant ciliary protein by this method. (K. Ohnishi, Y. Suzuki, and Y. Watanabe, personal communication.)

versal, designated "atalanta" and "spinner" have been reported in *P. tetraurelia* (Kung *et al.*, 1975).

Localization of calmodulin in such Ca^{2+}-sensitive apparatus seems to be important in relation to the possible role of calmodulin in ciliary reversal. Therefore, *Tetrahymena* cilia were fractionated into a membrane plus matrix fraction and an axoneme fraction, and the latter was further separated into a crude dynein fraction and an outer-doublet microtubule fraction according to Gibbons (1963). The amount of calmodulin in these fractions was investigated. As shown in Fig. 7, a considerable amount of calmodulin was associated with the outer-doublet microtubule fraction in a Ca^{2+}-dependent manner.

Calmodulin localization within the outer-doublet microtubule fraction was then demonstrated by immunoelectron microscopy using mono-

TABLE III

Effect of *Tetrahymena* Calmodulin Injected into Cells of *Paramecium caudatum* Stock 16D317[a-c]

No. of cells injected	Cells surviving after injection	Cells showing backward swimming	Cells showing trichocyst discharge
35	33	2	0
100[d]	100[d]	0	0

[a] A double mutant in terms of ciliary nonreversal and trichocyst nondischarge: *cnrC*/*cnrC*; *nd-2*/*nd-2*.

[b] Calmodulin was injected into each cell at 4.3 pg per cell. After 5–10 hr, trichocyst discharge and backward swimming were tested with a saturated picric acid solution and a Tris buffer containing 20 mM KCl and 1 mM CaCl$_2$, respectively.

[c] K. Hiwatashi and Y. Watanabe (personal communication).

[d] Control, no injection.

specific rabbit anti-*Tetrahymena* calmodulin IgG and ferritin-conjugated goat anti-rabbit IgG. As shown in Fig. 8, ferritin particles clustered on the outer-doublet microtubules at regular intervals of about 90 nm. These particles appeared to be localized on the interdoublet (nexin) links, the lateral linkers between outer-doublet microtubules. On the other hand, a control sample prewashed with EGTA buffer or treated with normal rabbit IgG in place of rabbit anti-calmodulin IgG showed random distribution of a very few ferritin particles (Ohnishi *et al.*, 1981). The specific localization of calmodulin on the outer-doublet microtubules may explain how this Ca^{2+}-dependent regulator is involved in ciliary reversal.

C. Effects of Trifluoperazine on Cellular Functions

Levin and Weiss (1976, 1977) have shown that, in the presence of Ca^{2+}, brain calmodulin binds 2 moles of trifluoperazine with a K_d of $10^{-6} M$ and that trifluoperazine-bound calmodulin becomes biologically inactive. Ca^{2+}-dependent stimulation of guanylate cyclase (Table II) and of brain phosphodiesterase (Kumagai *et al.*, 1980; Suzuki, 1980) by *Tetrahymena* calmodulin was also blocked by trifluoperazine. *Tetrahymena* calmodulin–lysozyme binding as an example of Ca^{2+}-dependent binding was inhibited completely (Fig. 9). However, the binding of *Tetrahymena* calmodulin to troponin I or certain *Tetrahymena* protein(s) was not inhibited even by 100 μM trifluoperazine (Fig. 9). These results could be interpreted to indicate that the drug binds to calmodulin in the presence of Ca^{2+} and that the diverse functions of *Tetrahymena* calmodulin may be influenced more or less seriously by the drug.

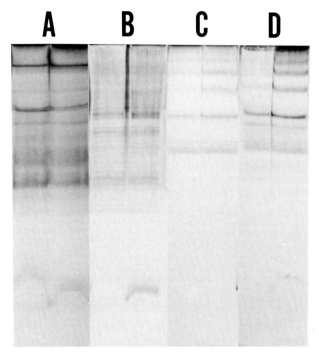

Fig. 7. Existence of calmodulin and Ca^{2+}-dependent calmodulin-binding protein(s) in ciliary fractions. A membrane plus matrix fraction (A), an axoneme fraction (B), a crude dynein fraction (C), and an outer-doublet microtubule fraction (D) were prepared from *Tetrahymena* cilia after Gibbons (1963). Each sample was subjected to alkali glycerol gel electrophoresis in the presence (left lane) or absence (right lane) of Ca^{2+}. Free calmodulin migrates as the fastest moving band. (Ohnishi *et al.*, 1981, reproduced by permission.)

The involvement of calmodulin in a certain function could be demonstrable if the function is seriously affected by trifluoperazine. Effects of phenothiazines on various cell metabolisms in *Tetrahymena* were first observed by Rogers (1966) from the pharmacological standpoint. The drugs appeared to enter the cells and exerted various influences on cell metabolism. As stated in Section IV,A, calmodulin was localized in the oral apparatus (Fig. 4), which is an important organelle for capture of food and for sustaining cell growth in the usual culture media (Rasmussen and Kludt, 1970). Moreover, development of the oral primordium is a prerequisite for the forthcoming division; for example, a setback response of the primordium development to a temperature shock is closely related to the temperature-induced division synchrony (Frankel, 1964a,b).

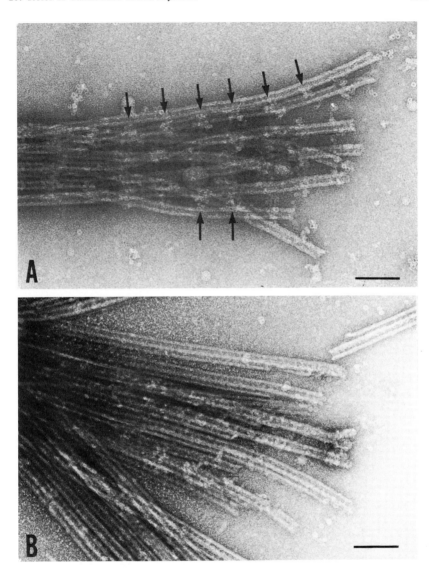

Fig. 8. Calmodulin localization in outer-doublet microtubules of *Tetrahymena* cilia by immunoelectron microscopy. The specimens were treated first with monospecific anti-*Tetrahymena* calmodulin IgG. The second antibody was ferritin-conjugated goat anti-rabbit IgG. A normal sample (A) and a sample prewashed with EGTA buffer (B) were treated as above. In (A), clustered ferritin particles are seen at regular intervals of 90 nm (arrows). Bars indicate 0.2 μm. (Ohnishi *et al.*, 1981, reproduced by permission.)

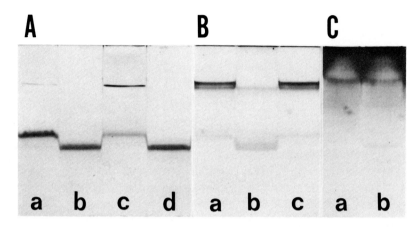

Fig. 9. Differential inhibitory effects of trifluoperazine on Ca^{2+}-dependent calmodulin–protein complexes. As examples of Ca^{2+}-dependent calmodulin–protein complexes, calmodulin–lysozyme (A), calmodulin–troponin I (B), and calmodulin–*Tetrahymena* undefined protein(s) (C) complexes were selected. In (A) a mixture of calmodulin, lysozyme, and 40 μM trifluoperazine was electrophoresed in the presence (a) or absence (b) of Ca^{2+}. Lanes (c) and (d) are same as (a) and (b) except that trifluoperazine was omitted. In (B) a mixture of calmodulin, troponin I, and 100 μM trifluoperazine was electrophoresed in the presence (a) or absence (b) of Ca^{2+}. Lane (c) is the same as (a) except that trifluoperazine was omitted. In (C) a crude *Tetrahymena* extract plus 100 μM trifluoperazine was electrophoresed in the presence (a) or absence (b) of Ca^{2+}. Note the difference in the appearance of the free calmodulin band in lane (a) of each sample. (From Suzuki, 1980, reproduced by permission.)

The oral apparatus is composed mainly of about 170 kinetosomes and interkinetosomal connectives (Nilsson and Williams, 1966) which consist of numerous microtubules, thick microfilaments, and latticed sheet structures (Numata *et al.*, 1980a,c); so that the oral apparatus includes tubulin and fiber-forming protein recently found (MW 38,000) to be major protein components (Numata *et al.*, 1980a–d). Ca^{2+} is required for *in vitro* polymerization of the fiber-forming protein into 14-nm filaments (Numata *et al.*, 1980a,d) which are components of the contractile ring of a dividing cell (Yasuda *et al.*, 1980; Numata *et al.*, 1980a,c,d). In the oral apparatus, calmodulin in ''crescent structure'' besides four membranelles was identified (Fig. 4). The structure is likely to be an area of membrane addition or presumably of endocytosis (N. E. Williams, The University of Iowa, personal communication). Thus analysis of the role of the Ca^{2+}–calmodulin system in the function of the oral apparatus assumes significance. As a preliminary approach, exponentially growing *Tetrahymena* cells were

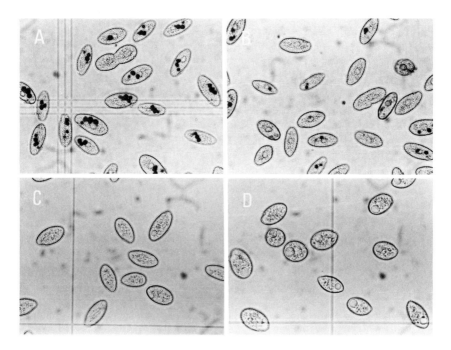

Fig. 10. Effect of trifluoperazine on food vacuole formation and excretion of contractile vacuole contents. Food vacuole formation in *Tetrahymena* was examined in a proteose–peptone medium containing 2% India ink. (A) Untreated cells; (B) cells treated with 40 μM trifluoperazine. For the investigation of excretory activity of the cells, they were transferred to a fivefold diluted proteose–peptone medium and the effect of trifluoperazine was evaluated from the size of the contractile vacuole. (C) Untreated cells; (D) cells treated with 40 μM trifluoperazine. Note that in (D) almost all the drug-treated cells have large vacuoles as a result of the inhibition of excretion. (From Suzuki, 1980, reproduced by permission.)

treated with various concentrations of trifluoperazine, and the cellular ability to take up carbon particles was investigated. As shown in Fig. 10A and B, trifluoperazine reduced the percentage of carbon-ingested cells. The response was dose-dependent. The results suggest that calmodulin may play an important role in nutrient uptake.

Calmodulin was also localized in contractile vacuole pores (Fig. 4) and seemed to bind tightly to the pore structures much like a structural protein; immunofluorescence remained in the vacuole pores of ethanol–acetone-fixed cells even after thorough washing with EGTA before fluorescent antibody staining (Suzuki, 1980). The effect of trifluoperazine on the excretion of contractile vacuole contents was examined in living

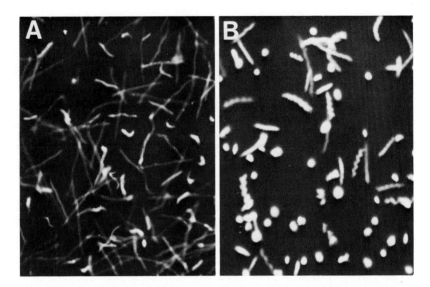

Fig. 11.   Effect of trifluoperazine on the swimming pattern of *Tetrahymena* cells. Living cells were photographed with dark-field illumination. The movement of each cell during a 2-sec exposure is shown as a white line. (A) Untreated cells; (B) cells treated with 40 μM trifluoperazine. In the micrographs, the swimming direction of each cell is unclear. However, it was directly observed that control cells swam forward, while most of the drug-treated cells swam backward. (From Suzuki, 1980, reproduced by permission.)

Tetrahymena cells (Fig. 10C and D). The results clearly indicated that trifluoperazine suppressed excretion, and the cells having a large contractile vacuole increased dramatically, suggesting that calmodulin may play a crucial role in excretion. Accumulation of liquor in a contractile vacuole and fusion between neighboring small contractile vacuoles are likely to be independent of the Ca^{2+}–calmodulin system. However, fusion between the contractile vacuole and the vacuole pore may be calmodulin-dependent.

Concerning the effects of trifluoperazine on ciliary movement and on swimming behavior, Suzuki (1980) observed that the addition of 40 μM trifluoperazine to the culture medium caused most *Tetrahymena* cells to swim backward, and the backward swimming (ciliary reversal) lasted for a long time (Fig. 11). This may indicate that trifluoperazine opens the membrane calcium channels and that the resulting Ca^{2+} influx shifts the ciliary gear to backward swimming. As suggested in the early part of this section, trifluoperazine seems to exert its influence more seriously upon the membrane calcium channels than upon Ca^{2+}-dependent reversal apparatus. From the results of Triton model experiments, forward swimming absolutely requires Mg-ATP but does not require Ca^{2+}. In the reversal of the swimming pattern Ca^{2+} serves as a transducer (Naitoh and Kaneko, 1973), presumably mediated through calmodulin.

V. CONCLUDING REMARKS

In this chapter we have described the cellular and subcellular localization of *Tetrahymena* calmodulin and several cell functions in which calmodulin may play a role.

The Ca^{2+}–calmodulin system presumably functions in multiple and diverse biological phenomena which constitute the *sine qua non* of *Tetrahymena* cell life, for example, nutrient ingestion (endocytosis), cell growth, morphogenesis, cell division, regulation of osmotic pressure, discharge of water-soluble excretion (exocytosis), and ciliary movement and reversal (cell behavior).

At the subcellular level, calmodulin may be involved in an ion transport membrane system (calcium channels), membrane fusion, excitatory signaling elicited by the binding of a chemical (or mechanical stimulus) to its membrane receptor, and the functions of such structures as 14-nm filaments, microtubules, dynein arms, and interdoublet links that are relevant to morphogenesis, cell division, and ciliary movement and reversal.

Ca^{2+}–calmodulin is likely to be functionally associated with guanylate cyclase and its product, cyclic GMP, since the distribution of both calmodulin (Fig. 4) and guanylate cyclase (Table I) in *Tetrahymena* cells parallel each other. A more detailed localization of Ca^{2+}, calmodulin, guanylate cyclase, and cyclic GMP in a certain organelle during its functional phase will be required for further understanding of the molecular basis underlying a biological phenomenon. In this respect, mutants affecting a certain biological phenomenon have easily been obtained in *T. thermophila*, and they should prove useful in future studies.

REFERENCES

Andersen, B., Osborn, M., and Weber, K. (1978). Specific visualization of the distribution of the calcium dependent regulatory protein of cyclic nucleotide phosphodiesterase (modulator protein) in tissue culture cells by immunofluorescence microscopy: Mitosis and intercellular bridge. *Cytobiologie* **17**, 354–364.

Black, R. A., Hobson, A. C., and Adler, J. (1980). Involvement of cyclic GMP in intracellular signaling in the chemotactic response of *Escherichia coli*. *Proc. Natl. Acad. Sci. U.S.A.* **77**, 3879–3883.

Chafouleas, J. G., Dedman, J. R., Munjaal, R. P., and Means, A. R. (1979). Calmodulin: Development and application of a sensitive radioimmunoassay. *J. Biol. Chem.* **254**, 10262–10267.

Chang, S. Y., and Kung, C. (1973). Genetic analyses of heat-sensitive pawn mutants of *Paramecium aurelia*. *Genetics* **75**, 49–56.

Cheung, W. Y. (1980). Calmodulin plays a pivotal role in cellular regulation. *Science* **207**, 19–27.

Cheung, W. Y., Lynch, T. J., and Wallace, R. W. (1978). An endogenous Ca^{2+}-dependent activator protein of brain adenylate cyclase and cyclic nucleotide phosphodiesterase. *Adv. Cyclic Nucleotide Res.* **9**, 233–251.

Colfey, R. G., Hadden, E. M., Lopez, C., and Hadden, J. W. (1978). cGMP and calcium in the initiation of cellular proliferation. *Adv. Cyclic Nucleotide Res.* **9**, 661–676.

Dedman, J. R., Welsh, M. J., and Means, A. R. (1978). Ca^{2+}-dependent regulator: Production and characterization of a monospecific antibody. *J. Biol. Chem.* **253**, 7515–7521.

Eckert, R. (1972). Bioelectric control of ciliary activity. *Science* **176**, 473–481.

Frankel, J. (1964a). Cortical morphogenesis and synchronization in *Tetrahymena pyriformis* GL. *Exp. Cell Res.* **35**, 349–360.

Frankel, J. (1964b). The effects of high temperatures on the pattern of oral development in *Tetrahymena pyriformis* GL. *J. Exp. Zool.* **155**, 403–436.

Friedman, D. L., Johnson, G. S., and Zeilig, C. E. (1976). The role of cyclic nucleotides in the cell cycle. *Adv. Cyclic Nucleotide Res.* **7**, 69–114.

Garbers, D. L., and Gray, J. P. (1974). Guanylate cyclase from sperm of the sea urchin, *Strongylocentrotus purpuratus*. *In* "Methods in Enzymology" (J. G. Hardman and B. W. O'Malley, eds.), Vol. 38, pp. 196–199. Academic Press, New York.

Gibbons, I. R. (1963). Studies on the protein components of cilia from *Tetrahymena pyriformis*. *Proc. Natl. Acad. Sci. U.S.A.* **50**, 1002–1010.

Gomes, S. L., Mennucci, L., and Maia, J. C. C. (1979). A calcium-dependent protein activator of mammalian cyclic nucleotide phosphodiesterase from *Blastocladiella emersonii*. *FEBS Lett.* **99**, 39–42.

Gray, N. C. C., Dickinson, J. R., and Swoboda, B. E. P. (1977). Cyclic GMP metabolism in *Tetrahymena pyriformis* synchronized by a single hypoxic shock. *FEBS Lett.* **81**, 311–314.

Greengard, P. (1976). Possible role for cyclic nucleotides and phosphorylated membrane proteins in postsynaptic actions of neurotransmitters. *Nature* (*London*) **260**, 101–108.

Greengard, P. (1978). Phosphorylated proteins as physiological effectors: Protein phosphorylation may be a final common pathway for many biological regulatory agents. *Science* **199**, 146–152.

Haga, N., and Hiwatashi, K. (1980). Nature of the factor which restores membrane excitability in a behavioral mutant of *Paramecium caudatum*. *Genetics* **94**, s40.

Hardman, J. G., and Sutherland, E. W. (1969). Guanyl cyclase, an enzyme catalyzing the formation of guanosine $3',5'$-monophosphate from guanosine triphosphate. *J. Biol. Chem.* **244**, 6363–6370.

Head, J. F., and Perry, S. V. (1974). The interaction of the calcium-binding protein (troponin

C) with divalent cations and the inhibitory protein (troponin I). *Biochem. J.* **137**, 145–154.

Head, J. F., Mader, S., and Kaminer, B. (1979). Calcium-binding modulator protein from the unfertilized egg of the sea urchin *Arbacia punctulata*. *J. Cell Biol.* **80**, 211–218.

Jamieson, G. A., Jr., and Vanaman, T. C. (1979). Calcium-dependent affinity chromatography of calmodulin on an immobilized phenothiazine. *Biochem. Biophys. Res. Commun.* **90**, 1048–1056.

Jamieson, G. A., Jr., Vanaman, T. C., and Blum, J. J. (1979). Presence of calmodulin in *Tetrahymena*. *Proc. Natl. Acad. Sci. U.S.A.* **76**, 6471–6475.

Jones, H. P., Matthews, J. C., and Cormier, M. J. (1979). Isolation and characterization of Ca^{2+}-dependent modulator protein from the marine invertebrate *Renilla reniformis*. *Biochemistry* **18**, 55–60.

Kakiuchi, S., Sobue, K., Yamazaki, R., Nagao, S., Umeki, S., Nozawa, Y., Yazawa, M., and Yagi, K. (1981). Ca^{2+}-dependent modulator proteins from *Tetrahymena pyriformis*, sea anemone, and scallop and guanylate cyclase activation. *J. Biol. Chem.* **256**, 19–22.

Kretsinger, R. H. (1977). Evolution of the informational role of calcium in eukaryotes. *In* "Calcium-Binding Proteins and Calcium Function" (R. H. Wasserman, R. A. Corrinadis, E. Carafoli, R. H. Kretsinger, D. H. MacLennan, and F. L. Siegel, eds.), pp. 63–72. Elsevier/North-Holland Publ., Amsterdam and New York.

Kudo, S., Nagao, S., Kameyama, Y. and Nozawa, Y. (1981a). Growth-associated changes in cyclic nucleotide enzymes in *Tetrahymena*. Involvement of calmodulin. *Cell Differentiation* (in press).

Kudo, S., Ohnishi, K., Muto, Y., Watanabe, Y. and Nozawa, Y. (1981b). *Paramecium* calmodulin also can stimulate membrane-bound guanylate cyclase in *Tetrahymena*. *Biochem. Int.* (in press).

Kudo, S., Nakazawa, K., and Nozawa, Y. (1980). Studies on cyclic nucleotide metabolism in *Tetrahymena pyriformis*: Partial characterization of cyclic AMP- and cyclic GMP-dependent phosphodiesterase. *J. Protozool.* **27**, 342–345.

Kumagai, H., Nishida, E., Ishiguro, K., and Murofushi, H. (1980). Isolation of calmodulin from the protozoan, *Tetrahymena pyriformis*, by the use of a tubulin-Sepharose 4B affinity column. *J. Biochem.* (*Tokyo*) **87**, 667–670.

Kung, C., Chang, S. Y., and Satow, Y. (1975). Genetic dissection of behavior in *Paramecium*. *Science* **188**, 898–904.

Levin, R. M., and Weiss, B. (1976). Mechanism by which psychotropic drugs inhibit adenosine 3′,5′-monophosphate phosphodiesterase of brain. *Mol. Pharmacol.* **12**, 581–589.

Levin, R. M., and Weiss, B. (1977). Binding of trifluoperazine to the Ca^{2+}-dependent activator of cyclic nucleotide phosphodiesterase. *Mol. Pharmacol.* **13**, 690–697.

Lovely, J. R., and Threlfall, R. J. (1979). The activity of guanylate cyclase and cyclic GMP phosphodiesterase during synchronous growth of the acellular slime mould *Physarum polycephalum*. *Biochem. Biophys. Res. Commun.* **86**, 365–370.

Marcum, J. M., Dedman, J. R., Brinkley, B. R., and Means, A. R. (1978). Control of microtubule assembly-disassembly by calcium-dependent regulator protein. *Proc. Natl. Acad. Sci. U.S.A.* **75**, 3771–3775.

Nagao, S., Suzuki, Y., Watanabe, Y., and Nozawa, Y. (1979). Activation by a calcium-binding protein of guanylate cyclase in *Tetrahymena pyriformis*. *Biochem. Biophys. Res. Commun.* **90**, 261–268.

Naitoh, Y., and Eckert, R. (1969). Ionic mechanisms controlling behavioral responses in paramecium to mechanical stimulation. *Science* **164**, 963–965.

Naitoh, Y., and Kaneko, H. (1972). Reactivated Triton-extracted models of *Paramecium*: Modification of ciliary movement by calcium ions. *Science* **176**, 523–524.

Naitoh, Y., and Kaneko, H. (1973). Control of ciliary activities by adenosine triphosphate and divalent cations in Triton-extracted models of *Paramecium caudatum*. *J. Exp. Biol.* **58**, 657–676.

Nakazawa, K., Shimonaka, H., Nagao, S., Kudo, S., and Nozawa, Y. (1979). Magnesium-sensitive guanylate cyclase and its endogenous activating factor in *Tetrahymena pyriformis*. *J. Biochem. (Tokyo)* **86**, 321–324.

Nilsson, J. R., and Williams, N. E. (1966). An electron microscope study of the oral apparatus of *Tetrahymena pyriformis*. *C. R. Trav. Lab. Carlsberg* **35**, 119–141.

Nozawa, Y. (1975). Isolation of subcellular components from *Tetrahymena*. *Methods Cell Biol.* **10**, 105–133.

Nozawa, Y., and Thompson, G. A. (1971). Studies of membrane formation in *Tetrahymena pyriformis*. II. Isolation and lipid analysis of cell fractions. *J. Cell Biol.* **49**, 712–721.

Numata, O., Yasuda, T., Hirabayashi, T., and Watanabe, Y. (1980a). A new fiber-forming protein from *Tetrahymena pyriformis*. *Exp. Cell Res.* **129**, 223–230.

Numata, O., Yasuda, T., Hirabayashi, T., and Watanabe, Y. (1980b). Isolation and some properties of a new fiber-forming protein from *Tetrahymena pyriformis*. *J. Biochem. (Tokyo)* **88**, 1487–1498.

Numata, O., Yasuda, T., Hirabayashi, T., and Watanabe, Y. (1980c). Localization of a new fiber-forming protein within *Tetrahymena pyriformis*. *J. Biochem. (Tokyo)* **88**, 1499–1504.

Numata, O., Yasuda, T., Ohnishi, K., and Watanabe, Y. (1980d). *In vitro* filament formation of a new fiber-forming protein from *Tetrahymena pyriformis*. *J. Biochem. (Tokyo)* **88**, 1505–1514.

Ohnishi, K., Suzuki, Y., and Watanabe, Y. (1981). Studies on calmodulin isolated from *Tetrahymena* cilia and its localization within cilium. *Exp. Cell Res.* (in press).

Pastan, I., Johnson, G. S., and Anderson, W. B. (1975). Role of cyclic nucleotides in growth control. *Annu. Rev. Biochem.* **44**, 491–520.

Rasmussen, L., and Kludt, T. A. (1970). Particulate material—A prerequisite for rapid cell multiplication in *Tetrahymena* cultures. *Exp. Cell Res.* **59**, 457–463.

Rogers, C. G. (1966). Effects of phenothiazines on growth, glucose uptake, and cell composition in *Tetrahymena pyriformis*. *Can. J. Biochem.* **44**, 1493–1503.

Springer, M. S., Goy, M. E., and Adler, J. (1979). Protein methylation in behavioural control mechanisms and in signal transduction. *Nature (London)* **280**, 279–284.

Suzuki, Y. (1980). Studies on a calcium-binding protein, calmodulin, from *Tetrahymena pyriformis*. Doctoral Thesis of the University of Tsukuba. A part of this thesis will be published in Suzuki, Y., Ohnishi, K., and Watanabe, Y. (1981). *Tetrahymena* calmodulin. Characterization of an anti-*Tetrahymena* calmodulin and the immunofluorescent localization in *Tetrahymena*. *Exp. Cell Res.* (In press).

Suzuki, Y., Hirabayashi, T., and Watanabe, Y. (1979). Isolation and electrophoretic properties of a calcium-binding protein from the ciliate *Tetrahymena pyriformis*. *Biochem. Biophys. Res. Commun.* **90**, 253–260.

Suzuki, Y., Nagao, S., Abe, K., Hirabayashi, T., and Watanabe, Y. (1981). *Tetrahymena* calcium-binding protein is indeed a calmodulin. *J. Biochem. (Tokyo)* **89**, 333–336.

Takagi, T., Nemoto, T., Konishi, K., Yazawa, M. and Yagi, K. (1980). The amino acid sequence of the calmodulin obtained from sea anemone (*Metridium senile*) muscle. *Biochem. Biophys. Res. Commun.* **96**, 377–381.

Takahashi, M. (1979). Behavioral mutants in *Paramecium caudatum*. *Genetics* **91**, 393–408.

Takahashi, M., and Naitoh, Y. (1978). Behavioural mutants of *Paramecium caudatum* with the defective membrane electrogenesis. *Nature (London)* 271, 656–659.

Takahashi, M., Onimaru, H., and Naitoh, Y. (1980). A mutant of *Tetrahymena* with nonexcitable membrane. *Proc. Jpn. Acad. Ser. B* 56, 585–590.

Teo, T. S., and Wang, J. H. (1973). Mechanism of activation of a cyclic adenosine 3':5'-monophosphate phosphodiesterase from bovine heart by calcium ions. *J. Biol. Chem.* 248, 5950–5955.

Waisman, D. M., Stevens, F. C., and Wang, J. H. (1978). Purification and characterization of a Ca^{2+}-binding protein in *Lumbricus terrestris*. *J. Biol. Chem.* 253, 1106–1113.

Wallace, R. W., and Cheung, W. Y. (1979). Calmodulin: Production of an antibody in rabbit and development of a radioimmunoassay. *J. Biol. Chem.* 254, 6564–6571.

Walter, M. F. and Schultz, J. E. (1981). Calcium receptor protein calmodulin isolated from cilia and cells of *Paramecium tetraurelia*. *Eur. J. Cell Biol.* 24, 97–100.

Watterson, D. M., Sharief, F., and Vanaman, T. C. (1980). The complete amino acid sequence of the Ca^{2+}-dependent modulator protein (calmodulin) of bovine brain. *J. Biol. Chem.* 255, 962–975.

Weiss, B., and Levin, R. M. (1978). Mechanism for selectively inhibiting the activation of cyclic nucleotide phosphodiesterase and adenylate cyclase by antipsychotic agents. *Adv. Cyclic Nucleotide Res.* 9, 285–303.

Welsh, M. J., Dedman, J. R., Brinkley, B. R., and Means, A. R. (1978). Calcium-dependent regulator protein: Localization in mitotic apparatus of eukaryotic cells. *Proc. Natl. Acad. Sci. U.S.A.* 75, 1867–1871.

White, A. A., and Aurbach, G. D. (1969). Detection of guanyl cyclase in mammalian tissues. *Biochim. Biophys. Acta* 191, 686–697.

Yasuda, T., Numata, O., Ohnishi, K., and Watanabe, Y. (1980). A contractile ring and cortical changes found in the dividing *Tetrahymena pyriformis*. *Exp. Cell Res.* 128, 407–417.

Yazawa, M., Sakuma, M., and Yagi. K. (1980). Calmodulins from muscles of marine invertebrates, scallop and sea anemone. *J. Biochem. (Tokyo)* 87, 1313–1320.

Yazawa, M., Yagi, K., Toda, H. Kondo, K., Narita, K. Yamazaki, R., Sobue, K., Kakiuchi, S., Nagao, S. and Nozawa, Y. (1981). The amino acid sequence of the *Tetrahymena* calmodulin which specifically interacts with guanylate cyclase. *Biochem. Biophys. Res. Commun.* 99, 1051–1057.

Zeuthen, E. (1964). The temperature-induced division synchrony in *Tetrahymena*. *In* "Synchrony in Cell Division and Growth" (E. Zeuthen, ed.), pp. 99–158. Wiley (Interscience), New York.

Chapter 11

Calcium Control of Actin Network Structure by Gelsolin

HELEN L. YIN
THOMAS P. STOSSEL

I.	Introduction	325
II.	Structure of the Cortical Cytoplasm	326
III.	Regulation of the Actin Gel–Sol Transformation	327
IV.	Calcium Regulation of Actin Filament by Gelsolin	328
V.	Mechanism of Action of Gelsolin	330
VI.	Gelsolin Is an Important Physiological Regulator	331
VII.	Effect of Other Calcium-Dependent Proteins on Actin	333
VIII.	Discussion	333
	References	335

I. INTRODUCTION

An increasing body of evidence implicates the protein actin as one of the most important constituents in the architecture and movement of cytoplasm in nonmuscle as well as muscle cells (Korn, 1978). Actin molecules, the primary structure of which is highly conserved throughout phylogeny (Vanderkeckhove and Weber, 1978), can assemble from monomers into long, semiflexible, double-helical polymers of variable length (Oosawa and Kasai, 1971) and, when so organized, bind reversibly to a variety of other proteins. The polymeric structures thus formed can be further organized at another level to suit the needs of diverse organisms depending on whether the filaments are aligned in parallel, randomly entangled or branched in some regular way, or attached to membranes or other cell components.

CALCIUM AND CELL FUNCTION, VOL. II

This chapter reviews a system which depends on calcium for the control of actin structure. The general principle behind this system is that calcium regulates the *average length* of actin filaments and can thereby control their viscosity and rigidity. The components of this system were first identified in rabbit lung macrophages which can be considered a prototype of highly motile cells. Therefore, this principle of the regulation of cell motility is likely to be applicable in general to other motile eukaryotic cells.

II. STRUCTURE OF THE CORTICAL CYTOPLASM

The motor of the macrophage appears to reside in the peripheral cytoplasm beneath the plasma membrane. Under light microscopy, the cortical region excludes organelles contributing to the gel-like appearance of the cytoplasm (Lewis, 1939). The thickness of this cortical region changes during various cell activities. It forms pseudopods which extend when the cell spreads or moves on a surface and engulf objects by phagocytosis, and it appears to squeeze the cytoplasm to produce the equatorial narrowing characteristic of cell division. Transmission electron microscopy of thin sections of fixed macrophages reveals that the cortical cytoplasm contains primarily overlapping filamentous material with the dimensions of actin polymers (Reaven and Axline, 1973). Electron microscopy of these meshworks after critical-point drying shows an anastomosing network of filaments (Trotter, 1979) identifiable by histochemical techniques as actin polymers (Allison *et al.*, 1971; Berlin and Oliver, 1978). Actin constitutes about 10% of the total protein of the macrophage, and it is very similar to skeletal muscle actin in structure and function (Hartwig and Stossel, 1975). The concentration of actin in the cortical region is estimated to be very high, possibly as high as 18 mg/ml (Davies and Stossel, 1977; Hartwig *et al.*, 1977). As mentioned above, morphological work indicates that some of this actin is in the filamentous state, although a large proportion of it may also exist as monomers. Some of the actin filaments may be cross-linked by other proteins to form a three-dimensional network. With such a highly concentrated actin polymer solution, it is obvious that there should be sufficient rigidity, particularly if the actin filaments are cross-linked, to produce the gel-like consistency inferred to exist in the cortical cytoplasm and to maintain the shape of this region. However, it is necessary to explain how this structure can also be rendered more fluid to bring about changes in cell shape and directional movement of the peripheral cytoplasm.

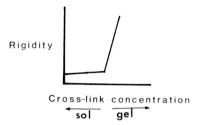

Fig. 1. Relation between the concentration of cross-linkers and rigidity of a solution of linear polymers. The sol-to-gel transition is marked by a large increase in the rigidity of the polymer solution. The relation between the critical concentration of cross-links required for gelatin to the concentration of monomers in filaments and filament length is defined in Eq. (1) in the text.

III. REGULATION OF ACTIN GEL–SOL TRANSFORMATION

Calcium regulation of actin filament length and consequently cytoplasmic rigidity was first demonstrated *in vitro* actin solutions containing actin-binding protein, a macromolecule from rabbit lung macrophages that cross-links actin filaments into a relatively isotropic gel (Stossel and Hartwig, 1975, 1976; Brotschi *et al.*, 1978; Hartwig and Stossel, 1979, 1980). The formation of an actin network by actin-binding protein appears to obey classic network theory (Brotschi *et al.*, 1978; Hartwig and Stossel, 1979) which predicts that a solution of polymers changes abruptly from the sol to the gel state at a critical concentration of cross-linker, defined as the ratio of the critical number of actin-binding protein cross-links (V_c) to the number of actin monomers in the polymer being cross-linked (N_0) (Flory, 1946) (Fig. 1). The abruptness of the sol-to-gel transition offers an excellent mechanism for controlling cytoplasmic consistency. In principle, such a transition can be achieved by one of several mechanisms (Fig. 2): (1) by removal of the actin-binding protein cross-links from the actin filaments, (2) by depolymerizing actin polymers, or (3) by decreasing the actin filament length distribution without altering the number of actin monomers and polymers. According to Flory's network theory, the critical actin cross-linker concentration is very sensitive to the length distribution of the polymers and is inversely proportional to the weight average degree of polymerization of the polymer (X_w):

$$V_c = N_0/X_w \qquad (1)$$

Therefore, the critical cross-linker density for incipient gelation can be

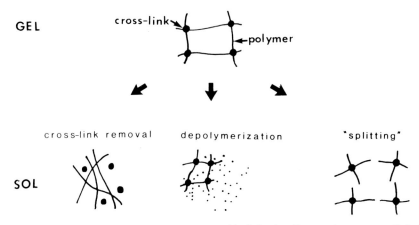

Fig. 2. Regulation of network structure. A gel is defined as linear polymers cross-linked into a network. Solation of the gel may be achieved by one of three mechanisms: (1) removal of cross-links, (2) depolymerization of polymers, (3) fragmentation of filaments between cross-links without net depolymerization.

changed by altering the filament length distribution, providing that the number of monomers and polymers remains unchanged.

At present, there is little compelling evidence for regulation of the network structure through the reversible addition and removal of cross-linking proteins. Reversible polymerization of actin between the monomeric and polymeric states can be predicted from Eq. (1) to be relatively ineffective, because a decrease in the concentration of monomers in polymers (N_0) would also decrease the weight average degree of polymerization (X_w) and therefore would be counterproductive in changing the critical actin cross-linker concentration. The third mechanism, which is equivalent to the reversible fragmentation of actin filaments, provides an extremely efficient means of altering V_c and appears to be the basis of the regulation of actin gel structure by a calcium-dependent regulatory protein isolated from rabbit lung macrophages.

IV. CALCIUM REGULATION OF ACTIN FILAMENT BY GELSOLIN

This regulator is a 91,000-dalton, heat-labile, globular protein. It shortens actin filaments when activated by calcium in the micromolar concentration range, contributing to the collapse of their three-dimensional lattice. When the ambient calcium concentration is decreased, the regulator is inactivated and the actin gel reforms. Since this protein causes an

TABLE I

Physicochemical Properties of Gelsolin

Parameter	Value
Stokes radius, a	44 Å
Sedimentation coefficient, s	4.9 S
Partial specific volume, $\overline{V}$	0.73 cm^3 g^{-1}
Isoelectric point (calcium-free)	6.1
Molecular weight	
Gel electrophoresis in the presence of sodium dodecyl sulfate	91,000
From s^0 and a	95,300
f/f_0 (from M_r, $\overline{V}$, and a)	1.43
Calcium binding	
K_a	$1.09 \times 10^6 \ \mu M^{-1}$
Capacity	1.7 moles Ca^{2+} mole^{-1} gelsolin

irreversible gel–sol transformation of the actin network, we have called it gelsolin (Yin and Stossel, 1979).

The physicochemical properties of gelsolin are summarized in Table I. Gelsolin binds 2 moles calcium mole^{-1} of gelsolin with high affinity ($K_a = 1.09 \times 10^6 \ M^{-1}$) and can bind to actin both in the monomeric or polymeric form. Ca^{2+}–gelsolin shortens preexisting actin filaments, as demonstrated directly by electron microscopy and indirectly by viscosity and flow birefringence measurements. Significant shortening of actin by Ca^{2+}–gelsolin occurs without a decrease in the sedimentability and turbidity of actin solutions, indicating that shortening does not occur primarily through depolymerization of actin (Yin et al., 1980). When added directly to G-actin, Ca^{2+}–gelsolin facilitates the nucleation step in the polymerization of monomeric actin prior to assembly, although the rate of elongation of the filaments following nucleation is decreased (Yin et al., 1981). In either case, filaments with shorter contour lengths are produced. The fact that gelsolin has similar effects on the length distribution of actin filaments whether it acts on preformed filaments or during actin assembly, or whether the actin is cross-linked or not, means that it can regulate the ultrastructure of actin filaments irrespective of the different states actin might manifest in the cell.

After shortening, it follows from the network theory that V_c increases and the gel will be solubilized at a given actin cross-linker concentration. Inhibition of gelation by gelsolin can be overcome by increasing the concentration of the cross-linker. The experimentally determined critical cross-linker concentration for gelation correlates well with the theoretical value derived from Flory's network theory for a given length of actin filament, confirming that filament shortening is indeed the basis of the solation of actin gels by gelsolin (Yin et al., 1980). The relation between filament shortening and gel solation is further substantiated by the parallel decrease in the gelation of actin by cross-linking proteins and by decreases in the viscosity of actin solutions in the absence of cross-linking proteins at increasing gelsolin concentrations (Yin et al., 1980). Furthermore, gelsolin does not decrease the binding of actin cross-linking proteins to actin filaments, so that solation of the gel is not likely to be due to a direct effect of gelsolin on the actin cross-linking protein (Yin and Stossel, 1979).

The activity of gelsolin is dependent on the ambient calcium concentration. In the presence of gelsolin, the viscosity of actin decreases slightly as the free calcium concentration rises from 10^{-8} to 10^{-7} M and falls sharply when the calcium concentration is above 10^{-7} M (Yin et al., 1980). Likewise, inhibition of actin gelation by gelsolin is also dependent on the calcium concentration (Yin and Stossel, 1979). When the calcium concentration is lowered by the addition of excess EGTA, both the viscosity of actin and its gelation are restored, suggesting that the effects of gelsolin are reversible.

V. MECHANISM OF ACTION OF GELSOLIN

The mechanism by which gelsolin shortens actin filaments has yet to be determined. Ca^{2+}–gelsolin acts rapidly on actin filaments, and a maximal decrease in the viscosity of actin is observed at 30 sec, the earliest time point measured. It is equally effective at 25° or 4°C. From the degree of shortening of actin filaments by gelsolin, we estimate that each of the added gelsolin molecules breaks 0.8 bond between the actin monomers in a filament. All these characteristics are consistent with a stoichiometric interaction between gelsolin and calcium. Because gelsolin remains attached to actin filaments after shortening them (Yin and Stossel, 1979) and the filaments remain short as long as the concentration of calcium is maintained above the threshold for activation of gelsolin, gelsolin probably blocks one or both ends of the actin filaments to prevent their reannealing. Electron microscopic studies on the growth of actin filaments off of actin nuclei formed in the presence of Ca^{2+}–gelsolin and decorated with heavy meromyosin suggest that gelsolin binds to the end of the actin filament normally preferred for elongation, blocking addition of actin monomers to that end (Yin et al., 1981a).

Theoretically, shortening of actin filaments by Ca^{2+}–gelsolin may arise by one or more of three mechanisms. First, gelsolin may rupture actin filaments in a stoichiometric manner after binding to actin filaments by directly breaking bonds between actin monomers in the polymers. Second, actin filaments have a tendency to break in response to thermal motion or imposed shear (Arisaka *et al.*, 1975) and reanneal spontaneously. Gelsolin may bind to one end of these fragmented filaments to prevent their reassociation. Third, gelsolin may operate through the recycling of actin between monomers in solution and in the filaments. In the presence of ATP, actin molecules in polymers exchange rapidly with those free in solution (Wegner, 1976; Hill, 1980). Gelsolin may shift the distribution of actin monomers from preexisting filaments to new filaments, and filament length may decrease as the number of filaments increases. At present, we favor shortening of actin filaments by direct breakage of the filaments as the mechanism of action of gelsolin.

VI. GELSOLIN IS AN IMPORTANT PHYSIOLOGICAL REGULATOR

Several lines of evidence suggest that this calcium-dependent system of the regulation of actin structure characterized *in vitro* is relevant to actin gel–sol transformation in living macrophages. First, actin, actin-binding protein, and gelsolin have been demonstrated by immunofluorescent techniques to be present in the cortical cytoplasm of macrophages (Stendahl *et al.*, 1980; Yin *et al.*, 1981b). Second, extracts of macrophage cytoplasm form a solid gel under conditions favoring actin polymerization and in the presence of excess EGTA (Stossel and Hartwig, 1976). It has been shown that actin-binding protein is responsible for at least 70% of the actin cross-linking activity in these extracts (Brotschi *et al.*, 1978). The extracts do not gel when micromolar calcium is present, and the gelled extract dissolves when calcium is added. From the yield of activity during purification of gelsolin from macrophage extracts to homogeneity, it can be concluded that gelsolin accounts for the bulk of the calcium-sensitive solation activity in the macrophage (Yin and Stossel, 1980). Third, expression of the regulatory function of gelsolin depends on the variation in free calcium concentration likely to occur in living cells, and its effects are reversible.

The ability of the cytoplasm to undergo reversible gel–sol transformation should confer unique opportunities for regulating cell functions. The gel state provides sufficient rigidity for maintaining cell shape and stabilizing cytoplasmic structures. The sol state provides fluidity, allowing dynamic movements within the cytoplasm. Because gelation is very sensitive to changes in the length of actin filaments, the cytoplasm can be considered to be "poised" in a balance between the two states, ready to be interconverted rapidly from gel to sol, or vice versa, should the need arise. Because gel–sol transformation can occur merely as the result of a severing of the actin filament network at crucial points and does not necessarily involve net actin depolymerization, the transition should be extremely efficient. This mechanism of dismantling and rebuilding the actin network allows for rapid, reversible remodeling of cytoplasmic structure. It also can restrict and/or direct the flow of organelles, macromolecules, and ions within the cytoplasm.

Besides the obvious importance of gel–sol transformation in remodeling of the cytoskeleton, regulating shape changes, and intracellular traffic, this process may also generate directional movement of the cytoplasm. From the organization of the actin network in macrophages, it is not immediately apparent how directional motion crucial to many cell functions is generated. As explained previously, actin filaments in the cytoplasm of the macrophage are very likely cross-linked in a three-dimensional network by actin-binding protein. Myosin is dispersed throughout the network and, in the presence of ATP, Mg^{2+}, and an as yet not completely defined activator (see below), interacts with actin filaments to generate tension. Gelsolin is also present and can be activated by calcium. When these proteins are combined *in vitro* and placed in a horizontal capillary tube, a crude motile system capable of directional motion is reconstructed. Calcium added to one end of the capillary tube causes the actin network to move away (Stendahl and Stossel, 1980). This directional movement can be explained solely on the basis of a localized change in actin network structure by gelsolin in response to calcium. We postulate that, under conditions of uniformly low calcium concentration in the capillary system, no net movement is produced because myosin within the gel lattice exerts equal and opposite tension on the uniformly rigid cytoplasm. With a local rise in calcium concentration, gelsolin is activated to dissolve actin gel. Since actin filaments in the solated region are no longer held together in a network, they have a tendency to slip past one another during myosin contraction, decreasing the effectiveness of myosin to exert tension. Amplification of contraction by cross-linking actin filaments is demonstrated *in vitro* by the fact that increasing concentrations of actin-binding protein added to actin filaments reduce the amount of myosin required to initiate the contraction of a fixed amount of actin (Stendahl and Stossel, 1980). An imbalance in the tension within the gel and sol portions of the actin network causes a net movement of actin fibers from the less gelled toward the more gelled domain. Likewise, the calcium-dependent reversible gel–sol transformation of actin gel networks by gelsolin can produce directional movement in the cytoplasm if a localized change in calcium concentration can be created and maintained.

There is now increasing evidence that gel–sol transformation of the cytoplasmic actin network is also involved in regulating the motility of a large number of other eukaryotic cells. Reversible, calcium-dependent gel–sol transformation has been demonstrated in extracts of many types of cells and, in the majority of cases, gelation of actin is attributable to the cross-linking of actin filaments by high-molecular-weight actin-binding proteins. Gelsolin can be identified by an immunohistochemical technique in a wide variety of cell types, including human polymorphonuclear leukocytes and platelets, rabbit smooth muscle, rabbit cardiac muscle, rabbit skeletal muscle, rabbit brain, bladder, liver, spleen, and thyroid (Yin et al., 1981b). Therefore, the system of calcium-dependent control of cytoplasmic structure and motility by gelsolin characterized in macrophages is likely to be applicable in general to other motile cells.

VII. EFFECT OF OTHER CALCIUM-DEPENDENT PROTEINS ON ACTIN

Interestingly, two other calcium-dependent proteins have also been shown to have similar effect on actin. Villin is a 95,000-dalton protein located exclusively in the microvilli of the brush border of intestinal epithelial cells of chickens (Bretscher and Weber, 1979, 1980). It is very similar to gelsolin in terms of its amino acid composition, calcium-binding affinity, and functional properties (Bretscher and Weber, 1980; Craig and Power, 1980; Mooseker et al., 1980). Further comparative studies should clarify whether the small difference in molecular weight between villin and gelsolin reflects important structural differences pertinent to the apparent localization of villin only in microvilli. Fragmin, a protein of *Physarum polycephalum* (Hasegawa et al., 1980), is also functionally similar, although it has a lower molecular weight (40,000).

VIII. DISCUSSION

A number of important questions concerning the calcium regulation of actomyosin contraction remain to be answered. First, what controls calcium concentrations in the cortex of the macrophage? Substances which stimulate directional movement and alter the cortical morphology of leukocytes stimulate the efflux of cell-associated calcium ions (Naccache et al., 1977). Phagocytic vesicles purified from rabbit lungs, which can be considered inverted plasma membrane sacs, may have a Mg^{2+}-dependent, ATP-dependent, high-affinity calcium pump which transports calcium against an electrochemical gradient from the cytoplasmic to the lumen of the phagolysosome. Bovine brain calmodulin stimulates the activity of this calcium pump (Lew and Stossel, 1980). These findings indicate that a membrane calcium pump can control the concentration of cytoplasmic calcium at the cell periphery. However, it remains to be shown how the activity of these pumps is coupled to environmental signals acting on cell surfaces. Second, is there any relation between gelsolin and calmodulin function? Calmodulin is very abundant in the macrophage and has been purified from these cells by affinity chromatography with fluphenazine-coupled resins (Lew and Stossel, 1981). Calmodulin has been identified by immunofluorescence to be associated with microfilament bundles of cultured cells (Welsh *et al.*, 1978), and trifluoperazine, a calmodulin-specific inhibitor, alters cell shape (Osborn and Weber, 1980). However, thus far, there is no evidence that calmodulin binds directly to actin (Glenney *et al.*, 1980) and that trifluoperazine directly affects the polymerization of actin (H. L. Yin, unpublished data). It is therefore likely that any interaction between gelsolin and calmodulin would be indirect. One possibility is that calmodulin exerts an effect on gelsolin activity indirectly by modulating the intracellular calcium concentration. As discussed above, calmodulin stimulates the calcium-extruding membrane pump. Calmodulin, with its high calcium-binding affinity and concentration, can also serve as a calcium buffer in modulating the intracellular calcium concentration (Glenney *et al.*, 1980). Another possibility is that calmodulin affects the activity of myosin. In our model of directional cell movement, it is obvious that alteration of actomyosin activity alone can also effect directional movement within the actin lattice, and these changes can potentiate the changes in lattice structure by gelsolin. A calmodulin-dependent myosin light chain kinase has been purified from chicken gizzard and shown to increase the actin-activated myosin ATPase activity of smooth muscle (Sobieszek and Small, 1976) and nonmuscle cells (Yerna *et al.*, 1979). A similar calmodulin-dependent kinase has also been partially purified from platelets (Hathaway and Adelstein, 1979). On the other hand, although myosin from macrophages can be activated by an endogenous kinase, there is no evidence as yet that its activity is calcium-dependent (Trotter and Adelstein, 1979).

In summary, information currently available permits a biochemical explanation of calcium-regulated directional movements of the cytoplasm. Calcium activates a regulatory protein, gelsolin, shortening actin filaments. This causes a localized breakdown in actin network structure, creating a gradient of cytoplasmic rigidity, and directional movement follows from the resultant imbalance in tension generated by myosin.

ACKNOWLEDGMENT

Supported by grants from the USPHS, the American Tobacco Council, and the Edwin S. Webster Foundation, and by gifts from Mr. Edwin W. Hiam and Charles L. and Jane D. Kaufman.

REFERENCES

Allison, A. C., Davies, P., and DePetris, S. (1971). Role of contractile microfilaments in macrophage movement and endocytosis. *Nature (London) New Biol.* **232**, 153–155.

Arisaka, F., Noda, H., and Maruyama, K. (1975). Kinetic analysis of the polymerization process of actin. *Biochim. Biophys. Acta* **400**, 263–274.

Berlin, R. D., and Oliver, J. M. (1978). Analogous ultrastructure and surface properties during capping and phagocytosis in leukocytes. *J. Cell. Biol.* **77**, 789–804.

Bretscher, A., and Weber, K. (1979). Villin: The major microfilament-associated protein of the intestinal microvillus. *Proc. Natl. Acad. Sci. U.S.A.* **76**, 2321–2325.

Bretscher, A., and Weber, K. (1980). Villin is a major protein of the microvillus cytoskeleton which binds both G and F actin in a calcium-dependent manner. *Cell* **20**, 839–847.

Brotschi, E. A., Hartwig, J. H., and Stossel, T. P. (1978). The gelation of actin by actin-binding protein. *J. Biol. Chem.* **253**, 8988–8993.

Craig, S. W., and Powell, L. D. (1980). Regulation of actin polymerization by villin, a 95,000 dalton cytoskeletal component of intestinal brush borders. *Cell* **22**, 739–746.

Davies, W. A., and Stossel, T. P. (1977). Peripheral hyaline blebs (podosomes) of macrophages. *J. Cell Biol.* **75**, 941–955.

Flory, P. J. (1946). Fundamental principles of condensation polymerization. *Chem. Rev.* **39**, 137–197.

Glenney, J. R., Jr., Bretscher, A., and Weber, K. (1980). Calcium control of the intestinal microvillus cytoskeleton: Its implications for the regulation of microfilament organization. *Proc. Natl. Acad. Sci. U.S.A.* **77**, 6458–6462.

Graessley, W. A. (1974). The entanglement concept in polymer rheology. *Adv. Polym. Sci.* **9**, 3–178.

Hartwig, J. H., and Stossel, T. P. (1975). Isolation and properties of actin, myosin, and a new actin-binding protein on rabbit alveolar macrophages. *J. Biol. Chem.* **250**, 5696–5705.

Hartwig, J. H., and Stossel, T. P. (1979). Cytochalasin B and the structure of actin gels. *J. Mol. Biol.* **134**, 539–554.

Hartwig, J. H., and Stossel, T. P. (1980). Macrophage actin-binding protein promotes the bipolar and branching polymerization of actin. *J. Cell Biol.* **87**, 841–848.

Hartwig, J. H., Davies, W. A., and Stossel, T. P. (1977). Evidence for contractile protein translocation in macrophage spreading, phagocytosis and phagolysome formation. *J. Cell Biol.* **75**, 956–967.

Hasegawa, T., Takahashi, S., Hayashi, H., and Hatano, S. (1980). Fragmin: A calcium ion sensitive regulatory factor in the formation of actin filaments. *Biochemistry* **19**, 2679–2683.

Hathaway, D. R., and Adelstein, R. S. (1979). Human platelet myosin light chain kinase requires the calcium-binding protein calmodulin for activity. *Proc. Natl. Acad. Sci. U.S.A.* **76**, 1653–1657.

Hill, T. L. (1980). Bioenergetic aspects and polymer length distribution in steady-state head-to-tail polymerization of actin or microtubules. *Proc. Natl. Acad. Sci. U.S.A.* **77**, 4803–4807.

Korn, E. D. (1978). Biochemistry of actomyosin-dependent cell motility (A review). *Proc. Natl. Acad. Sci. U.S.A.* **75**, 588–599.

Lew, P. D., and Stossel, T. P. (1980). Calcium transport by macrophage plasma membranes. *J. Biol. Chem.* **255**, 5841–5846.

Lew, P. D., and Stossel, T. P. (1981). Effect of calcium on superoxide production by phagocytic vesicles from rabbit alveolar macrophages. *J. Clin. Invest.* **67**, 1–9.

Lewis, W. H. (1939). Some cultural and cytological characteristics of normal and malignant cells *in vitro*. *Arch. Exp. Zellforsch. Besonders Gewebezuecht.* **23**, 8–26.

Mooseker, M. S., Graves, T. A., Whaton, K. A., Falco, N., and Howe, C. L. (1980). Regulation of microvillus structure: Calcium-dependent solation and crosslinking of actin filaments in the microvilli of intestinal epithelial cells. *J. Cell Biol.* **87**, 809–822.

Naccache, P. H., Showell, H. J., Becker, E. L., and Sha'afi, R. I. (1977). Transport of sodium, potassium, and calcium across rabbit polymorphonuclear leukocyte membranes: Effect of chemotactic factor. *J. Cell Biol.* **73**, 428–444.

Oosawa, F., and Kasai, M. (1976). *In* "Subunits in Biological Systems" (S. N. Timasheff and G. D. Fasman, eds.), Part A, pp. 261–322. Dekker, New York.

Osborn, M., and Weber, K. (1980). Damage of cellular functions by trifluoperazine, a calmodulin-specific drug. *Exp. Cell Res.* **130**, 484–488.

Reaven, E. P., and Axline, S. G. (1973). Subplasmalemmal microfilaments and microtubules in resting and phagocytizing cultivated macrophages. *J. Cell Biol.* **59**, 12–27.

Sobieszek, A., and Small, J. V. (1976). Myosin-linked calcium regulation in vertebrate smooth muscle. *J. Mol. Biol.* **101**, 75–92.

Stendahl, O. I., and Stossel, T. P. (1980). Actin-binding protein amplifies actomyosin contraction, and gelsolin confers calcium control on the direction of contraction. *Biochem. Biophys. Res. Commun.* **92**, 675–681.

Stendahl, O. I., Hartwig, J. H., Brotschi, E. A., and Stossel, T. P. (1980). Distribution of actin-binding protein and myosin in macrophages during spreading and phagocytosis. *J. Cell Biol.* **84**, 215–224.

Stossel, T. P., and Hartwig, J. H. (1975). Interactions of actin, myosin and an actin-binding protein of rabbit alveolar macrophages: Macrophage myosin Mg^{++}-adenosine triphosphatase requires a cofactor for activation by actin. *J. Biol. Chem.* **250**, 5706–5712.

Stossel, T. P., and Hartwig, J. H. (1976). Interactions of actin, myosin and a new actin-binding protein of rabbit pulmonary macrophages. II. Role in cytoplasmic movement and phagocytosis. *J. Cell Biol.* **68**, 602–619.

Trotter, J. A. (1979). The cytoskeleton of macrophages. *J. Cell Biol.* **83**, 321a.

Trotter, J. A., and Adelstein, R. S. (1979). Macrophage myosin: Regulation of actin-activated ATPase activity by phosphorylation of the 20,000-dalton light chain. *J. Biol. Chem.* **254**, 8781–8785.

Vanderkeckhove, J., and Weber, K. (1978). Mammalian cytoplasmic actin are the products of at least two genes and differ in primary structure in at least 25 identified positions from skeletal muscle actins. *Proc. Natl. Acad. Sci. U.S.A.* **75**, 1106–1110.

Wegner, A. (1976). Head-to-tail polymerization of actin. *J. Mol. Biol.* **108**, 139–150.

Welsh, M. J., Dedman, J. R., Brinkley, B. R., and Means, A. R. (1978). Calcium-dependent regulator protein: Localization in mitotic apparatus of eukaryotic cells. *Proc. Natl. Acad. Sci. U.S.A.* **75**, 1867–1871.

Yerna, M.-J., Dabrowska, R., Hartshorne, D. J., and Goldman, R. D. (1979). Calcium-sensitive regulation of actin-myosin interactions in baby hamster kidney (BHK-21) cells. *Proc. Natl. Acad. Sci. U.S.A.* **76**, 184–188.

Yin, H. L., and Stossel, T. P. (1979). Control of cytoplasmic actin gel-sol transformation by gelsolin, a calcium-dependent regulatory protein. *Nature (London)* **281**, 583–586.

Yin, H. L., and Stossel, T. P. (1980). Purification and structural properties of gelsolin, a Ca^{2+}-activated regulatory protein of macrophages. *J. Biol. Chem.* **255**, 9490–9493.

Yin, H. L., Zaner, K. S., and Stossel, T. P. (1980). Calcium control of actin gelation. Interaction of gelsolin with actin filaments and regulation of actin gelation. *J. Biol. Chem.* **255**, 9494–9500.

Yin, H. L., Hartwig, J. H., Maruyama, K., and Stossel, T. P. (1981a). Ca^{2+} control of actin polymerization: Interaction of macrophages gelsolin with actin monomers and effects on actin polymerization. *J. Biol. Chem.* **256**, 9693–9697.

Yin, H. L., Albrecht, J. H., and A. Fattoum (1981b). Identification of gelsolin, a Ca^{2+}-dependent regulatory protein of actin gel–sol transformation and its intracellular distribution in a variety of cells and tissues. *J. Cell Biol.* **91**.

Chapter 12

Calcium and the Metabolic
Activation of Spermatozoa

ROBERT W. SCHACKMANN
BENNETT M. SHAPIRO

I. Introduction 339
II. Nature of the Activation Process 341
   A. Regulation of Motility 341
   B. The Acrosome Reaction 343
   C. Other Metabolic Activations 345
III. Triggers for Sperm Activation 346
IV. Ca²⁺ Functions in Sperm–Egg Association 348
V. Conclusions 348
   References 349

I. INTRODUCTION

Fertilization acts as the agent of the continuity and diversity of meta-zoan species. The fusion of two haploid cells to initiate development of a new organism involves a complex hierarchy of events beginning with gametogenesis and leading to metabolically arrested sperm and eggs. Both gametes become activated during fertilization in the period just prior to and following gamete membrane fusion. Sperm become motile, undergo changes that allow them to penetrate the egg coats, and bind to the egg surface. Upon fusion of the gamete plasma membranes, ignition of the metabolic machinery of the egg initiates development.

CALCIUM AND CELL FUNCTION, VOL. II
Copyright © 1982 by Academic Press, Inc.
All rights of reproduction in any form reserved.
ISBN 0-12-171402-0

Spermatogenesis occurs in the testes. In mammals the sperm pass from the testes into a duct, the epididymis, where biochemical and physiological modifications prepare them for meeting the egg. In many species sperm motility is affected during this passage (reviewed in Bedford, 1979; Hoskins *et al.*, 1979; Turner, 1979). For example, sperm isolated from the caput epididymis (the region closest to the testes) of bulls show little flagellar activity and no coordinated swimming, whereas those from the cauda (terminal) epididymis become motile upon dilution. Additional changes occur in sperm after they leave the cauda epididymis. Bovine sperm isolated from the cauda epididymis have a higher permeability to Ca^{2+} than those obtained after ejaculation, an alteration that is associated with the addition of a glycoprotein to the sperm surface (Babcock *et al.*, 1979).

Even after ejaculation the sperm of many mammals are not capable of fertilizing eggs but require a period of residence within the female reproductive tract before they become capable of fertilizing. This process, capacitation, is associated with a number of surface changes in sperm (O'Rand, 1979), but the molecular nature of the changes is undefined. There is no evidence for such a complex series of changes in the male reproductive tract of invertebrates, nor is there an event directly analogous to capacitation. Sperm become motile either upon spawning, with its attendant dilution, or when they come near the egg and encounter its surface components (Harvey, 1930).

In several echinoids, egg components [called variably "egg jelly" (Lillie, 1914) or "factors released from eggs" (Garbers and Hardman, 1975)] stimulate sperm respiration (Ohtake, 1976a,b), as well as alter sperm motility (Hathaway, 1963) and trigger the acrosome reaction (Dan *et al.*, 1964; Dan, 1952, 1967), a secretion from the sperm head required for fertilization. In many invertebrates, the exocytosis of the acrosome reaction is accompanied by polymerization of actin at the tip of the sperm head into an acrosomal filament extending through the egg coats to encounter the egg. In mammals a filament does not form, but exocytosis occurs with exposure of hydrolytic enzymes to aid in penetration through the egg coats.

Thus sperm undergo several changes in structure and function as they mature and encounter the egg. In this chapter we will examine what is known about the role of Ca^{2+} in this process, which we will call sperm metabolic activation, in order to describe the sum of the physiological and biochemical changes that occur in preparation for fertilization. This process includes more than the whiplike movement associated with capacitation and the acrosome reaction of rodent sperm, which has been termed activation (Yanagimachi and Usui, 1974).

II. NATURE OF THE ACTIVATION PROCESS

A. Regulation of Motility

The sperm of many animals are stored in the male in a quiescent state, without flagellar activity. In invertebrates such as echinoderms, sperm obtained directly from a spawning animal are densely packed and immotile prior to dilution (Harvey, 1930), as are those of teleost fishes (Morisawa and Suzuki, 1980). Similarly, sperm of two rodents, rat and hamster, are immotile when obtained from the cauda epididymis (Morton et al., 1979; Turner, 1979). In other mammals there is disagreement about whether epididymal sperm are quiescent, since some studies (e.g., Morton et al., 1978) suggest that epididymal sperm of the bull, rabbit, and human are motile, whereas others (Turner, 1979) suggest that this motility is caused by the manipulation involved in removing them from the epididymis.

Ca^{2+} affects both mammalian and invertebrate sperm. Ca^{2+} stimulates the motility of quiescent sperm obtained from the cauda epididymis of the hamster (Morton et al., 1974, 1979), rat (Davis, 1978), and mouse (Miyamoto and Ishibashi, 1975), optimally at ~2 mM with a decrease at higher Ca^{2+} concentrations. Decreased sperm motility also occurs with relatively high levels of extracellular Ca^{2+} in the bull (McGrady et al., 1974; Bredderman and Foote, 1971) and chimpanzee (McGrady et al., 1974).

In sea urchin sperm extracellular Ca^{2+} is not required for the onset of motility (Gibbons, 1980). Sperm suspended in a divalent cation-free medium containing 0.2 mM EGTA become motile within 1–2 min. In the presence of 2 mM Ca^{2+} sperm become periodically quiescent and assume a canelike form. The addition of 40–200 μM Ca^{2+} to sperm in the presence of the divalent ionophore A23187 suggests that increased intracellular Ca^{2+} causes this shape change. Mg^{2+} does not substitute for Ca^{2+}; in fact, Mg^{2+} stimulates motility in ionophore-treated sperm made quiescent with Ca^{2+}. Further evidence that increased intracellular Ca^{2+} mediates quiescence comes from the study of Triton X-100-demembranated sperm (Brokaw et al., 1974; Gibbons and Gibbons, 1980). In these cell models, 1 μM Ca^{2+} produces asymmetric distortions of ATP-induced flagellar movements, and 100–200 μM Ca^{2+} causes quiescence. The canelike form of the sperm models is similar to that of intact sperm.

Sea urchin sperm have considerable quantities of calmodulin (Jones et al., 1978). Chlorpromazine, one of several phenothiazine derivatives that interact with calmodulin–Ca^{2+} complexes, abolishes quiescence at 5–10 μM (Gibbons, 1980). Whether or not it is this interaction which directly affects motility remains to be shown.

Other ions are involved in the regulation of sperm motility. Na^+ stimulates motility in rat (Wong *et al.*, 1981) and sea urchin sperm (Nishioka and Cross, 1978), presumably by increasing the intracellular pH. Dry sperm (collected from a spawning animal without dilution) do not become motile when suspended in seawater in which choline has replaced Na^+. Upon addition of 10 mM Na^+, H^+ efflux occurs and the sperm begin to swim (Nishioka and Cross, 1978). Likewise, addition of NH_4^+ leads to the motility of sperm in choline-substituted seawater, as does addition of the monovalent ionophore nigericin to sperm in seawater in which K^+ has replaced Na^+ (Lee *et al.*, 1980). In support of this, both cause efflux of the accumulated weak bases [^{14}C]methylamine and 9-aminoacridine (Lee *et al.*, 1980; Christen *et al.*, 1980). The efflux of these probes indicates that the internal pH increases (Rottenberg, 1979; Gillies and Deamer, 1979).

Wong *et al.* (1981) have shown that Na^+ also stimulates H^+ efflux from rat sperm in a Na^+-free medium (choline substituted for Na^+). This recent finding is in apparent disagreement with earlier results of Turner and Howards (1978) who were unable to detect specific ion requirements for motility. The discrepancy possibly results from different measurements of motility. Wong *et al.* (1981) measured the forward motility index, while Turner and Howards used the more general measurement of sperm tail motion. In the ameboid sperm of the nematode, the monovalent ionophore monesin, which exchanges Na^+ and H^+, also elicits sperm motility (Nelson and Ward, 1980).

The addition of Na^+ to sea urchin sperm in choline-substituted seawater leads to an increase in Ca^{2+} permeability (Lee *et al.*, 1980) associated with membrane depolarization as estimated with the lipophilic cation tetraphenylphosphonium (TPP^+). Increased Ca^{2+} influx also occurs as the Na^+/choline ratio is decreased from that of normal seawater (zero choline) to that containing low levels of Na^+ (5–10 mM) (Shapiro *et al.*, 1980). Both data suggest that external Na^+ might be used by the sperm to regulate intracellular Ca^{2+}.

There are other indications that interrelationships among intracellular Ca^{2+}, intracellular pH, and membrane potential might function in regulating sperm motility. Decreasing the extracellular pH causes sea urchin sperm motility to cease (Goldstein, 1979) and evokes a decrease in the intracellular pH as estimated from ^{31}P NMR measurements (Yoshioka and Inoue, 1980) and decreased methylamine efflux (Christen *et al.*, 1980). Sperm suspended in seawater containing 80 mM K^+ assume the quiescent wave form described above (Gibbons, 1980). Increased seawater K^+ appears to decrease the internal pH as estimated by a decreased efflux of methylamine (Christen *et al.*, 1980) and by ^{31}P nuclear magnetic resonance (NMR) spectroscopy (Yoshioka and Inoue, 1980). Increased K^+ also depolarizes the sperm membrane potential (Christen *et al.*, 1980). Hence intracellular pH and membrane potential might be regulators of the effects of sperm motility and flagellar shape either directly or by regulating intracellular Ca^{2+}.

Special flagellar activation is associated with capacitation and the acrosome reaction. Extracellular Ca^{2+} is required for this change in the motility of hamster and guinea pig sperm (Yanagimachi and Usui, 1974; Talbot *et al.*, 1976); a vigorous whiplike tail movement is induced by A23187 with extracellular Ca^{2+} but not with Mg^{2+} (Babcock *et al.*, 1976; Talbot *et al.*, 1976). A23187 has several effects on sperm Ca^{2+} fluxes. It causes a release of $^{45}Ca^{2+}$ at low ionophore concentrations or an accumulation of Ca^{2+} at higher concentrations (Babcock *et al.*, 1976). The motility activation occurs only under the latter conditions, supporting the hypothesis that Ca^{2+} influx is involved in the motility change.

B. The Acrosome Reaction

The echinoderm acrosome reaction (exocytosis and filament extension) is rapid compared with that of mammals (exocytosis alone). In sea urchin sperm induced to undergo the acrosome reaction *in vitro,* the entire population reacts less than 20 sec after the addition of egg jelly coat material. The reaction of an individual sperm probably takes place in 1 sec or less. The reaction in echinoids requires extracellular Ca^{2+} (Dan, 1954) and is associated with an H^+ efflux distinct from the proton release associated with motility onset (Schackmann *et al.*, 1978; Tilney *et al.*, 1978). After induction, sperm change their membrane permeabilities to Na^+, Ca^+, and K^+ and undergo a net depolarization of the membrane potential (Schackmann *et al.*, 1978; Schackmann and Shapiro, 1981).

In mammals the ion requirements and fluxes for induction of the acrosome reaction have not been well defined. In part this is because the mammalian reaction occurs over a period of minutes or hours in a population of sperm, depending upon the experimental conditions and whether or not the sperm have been capacitated (Yanagimachi and Usui, 1974; Cornett and Meizel, 1978). Whereas the physiological trigger for echinoids is an egg surface component (jelly), in mammals it is not known. Recent work suggests that the reaction is induced after contact with a protein in the zona pellucida that surrounds the egg (Saling *et al.*, 1978; Wasserman and Bleil, 1981).

Alternate methods of inducing the acrosome reaction exist, most of which involve an alteration of specific ion gradients. In all cases the alternate triggering methods require extracellular Ca^{2+}. The ionophore A23187 elicits the reaction in the presence of extracellular Ca^{2+} in invertebrates and mammals (Decker et al., 1976; Talbot et al., 1976; Collins and Epel, 1977; Green, 1978; Russell et al., 1979). The ionophore nigericin is similarly effective in both sea urchins and starfish (Schackmann et al., 1978; Tilney et al., 1978). Elevating the pH from 8 to 9.5 (Dan, 1952; Gregg and Metz, 1976) or increasing the extracellular Ca^{2+} to 30 mM (Collins and Epel, 1977; Wada et al., 1956) triggers the reaction in invertebrate sperm. The addition of 30 mM Na^+ to sperm suspended in choline-substituted seawater elicits the acrosome reaction in the sea urchin Strongylocentrotus purpuratus (Schackmann and Shapiro, 1981).

In starfish sperm extracellular Ca^{2+} is necessary for exocytosis but not for polymerization of actin. Polymerization of actin occurs when nigericin is added to sperm suspended in isotonic KCl, conditions that should increase the intracellular pH (Tilney et al., 1978). Polymerization, induced in the absence of Ca^{2+} even when EGTA is in the incubation medium, is not well organized, and no exocytosis occurs. Since normal filament assembly does not take place, exocytosis is apparently associated with the generation of an appropriate geometry for actin filament growth.

The requirement for extracellular Ca^{2+} suggests that influx of this cation may play a regulatory role. Ca^{2+} influx is associated with the acrosome reaction of sea urchins as estimated by several techniques. Increased $^{45}Ca^{2+}$ permeability, as shown by $^{45}Ca^{2+}$ uptake, can be detected as soon as sperm react and continues long after the acrosomal reaction has been completed (Schackmann et al., 1978). Carbonyl cyanide p-trifluoro-methoxyphenylhydrazone (FCCP) and other uncouplers of oxidative phosphorylation inhibit up to 90% of the Ca^{2+} uptake. Most of the Ca^{2+} accumulation is intramitochondrial, as shown by X-ray microanalysis (M. Cantino, unpublished results).

Support for the Ca^{2+} influx hypothesis was obtained from inhibitor studies. Drugs which interfere with Ca^{2+} binding or those which block Ca^{2+} channels prevent the acrosome reaction of several echinoids. The local anesthetics xylocaine and procaine (Collins and Epel, 1977), La^{3+} (Decker et al., 1976), and the Ca^{2+} channel and "slow" Na^+ channel blockers verapamil and D600 (Schackmann et al., 1978; Kopf and Garbers, 1980) all inhibit the acrosome reaction induced by egg jelly.

Ca^{2+} may play two roles in the triggering process: to initiate the interaction of jelly with the sperm and to enter the cell, perhaps to permit exocytosis of the acrosomal vesicle. Under appropriate conditions (in 5–10 mM Na^+ in choline-substituted seawater) Ca^{2+} influx occurs without exocytosis; thus Ca^{2+} influx does not appear to be sufficient for the reaction, although it is necessary.

The possibility that Ca^{2+} acts via a calmodulin–Ca^{2+} complex has been suggested by the observation that 1–2 μM trifluoperazine (TFP) blocks the acrosome reaction of sea urchin sperm triggered by egg jelly (Shapiro *et al.*, 1980; Garbers, 1981). That TFP does not simply have a deleterious effect on the sperm is indicated by the lack of effect of 10 μM TFP on the other methods of triggering the acrosome reaction (by nigericin or with 30 mM Na$^+$ in choline-substituted seawater) (R. W. Schackmann, unpublished).

Ca^{2+} influx occurs during capacitation of guinea pig spermatozoa, either prior to or along with the acrosome reaction (Singh *et al.*, 1978, 1980). Capacitation itself does not require extracellular Ca^{2+} (Yanagimachi and Usui, 1974), but the metabolic or surface changes that occur may allow Ca^{2+} influx. As was the case for the Ca^{2+} influx found in sea urchin sperm, uncouplers of oxidative phosphorylation inhibit mammalian ^{45}Ca^{2+} uptake. Again, Ca^{2+} accumulation is principally into the mitochondria, as detected by microanalysis with bovine sperm (Babcock *et al.*, 1978). In contrast to the case with the sea urchin sperm, the local anesthetic nupercaine and the Ca^{2+} channel inhibitor D600 actually stimulate ^{45}Ca^{2+} influx and induce the acrosome reaction of guinea pig sperm (Singh *et al.*, 1980). This may indicate that Ca^{2+} crosses the mammalian sperm plasma membrane by a mechanism different from that in echinoid sperm, but the situations are not directly comparable. Verapamil or D600 prevents ^{45}Ca^{2+} uptake and the acrosome reaction in sea urchin sperm only when added before egg jelly; there is only a slight inhibitory effect on the long-term Ca^{2+} uptake when added after the reaction is complete. Thus, in neither case is the slow mitochondrial accumulation prevented. Since we do not know the first step of triggering in the mammal, the relationship of the early Ca^{2+} uptake to triggering cannot be defined. In both guinea pig and sea urchin sperm Mg^{2+} inhibits the acrosome reaction (Singh *et al.*, 1978; Collins and Epel, 1977).

C. Other Metabolic Activations

In addition to the changes already discussed, fluctuations in cyclic-nucleotide levels are associated with the changes in sperm physiology found during metabolic activation. For example, elevated intracellular cAMP has been correlated with the initiation and stimulation of sperm motility (Cascieri *et al.*, 1976; Mrsny and Meizel, 1980), with sperm maturation in the epididymis (Hoskins *et al.*, 1974), and with induction of the acrosome reaction (Garbers and Hardman, 1975). We will not discuss all these changes in cyclic-nucleotide metabolism but will concentrate on those associated with the effects of Ca^{2+}. For a comprehensive review of cyclic-nucleotide effects on sperm motility the reader is referred to articles by Hoskins and colleagues (1978; Hoskins and Casillas, 1975).

Garbers and colleagues (Garbers and Hardman, 1975; Garbers, 1981) have demonstrated that factors in the egg jelly increase cAMP and cGMP levels in sea urchin sperm and increase cAMP levels in starfish and horseshoe crab sperm (Tubb et al., 1979). In sea urchin sperm increased cGMP or cAMP is caused by a peptide (Hansbrough and Garbers, 1981) that stimulates the respiration of sperm at pH 6.5 (seawater pH is usually near 8) and the oxidation of exogenous long-chain ($>C_8$) fatty acids (Hansbrough et al., 1980). A separate factor that induces the acrosome reaction increases only cAMP levels (Kopf et al., 1979). The effect of the latter factor is dependent upon extracellular Ca^{2+} (Tubb et al., 1978); that of the former is not. D600 blocks the increase in cAMP levels initiated by the Ca^{2+}-dependent factor, and substitution of choline for Na^+ also prevents the cAMP increase (Kopf and Garbers, 1980) as well as the acrosome reaction (Schackmann et al., 1978). The ionophore nigericin, which leads to $^{45}Ca^{2+}$ influx and the acrosome reaction, similarly increases cAMP levels. All these data indicate a relationship between Ca^{2+} influx and increased cAMP levels. Both processes begin as soon as acrosomal filaments appear; which, if either, functions in a primary role is not known.

In mammals conflicting data exist concerning cyclic-nucleotide levels, Ca^{2+}, and the acrosome reaction. The cAMP analogs dibutyryl-cAMP and 8-bromo-cAMP, as well as several phosphodiesterase inhibitors, either stimulated (Mrsny and Meizel, 1980; Hyne and Garbers, 1979) or inhibited (Rogers and Garcia, 1979) the acrosome reaction. In addition, dibutyryl-cGMP and 9-bromo-cGMP were found to stimulate the acrosome reaction in guinea pig sperm (Santos-Sacchi and Gordon, 1980), but detectable levels of cGMP were not found in these sperm under slightly different conditions (Hyne and Garbers, 1979). Extracellular Ca^{2+} leads to an increase in intracellular cAMP levels, as well as to the acrosome reaction of the guinea pig sperm.

Cyclic nucleotides also seem to play a role in regulating sperm motility. Caffeine and theophylline stimulate motility in porcine (Garbers et al., 1973), bovine (Hoskins and Casillas, 1975), and equine (Tamblyn et al., 1979) sperm under conditions where they increase cAMP levels. Likewise, cAMP applied to demembranated sperm stimulates flagellar motion (Lindemann, 1978; Tamblyn et al., 1979). The interactions between these cAMP effects and Ca^{2+} levels have not been defined.

III. TRIGGERS FOR SPERM ACTIVATION

In the sea urchin at least two factors in the egg jelly have physiological effects upon sperm. The jelly, which surrounds the egg, can be solubilized by treatment with pH 5–6 seawater or by merely allowing eggs to stand at high concentrations for several hours. These treatments produce a solution of macromolecules containing both the Ca^{2+}-dependent and the Ca^{2+}-independent factors mentioned above. The respiratory-stimulating factor (Ca^{2+}-independent) is a 1900-dalton peptide called speract that is effective below nanomolar concentrations (Hansbrough and Garbers, 1981). A similar dialyzable, protease-sensitive factor was isolated several years earlier by Ohtake (1976a,b) from a Japanese sea urchin. The peptide may work by stimulating sperm respiration in the acidic jelly microenvironment surrounding the egg.

The Ca^{2+}-dependent factor has only 12% protein; the remainder is principally fucose and sulfate (SeGall and Lennarz, 1979). This factor triggers the acrosome reaction in the species-specific manner characteristic of fertilization. However, the fucose sulfate polymer requires elevated Ca^{2+} concentrations for triggering, suggesting that a complex between the polymer and Ca^{2+} is the reactive species. Sequential incubation studies also suggest that a complex of egg jelly and Ca^{2+} is the triggering agent (Schackmann and Shapiro, 1981).

In mammals controversy exists over the factors in the female tract or surrounding the egg that are responsible for the physiological changes leading to the acrosome reaction or alterations in sperm motility. Recent work by Cornett and Meizel (1978) and Meizel and Working (1980) suggested that the catecholamines (epinephrine and norepinephrine) and the adrenergic agonist phenylephrine stimulated the activation of motility (whiplike tail movement) and the acrosome reaction in hamster sperm. Likewise, the adrenergic antagonists phentolamine and propanolol prevented this stimulation, but whether this was a pharmacological or physiological effect is not known. A bovine follicular fluid protein later identified as serum albumin stimulated the acrosome reaction in hamster sperm (Lui et al., 1977; Lui and Meizel, 1977), but the significance in vivo of the observation remains in doubt. Other small molecules, in particular cholinergic amines, affect the behavior of some sperm. Acetylcholine causes concentration-dependent motility changes (enhancement or reduction) in sea urchin sperm (Nelson, 1978), and acetylcholine changes the Ca^{2+} permeability of hypotonically treated mammalian sperm which may contain cholinergic receptors (Stewart and Forrester, 1978, 1979).

More convincing are the recent data implicating a Ca^{2+} requirement for the binding of nonacrosome reacted mouse sperm to the zona pellucida (Saling et al., 1978). These data are supported by an analysis of the interaction of mouse sperm with specific proteins of the zona pellucida. One of these seems to initiate the acrosome reaction specifically and effectively (Wasserman and Bleil, 1981). Thus mammalian sperm may also undergo the reaction at the egg surface. In invertebrates, this proximate reactivity is important, since sperm that have undergone the acrosome reaction die quickly (Kinsey *et al.*, 1979; Vacquier *et al.*, 1979).

IV. Ca^{2+} FUNCTIONS IN SPERM–EGG ASSOCIATION

As discussed above, Ca^{2+} appears to play a role in the regulation of sperm motility and in inducing the acrosome reaction. It may also function in the subsequent event in fertilization, fusion between gamete plasma membranes. When sperm and eggs are mixed, 1–5 μM Ca^{2+} must be present for fertilization to occur. Originally it was suggested that Ca^{2+} was required only for sea urchin sperm to undergo the acrosome reaction (Takahashi and Sugiyami, 1973). When sperm, jelly, and Ca^{2+} were mixed and then rapidly diluted into an egg suspension to lower the Ca^{2+} levels below those needed for fertilization, the eggs were still fertilized, suggesting that Ca^{2+} was not required in the sperm–egg interaction process. Recently, however, Sano and Kanatani (1980) studied the level of Ca^{2+} that remained in such an experiment and that needed for gamete fusion. They found that ~ 10 μM Ca^{2+} was required for fertilization. Ca^{2+} may facilitate binding to the egg, and a subsequent step, either penetration of the vitelline layer or sperm–egg fusion. However, this role of Ca^{2+} remains controversial (Chambers and Angeloni, 1981; Schmidt and Epel, 1981).

Recent work by Yanagimachi (1978) has demonstrated that Ca^{2+} is also necessary for sperm penetration of the zona pellucida and plasma membrane of the mammalian egg in hamsters, guinea pigs, and humans. In the denuded hamster egg penetration system, no penetration occurs in the absence of Ca^{2+} even if sperm have undergone the acrosome reaction, although binding is found. This imples that Ca^{2+} is required not only for the mammalian acrosome reaction but also for the union of mammalian gametes. In the hamster system 250 μM Ca^{2+} is required to obtain high percentages of penetration, although Mg^{2+} can substitute at higher (5 mM) concentrations. Taken together, these preliminary data suggest that Ca^{2+} may be required for sperm and egg fusion as well as in interactions between sperm and the egg coats, but the mechanism of the multiple Ca^{2+}-dependent interactions is not clear.

V. CONCLUSIONS

The system of fertilization provides many examples of the regulation of complex cellular behavior by ionic mechanisms. Even though the interrelationships of different ions, such as protons, Ca^{2+}, and K^+ are only beginning to be uncovered, certain generalizations have already appeared. Ca^{2+}

may act to elicit the membrane fusion of the acrosome reaction and sperm–egg association. The events are modulated by alterations in intracellular pH and membrane potential, so that an overall process, like the exocytosis and filament assembly of the acrosome reaction, requires a balance among changes in the three ion-dependent parameters. Calmodulin is present in sperm, and phenothiazine inhibitors of its action block such processes as the acrosome reaction and fertilization. Thus, an intracellular role for Ca^{2+} in the regulation of these events is hypothesized with the same vague uneasiness that attends all such implications of guilt by association. Yet the relationships that underlie these physiological observations remain unclear. How intracellular pH alterations are coordinated with changes in Ca^{2+} levels and how specific ions determine flagellar wave form are but two of the problems that require resolution at a deeper level of analysis. A great advantage of the fertilization system, especially in marine invertebrates, is that one does not lack for biochemical material to extend the analysis. This, along with the rapid, synchronous nature of the sperm metabolic activation process and its well-defined spatial localizations make it an ideal candidate for probing the association between molecular mechanisms and cellular activities.

REFERENCES

Babcock, D. F., First, N. L., and Lardy, H. A. (1976). Action of ionophore A23187 at the cellular level: Separation of effects at the plasma and mitochondrial membranes. *J. Biol. Chem.* **251**, 3881–3886.

Babcock, D. F., Stamerjohn, D. M., and Hutchinson, T. (1978). Calcium redistribution in individual cells correlated with ionophore action on motility. *J. Exp. Zool.* **204**, 391–400.

Babcock, D. F., Singh, J. P., and Lardy, H.A. (1979). Alteration of membrane permeability to calcium ions during maturation of bovine spermatozoa. *Dev. Biol.* **69**, 85–93.

Bedford, J. M. (1979). Evolution of the sperm maturation and storage functions of the epididymis. *In* "The Spermatozoon: Maturation, Motility, Surface Properties and Comparative Aspects" (D. W. Fawcett and J. M. Bedford, eds.), pp. 7–21. Urban & Schwarzenberg, Baltimore, Maryland.

Bredderman, P. J., and Foote, R. H. (1971). The effect of Ca ions on cell volume and motility of bovine spermatozoa. *Proc. Soc. Exp. Biol. Med.* **137**, 1440–1443.

Brokaw, C. J., Josslin, R., and Bobrow, L. (1974). Calcium ion regulation of flagellar beat symmetry in reactivated sea urchin spermatozoa. *Biochem. Biophys. Res. Commun.* **58**, 795–800.

Cascieri, M., Amann, R. P., and Hammerstedt, R. H. (1976). Adenine nucleotide changes at initiation of bull sperm motility. *J. Biol. Chem.* **251**, 787–793.

Chambers, E. L., and Angeloni, S. V. (1981). Is external Ca^{2+} required for fertilization of sea urchin eggs by acrosome reacted sperm? *J. Cell Biol.* **91**, No. 2, Pt. 2, 181a.

Christen, R., Schackmann, R. W., and Shapiro, B. M. (1980). Regulation of viability and the acrosome reaction of *Strongylocentrotus purpuratus* sperm by pH and K^+. *J. Cell Biol.* **87**, 140a, No. 2, Pt. 2.

Collins, F., and Epel, D. (1977). The role of calcium ions in the acrosome reaction of sea urchin sperm: Regulation of exocytosis. *Exp. Cell Res.* **106**, 211–222.

Cornett, L. E., and Meizel, S. (1978). Stimulation of *in vitro* activation and the acrosome reaction of hamster spermatozoa by catecholamines. *Proc. Natl. Acad. Sci. U.S.A.* **75**, 4954–4958.

Dan, J. C. (1952). Studies on the acrosome. I. Reaction to egg-water and other stimuli. *Biol. Bull. (Woods Hole, Mass.)* **103**, 54–66.

Dan, J. C. (1954). Studies on the acrosome. III. Effect of calcium deficiency. *Biol. Bull. (Woods Hole, Mass.)* **107**, 335–349.

Dan, J. C. (1967). Acrosome reaction and lysis. *In* "Fertilization: Comparative Morphology, Biochemistry and Immunology" (C. B. Metz and A. Monroy, eds.), Vol. 1, pp. 237–293. Academic Press, New York.

Dan, J. C., Ohori, Y., and Kushida, H. (1964). Studies on the acrosome. VII. Formation of the acrosomal process in sea urchin spermatozoa. *J. Ultrastruct. Res.* **11**, 508–524.

Davis, B. K. (1978). Effects of calcium on motility and fertilization by rat spermatozoa *in vitro*. *Proc. Soc. Exp. Biol. Med.* **157**, 54–56.

Decker, G. L., Joseph, D. B., and Lennarz, W. J. (1976). A study of factors involved in induction of the acrosomal reaction in sperm of the sea urchin, *Arbacia punctulata*. *Dev. Biol.* **53**, 115–125.

Garbers, D. L. (1981). The elevation of cyclic AMP concentrations in flagella-less sea urchin sperm heads. *J. Biol. Chem.* **256**, 620–624.

Garbers, D. L., and Hardman, J. G. (1975). Factors released from sea urchin eggs affect cyclic nucleotide metabolism in sperm. *Nature (London)* **257**, 677–678.

Garbers, D. L., First, N. L., Gorman, S. K., and Lardy, H. A. (1973). The effects of cyclic nucleotide phosphodiesterase inhibitors on ejaculated porcine spermatozoan metabolism. *Biol. Reprod.* **8**, 599–606.

Gibbons, B. H. (1980). Intermittent swimming in live sea urchin sperm. *J. Cell Biol.* **84**, 1–12.

Gibbons, B. H., and Gibbons, I. R. (1980). Calcium induced quiescence in reactivated sea urchin sperm. *J. Cell Biol.* **84**, 13–27.

Gillies, R. J., and Deamer, D. W. (1979). Intracellular pH: Methods and applications. *Curr. Top. Bioenerg.* **9**, 63–87.

Goldstein, S. F. (1979). Starting transients in sea urchin sperm flagella. *J. Cell Biol.* **80**, 61–68.

Green, D. P. L. (1978). The induction of the acrosome reaction in guinea pig sperm by the divalent metal cation ionophore A23187. *J. Cell Sci.* **32**, 137–151.

Gregg, K. W., and Metz, C. B. (1976). Physiological parameters of the sea urchin acrosome reaction. *Biol. Reprod.* **14**, 405–411.

Hansbrough, J. R., and Garbers, D. L. (1981). Speract: Purification and characterization of a peptide associated with eggs that activates spermatozoa. *J. Biol. Chem.* **256**, 1447–1452.

Hansbrough, J. R., Kopf, G. S., and Garbers, D. L. (1980). The stimulation of sperm metabolism by a factor associated with eggs and by 8-bromo-guanosine 3′,5′-monophosphate. *Biochim. Biophys. Acta* **630**, 82–91.

Harvey, E. B. (1930). An effect of lack of oxygen on the sperm and unfertilized eggs of *Arbacia punctulata* and on fertilization. *Biol. Bull. (Woods Hole, Mass.)* **58**, 288–292.

Hathaway, R. R. (1963). Activation of respiration in sea urchin spermatozoa by egg water. *Biol. Bull. (Woods Hole, Mass.)* **125**, 486–498.

Hoskins, D. D., and Casillas, E. R. (1975). Function of cyclic nucleotides in mammalian spermatozoa. *In* "Handbook of Physiology" (S. Geiger, ed.), Sect. 7, Vol. 5, pp. 453–460. Williams & Wilkins, Baltimore, Maryland.

Hoskins, D. D., Stephens, D. T., and Hall, M. L. (1974). Cyclic adenosine 3',5'-monophosphate and protein kinase levels in developing bovine spermatozoa. *J. Reprod. Fertil.* **37**, 131–133.

Hoskins, D. D., Brandt, H., and Acott, T. S. (1978). Initiation of sperm motility in the mammalian epididymis. *Fed. Proc., Fed. Am. Soc. Exp. Biol.* **37**, 2534–2542.

Hoskins, D. D., Johnson, D., Brandt, H., and Acott, T. S. (1979). Evidence for a role for a forward motility protein in the epididymal development of sperm motility. *In* "The Spermatozoon: Maturation, Motility, Surface Properties and Comparative Aspects" (D. W. Fawcett and J. M. Bedford, eds.), pp. 43–53. Urban & Schwarzenberg, Baltimore, Maryland.

Hyne, R. V., and Garbers, D. L. (1979). Calcium-dependent increase in adenosine 3'-5'-monophosphate and induction of the acrosome reaction in guinea pig spermatozoa. *Proc. Natl. Acad. Sci. U.S.A.* **76**, 5699–5703.

Jones, H. P., Bradford, M. M., McRorie, R. A., and Cormier, M. J. (1978). High levels of a calcium-dependent modulator protein in spermatozoa and its similarity to brain modulator protein. *Biochem. Biophys. Res. Commun.* **82**, 1264–1272.

Kinsey, W., SeGall, G. K., and Lennarz, W. J. (1979). The effect of the acrosome reaction on the respiratory activity and fertilizing capacity of echinoid sperm. *Dev. Biol.* **71**, 49–59.

Kopf, G. S., and Garbers, D. L. (1980). Calcium and a fucose-sulfate rich polymer regulate sperm cyclic nucleotide metabolism and the acrosome reaction. *Biol. Reprod.* **22**, 1118–1126.

Kopf, G. S., Tubb, D. J., and Garbers, D. L. (1979). Activation of sperm respiration by a low molecular weight egg factor and by 8-bromo guanosine 3',5'-monophosphate. *J. Biol. Chem.* **254**, 8554–8560.

Lee, H. C., Schuldiner, S., Johnson, C., and Epel, D. (1980). Sperm motility initiation: Changes in intracellular pH, Ca^{2+} and membrane potential. *J. Cell Biol.* **87**, No. 2, Pt. 2, 39a.

Lillie, F. R. (1914). Studies of fertilization. VI. The mechanism of fertilization in *Arbacia*. *J. Exp. Zool.* **16**, 523–590.

Lindemann, C. B. (1978). A cAMP-induced increase in the motility of demembranated bull sperm models. *Cell* **13**, 9–18.

Lui, C. W., and Meizel, S. (1977). Biochemical studies of the *in vitro* acrosome reaction inducing activity of bovine serum albumin. *Differentiation* **9**, 59–66.

Lui, C. W., Cornett, L. E., and Meizel, S. (1977). Identification of the bovine follicular fluid protein involved in the *in vitro* induction of the hamster sperm acrosome reaction. *Biol. Reprod.* **17**, 34–41.

McGrady, A. V., Nelson, L., and Ireland, M. (1974). Ionic effects on the motility of bull and chimpanzee spermatozoa. *J. Reprod. Fertil.* **40**, 71–76.

Meizel, S., and Working, P. K. (1980). Further evidence suggesting the hormonal stimulation of hamster sperm acrosome reactions by catecholamines *in vitro*. *Biol. Reprod.* **22**, 211–216.

Miyamoto, H., and Ishibashi, T. (1975). The role of calcium ions in fertilization of mouse and rat eggs *in vitro*. *J. Reprod. Fertil.* **45**, 523–526.

Morisawa, M., and Suzuki, K. (1980). Osmolality and potassium ions: Their roles in initiation of sperm motility in teleosts. *Science* **210**, 1145–1147.

Morton, B. E., Harrigan-Lum, J., Albagli, L., and Jooss, T. (1974). The activation of motility in quiescent hamster sperm from the epididymis by calcium and cyclic nucleotides. *Biochem. Biophys. Res. Commun.* **56**, 372–379.

Morton, B. E., Sagadraca, R., and Fraser, C. (1978). Sperm motility within the mammalian epididymis: Species variation and correlation with free calcium levels in epididymal plasma. *Fertil. Steril.* **29**, 695–698.

Morton, B. E., Fraser, C. F., and Sagadraca, R. (1979). Initiation of hamster sperm motility from quiescence: Effect of conditions upon flagellation and respiration. *Fertil. Steril.* **32**, 222–227.

Mrsny, R. J., and Meizel, S. (1980). Evidence suggesting a role for cyclic nucleotides in acrosome reactions of hamster sperm *in vitro*. *J. Exp. Zool.* **211**, 153–157.

Nelson, G. A., and Ward, S. (1980). Vesicle fusion, pseudopod extension and amoeboid motility are induced in nematode spermatids by the ionophore monensin. *Cell* **19**, 457–464.

Nelson, L. (1978). Chemistry and neurochemistry of sperm motility control. *Fed. Proc., Fed. Am. Soc. Exp. Biol.* **37**, 2543–2547.

Nishioka, D., and Cross, N. (1978). The role of external sodium in sea urchin fertilization. *In* "Cell Reproduction" (E. R. Dirksen, D. M. Prescott, and C. F. Fox, eds.), pp. 403–413. Academic Press, New York.

Ohtake, H. (1976a). Repiratory behavior of sea urchin spermatozoa. I. Effect of pH and egg water on the respiratory rate. *J. Exp. Zool.* **198**, 303–312.

Ohtake, H. (1976b). Respiratory behavior of sea urchin spermatozoa. II. Sperm-activating substance obtained from jelly coat of sea urchin eggs. *J. Exp. Zool.* **198**, 313–322.

O'Rand, M. G. (1979). Changes in sperm surface properties correlated with capacitation. *In* "The Spermatozoon: Maturation, Motility, Surface Properties and Comparative Aspects" (D. W. Fawcett and J. M. Bedford, eds.), pp. 195–203. Urban and Schwarzenberg, Baltimore, Maryland.

Rogers, B. J., and Garcia, L. (1979). Effect of cAMP on acrosome reaction and fertilization. *Biol. Reprod.* **21**, 365–372.

Rottenberg, H. (1979). The measurement of membrane potential and ΔpH in cells, organelles, and vesicles. *In* "Methods in Enzymology" (S. Fleischer and L. Packer, eds.), Vol. 55, pp. 547–569. Academic Press, New York.

Russell, L., Peterson, R. N., and Freund, M. (1979). Morphologic characteristics of the chemically induced acrosome reaction in human spermatozoa. *Fertil. Steril.* **32**, 87–92.

Saling, P. M., Storey, B. T., and Wolf, D. P. (1978). Calcium-dependent binding of mouse epididymal spermatozoa to the zona pellucida. *Dev. Biol.* **65**, 515–525.

Sano, K., and Kanatani, H. (1980). External calcium ions are requisite for fertilization of sea urchin eggs by spermatozoa with reacted acrosomes. *Dev. Biol.* **78**, 242–246.

Santos-Sacchi, J., and Gordon, M. (1980). Induction of the acrosome reaction in guinea pig spermatozoa by cGMP analogues. *J. Cell Biol.* **85**, 798–803.

Schackmann, R. W., and Shapiro, B. M. (1981). A partial sequence of ionic events associated with the acrosome reaction of *Strongylocentrotus purpuratus*. *Dev. Biol.* **81**, 145–154.

Schackmann, R. W., Eddy, E. M., and Shapiro, B. M. (1978). The acrosome reaction of *Strongylocentrotus purpuratus* sperm: Ion requirements and movements. *Dev. Biol.* **65**, 483–495.

Schmidt, T. and Epel, D. (1981). Is there a role for Ca^{2+} influx during fertilization of the sea urchin egg? *J. Cell Biol.* **91**, No. 2, Pt. 2, 179a.

SeGall, G. K., and Lennarz, W. J. (1979). Chemical characterization of one component of the jelly coat from sea urchin eggs responsible for induction of the acrosome reaction. *Dev. Biol.* **71**, 33–48.

Shapiro, B. M., Schackmann, R. W., Gabel, C. A., Foerder, C. A., Farrance, M. L., Eddy,

E. M., and Klebanoff, S. J. (1980). Molecular alterations in gamete surfaces during fertilization and early development. *Symp. Soc. Cell Biol.* **38,** 127–150.

Singh, J. P., Babcock, D. F., and Lardy, H. A. (1978). Increased Ca²⁺ influx is a component of sperm capacitation. *Biochem. J.* **172,** 549–546.

Singh, J. P., Babcock, D. F., and Lardy, H. A. (1980). Induction of accelerated acrosome reaction in guinea pig sperm. *Biol. Reprod.* **22,** 566–570.

Stewart, T. A., and Forrester, I. T. (1978). Identification of a cholinergic receptor in ram spermatozoa. *Biol. Reprod.* **19,** 965–970.

Stewart, T. A., and Forrester, I. T. (1979). Acetylcholine-induced calcium movements in hypotonically washed ram spermatozoa. *Biol. Reprod.* **21,** 109–115.

Takahashi, Y. M., and Sugiyama, M. (1973). Relationship between the acrosome reaction and fertilization in the sea urchin. I. Fertilization in Ca-free seawater with egg-water treated spermatozoa. *Dev. Growth Differ.* **15,** 261–267.

Talbot, P., Summers, R. G., Hylander, B. L., Keough, E. M., and Franklin, L. E. (1976). The role of calcium in the acrosome reaction: An analysis using ionophore A23187. *J. Exp. Zool.* **198,** 383–392.

Tamblyn, T. M., Singh, J. P., Lorton, S. P., and First, N. L. (1979). Mechanisms controlling motility of stallion spermatozoa. *J. Reprod. Fertil., Suppl.* **27,** 31–37.

Tilney, L. G., Kiehart, D. P., Sardet, C., and Tilney, M. (1978). Polymerization of actin. IV. Role of Ca²⁺ and H⁺ in assembly of actin and in membrane fusion in the acrosomal reaction of echnioderm sperm. *J. Cell Biol.* **77,** 536–550.

Tubb, D. J., Kopf, G. S., and Garbers, D. L. (1978). The elevation of sperm adenosine 3′,5′-monophosphate concentrations by factors released from eggs require calcium. *Biol. Reprod.* **18,** 181–185.

Tubb, D. J., Kopf, G. S., and Garbers, D. L. (1979). Starfish and horseshoe crab egg factors cause elevations of cyclic nucleotide concentrations in spermatozoa from starfish and horseshoe crabs. *J. Reprod. Fertil.* **56,** 539–542.

Turner, T. T. (1979). On the epididymis and its function. *Invest. Urol.* **16,** 311–321.

Turner, T. T., and Howards, S. S. (1978). Factors involved in the initiation of sperm motility. *Biol. Reprod.* **18,** 571–578.

Vacquier, V. D., Brandriff, B., and Glabe, C. G. (1979). The effect of soluble egg jelly on the fertilizability of acid-dejellied sea urchin eggs. *Dev. Growth Differ.* **21,** 47–60.

Wada, S. K., Collier, J. R., and Dan, J. C. (1956). Studies of the acrosome. V. An egg-membrane lysin from the acrosomes of *Mytilus edulis* spermatozoa. *Exp. Cell Res.* **10,** 168–180.

Wasserman, P., and Bleil, J. D. (1981). The role of zona pellucida glycoproteins as regulators of sperm-egg interactions in the mouse. *J. Supramol. Struct., Suppl.* **5,** 245.

Wong, P. Y. D., Lee, W. M., and Tsang, A. Y. F. (1981). The effects of extracellular sodium on acid release and motility initiation in rat caudal epididymal spermatozoa *in vitro. Exp. Cell Res.* **131,** 97–104.

Yanagimachi, R. (1978). Calcium requirement for sperm-egg fusion in mammals. *Biol. Reprod.* **19,** 949–958.

Yanagimachi, R., and Usui, N. (1974). Calcium dependence of the acrosome reaction and activation of guinea pig spermatozoa. *Exp. Cell Res.* **89,** 161–174.

Yoshioka, T., and Inoue, H. (1980). High potassium effect on the mobility of sea urchin sperm. *Nagoya J. Med. Sci.* **42,** 82–84.

Chapter 13

The Physiology and Chemistry of Calcium during the Fertilization of Eggs

DAVID EPEL

	I.	Introduction	356
II.	Evidence that Free Calcium Content Changes at Fertilization	357
III.	Calcium Permeability at Fertilization	358
	A. Flux Studies	358
	B. Membrane Potential Studies	358
IV.	The Rise in Intracellular Calcium as the Cause of Activation of the Egg	360
V.	Egg Activation as a Result of the Release of Calcium from Intracellular Stores	361
VI.	What Is the Role of the Calcium Influx after Fertilization?	363
VII.	Extracellular Calcium as a Requirement for Activation in Eggs	365
VIII.	Nature of the Cytoplasmic Calcium Stores	367
IX.	Calcium-Binding Proteins and Calcium Buffers of the Egg	368
X.	Control of Metabolism by Calcium	369
	A. Possible Reactions Mediated by the Calcium Rise	369
	B. Cortical Reaction	371
	C. NAD Kinase	373
	D. Lipoxygenase	377
XI.	Summary and Overview	378
	A. How Does Calcium Increase?	378
	B. Role of the Transient Calcium Increase	379
	References	379

CALCIUM AND CELL FUNCTION, VOL. II
Copyright © 1982 by Academic Press, Inc.
ISBN 0-12-171402-0

I. INTRODUCTION

The problem of how the egg is activated at fertilization has long intrigued cell biologists. Important early statements on the problem of cell activation at fertilization were made by Otto Warburg and Jacques Loeb at the turn of the century. Warburg (1908) described some of the metabolic consequences, and Loeb (1913) studied in depth the phenomenon of artificial parthenogenesis in which the egg is activated to develop in the absence of sperm. The concept that developed from their work, which was carried out on sea urchin eggs, was that of a meta-stable system in which development is normally triggered by the sperm but can be artificially induced by chemical or physical treatments.

Work in the 1920s and 1930s led to the concept that changes in intracellular calcium might be a critical aspect of the activation process. One line of evidence came from studies on artificial parthenogenesis; extracellular calcium was usually required, and elevated calcium levels could by itself cause egg activation. A critical line of evidence was the finding by Mazia (1937) that there was an increase in free or cytosolic calcium following fertilization of sea urchin eggs.

Subsequent work on fertilization was greatly influenced by concepts derived from immunology, and an active school of thought in the 1940s and 1950s centered around the hypothesis that surface interactions between sperm and egg are analogous to antigen–antibody reactions. In this view, an antigen–antibody type of reaction between sperm and egg somehow caused the egg to begin development (see, e.g., Tyler, 1963). But interest in the early changes in fertilization subsequently waned, and most of the emphasis in the 1950s and 1960s centered on the consequences of fertilization, such as the large increase in protein synthesis which occurred after activation of the sea urchin egg and appeared to be controlled at the translational level (reviewed by Giudice, 1973).

Several important changes have occurred in the last 10–15 years which have resulted in a literal renaissance in studies on the initial interactions between sperm and egg and the role of calcium in activation. A trivial factor in this redirection is a by-product of the jet age—air freight. This has allowed more widespread use of model systems for studying fertilization, such as the gametes of marine organisms, as from sea urchins. Previously these were only available in goodly amounts at marine biology stations or at coastal universities. A major center, for example, was the Marine Biological Laboratory at Woods Hole, Massachusetts, but research there was primarily a summer activity. In general, protracted research efforts were precluded. This limitation was changed by the advent of jet airplane transport, the design and commercial manufacture of refrigerated seawater aquaria, and the founding of commercial companies which would send marine material anywhere in the world. Thus, a new factor was the availability, on a year-round basis, of good material for fertilization studies.

A related factor in the renaissance of interest in fertilization was the development of other model systems for studying this phenomenon. These now include good *in vitro* systems for the frog (Wolf and Hedrick, 1971) and mammal (see e.g., Gwatkin, 1977) and new procedures for obtaining gametes in large amounts from starfish and mollusks (see, e.g., Kanatani, 1973; Morse *et al.*, 1977).

A second important breakthrough was the development and application of new techniques for monitoring intracellular activity *in vivo*. These included monitoring of oxidation and reduction of pyridine nucleotides by *in situ* fluorescence, monitoring of ion permeability changes with membrane potential microelectrodes, and monitoring of intracellular calcium with the use of aequorin. These studies, combined with the application of biochemical and cell biological insights from other systems, have now provided several important paradigms concerning the triggering of development at fertilization.

II. EVIDENCE THAT FREE CALCIUM CONTENT CHANGES AT FERTILIZATION

It was noted above that the 1937 studies of Mazia indicated an increase in free or cytosolic calcium. The conclusions of these studies were reexamined many years later with the use of the calcium-monitoring protein aequorin, initially on fish (Ridgeway *et al.*, 1977) and sea urchin eggs (Steinhardt *et al.*, 1977). The technique was to inject the photoprotein directly into the egg and then to observe changes in luminescence upon fertilization. The measurements with the fish egg, facilitated by its large size and ability to receive large amounts of the calcium indicator, were made on single cells; also, with the use of image intensifier techniques cytological localization could be carried out (Gilkey *et al.*, 1978). Such studies were not possible with the much smaller sea urchin egg; in this case the technique used was to inject a population of 10–30 eggs and to monitor luminescence from the entire population of eggs when they were fertilized (Fig. 1, from Steinhardt *et al.*, 1977).

The results were similar in both cases. A low or nonmeasurable level of luminescence was seen in the unfertilized state. Shortly after fertilization, coincident with the beginning of a secretory phase (referred to as the cortical reaction or cortical exocytosis; see below), a large increase in

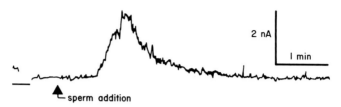

Fig. 1. Changes in free calcium concentration following fertilization of the sea urchin egg, as inferred from the luminescence of a group of aequorin-injected eggs. The luminescence, as measured by a photomultiplier, is related to the free calcium concentration. (Redrawn from Steinhardt *et al.*, 1977, and reprinted courtesy of Academic Press.)

luminescence began. This increase in luminescence (i.e., free calcium rise) lasted for several minutes, and its termination coincided approximately with termination of the aforementioned cortical reaction (Ridgeway *et al.*, 1977; Steinhardt *et al.*, 1977).

With the use of the image intensifier on fish eggs, it was found that the luminescence did not simultaneously spread over the entire egg surface; rather it began at the point of sperm entry and then moved through the egg as a narrow band of light (i.e., a band of calcium). These results indicated that free calcium appeared as a wave passing through the egg and was either sequestered or effluxed from the egg as the wave passed through it (Gilkey *et al.*, 1978). The important point, seen in both systems, was that the calcium rise was transient.

III. CALCIUM PERMEABILITY AT FERTILIZATION

A. Flux Studies

Fertilization is also accompanied by large changes in the influx and efflux of calcium, which are measured by isotope studies and direct chemical analysis. In sea urchin eggs, isotope studies show two phases of calcium uptake (Azarnia and Chambers, 1976; Paul and Johnston, 1978). The first begins almost as soon as sperm are added and ends sometime between 30 and 60 sec after insemination (Paul and Johnston, 1978). The second phase begins coincidentally with the formation of an extracellular coat referred to as the hyaline layer. This coat, which comes from the secretory products of cortical granule exocytosis, binds calcium with a high affinity; one possible explanation for the second phase of calcium uptake therefore is that this "uptake" is actually a binding of extracellular calcium to the layer. Such an hypothesis is supported by the finding that

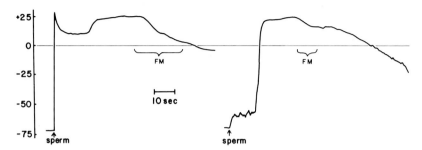

Fig. 2. Membrane potential changes following fertilization of *Lytechinus variegatus* eggs. (Right) A typical response in normal seawater. (Left) The response in low-calcium media. This experiment thus indicates the biphasic nature of the membrane potential changes, with the initial depolarization resulting from an influx of calcium ions. (Redrawn from Chambers and de Armendi, 1979, and reprinted courtesy of Academic Press.)

much of the calcium can be removed by washing the eggs in media which will remove the hyaline layer (Azarnia and Chambers, 1976). It thus appears that the major change in calcium permeability is an early one and is detected as a net influx of calcium into the egg. Superimposed on top of this is a calcium-binding phase in which calcium binds to the extracellular layers exposed by the cortical granule exocytosis. It is not clear whether this initial calcium uptake terminates at the time of exocytosis or continues and is obscured by the much larger magnitude of calcium binding to the hyaline layer.

Coincident with the increased uptake of calcium is an increased efflux. This was initially shown in experiments in which sea urchin eggs were preloaded with Ca[45] and the efflux then determined at various times before and after fertilization or artificial parthenogenesis (Steinhardt and Epel, 1974). Fertilization or activation with the ionophore A23187 resulted in a large efflux of calcium from the egg, which lasted for 30 min. This efflux might be of considerable magnitude; if one subtracts the loosely associated calcium (bound to the hyaline layer?), there actually is a net decrease in the total Ca^{2+} content of the egg (Azarnia and Chambers, 1976).

B. Membrane Potential Studies

Another way of detecting changes in ion permeability at fertilization is from measurements of the membrane potential of the egg and its changes at fertilization. For example, the sea urchin egg is rapidly depolarized at fertilization (Steinhardt *et al.*, 1971; Jaffe, 1976; Chambers and de Armendi, 1979) and, as shown in Fig. 2, the initial phase of this depolarization results from the aforementioned influx of calcium. This influx, therefore, is an electrogenic one.

Major studies on this calcium influx have been carried out on starfish and tunicate eggs, where the depolarization can be more properly labeled an action potential (reviewed by Hagiwara and Jaffe, 1979). Thus sufficient depolarization of the egg membrane results in the membrane undergoing a much more extensive depolarization analogous to an action potential. The evidence that the action potential results from a calcium influx has come from ion substitution studies which show that the major depolarization will not occur in the absence of extracellular calcium. These calcium-induced action potentials have also been seen in eggs of *Urechis* (Jaffe *et al.*, 1979) and sea urchins (Chambers and de Armendi, 1979).

The ability to remove extracellular calcium and to examine membrane potential changes allows one to assess the period of the calcium permeability change at fertilization. In the sea urchin egg, depolarization behavior is only affected for the first few seconds in the absence of calcium. This suggests that the electrogenic phase of calcium uptake only occurs in the first few seconds. This behavior is to be contrasted with the isotopic data which indicate that calcium influx lasts longer (up to 30 sec), and also with the efflux data which indicate that permeability changes might last for many more minutes. Perhaps one possible resolution of these differences is that there is a secondary calcium uptake which is nonelectrogenic, resulting either from calcium–cation exchange (such as calcium–calcium, calcium–sodium, or calcium–hydrogen) or from the opening of a voltage-gated calcium channel.

IV. THE RISE IN INTRACELLULAR CALCIUM AS THE CAUSE OF ACTIVATION OF THE EGG

Evidence that the rise in calcium is involved in activation comes from several types of studies. One major line of evidence is that experimental regimes which result in increased cytosolic calcium also result in egg activation. For example, incubation of eggs in an ionophore (A23187 or X537A) which increases free calcium results in activation of the egg (Chambers *et al.*, 1974; Steinhardt and Epel, 1974). This has been seen in a wide variety of organisms. Interestingly, these ionophores generally activate eggs in the absence of extracellular calcium, suggesting that these drugs act to release calcium from an intracellular store (Section V). The only exception to this rule is seen in the eggs of bivalve mollusks, where extracellular calcium is required (Schuetz, 1975). These findings suggest that the ionophore could act on calcium stores or on some membrane property of the egg which then nonspecifically results in activation of the egg. This latter argument is countered by the observation that in some species of eggs extracellular calcium is required for ionophore activation. Another argument is that all eggs so far tested are activated by calcium ionophores (see, e.g., Epel, 1978).

A second and extremely important line of evidence is that, if one injects a calcium chelator, such as EGTA, into eggs, the activation process is prevented. This has now been done in the sea urchin egg (Zucker and Steinhardt, 1980) and has also recently been reported for the fish egg (see review by Jaffe, 1980).

The third line of evidence is that frog eggs can be activated directly by the injection of calcium (Hollinger and Schuetz, 1976). Related to this is the aforementioned observation that eggs can be artificially activated by mechanical means, such as pricking with a pin, but only in the presence of extracellular calcium (see, e.g., Wolf, 1974). This could be because calcium is needed to prevent lysis of the egg or because the activation process requires the influx of calcium, which could occur when the egg is "insulted" by the opening up of a hole in its plasma membrane.

V. EGG ACTIVATION AS A RESULT OF THE RELEASE OF
CALCIUM FROM INTRACELLULAR STORES

The experiments with ionophore activation, showing that eggs from most organisms (except bivalve mollusks) can be activated in the absence of exogenous calcium, suggest that extracellular calcium is not required for the activation process, i.e., that the aforementioned influx of calcium that accompanies fertilization is not the Ca^{2+} detected intracellularly at fertilization. However, an argument against this hypothesis is that the ionophore might release calcium from some intracellular store not normally used during egg activation. For example, the ionophore could release calcium from membrane-bound stores in mitochondria, but this store may not be the source of intracellular calcium for activation.

Some evidence against this is offered by the studies of Steinhardt et al. (1977), who found that, if they allowed sperm to attach to aequorin-loaded sea urchin eggs and then quickly placed the eggs in calcium-free seawater, they still observed the rise in calcium at the time of the cortical reaction. Hence, in the absence of extracellular calcium, there was an increase in intracellular calcium. The problem with this experiment is that the eggs were inseminated in seawater containing calcium. Perhaps the influx of calcium that occurred immediately after sperm–egg binding might later result in the release of calcium from an intracellular store. In other words, the hypothesis is that the initial calcium influx at sperm–egg contact induces the further release of calcium and that this then results in activation of the egg.

One means of testing this would be to see if one could fertilize eggs in the absence of calcium. This, however, is not a simple experiment, since the acrosome reaction of the sperm (an exocytotic reaction which exposes the lysins necessary for the sperm to pass through the membrane and also exposes sperm–egg binding components) requires calcium (see, e.g., Tilney et al., 1978). This requirement can be bypassed. In sea urchins, for example, the acrosome reaction is normally induced by egg jelly, and a means of circumventing the problem of fertilizing the eggs in calcium-free seawater is first to induce the acrosome reaction in calcium-containing sea water and then add the acrosome-reacted sperm to eggs in calcium-free media.

The results obtained with this experimental protocol have been conflicting. The early experiments of Takahashi and Sugiyama (1973) suggested that calcium was not required. However, they did not carry out their experiments in the presence of EGTA, and the normal ambient levels of calcium in calcium-free seawater could have been at a level of 50 μM. In addition, experiments in which chelators were used have shown either a requirement for calcium (Sano and Kanatani, 1980) or no requirement (Chambers, 1980).

Recent work in our laboratory has essentially confirmed Chambers' finding; i.e., one can obtain a high percentage of fertilization in the complete absence of exogenous calcium (less than $10^{-8} M$ calcium as determined using calcium buffers). Our unpublished results (T. Schmidt, C. Patton, and D. Epel) suggest that these other workers most likely were not able to fertilize eggs in the absence of exogenous calcium because of the temperature at which they carried out fertilization. It appears that, although fertilization in normal seawater can occur at relatively high temperatures for these marine organisms ($\sim 20°C$), in the absence of calcium the eggs rapidly lose viability and then cannot be fertilized. As the temperature is lowered, the percentage of fertilization increases dramatically. These experiments, therefore, indicate that—at least in the sea urchin egg—extracellular calcium is not required and affirm the hypothesis that the sperm releases calcium in the egg from an intracellular store.

These results are puzzling, since they indicate that extracellular calcium is not required for sperm–egg fusion. This is a surprising result, since cell fusion normally requires calcium (see, e.g., Lucy, 1978; Poste and Pasternak, 1978). One possibility is that sperm–egg fusion is an exception to this rule. Another possibility is that the sperm carries a high concentration of calcium at the site specialized for sperm–egg fusion.

A third possibility is that the often accepted generalization that calcium is required for fusion is incorrect. The alternate possibility, suggested by our results on fertilization in calcium-free media, is that extracellular calcium is primarily required for cell stability and viability. We noted above that the eggs quickly lost their fertilizability in calcium-free media. This was also true for cell viability; within 5–10 min some eggs in the population begin to lyse. This is a common consequence of placing cells in calcium-free media, and perhaps the calcium requirement seen for cell fusion really represents this requirement for viability. To be sure, calcium is required for fusion in viral and model membrane systems (see Lucy, 1978; Poste and Pasternak, 1978), but in the case of living cells calcium might perhaps also be required for cell stability; stressed cells, as would occur in a calcium-free media, simply might be incapable of fusion.

VI. WHAT IS THE ROLE OF THE CALCIUM INFLUX AFTER FERTILIZATION?

We noted earlier the occurrence, in the sea urchin egg, of a transient and rapid accumulation of calcium in the period between the time of sperm–egg contact and the beginning of the cortical reaction. In terms of total calcium content the increase in calcium is small (the total calcium content of the egg is in the millimolar range, and the amount of calcium influx is about 1% of this, or about 10^{-5} M). However, if this calcium influx remained free in the cytosol, this would represent a marked increase in the free calcium level; indeed, it would be as much or greater than the release detected by aequorin measurements at the time of cortical exocytosis.

Is this calcium influx important to the physiology of the egg and the fertilization process? The aforementioned studies, showing that activation of the egg can occur in the complete absence of calcium, indicate that it is not. Indeed, we have found that, if eggs are fertilized and remain in calcium-free seawater, they will undergo pronuclear fusion and even pass through the first few cleavages (we have not followed it further than this). Thus the calcium influx is unnecessary for both fertilization and the events leading to at least the first division cycles.

One possibility is that this calcium influx is secondary to the movement of other ions. For example, it is known that sodium channels also open onto the egg surface (Steinhardt et al., 1971; Chambers and de Armendi, 1979), and perhaps calcium coincidentally passes through this sodium channel. Our recent studies on this calcium influx indicate that calcium transport occurs through a classic calcium channel. One line of evidence

364 David Epel

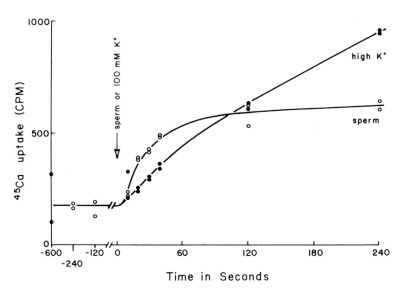

Fig. 3. Calcium influx following fertilization or incubation of unfertilized eggs in K⁺-supplemented seawater. Since the plasma membrane is partially depolarized by 100 mM K⁺, the increased Ca²⁺ influx suggests the presence of a voltage-gated Ca²⁺ channel. Regardless, the influx of Ca²⁺ is inadequate to activate the egg and, as described in the text, is also not required for activation by the sperm. (Data of D. Epel, T. Schmidt, and C. Patton.)

is that a K⁺-induced calcium uptake is inhibited by the calcium channel blocker verapamil.

The approach we have utilized in studying this calcium uptake is based on the possibility that it is carried by a voltage-dependent calcium channel. We have observed, as shown in Fig. 3, that when the egg plasma membrane is depolarized at elevated extracellular K⁺ levels, there is a corresponding influx of calcium ion. (This K⁺-induced calcium uptake, which is maximal at 100 mM potassium, corresponds to a membrane potential of about −20 mV; Jaffe and Robinson, 1978). The channel opened by this membrane depolarization is also sensitive to verapamil, suggesting that it is similar to the channel which opens at fertilization (D. Epel, unpublished observation).

The potassium-induced Ca²⁺ uptake is not affected by external sodium ion. If this channel is indeed identical to the one opened by fertilization, these observations argue against the hypothesis that the calcium influx is incidental and point to the concept of a voltage-dependent calcium channel, probably opened during the time when the membrane is depolarized.

Given that this calcium influx has some physiological basis and is not secondary to the influx of another ion, its role becomes even more mysterious. As noted above, the calcium influx does not appear to be necessary for egg activation. Also, the potassium-induced influx of calcium ion into the egg does not result in any obvious activation of egg metabolism. For example, there is no indication of cortical exocytosis and no activation of at least one of the calcium-dependent enzymes normally activated at fertilization, NAD kinase (see below).

Perhaps the only role of the calcium influx is to induce a blockage of polyspermy, a mechanism which allows the fusion of only one sperm with the egg and prevents supernumerary sperm from fusing (Jaffe, 1976; Gould-Somero et al., 1979). We noted earlier that the initial depolarization of the plasma membrane appears to result from the influx of calcium ion (Chambers and deArmendi, 1979; Cross, 1981). Also, there is a growing body of evidence that this initial electrical depolarization is part of a mechanism that prevents polyspermy. Thus the initial calcium influx could result in the voltage-dependent blockage of polyspermy. Since calcium uptake is also voltage-dependent, it might be that the second phase of calcium uptake is sustained by the subsequent and longer-lasting depolarization of the membrane, which results from an influx of sodium ion. If the above interpretation is correct, it would provide a role for the initial and electrogenic component of the calcium influx. Perhaps this secondary calcium influx is involved in some sort of later surface change which is also part of polyspermy blockage.

VII. EXTRACELLULAR CALCIUM AS A REQUIREMENT FOR ACTIVATION IN EGGS

The extensive studies summarized above on the sea urchin egg indicate that extracellular calcium is not required for activation by sperm or an ionophore. Although similar, intensive studies have not been done on eggs of other species, the absence of a requirement for extracellular calcium for ionophore activation of most eggs suggests that, in these cases also, activation ensues from the release of calcium from an intracellular store.

The exception to this rule, as noted above, appears to be the eggs of bivalve mollusks. Some recent unpublished studies suggest that these eggs also differ in the physiology of activation in a requirement for extracellular calcium ions. These studies (F. Dubé, unpublished) suggest that sperm–egg interaction in these species might not induce the release of calcium from an intracellular store but rather that fertilization induces the influx of calcium from seawater and that this influx then causes activation of the egg.

The evidence for this idea comes from studies on the phenomenon of activation by increased potassium ion; these eggs, unlike those of sea urchins, can be activated simply by raising the extracellular potassium level. The recent work of Dubé indicates that this activation also requires extracellular calcium, and isotope studies reveal that there is a calcium influx which is dependent on the extracellular potassium. Thus these eggs are similar to sea urchin eggs in having a potassium-induced calcium influx but are dissimilar in that the calcium influx is adequate to activate these eggs.

If a comparative conclusion is to be drawn from these types of studies on sea urchins and bivalve mollusks, it is that eggs might possess two calcium transport systems and calcium stores involved in activation. The simplest form appears to be that seen in mollusks (and possibly annelids, based on the reported activation of annelid eggs by high potassium; Lilly, 1902). In these situations one might imagine that the sperm induces depolarization of the egg plasma membrane and that this opens voltage-dependent channels through which calcium flows, resulting in egg activation. Membrane depolarization might also be involved in altering membrane proteins so as to establish the blockage of polyspermy.

The second and more complex case, as exemplified by the sea urchin egg, also involves an initial calcium influx. The electrogenic component of this calcium influx most probably induces the rapid blockage of polyspermy, and the secondary voltage-dependent influx might be involved in sustaining this depolarization or in some other yet to be described aspect of egg activation. An added aspect of activation, seen in these eggs, is the sperm-induced release of calcium from intracellular stores. This calcium influx then moves explosively through the egg, initiating cortical exocytosis and the other major features of egg activation. It is supported by the finding of large amounts of membrane-bound calcium in cortical vesicles, as visualized by chlortetracycline fluorescence (Schatten and Hemmer, 1979).

A problem with a cortical granule store hypothesis is that the granules do not break down at the site of sperm–egg entry; exocytosis in fact begins in an area somewhat removed from the site of sperm–egg fusion (D. Epel, unpublished). A second and more serious problem is that there are situations when nonpropagated cortical exocytosis occurs; under these conditions cortical granule fusion does not proceed completely around the egg (e.g., Sugiyama, 1955; Epel, 1980; Chambers and Hinkley, 1979). This argues against the idea that the release of calcium from one granule induces the fusion of adjacent granules, release of their contained calcium, and subsequent induction of a self-propagated cortical reaction.

Another possibility is that calcium is contained in a type of ooplasmic reticulum analogous to the sarcoplasmic reticulum of muscle. Eggs do have a reticulum which is present in the area of the cortical granules (see, e.g., Epel, 1978). An especially dramatic example of this is seen in the eggs of some amphibians which have a specific site of sperm–egg fusion; at this site there is a highly specialized reticulum (Campanella and Andreucetti, 1977).

A third candidate for calcium stores are vesicles which have been described as major components of the isolated mitotic apparatus of sea urchin eggs. The recent studies of Silver *et al.* (1980) have centered on the calcium-sequestering ability of these vesicles which concentrate calcium via an ATP-dependent mechanism. It is not known whether these vesicles are the same ones used in fertilization and later in mitosis.

There also is evidence of membrane-bound calcium in eggs. Mentioned earlier was the chlortetracycline fluorescence indicating a large amount of membrane-bound calcium in cortical granules (Schatten and Hemmer, 1979). Using the pyroantimonate technique for localizing calcium in eggs, Cardasis *et al.* (1978) found evidence of more widely distributed membrane-bound calcium. The results were somewhat disappointing in the sense of localizing a *single* calcium store; rather, it appeared that there were many sites of heavy calcium concentrations distributed in the plasma membrane, mitochondrial, and yolk granule membranes. However, under the fixation conditions used by these workers, much of the calcium could have been lost during the fixation process itself.

VIII. NATURE OF THE CYTOPLASMIC CALCIUM STORES

We noted above that sperm or an ionophore causes the release of calcium from an intracellular store in the sea urchin egg; although the identity of these stores has not yet been defined, there is evidence that they are completely discharged at fertilization. Utilizing aequorin-loaded eggs, Steinhardt *et al.* (1977) found that, if eggs were first fertilized and then ionophore added to them, there was no increased light output, indicating that no release of calcium was now induced by the ionophore (in contrast to the increased light output at fertilization). However, if they waited varying lengths of time, they observed that at about 40 min after fertilization a store of calcium appeared which could be released by the application of ionophore.

These results are extremely interesting; either a calcium store is "filled" beginning at about 40 min after fertilization or the calcium-sequestering system of the egg becomes more "relaxed" at this time. Unresolved by these experiments is the identity or nonidentity of these two stores; one would like to know if the store released at fertilization is the same one that is filled at 40 min after fertilization.

Some workers have proposed that the cortical granules contain calcium stores and that their release induces the breakdown of adjacent cortical granules (Vacquier, 1975). This elegant hypothesis would account for the autocatalytic propagation of cortical exocytosis and for further release of calcium; problems with this idea were noted earlier (p. 366).

IX. CALCIUM-BINDING PROTEINS AND CALCIUM BUFFERS OF THE EGG

Early studies on calcium-binding molecules in the egg indicated several calcium-binding factors of low affinity and at least one protein of high affinity. The low-affinity ones, which are probably not involved in the regulation of cytosolic calcium per se, are a binding factor detected by dialysis measurements (Nakamura and Yasumasu, 1974) with a K_d in the area of $10^{-4} M$ and a calcium-binding pigment (Perry and Epel, 1981a) with a K_d in the millimolar range. The former substance is unidentified. The pigmented Ca^{2+} complexer is a naphthoquinone pigment found in the eggs of some species and contained within pigment granules. Recent studies on *Arbacia* eggs by Perry and Epel (1981b) reveal that these naphthoquinone pigments can interact with calcium with the resultant production of H_2O_2. As naphthoquinone is released extracellularly at fertilization (Shapiro, 1946), it is possible that the naphthoquinone release and subsequent Ca^{2+}-induced H_2O_2 production is responsible for the production of peroxide which also accompanies egg activation. As the affinity of naphthoquinone for calcium is so low, it does not appear to function intracellularly. If this hypothesis is correct, its primary role is outside the cell where it can interact with the millimolar concentrations of calcium present in seawater to produce H_2O_2.

Calmodulin has been found both in sea urchin eggs and sperm and in mammalian sperm (Head *et al.*, 1979; Nishida and Kumagai, 1980; Jones *et al.*, 1978; Epel *et al.*, 1981). Coidentity with calmodulin has been confirmed on the basis of coelectrophoretic mobility with beef brain calmodulin, cross-reactivity of antibodies to bovine brain calmodulin in radioimmunoassays (Fig. 4), and its activation of bovine brain phosphodiesterase. There is also some evidence that there may be at least two different types of calmodulin in eggs (W. Burgess, University of Virginia, personal communication).

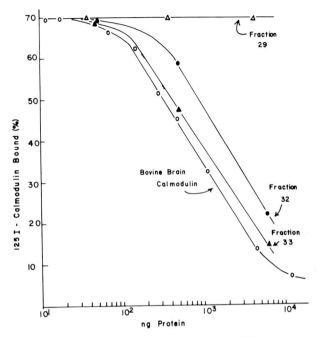

Fig. 4. Radioimmunoassay showing the presence of calmodulin in various column fractions prepared from a sea urchin egg homogenate. Details of the elution of the DEAE column are given in Epel *et al.* (1981), and the fractions noted correspond to the experiment depicted in Fig. 8. The figure compares calmodulin activity in these fractions with authentic calmodulin from bovine brain. (From Epel *et al.,* 1981, and reprinted courtesy of MIT Press.)

The role of calmodulin in egg metabolism is just beginning to be worked out. As indicated below, it may be involved in cortical exocytosis and also in the activation of at least one enzyme. There are as yet no studies on the specific localization of calmodulin in eggs and embryos other than a very intriguing cell homogenization study which indicated that more calmodulin was bound to particulate fractions in unfertilized eggs than in fertilized eggs (Carroll and Longo, 1979).

X. CONTROL OF METABOLISM BY CALCIUM

A. Possible Reactions Mediated by the Calcium Rise

Given that the increase in cytosolic calcium, as measured by the aequorin procedure, is temporally limited to a brief period of time after fertilization (between 30 sec and 1–2 min), one would surmise that the

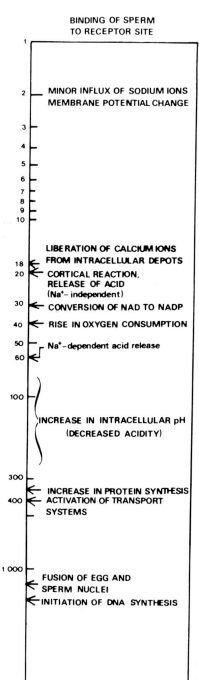

Fig. 5. Timetable of some of the events initiated at fertilization of the sea urchin egg. The ordinate is time, depicted on a logarithmic scale. (Reprinted with modifications from Epel, 1977, and with permission of *Scientific American*.)

effects of calcium must also be limited to this same time period. There-fore, if one were to look for the targets of calcium responsible for egg activation, one would look at the events that occur during this limited time period. A listing of these appears in Fig. 5. They include the aforemen-tioned cortical reaction or exocytosis of granules situated in the egg cortex (hence the term "cortical reaction"), the activation of the enzyme NAD kinase which converts part of the cells' NAD to NADP (Epel, 1964), the activation of a lipoxygenase which oxidizes arachidonic acid to peroxy fatty acids and hydroxy fatty acids (Perry and Epel, 1981a), the produc-tion of extracellular H_2O_2 (Foerder and Shapiro, 1977), a burst in oxygen consumption probably related to both lipoxygenase and H_2O_2 formation (Foerder et al., 1978; Perry and Epel, 1981a), a release of protons in a sodium-dependent fashion with a concomitant rise in intracellular pH (Johnson et al., 1976; Shen and Steinhardt, 1979), an apparent change in the location of glucose-6-phosphate dehydrogenase from a bound to a soluble form (Aune and Epel, 1981), and possibly other changes which have not yet been described.

Subsequent to these changes, all of which occur within the first 1–2 min, hence correspond to the time of elevated calcium, there are changes in the transport properties of the plasma membrane (e.g., Epel, 1972; Shen and Steinhardt, 1980) and a large increase in protein synthesis which occurs through the recruitment of preexisting mRNA (therefore a change at some level of the translational machinery; see, e.g., Humphreys, 1971).

The changes which occur during the period between 30 sec and 1–2 min after fertilization are presumably a direct result of the calcium rise. The later transport and protein synthesis changes could result secondarily from calcium effects; for example, pH changes could be involved (see, e.g., Shen and Steinhardt, 1980; Winkler and Steinhardt, 1981). Alterna-tively, a calcium-activated protein kinase might initiate these later changes, with the kinase activity requiring time for its action to be evident (see, e.g., Keller et al., 1980). In the sections below I shall go over each of the individual changes in which a role for calcium has been found or surmised.

B. Cortical Reaction

The cortical reaction refers to a massive exocytotic event centering on thousands of granules embedded in the egg cortex (see Schuel, 1978). These granules fuse with the egg plasma membrane and release their contents outward at the time of fertilization. The exocytosis begins near (but not at) the site of sperm–egg fusion and propagates over the egg cortex with a time course of about 30 sec (in sea urchin eggs) to several minutes (in fish and mammalian eggs). The products of the secretion include enzymes and structural proteins involved in altering the extracellular membranes around the egg, like the fertilization membrane in sea urchin eggs or the zona pellucida in mammals. The action of these enzymes serves as a fail-safe mechanism preventing polyspermy, most likely by keeping sperm away from the plasma membrane to prevent further sperm–egg fusions once the membrane potential has returned to its original depolarized state (reviewed by Epel, 1978; Schuel, 1978).

A role of calcium in cortical exocytosis might be assumed by the aforementioned common role of calcium in cell fusion systems. Correlative evidence that this might be the case has come from studies on aequorin-loaded fish eggs. In these large cells, the Ca^{2+}-induced luminescence is sufficiently intense to be followed with image intensifier techniques, and one can see exactly where the calcium is released in the egg and its relationship to various cytological events. As noted, these studies reveal that the cytosolic calcium passes through the egg in a wave form beginning at the site of sperm–egg fusion (Gilkey *et al.*, 1978). The wave of elevated cytosolic calcium slightly precedes cortical granule breakdown, suggesting a relationship between calcium release and cortical exocytosis.

Direct evidence for calcium involvement has come from studies in which cortical granule–plasma membrane complexes are isolated. The initial studies involved the isolation of membrane complexes attached to positively charged polyelectrolyte coatings on glass slides. It was found that, when calcium was applied to these complexes, a fusion occurred between adjacent granules with a release of the products of the cortical granules (Vacquier, 1975). With the use of calcium buffers, the concentration of calcium required for exocytosis of these systems was estimated to be ~10 μM (Steinhardt *et al.*, 1977).

Another approach in estimating the calcium requirement for cortical exocytosis was to induce the lysis of eggs in high electric fields, placing the eggs in media similar to the intracellular milieu as regards content of cations and anions, etc., and in a suitable calcium-buffering system (Baker *et al.*, 1980). These studies have indicated a lower calcium level, on the order of 1 μM, and also revealed a requirement for ATP in order to maintain this low calcium sensitivity.

A third approach, developed by H. Sasaki (unpublished), utilizes a procedure in which the cortical granule–plasma membrane complex is isolated as a complete structure in suspension, somewhat analogous to the preparation of a red blood cell ghost. When calcium is added to these suspensions, cortical exocytosis occurs and the kinetics and extent of the reaction can be monitored by changes in turbidity. Sasaki has studied the

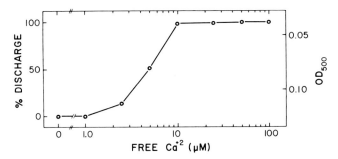

Fig. 6. Calcium dependency of cortical granule discharge in isolated cortexes of sea urchin eggs (*S. purpuratus*). The extent of the reaction is monitored by light scattering, and the calcium level is set with Ca^{2+}-EGTA buffers; as seen, the concentration for 50% discharge is $5 \times 10^{-6}\ M$ (unpublished data of H. Sasaki).

calcium dependency of exocytosis and found a K_d of 6 μM (Fig. 6). Also, in agreement with the results of Baker *et al.* (1980), the sensitivity to calcium is rapidly lost unless the cortexes are stored in the presence of ATP.

The nature of this calcium requirement is still unknown. Calmodulin appears to be involved, since low concentrations of trifluoperazine or other related drugs retard or inhibit the cortical reaction (Baker and Whitaker, 1980). It would be interesting to discern whether this calmodulin-sensitive process is a direct structural interaction between calcium–calmodulin and membrane proteins or via an enzymatic event (e.g., through the action of a protein kinase or phospholipase).

C. NAD Kinase

Evidence that this enzyme is activated came from studies on the levels of pyridine nucleotides after fertilization. Initial studies showed a dramatic alteration in the levels of the different pyridine nucleotides, with a large increase in NADP and NADPH and a concomitant decrease in the level of NAD (Krane and Crane, 1960). This interconversion of pyridine nucleotides indicated activity of the enzyme NAD kinase. This conversion was a very early event in fertilization, as discovered when the timing of the change was studied by determining the kinetics of interconversion of the pyridine nucleotides and by determining the *in situ* fluorescence of reduced NADPH and NADH (Epel, 1964).

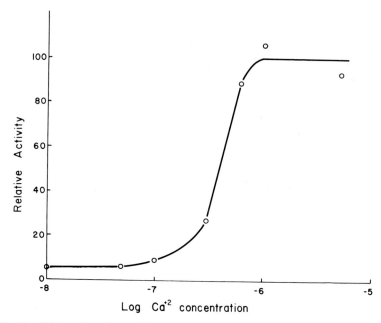

Fig. 7. Calcium dependency of NAD kinase. As seen, the concentration for 50% activity is $4 \times 10^{-7} M$, which makes this enzyme more than 10 times as sensitive to calcium as the cortical reaction. (From Epel *et al.*, 1981, and reprinted with permission of MIT Press.)

Subsequent work on the timing of the calcium rise indicated that the two changes were coincident, suggesting that calcium might be involved in the activation of NAD kinase. Also, work on plant NAD kinase has revealed that this enzyme is also calcium-activated and is controlled by its interaction with calmodulin (reviewed by Cormier *et al.*, 1980).

1. Regulation by Calmodulin

Recent work (Epel *et al.*, 1981) has shown that the NAD kinase of sea urchin eggs is also controlled by calcium via its interaction with calmodulin. This finding provides a clear example of the regulation of an enzyme by calcium both *in vitro* and *in vivo*.

The pertinent evidence that NAD kinase is activated by calcium via calmodulin comes from several different lines of evidence (see Epel *et al.*, 1981). The first is the aforementioned temporal coincidence of the calcium rise and the *in vivo* activity of NAD kinase (as evidenced by the NAD conversion to NADP). Second is the calcium sensitivity of the enzyme (Fig. 7). When enzyme activity is measured at different calcium concentrations with an EGTA buffer, the activity is seen to be highly calcium-dependent, with a K_d of $4 \times 10^{-7}\, M$. Third, one can separate an activator fraction from the NAD kinase by passing the enzyme fraction over a DEAE column in the presence of EGTA. As seen in Fig. 8, the enzyme loses its calcium sensitivity and this sensitivity is restored by a different fraction eluted from the DEAE column. This activator fraction contains calmodulin on the basis of its coelectrophoresis with calmodulin, activation of phosphodiesterase, and cross-reactivity with an antibody to bovine brain calmodulin in a radioimmunoassay. The calmodulin-like activity is not seen in adjacent column fractions, which also do not contain the NAD kinase activator. Finally, one attains NAD conversion *in vivo* when one activates the eggs with the calcium ionophore A23187.

Other *in vivo* experiments show that the activation of NAD kinase is not related to the intracellular pH increase after fertilization but solely to the calcium rise. This is seen in experiments where one activates eggs with ionophore in sodium-free seawater. Under these conditions there is a calcium rise but no pH_i increase (Shen and Steinhardt, 1979); yet, NAD conversion occurs. A final piece of evidence is that the enzyme is inhibited by a typical calmodulin-type inhibitor such as trifluoperazine (Epel *et al.*, 1981).

The role of the NAD conversion is not clear at this time. It is of interest that NAD kinase activation occurs in germinating seeds (Reed, 1970) and in several other animal eggs at fertilization (Krane and Crane, 1960). This suggests a common mechanism for inducing metabolic dormancy, which might involve the lowering of NADPH levels and the NADPH/NADP ratio so that synthesis involving NADPH-linked reductions cannot occur. As a part of the activation process, NADP-NADPH is increased, with the resulting activation or augmentation of reductive biosynthesis.

2. Other Consequences of NAD Kinase Activation

The increase in NADPH could have other consequences for embryonic metabolism. For example, the major source of reducing power for NADP is most probably the pentose phosphate shunt via glucose-6-phosphate dehydrogenase (Isono and Yasumasu, 1968). *In vitro* studies show that this enzyme is present as an inactive particulate enzyme in the unfertilized egg and as an active soluble enzyme in the fertilized embryo. The release of this enzyme appears to be mediated by both an intracellular pH increase and the increase in NADP (Aune and Epel, 1981).

There is also an augmentation of glycogen metabolism after fertilization in the eggs of many species (see, e.g., Yasumasu *et al.*, 1973). The exact mechanism of the increase in glycogenolysis has not been completely worked out but could well involve calcium acting through calmodulin on an enzyme such as phosophorylase kinase.

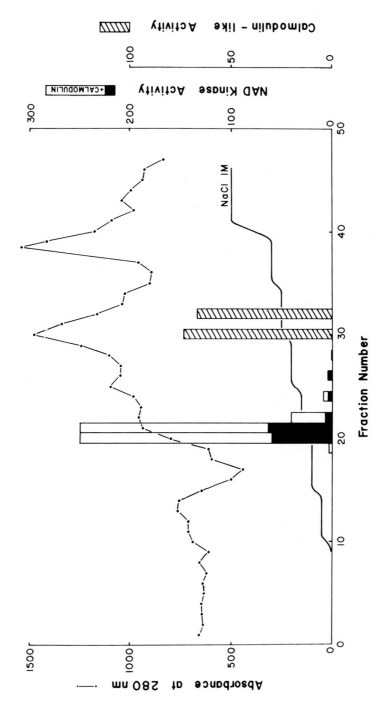

Fig. 8. Resolution of NAD kinase from its endogenous activator on a DEAE-cellulose column. The solid bars represent endogenous NAD kinase activity; the open bars represent the augmentation of activity when bovine brain calmodulin was added; the hatched bars represent NAD kinase activator activity (calmodulin-like activity) as assayed by adding various fractions to fraction 22 and assessing their effect on the endogenous NAD kinase activity of that fraction. (From Epel *et al.*, 1981, and reprinted with permission of MIT Press.)

D. Lipoxygenase

The evidence for activity of this enzyme came from the reexamination of an old observation (Hultin, 1950) that activation of oxygen consumption of sea urchin egg homogenates occurs upon the addition of calcium. A reexamination of this phenomenon revealed at least two mechanisms. Referred to in Section IX was the interaction of calcium with a naphthoquinone pigment, with the resultant production of H_2O_2 (as seen in *Arbacia punctulata* eggs). A second mechanism (seen in *Strongylocentotus purpuratus* eggs) was an oxidation system which was insensitive to mitochondrial inhibitors (such as cyanide, rotenone, and antimycin A) and considerably augmented by arachidonic acid (Perry, 1979; Perry and Epel, 1981a). Subsequent work showed that there was a calcium-mediated activation of arachidonic acid oxidation; rather than proceeding through the prostaglandin pathway, the major pathway in eggs was through the enzyme lipoxygenase, with the concomitant formation of hydroxy fatty acids such as [12-L-hydroxy-5,8,10,14-eicosatetraneoic acid (HETE)].

Perry and Epel (1981a) found virtually no arachidonic acid conversion in the absence of calcium, but oxidation was mediated by extremely low calcium levels, on the order of $1 \times 10^{-7} M$. It appears not to be mediated by calmodulin, since the enzyme activity is not inhibited by a calmodulin inhibitor such as trifluoperazine.

This enzyme activity is also seen *in vivo*. When eggs were preloaded with ^{14}C-labeled arachidonic acid and the conversion followed, a new product appeared after fertilization which comigrated with authentic HETE. Another indication that the enzyme might be modulated by calcium *in vivo* comes from kinetic studies on the arachidonic acid oxidation; it is restricted to the first 10 min after fertilization. This extends over a longer period than that of the calcium rise, at least as measured by the aequorin procedure, but the continued activity of the enzyme *in vivo* could be related to the extremely low level of calcium required for its activity; one might imagine that the calcium level drops too low to be detected by aequorin but still is above a basal level sufficient to stimulate lipoxygenase activity. An alternative possibility is that a calcium-regulated phospholipase is activated at fertilization (Perry and Epel, 1981a) and that the resultant products of this phospholipase action are then oxidized by the lipoxygenase. The lipoxygenase activity could then continue even after the phospholipase activity has ceased, since the calcium level is still sufficiently high for lipoxygenase activity.

The role of the lipoxygenase action is unknown. However, it is of interest that similar calcium-activated lipoxygenases have been seen in mammalian cells when these cells are activated, as in phagocytosis

(Borgeat and Samuellson, 1979; Siegel *et al.*, 1979). It is also possible that the hydroxy fatty acid is not the active component of the lipoxygenase action but rather that peroxy fatty acids, which are the immediate products of lipoxygenase, are the agents having profound effects on metabolism. These have been described to activate other enzymes, such as guanyl cyclase (Asano and Hidaka, 1977).

The final alternate possibility is that these peroxy compounds represent variations on a general theme of utilizing peroxides to regulate cell metabolism. It has recently been found that insulin action results in the production of H_2O_2 in certain insulin target tissues. The actions of insulin can be mimicked by peroxides, indicating that this might be a form of second messenger (May and DeHaen, 1979). Perhaps inorganic peroxides and lipid peroxides serve a general role in cells in this regard and, if so, the production of peroxy compounds may be part of this previously unsuspected second-messenger system.

XI. SUMMARY AND OVERVIEW

It is clear from the information presented in this chapter that the transient rise in calcium at fertilization is the major trigger in the activation of the egg. Two major problems of future research are discerning, first, the means by which the sperm–egg interaction increases the calcium level and, second, the role of this transient calcium increase in the triggering of development.

A. How Does Calcium Increase?

As noted, there appear to be two phases in the calcium rise in sea urchin eggs. The first is a voltage-dependent calcium influx which may be involved in causing membrane changes which prevent any additional sperm from fusing with the egg. This calcium influx, by itself, is not responsible for activating the egg. The second phase, which apparently results from a massive release of calcium from intracellular stores, is responsible for egg activation. The nature of these stores is not known, nor is it known how the brief sperm–egg interaction causes the discharge of calcium from these stores.

The situation in other eggs has not been as extensively studied. Fragmentary evidence suggests that in mollusk eggs the calcium is not released from intracellular stores but enters from the extracellular milieu. If these preliminary findings are borne out, it would appear that the sperm–egg interaction depolarizes the membrane and that this depolarization directly causes calcium to enter the egg in sufficient amounts to cause egg activation.

B. Role of the Transient Calcium Increase

The second major problem is how the transient rise in calcium triggers the egg to begin development. In the sea urchin egg, part of the rise in calcium is mediated by calcium binding to calmodulin with its subsequent involvement in other enzymatic and structural changes. Inhibitor studies suggest that calmodulin is involved in cortical exocytosis, and more direct evidence indicates Ca^{2+} and calmodulin regulation of NAD kinase. A third target for Ca^{2+}, which does not appear to involve calmodulin, is the enzyme lipoxygenase. This activity is seen in the eggs of some sea urchins, but not all. However, all sea urchin eggs so far examined increase their level of peroxides—whether inorganic or organic—after insemination.

An unanswered question is whether these three changes represent the only targets for calcium. One would think not. It will be interesting to examine other possible targets of calcium, such as protein phosphorylation, phospholipase activity, and perhaps protein and lipid methylation.

I have not discussed the consequences of an intracellular pH rise that occurs after the fertilization of sea urchin eggs (Johnson *et al.*, 1976; Shen and Steinhardt, 1978). As noted, there is good evidence that this change is involved in the large increase in protein synthesis that occurs after fertilization of the sea urchin egg. A pH_i increase has also been seen in frog eggs (Nuccitelli *et al.*, 1981), but not in starfish eggs (Johnson and Epel, 1980), after fertilization. Finally, there are large changes in cortical actin after fertilization (Burgess and Schroeder, 1977; Spudich and Spudich, 1979), and it is thought that the pH increase might be involved in this polymerization (Begg and Rebhun, 1979). Is there a role for Ca^{2+} also? These and other questions will surely be answered in the near future and could provide important insights into the cascade of events initiated by the brief rise in calcium ion.

REFERENCES

Asano, T., and Hidaka, H. (1977). Purification of guanylate cylase from human platelets and effect of arachidonic acid peroxide. *Biochem. Biophys. Res. Commun.* **78,** 910–918.

Aune, T., and Epel, D. (1981). In preparation.

Azarnia, R., and Chambers, E. L. (1976). The role of divalent cations in activation of the sea urchin egg. I. Effect of fertilization on divalent cation content. *J. Exp. Zool.* **198,** 65–78.

Baker, P. F., and Whitaker, M. J. (1980). Trifluoperazine inhibits exocytosis in sea-urchin eggs. *J. Physiol. (London)* **298**, 55P.

Baker, P. F., Knight, D. E., and Whitaker, M. J. (1980). The relation between ionized calcium and cortical granule exocytosis in eggs of the sea urchin *Echinus esculentus. Proc. R. Soc. London, Ser. B* **207**, 149–161.

Begg, D. A., and Rebhun, L. I. (1979). pH regulates the polymerization of actin in the sea urchin egg cortex. *J. Cell Biol.* **83**, 241–248.

Borgeat, P., and Samuelsson, B. (1979). Arachidonic acid metabolism in polymorphonuclear leukocytes: effects of ionophore A23187. *Proc. Natl. Acad. Sci. U.S.A.* **76**, 2148–2152.

Burgess, D., and Schroeder, T. E. (1977). Polarized bundles of actin filaments within microvilli of fertilized sea urchin eggs. *J. Cell Biol.* **74**, 1932–1937.

Campanella, C., and Andreucetti, P. (1977). Ultrastructural observations on cortical endoplasmic reticulum and residual cortical granules in the egg of *Xenopus laevis. Dev. Biol.* **50**, 1–10.

Cardasis, C., Schuel, H., and Herman L. (1978). Ultrastructural localization of calcium in unfertilized sea urchin eggs. *J. Cell Sci.* **77**, 101–115.

Carroll, A. G., and Longo, F. M. (1979). Changes in activity of the calcium dependent regulator protein calmodulin upon fertilization in the sea urchin, *Lytechinus pictus. J. Cell Biol.* **83**, 212a.

Chambers, E. L. (1980). Fertilization and cleavage of eggs of the sea urchin, *Lytechinus variegatus* in calcium-free seawater. *Eur. J. Cell Biol.* **22**, 476.

Chambers, E. L., and de Armendi, J. (1979). Membrane potential, action potential and activation potential of eggs of the sea urchin, *Lytechinus variegatus. Exp. Cell Res.* **122**, 203–218.

Chambers, E. L., and Hinkley, R. E. (1979). Non-propagative cortical reactions induced by the divalent ionophore A23187 in eggs of the sea urchin *Lytechinus variegatus. Exp. Cell Res.* **124**, 441–446.

Chambers, E. L., Pressman, B. C., and Rose, B. (1974). The activation of sea urchin eggs by the divalent ionophores A23187 and X537A. *Biochem. Biophys. Res. Commun.* **60**, 126–132.

Cormier, M. J., Anderson, J. M., Charbonneau, H., Jones, H. P., and McCann, R. O. (1980). *In* "Calcium and Cell Function" (W. Y. Cheung, ed.), Vol. 1, pp. 201–219. Academic Press, New York.

Cross, N. L. (1981). Initiation of the activation potential by an increase in intracellular calcium in eggs of the frog, *Rana pipiens. Dev. Biol.* **85**, 380–384.

Epel, D. (1964). A primary metabolic change of fertilization: Interconversion of pyridine nucleotides. *Biochem. Biophys. Res. Commun.* **17**, 62–68.

Epel, D. (1972). Activation of an Na^+-dependent amino acid transport system upon fertilization of sea urchin eggs. *Exp. Cell Res.* **72**, 74–89.

Epel, D. (1977). The program of fertilization (1977). *Sci. Amer* **237**(5), 128–139.

Epel, D. (1978). Mechanisms of activation of sperm and egg during fertilization of sea urchin gametes. *Curr. Top. Dev. Biol.* **12**, 186–246.

Epel, D. (1980). Ionic triggers in the fertilization of sea urchin eggs. *Ann. N.Y. Acad. Sci.* **339**, 74–86.

Epel, D., Patton, C., Wallace, R. W., and Cheung, W. Y. (1981). Calmodulin activates NAD kinase of sea urchin eggs: An early event of fertilization. *Cell* **23**, 543–549.

Foerder, C. A. and Shapiro, B. M. (1977). Release of ovoperoxidase from sea urchin eggs hardens the fertilization membrane with tyrosine crosslinks. *Proc. Natl. Acad. Sci. U.S.A.* **74**, 4214–4218.

Foerder, C. A., Klebanoff, S. J., and Shapiro, B. M. (1978). Hydrogen peroxide production, chemiluminescence, and the respiratory burst of fertilization: Interrelated events in early sea urchin development. *Proc. Natl. Acad. Sci. U.S.A.* **75**, 3183–3187.

Gilkey, J. C., Jaffe, L. F., Ridgeway, E. B., and Reynolds, G. T. (1978). A free calcium wave traverses the activating egg of the medaka, *Oryzias latipes*. *J. Cell Biol.* **76**, 448–466.

Giudice, G. (1973). "Developmental Biology of the Sea Urchin Embryo." Academic Press, New York.

Gould-Somero, M., Jaffe, L. A., and Holland, L. Z. (1979). Electrically mediated fast polyspermy block in eggs of the marine worm, *Urechis caupo*. *J. Cell Biol.* **82**, 426–440.

Gwatkin, R. B. L. (1977). "Fertilization Mechanisms in Man and Mammals." Plenum, New York.

Hagiwara, S., and Jaffe, L. A. (1979). Electrical properties of egg cell membranes. *Annu. Rev. Biophys. Bioeng.* **8**, 385–416.

Head, J. F., Mader, S., and Kaminer, B. (1979). Calcium-binding modulator protein from the unfertilized egg of the sea urchin, *Arbacia punctulata*. *J. Cell Biol.* **80**, 211–218.

Hollinger, T. G., and Schuetz, A. W. (1976). Cleavage and cortical granule breakdown in *Rana pipiens* oocytes induced by direct microinjection of calcium. *J. Cell Biol.* **71**, 395–401.

Hultin, T. (1950). On the oxygen uptake of *Paracentrotus lividus* egg homogenates after the addition of calcium. *Exp. Cell Res* **1**, 159–168.

Humphreys, T. (1971). Measurements of messenger RNA entering polysomes upon fertilization of sea urchin eggs. *Dev. Biol.* **26**, 201–208.

Isono, N., and Yasamasu, I. (1968). Pathways of carbohydrate breakdown in sea urchin eggs. *Exp. Cell Res.* **50**, 616.

Jaffe, L. A. (1976). Fast block to polyspermy is electrically mediated. *Nature (London)* **261**, 68–71.

Jaffe, L. A., and Robinson, K. R. (1978). Membrane potential of the unfertilized sea urchin egg. *Dev. Biol.* **62**, 215–228.

Jaffe, L. A., Gould-Somero, M. and Holland, L. (1979). Ionic mechanism of the fertilization potential of the marine worm, *Urechis caupo*. *J. Gen. Physiol.* **73**, 469–492.

Jaffe, L. F. (1980). Calcium explosions as triggers of development. *Ann. N.Y. Acad. Sci.* **339**, 86–101.

Johnson, C. H., and Epel, D. (1980). Intracellular pH does not regulate protein synthesis in starfish oocytes. *J. Cell Biol.* **87**, 142a.

Johnson, J. D., Epel, D., and Paul, M. (1976). Intracellular pH and activation of sea urchin eggs after fertilization. *Nature (London)* **262**, 661–664.

Jones, H. P., Bradford, M. J., McRorie, R. A., and Cormier, M. J. (1978). High levels of a calcium-dependent modulator protein in spermatozoa and its similarity to brain modulator protein. *Biochem Biophys. Res. Commun.* **82**, 1264–1272.

Kanatani, H. (1973). Maturation-inducing substance in starfishes. *Int. Rev. Cytol.* **35**, 253–298.

Keller, C., Gunderson, G., and Shapiro, B. M. (1980). Altered *in vitro* phosphorylation of specific proteins accompanies fertilization of *Strongylocentrotus purpuratus* eggs. *Dev. Biol.* **74**, 86–100.

Krane, S. M., and Crane, R. K. (1960). Changes in levels of triphosphopyridine nucleotide in marine eggs subsequent to fertilization. *Biochim. Biophys. Acta* **43**, 369–373.

Lilly, F. R. (1902). Differentiation without cleavage in the egg of the annelid *Chaetopterus pergamentaceus*. *Arch. Entwicklungsmech. Org.* **14**, 477–499.

Loeb, J. (1913). "Artificial Parthenogenesis and Fertilization." Univ. of Chicago Press, Chicago, Illinois.

Lucy, J. A. (1978). Mechanisms of chemically induced cell fusion. *Cell Surf. Rev.* **5**, 268–305.

May, J. M., and DeHaen, C. (1979). The insulin-like effect of hydrogen peroxide on pathways of lipid synthesis in rat adipocytes. *J. Biol. Chem.* **254**, 9017–9021.

Mazia, D. (1937). The release of calcium in *Arbacia* eggs on fertilization. *J. Cell. Comp. Physiol.* **10**, 291–304.

Morse, D. E., Duncan, H., Hooper, N., and Morse, A. (1977). Hydrogen peroxide induces spawning in mollusks, with activation of prostaglandin endoperoxide synthetase. *Science* **196**, 298–300.

Nakamura, M., and Yasumasu, I. (1974). Mechanism for increase in intracellular concentration of free calcium in fertilized sea urchin eggs. *J. Gen. Physiol.* **63**, 374–388.

Nishida, E., and Kumagai, H. (1980). Calcium-sensitivity of sea urchin tubulin in *in vitro* assembly and the effects of calcium-dependent regulator (CDR) proteins isolated from sea urchin eggs and porcine brains. *J. Biochem. (Tokyo)* **87**, 143–151.

Nuccitelli, R., Webb, D. J., Lagier, S. T., and Matson, G. B. (1981). ^{31}P NMR reveals an increase in intracellular pH after fertilization in *Xenopus* eggs. *Proc. Natl. Acad. Sci. U.S.A.* **78**, 4421–4425.

Paul, M., and Johnston, R. N. (1978). Uptake of Ca^{+2} is one of the earliest responses to fertilization of sea urchin eggs. *J. Exp. Zool.* **203**, 143–149.

Perry, G. (1979). Studies on calcium-stimulated oxidation in the sea urchin egg. Ph.D. Thesis, University of California, San Diego.

Perry, G., and Epel, D. (1981a). Activation of a lipoxygenase-like activity at fertilization of the sea urchin egg (submitted for publication).

Perry, G., and Epel, D. (1981b). Ca^{+2}-stimulated production of H_2O_2 from naphthoquinone oxidation in *Arbacia* eggs. *Exp. Cell Res.* (in press).

Poste, G., and Pasternak, C. A. (1978). Virus-induced cell fusion. *Cell Surf. Rev.* **5**, 306–369.

Reed, J. (1970). On the control and activation of metabolism during germination of *Arachis hypogea*. Ph.D. Thesis, University of Pennsylvania, Philadelphia.

Ridgeway, E. B., Gilkey, J. C., and Jaffe, L. F. (1977). Free calcium increases explosively in activating medaka eggs. *Proc. Natl. Acad. Sci. U.S.A.* **74**, 623–627.

Sano, K., and Kanatani, H. (1980). External calcium ions are requisite for fertilization of sea urchin eggs by spermatozoa with reacted acrosomes. *Dev. Biol.* **78**, 242–246.

Schatten, G., and Hemmer, M. (1979). Localization of sequestered calcium in unfertilized sea urchin eggs: Discharge upon activation. *J. Cell Biol.* **83**, 199a.

Schuel, H. (1978). Secretory function of egg cortical granules in fertilization and development: A critical review. *Gamete Res.* **1**, 299–382.

Schuetz, A. (1975). Induction of nuclear breakdown and meiosis in *Spisula solidissima* oocytes by calcium ionophores. *J. Exp. Zool.* **191**, 443–446.

Shapiro, H. (1946). The extracellular release of echinochrome. *J. Gen. Physiol.* **29**, 267–275.

Shen, S. S., and Steinhardt, R. A. (1978). Direct measurement of intracellular pH during metabolic derepression of the sea urchin egg. *Nature (London)* **272**, 253–254.

Shen, S. S., and Steinhardt, R. A. (1979). Intracellular pH and the sodium requirement at fertilisation. *Nature (London)* **282**, 87–89.

Shen, S. S., and Steinhardt, R. A. (1980). Intracellular pH controls the development of new potassium conductance after fertilization of the sea urchin egg. *Exp. Cell Res.* **125**, 55–61.

Siegel, M. I., McConnell, R. T., Abrams, S. L., Porter, N. A., and Cuatrecasas, P. (1979). Regulation of arachidonate metabolism via lipoxygenase and cyclo-oxygenase by

12-HPETE, the product of human platelet lipoxygenase. *Biochem. Biophys. Res. Commun.* **89,** 1273–1280.

Silver, R. B., Cole, R. D., and Cande, W. Z. (1980). Isolation of mitotic apparatus containing vesicles with calcium sequestration activity. *Cell* **19,** 505–516.

Spudich, A., and Spudich, J. A. (1979). Actin in Triton-treated cortical preparations of unfertilized and fertilized sea urchin eggs. *J. Cell Biol.* **82,** 212–226.

Steinhardt, R. A., and Epel, D. (1974). Activation of sea urchin eggs by a Ca^{+2}-ionophore. *Proc. Natl. Acad. Sci. U.S.A.* **71,** 1915–1919.

Steinhardt, R. A., Lundin, L., and Mazia, D. (1971). Bioelectric responses of the echinoderm egg to fertilization. *Proc. Natl. Acad. Sci. U.S.A.* **68,** 2426–2430.

Steinhardt, R. A., Zucker, R., and Schatten, G. (1977). Intracellular calcium release at fertilization in the sea urchin egg. *Dev. Biol.* **58,** 185–196.

Sugiyama, M. (1955). Physiological analysis of the cortical response of the sea urchin egg. *Exp. Cell Res.* **10,** 370–376.

Takahashi, Y., and Sugiyama, M. (1973). Relation between the acrosome reaction and fertilization in the sea urchin. I. Fertilization in Ca^{+2}-free sea water with egg-water treated spermatozoa. *Dev. Growth Differ.* **15,** 261–267.

Tilney, L. G., Kiehart, D. P., Sardet, C., and Tilney, M. (1978). Polymerization of actin. IV. Role of Ca^{+2} and H^+ in the assembly of actin and in membrane fusion in the acrosomal reaction of echinoderm sperm. *J. Cell Biol.* **77,** 536–550.

Tyler, A. (1963). The manipulations of macromolecular substances during fertilization and early development of animal eggs. *Am. Zool.* **3,** 109–126.

Vacquier, V. D. (1975). The isolation of intact cortical granules from sea urchin eggs: Calcium ions trigger granule discharge. *Dev. Biol.* **43,** 62–74.

Warburg, O. (1908). Beobachtungen uber die Oxydation prozesse im Seeigelei. *Hoppe-Seyler's Z. Physiol. Chem.* **66,** 305–340.

Winkler, M. M., and Steinhardt, R. A. (1981). Activation of protein synthesis in a sea urchin cell free system. *Dev. Biol.* **84,** 432–439.

Wolf, D. P. (1974). The cortical response in *Xenopus laevis* ova. *Dev. Biol.* **40,** 102–115.

Wolf, D. P., and Hedrick, J. L. (1971). A molecular approach to fertilization. II. Viability and artificial fertilization of *Xenopus laevis* gametes. *Dev. Biol.* **25,** 348–359.

Zucker, R., and Steinhardt, R. A. (1978). Prevention of the cortical reactions in fertilized sea urchin eggs by injection of calcium-chelating ligands. *Biochim. Biophys. Acta* **541,** 459–466.

Chapter 14

Calcium and Phospholipid Turnover as Transmembrane Signaling for Protein Phosphorylation

YOSHIMI TAKAI
AKIRA KISHIMOTO
YASUTOMI NISHIZUKA

I. Introduction 386
II. Enzymology of Calcium-Activated, Phospholipid-Dependent
Protein Kinase 387
A. Tissue Distribution 387
B. Physical and Kinetic Properties 388
III. Mode of Enzyme Activation 389
A. Outline of Activation 389
B. Specificity of Diacylglycerol and Phospholipid 391
C. Reversibility and Selective Inhibition 393
D. Proteolytic Activation 394
IV. Phospholipid Metabolism and Receptor Function 396
A. Phosphatidylinositol Turnover for Enzyme Activation 396
B. Role in Cellular Activation 398
V. Physiological Implication in Transmembrane Control 401
A. Possible Cascade Mechanism for Protein Phosphorylation . . . 401
B. Recognition of Substrate Proteins 403
VI. Coda and Prospectives 405
References 406

I. INTRODUCTION

Cellular function and proliferation are frequently activated by interaction of extracellular messengers with specific cell surface receptors, and the mechanism of such activation, particularly of transmission of information across cell membranes, has attracted much attention. It is now clear that enzymatic phosphorylation and dephosphorylation of proteins play crucially important roles in such transmembrane control of various cellular processes, and that cyclic AMP, cyclic GMP, and Ca^{2+} serve as nearly universal, interrelated intracellular messengers which may regulate protein phosphorylation reactions. Nevertheless, there seem to be dramatic variations on this universal theme of cell activation, and several species of protein kinases which depend upon Ca^{2+} for enzymatic activity have been described. For instance, the enzymes which require Ca^{2+}-dependent modulator protein, that is, calmodulin, include muscle phosphorylase kinase (Cohen *et al.*, 1978), myosin light chain kinase (Dabrowska *et al.*, 1978; Yagi *et al.*, 1978), muscle glycogen synthase (Payne and Soderling, 1980), and poorly defined enzymes found in brain (Schulman and Greengard, 1978; DeLorenzo *et al.*, 1979; Yamauchi and Fujisawa, 1979), heart (Le Peuch *et al.*, 1979), and adipose tissue (Landt and McDonald, 1980). In general, these enzymes appear to be relatively specific for their respective phosphate acceptor proteins.

A series of recent studies in this laboratory have found another species of Ca^{2+}-dependent protein kinase in various tissues and organs, which apparently shows multifunctional properties (Takai *et al.*, 1979a,b,c; Kishimoto *et al.*, 1980; for review, see Nishizuka *et al.*, 1979; Nishizuka, 1980; Nishizuka and Takai, 1981; Takai *et al.*, 1981). The enzyme absolutely requires phospholipid in addition to Ca^{2+}; cyclic nucleotides show no effect. The activation process seems to be directly coupled with phosphatidylinositol turnover provoked by a variety of extracellular messengers which do not normally produce cyclic AMP in their target tissues. Such phosphatidylinositol turnover was first described by M. R. Hokin and Hokin (1953, 1954; L. E. Hokin and Hokin, 1955a,b) almost three decades ago in acetylcholine-sensitive tissues and subsequently found by many investigators in a large number of tissues that are stimulated by various extracellular messengers including neurotransmitters, peptide hormones, mitogenic substances, and oncogenic viruses (for review, see Hawthorne, 1960; Hawthorne and White, 1975; Michell, 1975, 1979). A signal of such extracellular messengers induces specific hydrolysis of phosphatidylinositol in a phospholipase C-type reaction to produce diacylglycerol which markedly increases the affinity of protein kinase for Ca^{2+} as well as for phospholipid, thereby serving as a second messenger in

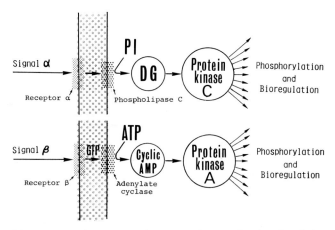

Fig. 1. Two transmembrane control systems for protein phosphorylation. PI, Phosphatidylinositol; DG, diacylglycerol.

the selective activation of enzyme in a manner analogous to the behavior of cyclic AMP as schematically given in Fig. 1. The protein kinase activated in this way appears to play important roles in transmembrane control of protein phosphorylation in a variety of mammalian tissues. In this chapter the mode of activation, physical and kinetic properties, and possible roles of this enzyme in receptor functions will be briefly described. For convenience, the Ca^{2+}-activated, phospholipid-dependent protein kinase described here will be referred to tentatively as protein kinase C. Cyclic AMP-dependent and cyclic GMP-dependent protein kinases will be designated hereafter as protein kinases A and G, respectively.

II. ENZYMOLOGY OF CALCIUM-ACTIVATED, PHOSPHOLIPID-DEPENDENT PROTEIN KINASE

A. Tissue Distribution

It has been well known that protein kinases A and G are found widely in eukaryotic organisms including various vertebral and invertebral tissues (Kuo and Greengard, 1969; Kuo, 1974). In general, protein kinase A appears to be present at much higher levels than protein kinase G, and only trace amounts of the latter class of enzyme are detected in most mammalian tissues. In contrast, protein kinase C is distributed in all tissues and organs so far tested in amounts comparable to those of protein kinase A (Minakuchi *et al.*, 1981). Table I shows the relative activities of the three

TABLE I

Tissue Distribution and Relative Activities of Protein Kinases C, A, and G[a]

Tissue	Protein kinase C	Protein kinase A	Protein kinase G
Platelets	6300	340	—
Brain	3270	250	6
Lymphocytes	1060	320	<1
Granulocytes	530	100	—
Small intestine smooth muscle	770	560	—
Lung	360	290	13
Kidney	280	150	—
Liver	180	130	2
Adipocyte	170	270	2
Heart	110	230	7
Skeletal muscle	80	110	—

[a] Rat tissues were employed except for platelets, lymphocytes, and granulocytes which were obtained from human volunteers. The enzymes were assayed under comparable conditions with calf thymus H1 histone as a model substrate. Detailed conditions will be described elsewhere (Minakuchi *et al.*, 1981). Values are units per milligram of protein. One unit of protein kinase activity was defined as the amount of enzyme that incorporated 1 pmole of phosphate from ATP into H1 histone per minute.

species of protein kinases when assayed under similar conditions with calf thymus H1 histone as a model phosphate acceptor. It seems to be a difficult task at present to estimate the exact amounts of these enzymes, since they may utilize different substrates and the molecular activities probably differ from one another. Nevertheless, the numbers given in this table may be directly comparable within these limits. It is clear that protein kinase C is found in large quantities, particularly in brain and some blood elements such as platelets and lymphocytes.

B. Physical and Kinetic Properties

In most tissues protein kinase C is present as an inactive soluble form under normal conditions and is extracted with isotonic solutions containing ethylene glycol bis(β-aminoethyl ether)-N,N,N',N' tetraacetic acid (EGTA). The enzyme has been partially purified from mammalian tissues such as rat liver (Takai *et al.*, 1977a), rat brain (Inoue *et al.*, 1977), human platelets (Kawahara *et al.*, 1980a), and human peripheral lymphocytes (Ogawa *et al.*, 1981) by DEAE-cellulose column chromatography and gel filtration followed by isoelectrofocusing electrophoresis. The enzymes obtained from several tissues are indistinguishable from one another and

TABLE II

Physical and Kinetic Properties of Protein Kinase C

Property	Value
Molecular weight	7.7×10^4
Sedimentation coefficient	5.1 S
Stokes radius	42 Å
Isoelectric point	pH 5.6
K_m for ATP	$6.6 \times 10^{-6}\ M$
K_m for H1 histone	30 μg/ml
Optimum pH	7.5–8.0
Optimum Mg^{2+}	5–10 mM

appear to show no species and tissue specificities, at least in their physical and kinetic properties. The molecular weight is estimated to be 7.7×10^4, based on a sedimentation coefficient of 5.1 S. The Stokes radius is 42 Å. The subunit structure of the enzyme is not known, and GTP is unable to substitute for ATP as a phosphate donor. Some of the physical and kinetic properties are shown in Table II.

III. MODE OF ENZYME ACTIVATION

A. Outline of Activation

Protein kinase C is routinely assayed by measuring the incorporation of radioactive phosphate of ATP into calf thymus H1 histone as a model substrate. The enzyme per se is inactive with histone unless membrane and Ca^{2+} are added. Cyclic nucleotides are unable to activate the enzyme. Initially, the enzyme was found as an undefined protein kinase which could be activated in an irreversible manner by limited proteolysis with Ca^{2+}-dependent neutral protease (Section III,D) (Inoue et al., 1977; Takai et al., 1977a). However, it soon became clear that in the presence of Ca^{2+} the enzyme was activated without proteolysis by reversible association with membranes obtained from various tissues such as brain synapses and erythrocyte ghosts (Takai et al., 1979a,b). Subsequent analysis has revealed that the active components in membranes are phospholipid and diacylglycerol (Takai et al., 1979c; Kishimoto et al., 1980). However, the enzyme alone shows catalytic activity when protamine is used as a phosphate acceptor. A series of analyses have indicated that the enzymatic activities with histone and protamine seem to originate from a signal

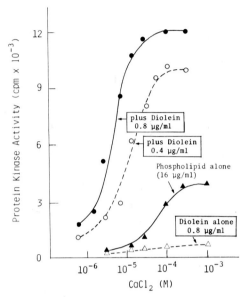

Fig. 2.   Activation of protein kinase C by unsaturated diacylglycerol. Protein kinase C was assayed in the presence of various concentrations of Ca^{2+} as indicated. Synthetic diolein and a mixture of various phospholipids prepared from human erythrocyte ghosts were employed. All reagents were prepared with Ca^{2+}-free water which was made with a double-distillation apparatus followed by passing through a Chelex-100 column. Other conditions were as described elsewhere (Kishimoto *et al.*, 1980).

protein, although the reason for this is not clear at present (Inoue *et al.*, 1977; Takai *et al.*, 1977a).

Figure 2 shows a typical result of kinetic analysis with calf thymus H1 histone as substrate. In this experiment the enzyme activity is plotted against the Ca^{2+} concentration on a logarithmic scale. In the presence of phospholipid alone the enzyme requires much higher concentrations of Ca^{2+} and is practically inactive at physiological concentrations of this divalent cation. However, when a small amount of diolein is added, the reaction velocity is greatly increased along with a concomitant decrease in Ca^{2+} concentration, giving rise to maximum enzyme activation. Diolein alone shows little or no effect. This activation process absolutely requires Ca^{2+}, and no other divalent cation can substitute for Ca^{2+}, except Sr^{2+}, which is less than 10% as active under comparable conditions. As will be described below, the apparent K_a value for Ca^{2+}, which is the concentration necessary for half-maximal activation, is less than the micromolar range in the presence of both phosphatidylserine and unsaturated diacylglycerol. In the physiological range of Ca^{2+} concentrations the en-

TABLE III

Specificity of Neutral Lipid for Activation of Protein Kinase C[a]

Experiment	Diacylglycerol added	K_a for Ca²⁺ ($10^{-6} M$)	Reaction velocity at $6 \times 10^{-6} M$ CaCl₂ (cpm)
1	Phospholipid alone	70	980
	Plus triolein	70	970
	Plus diolein	4	6500
	Plus monoolein	70	990
	Plus tripalmitin	70	980
	Plus dipalmitin	70	970
	Plus monopalmitin	70	980
2	Phospholipid alone	70	980
	Plus diolein	4	6500
	Plus dilinolein	7	5510
	Plus diarachidonin	4	6220
	Plus 1-stearoyl-2-oleoyl diglyceride	8	4610
	Plus 1-oleoyl-2-stearoyl diglyceride	7	4820
	Plus distearin	80	950

[a] Protein kinase C was assayed with a fixed amount of phospholipid (16 μg/ml) and various synthetic neutral lipids (0.8 μg/ml each) as indicated. Other conditions were as described elsewhere (Kishimoto et al., 1980).

zymatic activity solely depends upon the presence of a small quantity of unsaturated diacylglycerol, since membrane phospholipid may not be limited. Thus the enzyme can be activated without a net increase in Ca²⁺ within the cell, and unsaturated diacylglycerol, which may be produced from phosphatidylinositol hydrolysis as discussed below (Section IV,A), serves as a second messenger for the selective activation of this unique protein kinase. Presumably, the preexisting membrane-bound Ca²⁺ concentration is enough to play an essential role in this enzyme activation process. However, it is also possible, as described above, that a large increase in intracellular Ca²⁺ also activates the enzyme in the absence of diacylglycerol.

B. Specificity of Diacylglycerol and Phospholipid

The unique effect of neutral lipid mentioned above is most remarkable for diacylglycerol containing unsaturated fatty acid. Table III compares the efficacy of some synthetic neutral lipids in supporting enzyme activation. Diolein, dilinolein, and diarachidonin are equally active, whereas diacylglycerol, possessing saturated fatty acids such as dipalmitin and distearin, is far less effective. A series of experiments have indicated that

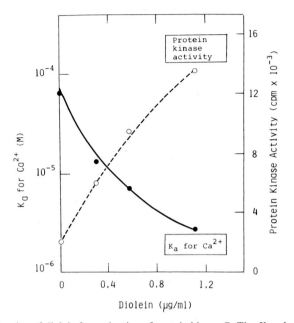

Fig. 3. Titration of diolein for activation of protein kinase C. The K_a value for Ca^{2+} was estimated in the presence of various concentrations of Ca^{2+} as described in Fig. 2. Protein kinase activity was assayed at $6 \times 10^{-6} M$ CaCl$_2$. Other conditions were as described elsewhere (Kishimoto *et al.*, 1980).

diacylglycerol with one unsaturated fatty acid at either position 1 or 2 is active enough irrespective of the chain length of the other fatty acyl moiety. Kinetically, such unsaturated diacylglycerols appear to increase the affinity of the enzyme for phospholipid as well as for Ca^{2+}. None of the triacylglycerols, monoacylglycerols, and free fatty acids thus far tested are able to substitute for unsaturated diacylglycerol. Cholesterol and glycolipid are totally ineffective. Figure 3 shows the titration of diolein for activating protein kinase C, and the results indicate that a very small quantity of diolein, less than 10% of phospholipid, shows significant effects.

Another set of experiments given in Table IV are designed to show the specificity of phospholipid for the activation of enzyme. In the presence of physiological concentrations of Ca^{2+}, phosphatidylserine is the most effective, and other phospholipids appear to be practically inactive. However, it is noted that, in the presence of phosphatidylserine, enzymatic activity is enhanced further by the addition of phosphatidylethanolamine but is diminished by the addition of phosphatidylcholine and sphingomyelin.

TABLE IV

Specificity of Phospholipid for Activation of Protein Kinase C[a]

	Reaction velocity (cpm)	
Phospholipid added	Without diolein	With diolein
None	80	70
Phosphatidylserine	1400	6940
Phosphatidylinositol	220	310
Phosphatidylethanolamine	60	30
Phosphatidic acid	100	200
Phosphatidylcholine	110	110
Sphingomyelin	60	80

[a] Protein kinase C was assayed with and without diolein (0.8 μg/ml) in the presence of $2 \times 10^{-6}\ M$ CaCl$_2$ and various phospholipids (8 μg/ml each) as indicated. Other conditions have been described elsewhere (Takai et al., 1979c).

Both phosphatidylinositol and phosphatidic acid do not affect the enzymatic reaction. It has been well established for red blood cells and platelets that most of the phosphatidylcholine and sphingomyelin is located in the outer monolayer, whereas phosphatidylserine, phosphatidylethanolamine, and phosphatidylinositol are largely located in the inner monolayer of membranes (for review, see Roelofsen and Zwaal, 1976; Rothman and Lenard, 1977; Op den Kamp, 1979). Although it is not known whether such asymmetric distribution of various phospholipids favorable for the activation of protein kinase C may be generalized for plasma membranes of other tissues and organs, it is possible to assume that various species of membrane phospholipids display specific biological roles with positive and negative cooperativities in this receptor function. A certain lipid bilayer structure may be necessary for rendering the enzyme fully active, and a better physiological picture will emerge from further investigations.

C. Reversibility and Selective Inhibition

The activation process of protein kinase C described above is experimentally reversible, and the enzyme once activated in this way is inactivated and recovered in an inactive soluble form by removing Ca^{2+} with EGTA. Such an activation–inactivation cycle may be repeated without loss of enzymatic activity. However, the association with membrane phospholipid does not necessarily mean activation of the enzyme, since in the presence of various local anesthetics, tranquilizers, and other phospholipid-interacting drugs the enzyme is profoundly inhibited but may still be associated with membranes (Mori *et al.*, 1980). Kinetic analysis has indicated that these drugs interact with the phospholipid, thereby competitively inhibiting the activation process of protein kinase C, and that none of these drugs interact with the catalytic site of this enzyme. These drugs include dibucaine, tetracaine, chlorpromazine, imipramine, phentolamine, and verapamil, and it may be noted that many such phospholipid-interacting drugs and compounds interact with membrane phospholipid and affect a variety of neuronal and nonneuronal cellular activities (for review, see Papahadjopoulos, 1972; Seeman, 1972). Neither protein kinase A nor G is susceptible to these drugs. Presumably, these drugs confer profound modification of specific hydrophobic lipid–protein interactions.

Calmodulin is unable to substitute for phospholipid or for diacylglycerol over a wide range of concentrations. Although Mg^{2+} or Mn^{2+} is essential for the enzymatic reaction, Ca^{2+} is dispensable for catalytic activity. The best evidence for this interpretation has been provided by the fact that a catalytically active fragment of protein kinase C, obtained by limited proteolysis as described below, is fully active without Ca^{2+}, phospholipid, or diacylglycerol and is no more susceptible to the phospholipid-interacting drugs listed above.

D. Proteolytic Activation

As briefly outlined above, the enzyme was uncovered initially as an inactive protein kinase which could be activated by limited proteolysis with Ca^{2+}-dependent neutral protease or trypsin (Inoue *et al.*, 1977; Kishimoto *et al.*, 1977; Takai *et al.*, 1977a,b; Yamamoto *et al.*, 1978). In fact, when the enzyme is incubated with either trypsin or Ca^{2+}-dependent protease, the catalytic activity for histone phosphorylation rapidly appears. When Ca^{2+}-dependent protease is employed, the catalytic activity soon reaches a plateau and remains unchanged for prolonged incubation, suggesting that limited proteolysis takes place. Upon gel filtration analysis a smaller component carrying catalytic activity is isolated, and the molecular weight is estimated to be about 51,000. When Ca^{2+}-dependent protease is replaced by trypsin, an active component is similarly produced but disappears rapidly, probably as a result of further proteolytic degradation. It is not known whether a single active component is generated during limited proteolysis. The catalytically active component thus produced is entirely independent of Ca^{2+}, phospholipid, and diacylglycerol.

Cyclic nucleotides again show no effect. So far, active protein kinases activated in either a reversible or an irreversible manner as described above show similar kinetic and catalytic properties.

The Ca^{2+}-dependent neutral protease responsible for the proteolytic activation of enzyme appears to be present in soluble fractions of various tissues and organs (Inoue et al., 1977). This class of neutral proteases, designated calpain (Murachi et al., 1981), was described first in rat brain (Guroff, 1964), and subsequently in mammalian and chicken skeletal muscles (Meyer et al., 1964; Drummond and Duncan, 1968; Huston and Krebs, 1968; Busch et al., 1972; Reddy et al., 1975; Dayton et al., 1976; Ishiura et al., 1978), rat uterus (Puca et al., 1977), human platelets (Phillips and Jakabova, 1977), and rat liver (Takai et al., 1977a; Nishiura et al., 1978), and all proteases presumably belong to the same entity. These proteases show rather broad substrate specificity and react with muscle phosphorylase kinase and glycogen synthase, resulting in activation and inactivation of the respective enzymes (Meyer et al., 1964; Belocopitow et al., 1965). However, one of the major drawbacks to this proteolytic activation of protein kinase C is that the protease requires millimolar Ca^{2+} for catalytic activity. Nevertheless, such a drawback has been recently overcome by the finding of Mellgren (1980), who reported a similar type of Ca^{2+}-dependent protease in dog heart which showed maximum activity at 10^{-5} M Ca^{2+}. The latter class of enzymes are also capable of activating protein kinase C and are also distributed in many other tissues including liver, skeletal muscle, brain, and kidney (Kishimoto et al., 1981). A typical result is shown in Fig. 4. Thus it is possible to assume that the proteolytic activation of protein kinase C may play a role in a large variety of irreversible biological processes linked with digestion, secretion, lactation, metamorphosis, fertilization, etc.

It is also noted that such an irreversible proteolytic activation process may be blocked by several thiol protease inhibitors including leupeptin, chymostatin, antipain, and E-64 {N-[N-(L-3-trans-carboxyoxiran-2-carbonyl)-1-leucyl]agmatine} (Takai et al., 1979a; Sugita et al., 1980). These protease inhibitors do not affect the reversible activation process or the already proteolytically activated protein kinase C. Nevertheless, it is not known whether these protease inhibitors actually work in in vivo systems. Some reports have appeared in the literature which describe the natural occurrence of macromolecular protease inhibitors (Nishiura et al., 1978; Waxman and Krebs, 1978). Although both protein kinases A and G may also be activated by trypsin in a proteolytic manner (Huang and Huang, 1975; Inoue et al., 1976), all attempts to activate these two protein kinases by Ca^{2+}-dependent protease have been thus far unsuccessful. The reason is not clear at present, but this fact is probably of physiological significance.

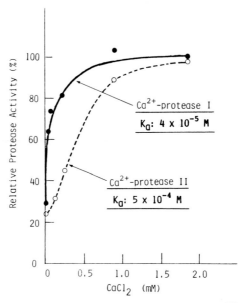

Fig. 4. Two types of Ca^{2+}-dependent proteases activated by different Ca^{2+} concentrations. Protease activity was assayed at various concentrations of Ca^{2+} by measuring the formation of a catalytically active fragment from native inactive protein kinase C. Detailed assay conditions were as described earlier (Takai *et al.*, 1977a). Ca^{2+} proteases I and II represent the newly and previously found Ca^{2+}-dependent neutral proteases, respectively. K_a indicates the Ca^{2+} concentration needed for half-maximal enzyme activation.

IV. PHOSPHOLIPID METABOLISM AND RECEPTOR FUNCTION

A. Phosphatidylinositol Turnover for Enzyme Activation

Based on the foregoing discussion it is reasonable to raise the question of whether the activation of protein kinase C is directly related to the phosphatidylinositol turnover which has been known for many years. M. R. Hokin and Hokin (1953, 1954; L. E. Hokin and Hokin, 1955a,b) described the first evidence that this particular phospholipid turns over very rapidly in response to acetylcholine and that this turnover may be blocked by atropine. Early work has found that such a phosphatidylinositol response may be observed in activated plasma membranes of various types of secretory tissues such as pancreas (M. R. Hokin and Hokin, 1953, 1954; L. E. Hokin and Hokin, 1955b), salivary gland (Hokin and Sherwin, 1957), and salt-excreting gland from seabirds (L. E. Hokin and Hokin, 1959), and that inorganic phosphate is rapidly incorporated into phos-

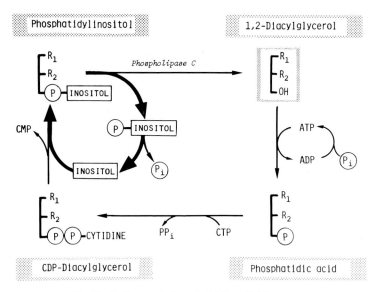

Fig. 5. Pathway of phosphatidylinositol turnover.

phatidic acid as well as into phosphatidylinositol with little or no labeling of other major phospholipids including phosphatidylethanolamine, phosphatidylcholine, and sphingomyelin. The pathway of such a labeling pattern, namely, the turnover pathway of phosphatidylinositol, has been clarified subsequently by many investigators as schematically shown in Fig. 5 (for review, see Hawthorne and White, 1975; Michell, 1975). The phospholipid turnover is initiated by the cleavage of a phosphodiester bond in the manner of a phospholipase C-type reaction, resulting in the formation of diacylglycerol and inositol phosphate. Since the latter compound is produced largely as D-myoinositol 1,2-cyclic phosphate (Dawson *et al.*, 1971), the possibility has been sometimes discussed that this inositol phosphate may serve as a second messenger in a manner analogous to that of cyclic AMP. Nevertheless, nearly all attempts to elucidate such a possible role of this compound have been unsuccessful. Instead, of various approaches in exploring the biological significance of phosphatidylinositol turnover, a most attractive and obvious possibility at this time is that diacylglycerol may occur in plasma membranes as an integral part of the receptor functions.

It has been well established that phosphatidylinositol obtained from mammalian tissues is not a single chemical entity but a mixture of different molecular species. That is, this phospholipid is composed of unsaturated fatty acids, most often arachidonic acid at position 2 (Holub *et al.,* 1970), hence one of the primary products of phosphatidylinositol response to extracellular messengers is expected to be unsaturated diacylglycerol. Therefore, it is tempting to assume that phosphatidylinositol turnover may be directly related to the selective activation of protein kinase C. In order to obtain such evidence, the following set of experiments were carried out using human platelets which are stimulated by thrombin (Kawahara *et al.,* 1980a,b).

B. Role in Cellular Activation

Platelets have been well known to respond to thrombin with an aggregation and release reaction. Since such activation reactions are analogous to hormonal control of cellular processes, numerous studies have been made to clarify the molecular basis of the mechanism, particularly transmembrane control of intracellular events leading to the aggregation and release reaction (for review, see Haslam *et al.,* 1978). Among these events attention has been paid to the phosphorylation of endogenous proteins having molecular weights of about 40,000 (40 K protein) and 20,000 (20 K protein). The phosphorylation of both proteins is observed specifically upon stimulation by thrombin. The 20 K protein has been identified as myosin light chain, and a calmodulin-dependent myosin light chain kinase has been proposed to participate in this reaction (Daniel *et al.,* 1977). In contrast, the enzyme responsible for the 40 K-protein phosphorylation has yet to be determined, although the reaction seems to be intimately related to the release reaction (Lyons *et al.,* 1975; Haslam and Lynham, 1977; Wallace and Bensusan, 1980). Rittenhouse-Simmons (1979) and Bell and Majerus (1980) have recently reported that diacylglycerol is rapidly generated at the expense of phosphatidylinositol upon stimulation of platelets with thrombin. In confirmation of these observations, Fig. 6 shows that diacylglycerol is rapidly produced when platelets are activated by thrombin. It is also noted that 40 K-protein phosphorylation and serotonin release proceed immediately after diacylglycerol formation. The rapid disappearance of diacylglycerol once it is produced is presumably due to conversion to phosphatidic acid and also partly to degradation to arachidonic acid (Section V,A). The three reactions given in this figure are concomitantly supressed by the addition of prostaglandin E_1 which is well known as a potent inhibitor of platelet activation, and diacylglycerol formation appears to be always associated with the 40 K-protein phosphorylation and release reaction. In fact, in another set of experiments given in

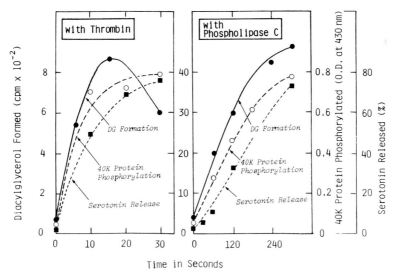

Fig. 6.   Stimulation of diacylglycerol (DG) formation, 40K-protein phosphorylation, and serotonin release by thrombin and phospholipase C. Washed human platelets were pre-labeled with [³H]arachidonic acid, ³²P$_i$, or [¹⁴C]serotonin as described by Rittenhouse-Simmons (1979), Lyons *et al.* (1975), and Haslam *et al.* (1975), respectively. The radioactive platelets were then stimulated by thrombin or phospholipase C. At the various times indicated, the radioactive diacylglycerol formed was extracted and analyzed by thin-layer chromatography. The radioactive proteins were subjected to sodium dodecyl sulfate–slab gel electrophoresis. The stained gel was dried on a filter paper and autoradiographed on x-ray film. The relative intensity of radioactive 40K protein was quantitated by tracing at 430 nm. Serotonin release was determined by measuring the release of radioactive serotonin from platelets. Detailed experimental procedures were as described elsewhere (Kawahara *et al.*, 1980a,b).

Fig. 6, the exogenous addition of *Clostridium perfringens* phospholipase C mimics thrombin action (Schick and Yu, 1974; Chap *et al.*, 1978) and diacylglycerol formation again accompanies the 40K-protein phosphorylation and release reaction (Kawahara *et al.*, 1980b).

Human platelets contain a large amount of protein kinase C, as briefly described in Section II,A, and the enzymatic activity is more than 20 times higher than that of protein kinase A when assayed with calf thymus H1 histone as a phosphate acceptor, as shown in Fig. 7. A series of experiments have revealed that, among a variety of endogenous proteins, protein kinase C rapidly and preferentially phosphorylates 40K protein when tested in *in vitro* systems. Although protein kinase A also reacts with

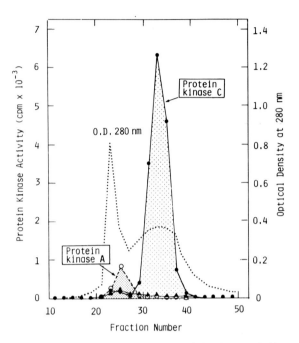

Fig. 7. Sephadex G-150 column chromatography of platelet protein kinases. The soluble fraction of platelets was prepared in the presence of EGTA and subjected to gel filtration on a Sephadex G-150 column (2 × 80 cm). Protein kinases of each fraction were assayed, and detailed experimental conditions were as described elsewhere (Kawahara *et al.*, 1980a,b). Triangles indicate blank values, and no peak of protein kinase G was detected under the standard conditions specified earlier (Hashimoto *et al.*, 1976).

40 K protein when tested in *in vitro* systems, cyclic AMP is rather inhibitory for platelet activation. Rather, it is now accepted that neither cyclic AMP nor protein kinase A is responsible for this 40 K-protein phosphorylation reaction *in vivo* (Lyons *et al.*, 1975). Another piece of evidence suggesting that protein kinase C is involved in the phosphorylation of 40 K protein has come from experiments using certain inhibitory drugs, as mentioned in Section III,C. When platelets are treated with dibucaine or chlorpromazine before being stimulated by thrombin, 40 K-protein phosphorylation is selectively and profoundly inhibited. The phosphorylation of other proteins, including 20 K protein, is far less susceptible to these drugs. The experimental results so far obtained do not appear to provide convincing evidence but strongly suggest that protein kinase C is *ipso facto* responsible for the phosphorylation of 40 K protein, which may be specifically observed in platelets stimulated by thrombin.

TABLE V

Phosphatidylinositol Turnover and Cyclic Nucleotides[a]

Stimulus	Tissue	Phosphatidylinositol turnover	Cyclic GMP	Cyclic AMP
Glucagon	Liver, fat cell	→	→	↑
Adrenergic (β)	Various	→	→	↑
Prostaglandin E_1	Various	→	→	↑
Adrenergic (α)	Various	↑	↑	→ or ↓
Cholinergic (muscarinic)	Various	↑	↑	→ or ↓
Vasopressin	Liver	↑	↑	→
Thrombin	Platelets	↑	↑	→ or ↓
fMet-Leu-Phe	Granulocytes	↑	↑	→ or ↓
Glucose	Langerhans' islets	↑	→	→
Plant lectin	Lymphocytes	↑	↑	→ or ↓

[a] ↑, →, and ↓ denote increased, unchanged, or decreased values, respectively.

V. PHYSIOLOGICAL IMPLICATION IN TRANSMEMBRANE CONTROL

A. Possible Cascade Mechanism for Protein Phosphorylation

Although the experimental support described above comes from studies with a limited number of tissues, it is conceivable that phosphatidylinositol turnover is directly coupled to the activation of protein kinase C and that the enzyme plays roles of crucial importance in the transmission of information from a large number of extracellular messengers. These messengers probably include α-adrenergic and muscarinic cholinergic stimulators, peptide hormones such as vasopressin, angiotensin II, and thyroid-stimulating hormone, and various mitogenic substances such as growth factors and plant lectins, oncogenic viruses, and many other biologically active substances known to provoke phospholipid turnover (for review, see Michell, 1975, 1979). As pointed out by Michell (1975), the wide variety of extracellular messengers also increase the intracellular cyclic GMP level in most tissues but do not induce cyclic AMP formation as exemplified in Table V. Conversely, the extracellular messengers known to produce cyclic AMP do not provoke phosphatidylinositol turnover or increase cyclic GMP levels in their target tissues. Although such a generalization may not always be possible, there appears to be an intimate correlation between phospholipid turnover and cellular cyclic GMP. As schematically outlined in Fig. 8, part of the

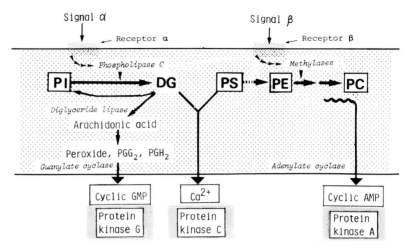

Fig. 8. Two possible major receptor functions. PI, Phosphatidylinositol; DG, diacylglycerol; PS, phosphatidylserine; PE, phosphatidylethanolamine; PC, phosphatidylcholine; PGG_2, prostaglandin G_2; PGH_2, prostaglandin H_2.

diacylglycerol intermediately produced during the turnover of phosphatidylinositol may undergo an alternative pathway to release free unsaturated fatty acids, mostly arachidonic acid. Probably, human platelets again share with other target tissues extensive studies on such lipid metabolism in response to various messengers including thrombin. Although a phospholipase A_2-type reaction may play a role in arachidonic acid release in such activated platelets (Bills *et al.*, 1976; Blackwell *et al.*, 1977), a line of plausible evidence recently presented indicates that phospholipase C and diacylglycerol lipase show sufficient activity to account for the rapid release of arachidonic acid (Bell *et al.*, 1979; Rittenhouse-Simmons, 1979, 1980; Bell and Majerus, 1980; Billah *et al.*, 1980). Such a diacylglycerol lipase has also been found recently in thyroid tissues (Igarashi and Kondo, 1981). Presumably, both phospholipases A_2 and C play roles in different ratios in different tissues in providing unsaturated free fatty acids which serve as precursors of various species of prostaglandins and other derivatives. Several reports have appeared in the literature indicating that fatty acids or their metabolites activate guanylate cyclase to produce cyclic GMP in various mammalian tissues. This activation was initially observed with unsaturated fatty acids (Asakawa *et al.*, 1976; Barber, 1976; Wallach and Pastan, 1976) but later shown with their metabolites, particularly fatty acid hydroperoxides and prostaglandin endoperoxides in various types of tissues (Hidaka and Asano, 1977; Graff *et al.*, 1978). Since extracellular messengers do not directly activate guanylate cyclase under any conditions so far described, it is likely that unsaturated fatty acid derivatives are possible candidates which serve as immediate stimulators for this enzyme.

Based on the foregoing discussion it is attractive to postulate a cascade mechanism, starting from the phosphatidylinositol breakdown, which leads eventually to the simultaneous activation of protein kinases C and G. Although there should be dramatic variations and heterogeneity from tissue to tissue in such receptor mechanisms, the experimental results described in this chapter may provide clues for elucidating a coupling between membrane phospholipid metabolism and the regulation of metabolic as well as other intracellular processes. Nevertheless, the mechanism by which the reaction of phospholipase C is triggered by extracellular messengers in a variety of target tissues is not known. It may be possible that some physicochemical state of lipid–lipid and lipid–protein interactions is modulated by binding of messengers to their specific receptor molecules. For instance, lateral movement or phase separation of various phospholipids, as well as the perturbation of cellular skeleton systems, may be related to such receptor functions. Axelrod and his coworkers (Hirata *et al.*, 1978, 1979; Hirata and Axelrod, 1978; for review, see Hirata and Axelrod, 1980) have developed a series of studies from which they have proposed that, in some mammalian tissues such as reticulocytes, glioma astrocytoma, and HeLa cells, another flow of phospholipids starting from phosphatidylethanolamine to phosphatidylcholine is intimately related to β-adrenergic receptors, which have been well known to stimulate adenylate cyclase, and ultimately leads to the activation of protein kinase A. It is likewise possible that such a methylation reaction of phospholipid is a result of membrane perturbation or an integral part of the cascade mechanism caused by extracellular messengers. The precise role of phospholipid methylation in adenylate cyclase activation will be clarified by further investigations.

B. Recognition of Substrate Proteins

Among the various species of protein kinases thus far described, protein kinases C, A, and G appear to have many similarities. These enzymes are widely distributed in a variety of tissues and directly related to the universal intracellular messengers, namely, Ca^{2+}, cyclic AMP, and cyclic GMP, respectively. These enzymes show relatively broad substrate specificities and thereby appear to play multifunctional roles in biological processes, although their physiological roles are expected to be distinctly different from one another. Nevertheless, the three enzymes show similar

TABLE VI

Relative Reaction Velocities of Protein Kinases C, A, and G toward Five Histone Fractions[a]

Histone	Reaction velocity (cpm)		
	Protein kinase C	Protein kinase A	Protein kinase G
H1 histone	17,000	3,830	4,500
H2A histone	2,850	2,470	5,560
H2B histone	4,380	17,000	17,000
H3 histone	1,210	468	670
H4 histone	240	260	450

[a] Protein kinases C, A, and G employed in this experiment were partially purified from rat brain, rabbit skeletal muscle, and bovine cerebellum, respectively. Detailed conditions have been described elsewhere (Takai *et al.*, 1977b; Iwasa *et al.*, 1980).

catalytic properties as far as they have been tested in *in vitro* systems (for review, see Nishizuka *et al.*, 1979; Nishizuka, 1980). It seems to be generally accepted that protein kinases A and G recognize the same specific seryl and threonyl residues of a variety of naturally occurring proteins, as well as of synthetic peptides, but the relative rates of reactions and affinities of enzymes for various substrates are markedly different (Hashimoto *et al.*, 1976; Chihara-Nakashima *et al.*, 1977; Edlund *et al.*, 1977; Khoo *et al.*, 1977; Lincoln and Corbin, 1977, 1978; Riou *et al.*, 1977; Yamamoto *et al.*, 1977; Blumenthal *et al.*, 1978). Apparently, in *in vitro* systems protein kinase C also favors a similar set of substrate proteins. Table VI shows reactions of the three protein kinases toward five fractions of calf thymus histone, although the reaction rates are greatly different. Protein kinase C is shown to react also with muscle phosphorylase kinase and glycogen synthase very slowly, resulting in activation and inactivation of the respective enzymes (Kishimoto *et al.*, 1977, 1978; Takai *et al.*, 1979a). More recently, it has been reported that protein kinase C may react with cardiac sarcoplasmic reticulum proteins and stimulate Ca^{2+}-dependent ATPase activity (Limas, 1980). However, it has not been clarified whether protein kinases C and A recognize identical amino acid residues in these phosphate acceptor proteins. In any case the evidence so far available in this laboratory has suggested that protein kinase C has a marked preference for seryl and threonyl residues that differ from those preferentially phosphorylated by protein kinase A, as well as from those phosphorylated by protein kinase G, and thereby plays its own specific role in physiological processes (Iwasa *et al.*, 1980; Kawahara *et al.*, 1980a,b). Obviously, the primary sequence around the seryl and threonyl residues seems to be important in the recognition of substrate. It is also noted that the shape and size of enzyme and substrate proteins, namely, the macromolecular interactions between enzyme and substrate, are equally or even more important in the selectivity of reactions. In addition, it has also been suggested that the topographical arrangement or subcellular localization of an enzyme as well as its substrate is another important determining factor for the functions and specificities of protein kinase reactions. Probably the best evidence is provided by a model reaction with protein kinase G which preferentially reacts with histone associated with DNA rather than with free histone (Hashimoto *et al.*, 1979). Along the same lines, protein kinase C appears to favor greatly some endogenous proteins associated tightly or loosely with membranes in most mammalian tissues. The physiological substrate proteins for this unique protein kinase will be clarified by further investigations in the next few years.

VI. CODA AND PROSPECTIVES

This chapter has briefly described a new species of protein kinase recently found in our laboratory. This enzyme, protein kinase C, appears to be an integral part of the receptor function which seems to be directly involved in the transmembrane control of protein phosphorylation. This function absolutely requires Ca^{2+} and is activated by a large number of extracellular messengers whose biological effects do not appear to be mediated through the action of cyclic AMP. Instead, the activation of this receptor function appears to be tightly coupled to phospholipid metabolism in plasma membranes, and the mechanism proposed in this chapter may provide a new means of elucidating the mode of action of a large number of hormones, neurotransmitters, and other biologically active materials. Nevertheless, the underlying principles and mechanisms described here may undoubtedly be modified by current knowledge which will be expanded rapidly through analysis of lipid–lipid and lipid–protein interactions in various target tissues.

It should be emphasized that there appear to be three species of closely similar protein kinases which may be responsible for the three intracellular messengers. Although the existence of protein kinase A responsible for cyclic AMP has been known for more than a decade, the exact role of this enzyme in regulating a large number of biological processes is not yet completely understood. Similarly, protein kinase G, responsible for cyclic GMP, has been known for many years, but virtually nothing has been clarified concerning its physiological functions in molecular terms. The roles of protein kinase C, responsible for Ca^{2+}, have just begun to be elucidated. One of the difficulties in approaching these problems comes from the fact that the three protein kinases show apparently similar, broad substrate specificities when tested in *in vitro* systems. Presumably, the topographical arrangement or compartment in the enzyme–substrate interaction is an important factor in obtaining a complete understanding of the chemical basis of transmembrane control through various receptor functions.

ACKNOWLEDGMENTS

The authors express indebtedness to their collaborators who have so effectively and actively carried out the experiments presented in this chapter. We are particularly grateful to Drs. H. Yamamura, Y. Kawahara, H. Tabuchi, R. Minakuchi, K. Sano, U. Kikkawa, T. Mori, B. Yu, K. Kaibuchi, N. Kajikawa, and T. Matsubara. The skillful secretarial assistance of Mrs. S. Nishiyama and Miss K. Yamasaki is also acknowledged. This investigation was supported in part by research grants from the Scientific Research Fund of the Ministry of Education, Science and Culture, the Intractable Diseases Division, Public Health Bureau, the Ministry of Health and Welfare, a Grant-in-Aid of New Drug Development from the Ministry of Health and Welfare, the Yamanouchi Foundation for Research on Medical Resources, and the Research Fund of Takeda Pharmaceutical Company.

REFERENCES

Asakawa, T., Scheinbaum, I., and Ho, R. (1976). Stimulation of guanylate cyclase activity by several fatty acids. *Biochem. Biophys. Res. Commun.* **73**, 141–148.
Barber, A. J. (1976). Cyclic nucleotides and platelet aggregation: Effect of aggregating agents on the activity of cyclic nucleotide-metabolizing enzymes. *Biochim. Biophys. Acta* **444**, 579–595.
Bell, R. L., and Majerus, P. W. (1980). Thrombin-induced hydrolysis of phosphatidylinositol in human platelets. *J. Biol. Chem.* **255**, 1790–1792.
Bell, R. L., Kennerly, D. A., Stanford, N., and Majerus, P. W. (1979). Diglyceride lipase: A pathway for arachidonate release from human platelets. *Proc. Natl. Acad. Sci. U.S.A.* **76**, 3238–3241.
Belocopitow, E., Appleman, M. M., and Torres, H. N. (1965). Factors affecting the activity of muscle glucogen synthase. II. The regulation by Ca^{2+}. *J. Biol. Chem.* **240**, 3473–3478.
Billah, M. M., Lapetina, E. G., and Cuatrecasas, P. (1980). Phospholipase A_2 and phospholipase C activities of platelets: Differential substrate specificity, Ca^{2+} requirement, pH dependence, and cellular localization. *J. Biol. Chem.* **255**, 10227–10231.
Bills, T. K., Smith, J. B., and Silver, M. J. (1976). Metabolism of [^{14}C]arachidonic acid by human platelets. *Biochim. Biophys. Acta* **424**, 303–314.
Blackwell, G. J., Duncombe, W. G., Flower, R. J., Parsons, M. F., and Vane, J. R. (1977).

The distribution and metabolism of arachidonic acid in rabbit platelets during aggregation and its modification by drugs. *Br. J. Pharmacol.* **59**, 353–366.

Blumenthal, D. K., Stull, J. T., and Gill, G. N. (1978). Phosphorylation of cardiac troponin by guanosine 3':5'-monophosphate-dependent protein kinase. *J. Biol. Chem.* **253**, 334–336.

Busch, W. A., Stromer, M. H., Goll, D. E., and Suzuki, A. (1972). Ca^{2+}-specific removal of Z lines from rabbit skeletal muscle. *J. Cell Biol.* **52**, 367–381.

Chap, H., Mauco, G., Perret, B., Simon, M. F., and Douste-Blazy, L. (1978). Role of exogenous phospholipases in triggering platelet aggregation. *Adv. Prostaglandin Thromboxane Res.* **3**, 97–104.

Chihara-Nakashima, M., Hashimoto, E., and Nishizuka, Y. (1977). Intrinsic activity of guanosine 3'-5'-monophosphate-dependent protein kinase similar to adenosine 3',5'-monophosphate-dependent protein kinase. II. Phosphorylation of ribosomal proteins. *J. Biochem. (Tokyo)* **81**, 1863–1867.

Cohen, P., Burchell, A., Foulkes, J. G., Cohen, P. T. W., Vanaman, T. C., and Nairn, A. C. (1978). Identification of the Ca^{2+}-dependent modulator protein as the fourth subunit of rabbit skeletal muscle phosphorylase kinase. *FEBS Lett.* **92**, 287–293.

Dabrowska, R., Sherry, J. M. F., Aromatorio, D. K., and Hartshorne, D. J. (1978). Modulator protein as a component of the myosin light chain kinase from chicken gizzard. *Biochemistry* **17**, 253–258.

Daniel, J. L., Holmsen, H., and Adelstein, R. S. (1977). Thrombin-stimulated myosin phosphorylation in intact platelets and its possible involvement secretion. *Thromb. Haemostasis* **38**, 984–989.

Dawson, R. M. C., Freinkel, N., Jungalwala, F. B., and Clarke, N. (1971). The enzymic formation of *myo*-inositol 1:2-cyclic phosphate from phosphatidylinositol. *Biochem. J.* **122**, 605–607.

Dayton, W. R., Goll, D. E., Zeece, M. G., Robson, R. M., and Reville, W. J. (1976). A Ca^{2+}-activated protease possibly involved in myofibrillar protein turnover: Purification from porcine muscle. *Biochemistry* **15**, 2150–2158.

DeLorenzo, R. J., Freedman, S. D., Yohe, W. B., and Maurer, S. C. (1979). Stimulation of Ca^{2+}-dependent neurotransmitter release and presynaptic nerve terminal protein phosphorylation by calmodulin and a calmodulin-like protein isolated from synaptic vesicles. *Proc. Natl. Acad. Sci. U.S.A.* **76**, 1838–1842.

Drummond, G. I., and Duncan, L. (1968). On the mechanism of activation of phosphorylase *b* kinase by calcium. *J. Biol. Chem.* **243**, 5532–5538.

Edlund, B., Zetterqvist, O., Ragnarsson, U., and Engström, L. (1977). Phosphorylation of synthetic peptides by [^{32}P]ATP and cyclic GMP-stimulated protein kinase. *Biochem. Biophys. Res. Commun.* **79**, 139–144.

Graff, G., Stephenson, J. H., Glass, D. B., Haddox, M. K., and Goldberg, N. D. (1978). Activation of soluble splenic cell guanylate cyclase by prostaglandin endoperoxides and fatty acid hydroperoxides. *J. Biol. Chem.* **253**, 7662–7676.

Guroff, G. (1964). A neutral, calcium-activated proteinase from the soluble fraction of rat brain. *J. Biol. Chem.* **239**, 149–155.

Hashimoto, E., Takeda, M., Nishizuka, Y., Hamana, K., and Iwai, K. (1976). Studies on the sites in histones phosphorylated by adenosine 3',5'-monophosphate-dependent and guanosine 3',5'-monophosphate-dependent protein kinases. *J. Biol. Chem.* **251**, 6287–6293.

Hashimoto, E., Kuroda, Y., Ku, Y., and Nishizuka, Y. (1979). Stimulation by polydeoxyribonucleotide of histone phosphorylation by guanosine 3':5'-

monophosphate-dependent protein kinase. *Biochem. Biophys. Res. Commun.* **87**, 200–206.

Haslam, R. J., and Lynham, J. A. (1977). Relationship between phosphorylation of blood platelet proteins and secretion of platelet granule constituents. I. Effects of different aggregating agents. *Biochem. Biophys. Res. Commun.* **77**, 714–722.

Haslam, R. J., Davidson, M. M. L., and McClenaghan, M. D. (1975). Cytochalasin B, the blood platelet release reaction and cyclic GMP. *Nature (London)* **253**, 455–457.

Haslam, R. J., Davidson, M. M. L., Davies, T., Lynham, J. A., and McClenaghan, M. D. (1978). Regulation of blood platelet function by cyclic nucleotides. *Adv. Cyclic Nucleotide Res.* **9**, 533–552.

Hawthorne, J. N. (1960). The inositol phospholipids. *J. Lipid Res.* **1**, 255–280.

Hawthorne, J. N., and White, D. A. (1975). Myo-inositol lipids. *Vitam. Horm. (N.Y.)* **33**, 529–573.

Hidaka, H., and Asano, T. (1977). Stimulation of human platelet guanylate cyclase by unsaturated fatty acid peroxides. *Proc. Natl. Acad. Sci. U.S.A.* **74**, 3657–3661.

Hirata, F., and Axelrod, J. (1978). Enzymatic methylation of phosphatidylethanolamine increases erythrocyte membrane fluidity. *Nature (London)* **275**, 219–220.

Hirata, F., and Axelrod, J. (1980). Phospholipid methylation and biological signal transmission. *Science* **209**, 1082–1090.

Hirata, F., Viveros, O. H., Diliberto, E. J., Jr., and Axelrod, J. (1978). Identification and properties of two methyltransferases in conversion of phosphatidylethanolamine to phosphatidylcholine. *Proc. Natl. Acad. Sci. U.S.A.* **75**, 1718–1721.

Hirata, F., Strittmatter, W. J., and Axelrod, J. (1979). β-Adrenergic receptor agonists increase phospholipid methylation, membrane fluidity, and β-adrenergic receptor-adenylate cyclase coupling. *Proc. Natl. Acad. Sci. U.S.A.* **76**, 368–372.

Hokin, L. E., and Hokin, M. R. (1955a). Effects of acetylcholine on phosphate turnover in phospholipids of brain cortex *in vitro*. *Biochim. Biophys. Acta* **16**, 229–237.

Hokin, L. E., and Hokin, M. R. (1955b). Effects of actylcholine on the turnover of phosphoryl units in individual phospholipids of pancreas slices and brain cortex slices. *Biochim. Biophys. Acta* **18**, 102–110.

Hokin, L. E., and Hokin, M. R. (1959). Evidence for phosphatidic acid as the sodium carrier. *Nature (London)* **184**, 1068–1069.

Hokin, L. E., and Sherwin, A. L. (1957). Protein secretion and phosphate turnover in the phospholipids in salivary glands *in vitro*. *J. Physiol. (London)* **135**, 18–29.

Hokin, M. R., and Hokin, L. E. (1953). Enzyme secretion and the incorporation of P^{32} into phospholipids of pancreas slices. *J. Biol. Chem.* **203**, 967–977.

Hokin, M. R., and Hokin, L. E. (1954). Effects of acetylcholine on phospholipids in the pancreas. *J. Biol. Chem.* **209**, 549–558.

Holub, B. J., Kuksis, A., and Thompson, W. (1970). Molecular species of mono-, di-, and triphosphoinositides of bovine brain. *J. Lipid Res.* **11**, 558–564.

Huang, L. C., and Huang, C. (1975). Rabbit skeletal muscle protein kinase: Conversion from cAMP dependent to independent form by chemical perturbations. *Biochemistry* **14**, 18–24.

Huston, R. B., and Krebs, E. G. (1968). Activation of skeletal muscle phosphorylase kinase by Ca^{2+}. II. Identification of the kinase activating factor as a proteolytic enzyme. *Biochemistry* **7**, 2116–2122.

Igarashi, Y., and Kondo, Y. (1981). Demonstration and characterization of partial glyceride specific lipases in pig thyroid plasma membranes. *Biochem. Biophys. Res. Commun.* **97**, 766–771.

Inoue, M., Kishimoto, A., Takai, Y., and Nishizuka, Y. (1976). Guanosine 3':5'-

monophosphate-dependent protein kinase from silkworm: Properties of a catalytic fragment obtained by limited proteolysis. *J. Biol. Chem.* **251**, 4476–4478.

Inoue, M., Kishimoto, A., Takai, Y., and Nishizuka, Y. (1977). Studies on a cyclic nucleotide-independent protein kinase and its proenzyme in mammalian tissues. II. Proenzyme and its activation by calcium-dependent protease from rat brain. *J. Biol. Chem.* **252**, 7610–7616.

Ishiura, S., Murofushi, H., Suzuki, K., and Imahori, K. (1978). Studies of a calcium-activated neutral protease from chicken skeletal muscle. I. Purification and characterization. *J. Biochem. (Tokyo)* **84**, 225–230.

Iwasa, Y., Takai, Y., Kikkawa, U., and Nishizuka, Y. (1980). Phosphorylation of calf thymus H1 histone by calcium-activated, phospholipid-dependent protein kinase. *Biochem. Biophys. Res. Commun.* **96**, 180–187.

Kawahara, Y., Takai, Y., Minakuchi, R., Sano, K., and Nishizuka, Y. (1980a). Possible involvement of Ca^{2+}-activated, phospholipid-dependent protein kinase in platelet activation. *J. Biochem. (Tokyo)* **88**, 913–916.

Kawahara, Y., Takai, Y., Minakuchi, R., Sano, K., and Nishizuka, Y. (1980b). Phospholipid turnover as a possible transmembrane signal for protein phosphorylation during human platelet activation by thrombin. *Biochem. Biophys. Res. Commun.* **97**, 309–317.

Khoo, J. C., Sperry, P. J., Gill, G. N., and Steinberg, D. (1977). Activation of hormone-sensitive lipase and phosphorylase kinase by purified cyclic GMP-dependent protein kinase. *Proc. Natl. Acad. Sci. U.S.A.* **74**, 4843–4847.

Kishimoto, A., Takai, Y., and Nishizuka, Y. (1977). Activation of glycogen phosphorylase kinase by a calcium-activated, cyclic nucleotide-independent protein kinase system. *J. Biol. Chem.* **252**, 7449–7452.

Kishimoto, A., Mori, T., Takai, Y., and Nishizuka, Y. (1978). Comparison of calcium-activated, cyclic nucleotide-independent protein kinase and adenosine 3′:5′-monophosphate-dependent protein kinase as regards the ability to stimulate glycogen breakdown *in vitro*. *J. Biochem. (Tokyo)* **84**, 47–53.

Kishimoto, A., Takai, Y., Mori, T., Kikkawa, U., and Nishizuka, Y. (1980). Activation of calcium and phospholipid-dependent protein kinase by diacylglycerol: Its possible relation to phosphatidylinositol turnover. *J. Biol. Chem.* **255**, 2273–2276.

Kishimoto, A., Kajikawa, N., Tabuchi, H., Shiota, M., and Nishizuka, Y. (1981). Calcium-dependent neutral proteases, widespread occurrence of a species of protease active at lower concentrations of calcium. *J. Biochem. (Tokyo)* **90**, 889–892.

Kuo, J. F. (1974). Guanosine 3′:5′-monophosphate-dependent protein kinases in mammalian tissues. *Proc. Natl. Acad. Sci. U.S.A.* **71**, 4037–4041.

Kuo, J. F., and Greengard, P. (1969). Cyclic nucleotide-dependent protein kinases. IV. Widespread occurrence of adenosine 3′,5′-monophosphate-dependent protein kinase in various tissues and phyla of the animal kingdom. *Proc. Natl. Acad. Sci. U.S.A.* **64**, 1349–1355.

Landt, M., and McDonald, J. M. (1980). Calmodulin-activated protein kinase activity of adipocyte microsomes. *Biochem. Biophys. Res. Commun.* **93**, 881–888.

Le Peuch, C. J., Haich, J., and Demaille, J. G. (1979). Concerted regulation of cardiac sarcoplasmic reticulum calcium transport by cyclic adenosine monophosphate-dependent and calcium-calmodulin-dependent phosphorylations. *Biochemistry* **18**, 5150–5157.

Limas, C. J. (1980). Phosphorylation of cardiac sarcoplasmic reticulum by a calcium-activated, phospholipid-dependent protein kinase. *Biochem. Biophys. Res. Commun.* **96**, 1378–1383.

Lincoln, T. M., and Corbin, J. D. (1977). Adenosine 3':5'-cyclic monophosphate- and guanosine 3':5'-cyclic monophosphate-dependent protein kinases: Possible homologous proteins. *Proc. Natl. Acad. Sci. U.S.A.* **74**, 3239–3243.

Lincoln, T. M., and Corbin, J. D. (1978). Purified cyclic GMP-dependent protein kinase catalyzes the phosphorylation of cardiac troponin inhibitory subunit (TN-I). *J. Biol. Chem.* **253**, 337–339.

Lyons, R. M., Stanford, N., and Majerus, P. W. (1975). Thrombin-induced protein phosphorylation in human platelets. *J. Clin. Invest.* **56**, 924–936.

Mellgren, R. L. (1980). Canine cardiac calcium-dependent proteases: Resolution of two forms with different requirements for calcium. *FEBS Lett.* **109**, 129–133.

Meyer, W. L., Fischer, E. H., and Krebs, E. G. (1964). Activation of skeletal muscle phosphorylase *b* kinase by Ca^{2+}. *Biochemistry* **3**, 1033–1039.

Michell, R. H. (1975). Inositol phospholipids and cell surface receptor function. *Biochim. Biophys. Acta* **415**, 81–147.

Michell, R. H. (1979). Inositol phospholipids in membrane function. *Trends Biochem. Sci.* **4**, 128–131.

Minakuchi, R., Takai, Y., Yu, B., and Nishizuka, Y. (1981). Widespread occurrence of calcium-activated, phospholipid-dependent protein kinase in mammalian tissues. *J. Biochem. (Tokyo)* **89**, 1651–1654.

Mori, T., Takai, Y., Minakuchi, R., Yu, B., and Nishizuka, Y. (1980). Inhibitory action of chlorpromazine, dibucaine, and other phospholipid-interacting drugs on calcium-activated, phospholipid-dependent protein kinase. *J. Biol. Chem.* **255**, 8378–8380.

Murachi, T., Tanaka, K., Hatanaka, M., and Murakami, T. (1981). Intracellular Ca^{2+}-dependent protease (calpain) and its high-molecular-weight endogenous inhibitor (calpastatin). *Adv. Enzyme Regul.* **19**, 407–424.

Nishiura, I., Tanaka, K., Yamamoto, S., and Murachi, T. (1978). The occurrence of an inhibitor of Ca^{2+}-dependent neutral protease in rat liver. *J. Biochem. (Tokyo)* **84**, 1657–1659.

Nishizuka, Y. (1980). Three multifunctional protein kinase systems in transmembrane control. *Mol. Biol. Biochem. Biophys.* **32**, 113–135.

Nishizuka, Y., and Takai, Y. (1981). Calcium and phospholipid turnover in a new receptor function for protein phosphorylation. *Cold Spring Harbor Conf. Cell Proliferation* **8**, 237–249.

Nishizuka, Y., Takai, Y., Hashimoto, E., Kishimoto, A., Kuroda, Y., Sakai, K., and Yamamura, H. (1979). Regulatory and functional compartment of three multifunctional protein kinase systems. *Mol. Cell. Biochem.* **23**, 153–165.

Ogawa, Y., Takai, Y., Kawahara, Y., Kimura, S., and Nishizuka, Y. (1981). A new possible regulatory system for protein phosphorylation in human peripheral lymphocytes. I. Characterization of a calcium-activated, phospholipid-dependent protein kinase. *J. Immunol.* **127**, 1369–1374.

Op den Kamp, J. A. F. (1979). Lipid asymmetry in membranes. *Annu. Rev. Biochem.* **48**, 47–71.

Papahadjopoulos, D. (1972). Studies on the mechanism of action of local anesthetics with phospholipid model membranes. *Biochim. Biophys. Acta* **265**, 169–186.

Payne, M. E., and Soderling, T. R. (1980). Calmodulin-dependent glycogen synthase kinase. *J. Biol. Chem.* **255**, 8054–8056.

Phillips, D. R., and Jakabova, M. (1977). Ca^{2+}-dependent protease in human platelets: Specific cleavage of platelet polypeptides in the presence of added Ca^{2+}. *J. Biol. Chem.* **252**, 5602–5605.

Puca, G. A., Nola, E., Sica, V., and Bresciani, F. (1977). Estrogen binding proteins of calf uterus: Molecular and functional characterization of the receptor transforming factor—A Ca^{2+}-activated protease. *J. Biol. Chem.* **252**, 1358–1366.

Reddy, M. K., Etlinger, J. D., Rabinowitz, M., Fischman, D. A., and Zak, R. (1975). Removal of Z-lines and α-actinin from isolated myofibrils by a calcium-activated neutral protease. *J. Biol. Chem.* **250**, 4278–4284.

Riou, J. P., Claus, T. H., Flockhart, D. A., Corbin, J. D., and Pilkis, S. J. (1977). *In vivo* and *in vitro* phosphorylation of rat liver fructose-1,6-bisphosphatase. *Proc. Natl. Acad. Sci. U.S.A.* **74**, 4615–4619.

Rittenhouse-Simmons, S. (1979). Production of diglyceride from phosphatidylinositol in activated human platelets. *J. Clin. Invest.* **63**, 580–587.

Rittenhouse-Simmons, S. (1980). Indomethacin-induced accumulation of diglyceride in activated human platelets: The role of diglyceride lipase. *J. Biol. Chem.* **255**, 2259–2262.

Roelofsen, B., and Zwaal, R. F. A. (1976). The use of phospholipases in the determination of asymmetric phospholipid distribution in membranes. *Methods Membr. Biol.* **7**, 147–177.

Rothman, J. E., and Lenard, J. (1977). Membrane assymmetry. *Science* **195**, 743–753.

Schick, P. K., and Yu, B. P. (1974). The role of platelet membrane phospholipids in the platelet release reaction. *J. Clin. Invest.* **54**, 1032–1039.

Schulman, H., and Greengard, P. (1978). Ca^{2+}-dependent protein phosphorylation system in membranes from various tissues, and its activation by "calcium-dependent regulator." *Proc. Natl. Acad. Sci. U.S.A.* **75**, 5432–5436.

Seeman, P. (1972). The membrane actions of anesthetics and tranquilizers. *Pharmacol. Rev.* **24**, 583–655.

Sugita, H., Ishiura, S., Suzuki, K., and Imahori, K. (1980). Inhibition of epoxide derivatives on chicken calcium-activated neutral protease (CANP) *in vitro* and *in vivo*. *J. Biochem. (Tokyo)* **87**, 339–341.

Takai, Y., Yamamoto, M., Inoue, M., Kishimoto, A., and Nishizuka, Y. (1977a). A proenzyme of cyclic nucleotide-independent protein kinase and its activation by calcium-dependent neutral protease from rat liver. *Biochem. Biophys. Res. Commun.* **77**, 542–550.

Takai, Y., Kishimoto, A., Inoue, M., and Nishizuka, Y. (1977b). Studies on a cyclic nucleotide-independent protein kinase and its proenzyme in mammalian tissues. I. Purification and characterization of an active enzyme from bovine cerebellum. *J. Biol. Chem.* **252**, 7603–7609.

Takai, Y., Kishimoto, A., Iwasa, Y., Kawahara, Y., Mori, T., and Nishizuka, Y. (1979a). Calcium-dependent activation of a multifunctional protein kinase by membrane phospholipids. *J. Biol. Chem.* **254**, 3692–3695.

Takai, Y., Kishimoto, A., Iwasa, Y., Kawahara, Y., Mori, T., Nishizuka, Y., Tamura, A., and Fujii, T. (1979b). A role of membranes in the activation of a new multifunctional protein kinase system. *J. Biochem. (Tokyo)* **86**, 575–578.

Takai, Y., Kishimoto, A., Kikkawa, U., Mori, T., and Nishizuka, Y. (1979c). Unsaturated diacylglycerol as a possible messenger for the activation of calcium-activated, phospholipid-dependent protein kinase system. *Biochem. Biophys. Res. Commun.* **91**, 1218–1224.

Takai, Y., Kishimoto, A., Kawahara, Y., Minakuchi, R., Sano, K., Kikkawa, U., Mori, T., Yu, B., Kaibuchi, K., and Nishizuka, Y. (1981). Calcium and phosphatidylinositol turnover as signalling for transmembrane control of protein phosphorylation. *Adv. Cyclic Nucleotide Res.* **14**, 301–313.

Wallace, W. C., and Bensusan, H. B. (1980). Protein phosphorylation in platelets stimulated by immobilized thrombin at 37° and 4°C. *J. Biol. Chem.* **255**, 1932–1937.

Wallach, D., and Pastan, I. (1976). Stimulation of guanylate cyclase of fibroblasts by free fatty acids. *J. Biol. Chem.* **251**, 5802–5809.

Waxman, L., and Krebs, E. G. (1978). Identification of two protease inhibitors from bovine cardiac muscle. *J. Biol. Chem.* **253**, 5888–5891.

Yagi, K., Yazawa, M., Kakiuchi, S., Ohshima, M., and Uenishi, K. (1978). Identification of an activator protein for myosin light chain kinase as the Ca^{2+}-dependent modulator protein. *J. Biol. Chem.* **253**, 1338–1340.

Yamamoto, M., Takai, Y., Hashimoto, E., and Nishizuka, Y. (1977). Intrinsic activity of guanosine 3′,5′-monophosphate-dependent protein kinase similar to adenosine 3′,5′-monophosphate-dependent protein kinase. I. Phosphorylation of histone fractions. *J. Biochem.* (*Tokyo*) **81**, 1857–1862.

Yamamoto, M., Takai, Y., Inoue, M., Kishimoto, A., and Nishizuka, Y. (1978). Characterization of cyclic nucleotide-independent protein kinase produced enzymatically from its proenzyme by calcium-dependent neutral protease from rat liver. *J. Biochem.* (*Tokyo*) **83**, 207–212.

Yamauchi, T., and Fujisawa, H. (1979). Activation of tryptophan 5-monooxygenase by calcium-dependent regulator protein. *Biochem. Biophys. Res. Commun.* **90**, 28–35.

Index

A

Abnormal prothrombin, 218
Acidic amino acids, in solid-state proteins, 17
Acrosome reaction, 343–345
Actin
  calcium-dependent proteins and, 333–334
  myosin-binding site on, 163
  skeletal and cardiac troponin subunit interactions with, 152
F-Actin, fluorescent energy transfer for ϵ-ADP, 164
Actin filament length
  average, 326
  calcium regulation of, 327
Actin gel-sol transformation, regulation of, 327–328
Actin network structure, calcium control and regulation of, 325–335
Actin polymerization, in starfish sperm, 344
Actin polymers, depolymerizing, 327
Actomyosin, calcium sensitivity on, 146
Actomyosin ATPase inhibitory subunit, 146
Actomycin-Tn, Tm, and Ca^{2+} interactions, model for regulation of, 154
Adenosine triphosphate, *see* ATP
Adenylate cyclase, intracellular distribution of, in *Tetrahymena pyriformis*, 302

ϵ-ADP, fluorescence energy transport from, 164
Adrenal glucocorticoids, in mitochondrial Ca^{2+} uptake, 48
Aequorin, 7
Aging, CaBP and, 190–192
Alkali gel electrophoresis, calmodulin localization in cilia as revealed by, 308–313
Allosteric control, by coenzyme, 12
Allosteric effect, protein motion and, 12
Allosteric switch, two-state, 31
Ammonium sulfate fractionation, SCP purification by, 261
Amphioxus, SCPs in, 249, 270
Amylase, calcium-binding site for, 9
Ancestral one-domain polypeptide, reconstruction of, 117
1-Anilinonaphthalene-8-sulfonate, 181
Anion exchange, SCP purification by, 261
Anionic polymers, calcium ions in binding of, 17–18
Annelida, SCPs in, 247
Antipsychotic drugs, as calmodulin inhibitors, 300
Arachidonic acid
  egg preloading with, 377
  phosphatidylinositol turnover and, 402
Arbacia punctulata, 377
Arg-Ala peptide bond, cleavage of in factor IX, 232
Arginine residue, in porcine CaBP, 178
Arthropoda, SCPs in, 247

413

Aspartate resonance, in Ca^{2+} binding, 158
ATP (adenosine triphosphate)
    changing concentrations of, for light-chain kinase-phosphatase system, 291–292
    in protein kinase C assays, 389
    Ca^{2+}-transporting, 40
ATPase
    in cholinergic synaptic vessels, 88
    of endoplasmic reticulum and plasmalemma, 66
    membrane channel for, 34
ATPase pump, 7
ATP concentrations, for light-chain kinase-phosphatase system test, 291–292
ATP-dependent mechanism, calcium sequestration by, 89
ATP hydrolysis
    in calcium influx, 41
    for calcium proteins in membranes, 34
    by myofibrils, 146
ATPγS
    pretreatment with, 287
    skinned fiber preparations and, 288
ATTσS, irreversible thiophosphorylation of myosin light chains using, 286–287

B

Bacterial proteases, 7
Binding layers, types of, 18, *see also* Calcium binding
Biological anion centers, calcium binding to, 1–35
Birds, calcium-binding protein vs. gonadal hormones in, 195–197
Bivalent metal ion specificity, in mitochondrial Ca^{2+} transport, 43–44
Blood, calcium-transport proteins in, 21–22
Blood coagulation
    factor X in, 229
    protein C inhibition of, 235
Bone
    calcium activity and, 21
    γ-carboxyglutamic acid-containing protein of, 236–237
Bone proteins, 7, 236–237
    disulfide bond reduction in, 237
Bone solubility products, 30

Bovine bone, γ-carboxyglutamic-acid containing protein of, 236–237
Bovine brain calmodulin, rabbit skeletal muscle troponin C and, 132
Bovine calcium-binding protein, structural changes to, in calcium-binding state, 205
Bovine follicular fluid protein, in hamster sperm, 347
Bovine intestinal CaBP, 177–179
    amino acid sequences in, 184–185, 227
    Ca^{2+} binding by, 182
    three distinct proteins in, 183
Bovine prothrombin
    intrinsic fluorescence of, 224
    Kringle structure of fragment 1 of, 15
Brain mitochondrin, Ca^{2+} sequestration in, 85–86
Bridge swivel, calcium role of, 15–16

C

Ca^{2+}, *see* Calcium; Extracellular calcium; Intracellular calcium
CaBP, *see* Calcium-binding protein
CaBP-coded nuclear mRNA, 200
"CaBPology," defined, 176
CaBP-specific fluorescence, globlet cells and, 187–188
Calcisomes, defined, 137
Calcitonin, in calcium precipitation, 21
Calcium
    arachidonic acid conversion and, 377
    buffering of by intraterminal organelles, 91–92
    in control and substrate binding, 12
    in cortical exocytosis, 372
    as cross-link between proteins, 13, 22
    in egg fertilization, 355–379
    free, 357–358
    in frog eggs activation, 361
    in heart mitochondria, 63
    intracellular, *see* Intracellular calcium
    intraterminal sequestration of, 84–89
    in isolated mitochondria, 62–64
    lipid binding and, 19
    vs. magnesium in protein binding, 6–7
    metabolism control by, 369–378
    mitochondrial high affinity for, 85
    mobilization of, 19

in phosphorylation regulation, 386
in polyspermy blockage, 365
precursor proenzymes and, 9
at presynaptic nerve terminals, 82–83
in secretory mechanisms, 102
in serum function, 21
spermatozoa metabolic activation and, 339–349
in sperm-egg association, 348
spontaneous release of in mitochondrial transport, 53
in vesicle or plasma membrane binding, 84
in viral invasion of cells, 21
Calcium absorption, efficiency of, 189
Calcium-activated phospholipid-dependent protein hinge, enzymology of, 387–389
Calcium activity, extracellular, *see* Extracellular calcium
Calcium barrier proteins, in fluids connecting calcium stores, 19
Calcium binding
aspartate resonance in, 158
by calmodulin, 29
to check intestinal CaBP, 181
evolutionary changes in, 119
to factor X, 230
glutamate resonance in, 158
layers of, 18
microcalorimetry in, 157
in muscle contraction activation and regulation, 145–147
in myofibrillar ATPase activation, 150
by parvalbumin, 259–261
phenylalanine resonance in, 158
polypeptide backbone tightening in, 158
to proteins and biological anion centers, 1–35
rates of, 31
structural changes following, 158
transmission of information in, 29
trigger proteins in, 24–30
to troponin, 147
tyrosine fluorescence and, 159
and ultraviolet circular dichroic spectra of CTnC and STnC, 159
Calcium-binding constants, 22, 27
Calcium-binding intestinal proteins, function of, 32–34

Calcium-binding loops
evolutionary rates in, 118–120
evolution of middomain positions in, 118–119
Calcium-binding protein family
ancestral one-domain polypeptide in, 116–117
evolution of, 113–120
nucleotide replacement rates in, 118
structural and functional features of, 114
Calcium-binding proteins, *see also* Calmodulin
aging and, 190–194
binding constants for, 24
as bona fide transport proteins, 206
bovine intestine in, 177
calcium absorption and, 203
and calcium buffers of egg, 368–369
in calcium reabsorption in kidney, 202
in calcium "shuttle" activity, 207
γ-carboxyglutamic acid-containing, *see* γ-Carboxyglutamic acid-containing calcium-binding proteins
cellular localization of, 187–188
chick intestinal, 176
cortisol and, 194–195
diabetes and, 197–198
in embryonic intestine development, 201–202
evolution of, 118–119, 245–246
evolutionary diversification of structure and function in, 111–138
extracellular, 8–17
functions of, 7–8
and gonadal hormones in birds, 197
heat treatment and, 181
hen uterine, 176–177
from human jejenum, 181
human kidney, 178
immunoreactive, 177–178
in intestinal mucosa, 203, 206
in vitro synthesis of, 198–201
in laying birds, 195–197
lysolecithin and, 206
in maturation of female quail, 196
molecular biology of, 201
rat kidney, 177
sarcoplasmic, *see* Sarcoplasmic calcium-binding proteins
physiological factors affecting, 188–189

phosphate deficiency and, 189
properties of, 180–186
temporal responses of to acute doses of
1,25(OH)$_2$D$_3$, 203–205
vitamin D-induced, 175–207
Calcium binding rates, to trigger products, 31
Calcium binding specific fluorescence, cell
cytoplasm and, 188
Calcium bone proteins, 16–17
Calcium-calmodulin-dependent enzymes,
minimal kinetic scheme for activation
of, 135–136
Calcium-calmodulin-light chain-kinase-
phosphatase system, model for, 282
Calcium channels
blocking of by polyvalent cations, 94
"fast" and "slow" types of, in synaptosomes, 95–100
heterogeneity of, 93–95
properties in, 96
sensitivities of, 94
voltage-regulated, 98
Calcium complexes, structure of, 3
Calcium concentration, cytosolic free, 39
Calcium containing mineralized tissue, 237
Calcium cross-links, functions of, 16
Calcium-dependent calmodulin-binding
protein, 311
Calcium-dependent enzymes, 12–14
sequential activation in deactivation of,
135–138
Calcium-dependent transmitter release, in
squid and frog nerve endings, 92–93
Calcium distribution, in nerve terminals,
83–84, 92–93
Calcium efflux, 50–62
isolation and reinstitution of Ca^{2+}–Na$^+$
antiport activity, 52
kinetics of, 51–52
mechanism of, 50–51
from nerve terminals, 100–101
regulation of, 53–62
Calcium entry mechanisms, 92–100
Calcium exchange rates, troponin C
structural changes and, 164–167
Calcium extrusion, sodium-calcium exchange in, 100–101
Calcium increase
in sea urchin egg fertilization, 378
transient, 379

Calcium-induced structural reorganization,
diagrammatic representation of, 27
Calcium-induced structural troponin C
changes, propagation to thin-filament
proteins, 163–164
Calcium influx
inhibitors of, 45
internal Na$^+$-dependent, 99
in mitochondrial Ca^{2+} transport, 41–50
role of, after fertilization, 363
Calcium ion binding, *see also* Calcium
binding
to large particles, membranes, and surfaces, 17–19
physiological concentrations in, 28
to prothrombin fragment 1, 225
Calcium ions
in actin-myosin interaction, 243
bone and, 21
as cofactor in conversion of vitamin K-
dependent blood coagulation
zymogens, 221
in extracellular fluids, 21
in muscle contraction, 243
as second messengers, 112–113
short-term buffering of, 137
Calcium level, precipitating anion levels
and, 20
Calcium metabolism, aging and, 191–192
Calcium movement, in presynaptic nerve
terminals, 81–102
Calcium permeability, at fertilization, 358–
360
Calcium-polysaccharide interactions, 18–19
Calcium precipitation, equilibria involved
in, 20
Calcium protection, from external proteolysis, 11
Calcium proteinate, 20
Calcium protein binding, in muscle contraction activation, 145
Calcium proteins, 30–31
classes of, 7
in membranes, 34
Scatchard plot of binding in, 235
Calcium reflux, intracellular factors in, 53–
59
Calcium release
from FCCP, 83
ruthenium red-induced, 50–52

Calcium-requiring extracellular triggers, 15
Calcium sequestration
  by ATP-dependent mechanisms, 89
  in brain mitochondria, 85
  in liver mitochondria, 85
  in smooth endoplasmic reticulum, 87
Calcium sites, in proteins, 13
Calcium storage, in synaptic vessels, 88–89
Calcium stores
  cytoplasmic, 367–368
  vesicles and, 23
Calcium-storing organelles, morphology of, 89–92
Calcium tooth proteins, 17
Calcium translocation, intestinal calcium-binding protein in, 33
Calcium transport, 19–23
  across endothelial cells, 19
  mitochondrial, *see* Mitochondrial calcium uptake
Calcium trigger cells, 23–28
Calcium turnover, as transmembrane signaling for protein phosphorylation, 385–406
Calcium uptake, *see also* Mitochondrial calcium uptake
  mitochondrial energy coupling and, 64
  phosphate stimulation of, 44–45
  respiration-dependent, 42
Callinectes sapidus, 45
Calmodulin, *see also* Sarcoplasmic calcium-binding proteins
  calcium binding by, 29
  calcium titration of, 125
  conformational changes in, 29
  in contractile vacuole pores, 317
  in egg metabolism, 269
  evolution of, 249
  evolutionary diversification of structure and function in, 111–138
  exogenous, 289
  and four-domain ancestral protein, 129
  glycogen phosphorylase *b* kinase activation by, 130
  invertebrate, 133
  ion-binding constants of, 124
  octopus, *see* Octopus calmodulin
  localization of in cilia, 308–313
  NMR studies of, 24

  in outer-doublet microtubules of *Tetrahymena* cilia, 315
  in *Paramecium* and *Tetrahymena* cilia, 308–315
  phenothiazine drug binding to, 30
  phenothiazine agents as inhibitors of, 305
  as prototype of sensor molecules, 124–135
  replacement of troponin C and leiotinin C by, 130
  in sea urchin eggs and sperm, 341, 368–369
  sequence of, 13, 124–129
  specificity of in activation of target enzymes, 129
  structural changes to, in calcium-binding state, 205
  terbium binding to, 127
  in *Tetrahymena pyriformis*, 297–319
  trifluoperazine binding to two domains of, 31
  as trigger proteins, 23–30
Calmodulin activity, inhibitors of, 288–289
Calmodulin binding, in ancestral one-domain polypeptide, 116
Calmodulin-binding proteins, in sequential deactivation of calmodulin-dependent enzymes, 137
Calmodulin troponin, 7
Calmodulin/troponin C, as biheaded protein, 134
Calsequestrin, 7, 23
CaMBP, *see* Calcium-dependent calmodulin-binding protein
cAMP (cyclic adenosine 3′,5′-monophosphate)
  as interrelated intracellular messenger in phosphorylation reactions, 112, 386–387
  phospatidylinositol turnover and, 401
cAMP-dependent protein kinases, 112, 290
  phosphorylation of calmodulin-dependent enzymes by, 138
Carbonic anhydrases, movements of, 10
γ-Carboxyglutamic acid
  as calcium ion ligand in vitamin K-dependent proteins, 218
  chemical properties of, 219
  discovery of, 217–218

factor X and, 229–232
history of, 218–219
as metal ligand, 219–220
in prothrombin, 218
protein C and, 235
γ-Carboxyglutamic acid-containing peptides, as models for metal-binding studies, 270–271
γ-Carboxyglutamic acid-containing proteins, 217–238
of blood plasma, 221–235
of bone, 236–237
in calcified tissues, 237–238
of factor IX, 232
γ-Carboxyglutamic-acid residues, and metal-liganding properties of proteins, 238
Carboxypeptidase
calcium-binding site for, 9
movements of, 10
Carboxylate, binding of calcium through, 2
Cardiac muscle contraction, troponin in regulation of, 168
Cardiac troponin C, Ca^{2+} and Mg^{2+} binding properties of, 148, 162
Cardiac troponin C_{IA}, complexing of with troponin I, 163
Cardiac troponin subunits, Ca^{2+} and Mg^{2+} binding properties of, 148
Carp parvalbumin
as globular molecule, 257
peptide backbone of, 258
Carp protein, secondary structure of, 254–257
Catalytic site control, by coenzyme, 12
CD spectra, see Circular dichroic spectra
Cellular Ca^{2+} fluxes, mitochondrial involvement in, 64–66
Cellular factors, in mitochondrial Ca^{2+} efflux regulation, 61–62
Cellular functions, trifluoperazine effects on, 313–319
Cephalochordata, SCPs, 247
Cestrum diurnum, 190
cGMP (cyclic guanosine monophosphate)
eukaryotic cells and, 112
phosphatidylinositol turnover and, 401
phosphorylation reactions in, 386
protein kinase G and, 405–406
Chick cerebellum, CaBP in Purkinje cell of, 188

Chick chorioallantoic membrane, vitamin K-dependent carboxylation systems in, 238
Chick heart muscle, parvalbumin in, 123
Chick embryo leg, parvalbumin levels in, 121
Chick embryonic intestine development, CaBP in, 201–202
Chick intestinal CaBP, 176–179, 203
calcium binding to, 181
isolated polyribosomes and, 199
Chick intestinal nuclear mRNA, 200
Chick kidney
1α-hydroxylase activity in, 190
polyribosomes from, 200
Chironomus salivary gland cells, cytosolic free Ca^{2+} concentration in, 65
Chlorpromazine
calmodulin activity and, 292
as calmodulin inhibitor, 300
sea urchin sperm and, 341
Chlortetracycline, intracellular Ca^{2+} distribution changes and, 65
Cholecalciferol, in calcium precipitation, 20–21
Circular dichroism
Ca^{2+} binding changes and, 159
Clam, SCP for, 249
Clam-amphioxus isologies, for SCPs, 249
Clam-lobster isologies, for SCPs, 249
Clam-sandworm isologies, for SCPs, 249
Clostridium perfringens, 399
^{13}CNMR spectroscopy, 226–227, see also Nuclear magnetic spectroscopy
Coelamates, SCPs for, 249
Coenzymes, allosteric control by, 12
Con A, sequence in, 13
Contraction, skinned fibers as models for, 283–285
Cortical cytoplasm, 326
Cortical exocytosis, calcium role in, 372
Cortisol, CaBP and, 194–195
Cow calcium binding protein, 178, 205, see also Bovine (adj.)
Crayfish SCPs, 266–269
antibodies for, 255
localization of, 271
Crustacean SCPs, molecular weights of, 262
CTnC, see Cardiac troponim C

Cyclic adenosine 3',5'-monophosphate, *see* cAMP
Cyclic AMP-dependent protein kinase, skinned muscle fiber preparations and, 289–290
Cyclic nucleotides, phosphatidylinositol turnover and, 401
Cytochrome C, movements of, 10
Cytoplasmic calcium stores, nature of, 367–368
Cytosol
  calcium uptake from, 40
  phosphate concentrations in, 44
Cytosolic calcium, low-level maintenance of, 40
Cytosolic calcium homeostasis, dynamics of, 67

D

Dansylaziridine, as fluorescent probe, 160–161
Dansylaziridine-labeled STnC, 149
Decarboxy-factor X, activation of by RVV, 230–231
Depolarization-neurotransmitter release coupling, Ca^{2+} role in, 102
Depolarization-secretion coupling, Ca^{2+} role in, 102
Depolarizing potential, calcium channel inactivation in presence of, 94
Diabetes, CaBP and, 197–198
Diacylglycerol
  calmodulin as substitute for, 394
  specificity of, 391–393
  stimulation of, 399
Diarachidonia, 391
Digestive enzymes, calcium vs. magnesium selectivity for, 9
Dihydrotachysterol, 190
Dilinolein, 391
Diolein, 391–392
Dipalmitin, 391
Direct binding, 18
Distearin, 391

E

Echinoderm acrosome reaction, 343–345
EDTA (ethylenediaminetetraacetic acid), in calcium-binding studies, 159, 165

EDTA buffer system, 267
Egg activation
  extracellular calcium in, 365–367
  intracellular calcium stores and, 361–362
Egg fertilization, calcium physiology and chemistry in, 355–379
Egg jelly, 340
Egg metabolism, calcium in control of, 369–378
EGTA [ethyleneglycol bis(β-aminoethyl ether)-N,N'-tetraacetic acid]
  buffer system of, 267
  in calcium binding studies, 5–6
  in calmodulin localization via alkali gel electrophoresis, 310
  egg fertilization and, 361
  in endogenous Ca^{2+} removal, 159
  free Ca^{2+} concentration and, 68
  gelsolin reaction in, 331
  in protein kinase C reversibility, 393
Electrophoretic Ca^{2+} efflux system, 50–62
Endoplasmic reticulum
  calcium reflux from, 40
  calcium-transporting ATPases of, 66
  calcium uptake into cisternae of, 40
Enzymes
  active site residues of, 10
  calcium-dependent, 12–14
  hydrophobic, 14
  mobility in, 10–11
  mobility maintenance inside, 11
  movements of active site residues of, 10
Epididymis, sperm passage into, 340
Epithelial cells, from rat duodenum, 199
Equilibria-allosteric switches, mobility and, 11
Erythrocruorin, 17
Escherichia coli, 298
N-Ethylmaleimide, in Ca^{2+} binding, 181
Eukaryotic cells, cAMP in, 112
Exocytotic event, cortical reactions to, 371
Exogenous calmodulin, *see also* Calmodulin
  calcium sensitivity and, 290
  skinned muscle fiber preparations and, 289
External proteolysis, calcium protection from, 11
Extracellular calcium
  activities of, 14–15

in egg activation, 365–367
protein-bound, 15
Extracellular calcium proteins, general
summary of, 22
Extracellular enzymes
calcium-containing, 8
calcium-requiring, 9
Extracellular factors
in mitochondrial Ca^{2+} efflux regulation,
59
in mitochondrial Ca^{2+} uptake, 48–49
Extracellular fluids, calcium concentration
in, 20–22
Extracellular messengers, in cellular function and proliferation, 386

F

Factor VII, 218
metal binding data of, 234
structure of, 234
Factor IX, 218
calcium binding to, 233
conversion to factor IXaα, 233–234
molecular weight and structure of, 232–
234
N-terminal region of, 232
Factor IXaβ, conversion of factor IX to,
234
Factor X, 218
activated factors VIII and IX in relation
to, 229
activation of by RVV, 231
metal ion binding to, 231
molecular weight and structure of, 229
Russell's viper venom and, 229
Fast Ca^{2+} channels, voltage-regulated, 98
Fast-muscle fibers, calcium cycle in, 123
FCCP (carbonyl cyanide p-
trifluoromethoxyphenylhydrazone)
in Ca^{2+} buffering by intraterminal
organelles, 91
calcium release and, 83
Fertilization
as agent of species continuity, 339
calcium influx and efflux in, 358–359,
363–364
calcium permeability at, 358–360
Fibrin, as calcium protein, 7
Fibrinogen, calcium-binding site of, 16

Free calcium content, changes in at fertilization, 357–358
Frog eggs, calcium activation of, 361

G

Gating mechanism, for calcium protein in
membranes, 34
Gel filtration, SCP purification by, 261
Gelsolin
amino acid composition of, 333
calcium control of action network structure by, 325–335
defined, 328
lattice structure changes by, 334
mechanism of action of, 330–331
physicochemical properties of, 329
as physiological regulator, 331–333
Gla-containing fragment, changes in on
binding of calcium ion, 14
Glucagon, in mitochondrial Ca^{2+} uptake,
49
Glucocorticoids, inhibiting effect of, in calcium intestinal absorption, 194
Glutamate resonance, in Ca^{2+} binding, 158
Glycogen phosphorylase b kinase, activation of by troponin C and calmodulin,
130–131
Goblet cells, CaBP and, 187–188
Gonadal hormones, $1,25(OH)_2D_3$ formation
and, 197
Gouy-Chapman diffuse layer, 18
GTPγS, in skinned fiber treatment, 291
Guanosine triphosphate, in skinned fiber
treatment, 291
Guanylate cyclase
intracellular distribution of, in *Tetrahymena pyriformis,* 302
membrane-bound, 303–307
Guanylate cyclase activity, cell cycle-
associated changes in, 306–307
Guanylate cyclase-cyclic guanosine 3′,5′-
monophosphate, *see* cGMP
Guinea pig spermatozoa, Ca^{2+} influx during capacitation of, 345

H

Heart mitochondria, Ca^{2+} content of, 63
Hemocyanin, 17
Hen uterine CaBP, 176–177

Hill coefficient, 44

Histone fractions, protein kinase reaction velocities and, 404

Homeostasis, in controlled calcium ion buffering, 21

Human CaBP, human kidney and, 179

Human jejenum CaBP, 181

Human kidney CaBP, 178

Human prothrombin, intrinsic fluorescence of, 224

Hydrophilic protein, defined, 14

Hydrophobic enzyme, 14

Hydrophobic-to-charged amino acid ratio, 25

Hydroxyapatites, binding of in osteocalcium of bone, 237

1,25-Hydroxylated vitamin D_3, 20–21, 195–196, 199, 202–203

I

IAANS [2-(4'-iodoacetamidoanilino) naphthalene-6-sulfonic acid], 160, 163

CYs-98 fluorescent decrease and, 165

Immunoreactive CaBP, 177–178

Inosine triphosphate, as test for light-chain kinase, 291

Insulin, CaBP and, 197–198

Internal Na^+-dependent Ca^{2+} influx, stoichiometry of, 99–100

Intestinal calcium-binding protein, see also Calcium-binding protein

function of, 32–33

in intestinal calcium transport, 193

in proximal small intestine, 178

in vitamin-D-mediated calcium metabolism, 187–188

Intestinal calcium transport, vs. intestinal CaBP, 193

Intracellular calcium

egg activation from release of, 361–363

mitochondrial regulation of, 39–71

role of, 35

transport-mediated regulation of, 40

Intracellular calcium-binding protein, ion-binding properties of, 113

Intracellular calcium distribution, chlortetracycline monitoring of, 65

Intracellular calcium rise, egg activation and, 360–363

Intracellular factors

in calcium reflux, 53–58

in mitochondrial Ca^{2+} uptake, 49–50

Intracellular proteins, 23–31

Intramitochondrial pyridine nucleotides, in mitochondrial Ca^{2+} efflux regulation, 55–56

Intraterminal calcium distribution, calcium content and, 82–92

Intraterminal organelles, calcium buffering by, 91–92

Invertebrate calmodulin, trimethylline residue in, 133

Invertebrates, sarcoplasmic calciumbinding proteins from, 247, 261–270

Iodoacetate, in Ca^{2+} binding, 181

Ionophore activation, in absence of exogenous calcium, 361

Isolated mitochondria, Ca^{2+} content of, 62–64

ITP, as ATP analogue, 291

ITPγS, 291, see also Inosine triphosphate

K

Kidney 1,25(OH)D_3-1α hydroxylase increased activity following uptake of, 197

strontium intake and, 189–190

Kringle structure, of bovine prothrombin fragment 1, 15

K^+-stimulated Ca^{2+} uptake, in synaptosomes, 97

L

Large particles, calcium ion binding to, 17–19

Leiotonin A and C, 283

Ligand protein binding, Scatchard analysis of, 222

Ligands, proteins and small molecules as, 5–6

Light chain hormone-phosphatase activity, 286–292

Light chain hormone phosphatase system

evidence for, in skinned fibers, 285–292

in vitro and in vivo evidence for, 280–281

postulated physiological role of, 281–283

Lipids, calcium binding to, 19

Lipoxygenase
 arachidonic acid oxidation and, 371
 sea urchin egg oxygen consumption and,
  377
Liver mitochondria, Ca²⁺ sequestration in,
 85
Lobster-amphioxus, from SCPs, 249
Lobster-crayfish isologies, from SCPs, 249
Low-calcium diet, calcium absorption and,
 189
Low cystolic calcium, maintenance of, 40
Low phosphorus diet, calcium absorption
 and, 189
Lysolecithin, CaBP and, 206
Lysozyme, movements of, 10
Lytechinus varietatus, membrane potential
 changes following fertilization of, 359

M

Magnesium binding
 by parvalbumin, 259–261
 physiological concentration in, 28
 protein binding and, 6–7
Magnesium ions, binding to membranes in
 fields, 19
Mammalian calmodulin, tyrosine residue
 and, 126, *see also* Calmodulin
Membrane-bound guanylate cyclase,
 activation of in *Tetrahymena*
Membrane potential studies, fertilization
 and, 359–360
Membranes
 calcium and magnesium binding to, 19
 calcium proteins in, 34
β-Mercaptoethanol, in Ca²⁺ binding, 181
Messenger molecules, in metazoan
 homeostasis, 111–112
Metal-γ-carboxyglutamate complex, solution geometry of, 219
Metazoans, homeostasis of, 111
Microcalorimetry, in Ca²⁺ binding
 measurements, 157
Mitochondria
 calcium accumulation in, 85
 calcium content in, 62–64
 calcium sequestration in, 84–86
 "high affinity" of, for Ca²⁺, 85
 in intracellular calcium regulation, 39–71
 steady-state buffering of free calcium by,
  67–71

Mitochondrial calcium, endogenous, 63–64
Mitochondrial calcium accumulation, release of by uncoupling agents, 50
Mitochondrial calcium buffering, steady-
 state, 69
Mitochondrial calcium efflux
 cellular factors in regulation of, 61–62
 extracellular factors in, 59
 intracellular factors in, 54
 oxidation-reduction state of intramitochondrial pyridine nucleotides in, 55
 and pH of extramitochondrial milieu, 57
 triiodothyronine in, 60–61
Mitochondrial calcium level, influx-efflux
 cycling and, 70
Mitochondrial calcium "set points," 67–70
Mitochondrial calcium transport
 calcium influx in, 41–50
 mechanics, kinetics, and regulation of,
  41–62
Mitochondrial calcium uptake
 adrenal glucocorticoids in, 48
 biological factors in, 46
 cellular factors in, 47–48
 extracellular factors in, 48–49
 glucagon and, 49
 inhibition of, 45
 intracellular factors in, 49–50
 kinetic properties and, 44
 respiration-dependent, 42
 sarcoplasmic Na⁺ changes and, 58
 translocase isolation in, 45–46
Mitochondrial regulation of cellular calcium, evidence for, 62–67, *see also*
 Mitochondrial calcium efflux
Mitochondrial transport, bivalent metal ion
 specificity in, 43–44
Mobile polymer, states open to, 11
Mobility
 of enzyme or protein, 10–11
 vs. two-state equilibria in allosteric
  switches, 11
Molecular tumbling, 5
Mollusca, SCPs in, 247, 270
mRNA, from chick intestine, 200
Muscle, calcium trigger proteins in, 23
Muscle contraction
 initial event in activation of, 145
 thin-filament protein interactions in,
  151–155

Multidentate chelation, binding selectivity and, 22
Myofibrillar ATPase, Ca^{2+} binding to Tn and, 150–151
Myofibrils, parvalbumin as relaxing factor within, 120
Myosin light-chain kinase
  calmodulin activation of, 135
  calmodulin regulation of, 288–289
  irreversible thiophosphorylation of, 288
  isolation for skeletal muscle, 280
  in skinned fibers, 279–293
  troponin C in activation of, 131
Myosin light chain phosphorylation and contraction, correlations between, 285–287

N

NAD hormone activation
  embryonic metabolism and, 375
  by pyridine nucleotides after fertilization, 373
NAD hormone resolution, for endogenous activator, 376
NADP, increase of following fertilization, 373
NADPH, *in situ* fluorescence of, 373
NADPH increase, embryonic metabolism and, 375
NADPH oxidation, mitochondrial Ca^{2+} efflux regulation and, 60
Nerve terminals
  Ca^{2+} distribution in, 83–84
  Na^+/Ca^{2+} exchange in, 99–100
  $[Ca^{2+}]_{in}$ and transmitter release relationships in, 101–102
  presynaptic, 81–102
NMR spectrometry, *see* Nuclear magnetic resonance spectroscopy
NOEs, *see* Nuclear Overhauser effects
Nonacrosome reacted mouse sperm, binding of to zona pellucida, 347–348
Nonmitochondrial Ca^{2+} sequestering organelle, 87–88
Nuclease, movements of, 10
Nuclear magnetic resonance spectroscopy
  in bovine intestinal CaBP, 182
  in calmodulin and troponin studies, 24
  in protein studies, 3–5, 11
  in sperm motility studies, 342

Nuclear Overhauser effects, in NMR spectroscopy, 5
Nucleotide replacement rates, in evolutionary periods of calcium-binding protein family, 118
Nucleotide replacements and evolutionary changes, in calcium-binding activity, 119

O

Octopus calmodulin
  activation of turkey gizzard myosin light-chain kinases by, 135
  titration of by Tb^{3+} ions, 131
  trimethyllsine residue in, 133–134
  tyrosine fluorescence quantum yield for, 127
  ultraviolet absorption spectrum of, 128
$1,25(OH)_2D_3$ (25 hydroxycholecalciferol-1α-hydroxylase)
  in CaBP synthesis, 199
  intestinal concentration of in laying birds, 195–196
  stimulation of calcium absorption by, 202–203
Organelles, calcium sequestration by, 90
Osteocalcin, hydroxyapatite binding to, 237

P

Paramecium spp.
  calmodulin in cilia of, 308–313
  cilia reversal in, 298
Paramecium caudatum, 310
  calmodulin injection in, 313
Paramecium tetraurelia, 310
Parathyroid hormone
  in calcium precipitation, 21
  $1,25(OH)_2D_3$ and, 189
Particulate guanylate cyclase, activation of by *Tetrahymena* calmodulin, 303–306
Parvalbumin(s), 7, 243–273, *see also* Sarcoplasmic calcium-binding proteins
  amino acid composition of, 251
  in amphibians, 246
  in birds, 246
  in body fishes, 246
  calcium and magnesium binding of, 259–261

calcium binding constant in, 27
in coelecanth tissue, 253
in carp tissue, 253
in cartilage fishes, 246
characterization of, 250–251
in chick breast muscle, 123–124
in chick embryo leg, 121
in chicken tissue, 253
conformational states of, 123, 261
defined, 120, 251
in fast-muscle fibers, 123
in frog tissue, 253
in hake tissue, 253
helix-loop-helix configuration of calcium-
    binding site of, 26
high-affinity Ca^{2+}–Mg^{2+} sites in, 118
in human tissue, 253
isotypes of, 250
magnesium and calcium binding to, 122
in mammals, 246
molecular weights of, 252
as muscle proteins, 251–254
physicochemical properties of, 252
purification of, 254
in rabbit cardiac muscle, 124
in rabbit tissue, 253
as relaxing factor with myofibrils, 120
in reptiles, 246
as sarcoplasmic proteins, 254
sequence in, 13
in shrew tissue, 253
as soluble relaxing factor, 272
structure of, 254–259
tissue and intracellular distribution of,
    251–254
tissue content of, 253
troponin C and, 250
in turtle tissue, 253
vertebrate classes containing, 246
Parvalbumin-like protein, in shrew heart
    muscle, 124
Peptides, calcium binding constants in, 27
Perch parvalbumin, immunofluorescence
    localization in, 255
pH, in mitochondrial Ca^{2+} efflux regula-
    tion, 57
Phenothiazine psychotropic agents, as cal-
    modulin inhibitors, 305
Phenylalanine resonance, in Ca^{2+} binding,
    158

Phenylephrine, mobility activation of, 347
Phosphate, Ca^{2+} uptake stimulation of, 44
Phosphate deficiency, CaBP synthesis and,
    189
Phosphatidylcholine, 397
Phosphatidylethanolamine, 392, 397
Phosphatidylinositol, 397
Phosphatidylinositol turnover
    cyclic nucleotides and, 401
    for enzyme activation, 396–398
Phosphatidylserine, in enzyme activation,
    392
Phospholipase A_2
    calcium binding site for, 9
    as calcium protein, 7
    carbonic anhydrase and, 12
    NMR resonance energies of groups in, 5
    sequence in, 13
Phospholipase C-type reaction, 397
Phospholipid(s)
    calmodulin as substitute for, 394
    specificity of, 391–393
Phospholipid metabolism
    in cellular activation, 398
    receptor function and, 396–400
    as transmembrane signaling for protein
        phosphorylation, 385–406
Phosphorylase kinase calmodulin subunit,
    myosin light-chain kinase and, 135
Phosphorylated serine, in solid-state pro-
    tein, 17
Phosphorylation
    Ca^{2+} binding control system and, 283
    irreversible, 282
Phosphorylation reactions, cAMP and
    cGMP as regulators of, 386–387
Phosphorus intake, low dietary calcium
    and, 189–190
Physarum polycephalum, 303, 333
Pig, calcium absorption in, 189
Polyanionic surface, negative potential
    generation of, 18
Polypeptide backbone, tightening of, 158–
    159
Polyribosomes, in *in vitro* synthesis of
    chick CaBP, 199
Polysaccharide-calcium interactions, 18–19
Polyspermy, calcium blockage of, 365
Polyvalent cations, blocking of Ca^{2+} chan-
    nels by, 94

Porcine calcium-binding protein, 178, *see also* Calcium-binding protein
  amino acide sequences in, 183–186
Potassium-stimulated influx, time courses of, 96–97
Precursor proenzymes, calcium and, 9
Presynaptic nerve terminals
  calcium movement and regulation in, 81–102
  total calcium at, 82–83
Protein(s)
  calcium binding to, 1–35, *see also* Calcium-binding proteins
  γ-carboxyglutamic acid as constituent of, 218
  hydrophilic, 14
  as internal triggers, 23–28
  intracellular, *see* Intracellular proteins
  mobility of, 11
  rhetatic, 30
  vs. small molecules as ligands, 5–6
  solid-state, 17
  specific features as ligands, 5
Protein C
  blood coagulation inhibition by, 235
  sequence homology of, 235
Protein classification, hydrophobic/charged amino acid ratio in, 25
Protein kinase(s)
  intracellular messengers and, 405
  physical and kinetic properties of, 388–389
  proteolytic activation of, 394–396
  relative reaction velocities of, 404
  tissue distribution and relative activities of, 388
Protein kinase C
  activation of by unsaturated deacyl-glycerol, 390
  enzyme activation for, 389–391
  as receptor function in transmembrane control of protein phosphorylation, 405
  reversibility and selective inhibition in, 393–394
  soluble form of, 388
  specificity of neutral lipid for activation of, 391
  specificity of phospholipid for activation of, 393

Protein kinase G, cGMP and, 405–406
Protein motion, allosteric effect and, 12
Protein phosphorylation
  calcium and phospholipid turnover as transmembrane signaling for, 385–406
  possible cascade mechanism for, 401–403
  transmembrane control systems for, 387
Proteolytic activation, of protein kinase C, 394–395
Prothrombin
  abnormal, 218
  amino acid sequence in, 227
  antisera raised in rabbits against, 227–229
  as calcium protein, 7
  γ-carboxyglutamic acid in, 219, 221–229
  generation of in presence of activated factor X, 223
  human, 224
  intrinsic fluorescence of, 224
  metal ion binding to, 224
  as most abundant vitamin K-dependent protein in plasma, 221
Prothrombin Gla units, 14
Prothrombin fragment, 12–44, 226, 228
Prothrombin molecule, prethrombin-1 portion of, 223
Proton magnetic resonance, *see also* Nuclear magnetic resonance spectroscopy
  circular dichroism and, 159
  in troponin C structure studies, 157–159
Pseudo vitamin D hormone, 190
Pyridine nucleotides, in mitochondrial Ca^{2+} efflux regulation, 55–57

R

Rabbit cardiac muscle, parvalbumin in, 124, 254–256
Rabbit skeletal muscle troponin C
  amino acid sequence in, 265
  bovine brain calmodulin and, 132
Rachitic chicks, response of to intravenous $1,25(OH)_2D_3$, 204
Ram testis calmodulin
  activation of turkey gizzard myosin light-chain kinases by, 135
  cation dissociation constants of, 125

in myosin light-chain kinase activation, 134

Rat duodenum, epithelial cells from, 199

Rat intestinal CaBP, molecular weight of, 182

Rat intestinal proteins, porcine kidney and, 179

Rat kidney CaBP, 177

Rat placental CaBP, electrophoretic protein bands in, 185

Receptor protein, transmission of information upon calcium binding to, 29

Renilla reniformis, 133

Rhetatic protein, 30

Russell's viper venom, in factor X activation, 229

Ruthenium red, mitochondrial Ca^{2+} uptake inhibition by, 45

Ruthenium red-sensitive Ca^{2+} influx system, 50

RVV, *see* Russell's viper venom

S

Sandworm SCPs, 249, 268–270

Saponin, in calcium sequestration studies, 90

Sarcoplasmic Na^+, transient changes in, 58

Sarcoplasmic calcium binding proteins, 243–273
  absence of in sedentary animals, 250
  absence of in various organisms, 248
  amino acid composition of, 262–264
  in amphioxus, 249
  binding of calcium to, 267
  of crayfish, 264–268
  calcium and magnesium binding in, 266–270
  defined, 244
  distribution of, 245–246
  "EF hand" arrangement in, 265
  historical review of, 244–245
  homogeneity of, 263
  in invertebrate phyla, 247
  molecular weight of, 262–263
  physiological implications of, 270–272
  pI values of, 262
  purification of, 261
  structure of, 265–266
  in various muscles, 265

Sarcoplasmic reticulum, calcium uptake into cisternae of, 40

Scallop EDTA light chain myosin, 249

SCPs, *see* Sarcoplasmic calcium binding proteins

SDS, *see* Sodium dodecyl sulfate

Sea urchin eggs
  aequorin-loaded, 361
  calcium regulation of NAD kinase of, 374
  calcium rise in, 378–379
  depolarization behavior in, 360
  events initiated at fertilization of, 370

Sea urchin sperm
  calmodulin in, 341
  egg jelly effect on, 346–347
  Na^+ addition to, 342
  TFP blocking of acrosome reaction in, 344–345

Sedentary animals, SCP absence in, 250

SER, *see* Smooth endoplasmic reticulum

Serum function, calcium and, 21

"Set points," mitochondrial, 67–70

Shrew heart muscle, parvalbumin-like protein in, 124

Skeletal muscle, parvalbumins in, 121–123

Skeletal muscle contraction, troponin in regulation of, 168

Skeletal myofibrillar light-chain kinase, activation or inhibition of by calmodulin and troponin C, 133–134

Skeletal troponin C
  Ca^{2+}-coordinating amino acid replacements and, 155
  Ca^{2+}-induced changes in fluorescence of, 161
  Ca^{2+} and Mg^{2+} binding properties of, 148
  dansylaziridine-labeled, 149
  molar ratio complexes of, 149

Skeletal troponin C_{DANZ}
  Ca^{2+}-induced changes in fluorescence of, 161
  production of highly fluorescent transient state in, 166

Skeletal troponin subunits, Ca^{2+} and Mg^{2+} binding properties of, 148

Skinned muscle fibers
  exogenous calmodulin and, 289
  as model for contraction, 283–285

myosin light chain kinase in, 279–293
  preparation of, 280, 284
Slow Ca^{2+} channels, voltage-regulated, 98
Smooth endoplasmic reticulum
  Ca^{2+} sequestration in, 87–88
  cisternal profile of, 86
  intraterminal, 91
  regulation of, 282
Sodium-calcium exchange, 99–101
Sodium dodecyl sulfate, 146, 251
Solanum malacoxylon, 190–191
Speract, defined, 347
Sperm
  changes in, 340
  storage of in animal, 341
Sperm activation, triggers for, 346–348
Spermatogenesis, in mammals, 340
Spermatozoa, calcium in metabolic activity
  of, 339–349
Spermatozoa activation process, 341–346
Sperm-egg association, Ca^{2+} functions in,
  348
Sperm-egg fusion, extracellular calcium
  absence in, 362
Sperm motility
  other ions in regulation of, 342
  phenylephrine activation of, 347
  regulation of, 341–343
Sperm physiology, changes in during meta-
  bolic activation, 345–346
Sphingomyelin, 392, 397
Squid axons, ATP effects on, 100–101
S-S bridges, calcium predating of, 13
Staphylococcal nuclease
  calcium binding site for, 9
  sequence in, 13
Starfish sperm, extracellular Ca^{2+} in, 344
Steady-state mitochondrial Ca^{2+} buffering
  studies, 69
Stern layer, 18
STnC, *see* Skeletal troponin C
Streptococcus griseus, calcium binding site
  for, 9
Strongylocentrotus purpuratus, 344, 377
Strontium ions, in factor IX activation by
  activated factor XI, 233
Strontium salt diet, calcium absorption
  and, 190
Submitochondrial vesicles, "inside-out,"
  50

Subtilisin Carlsberg, calcium binding site
  for, 9
Sugar oxygen, as neutral donor, 19
Substrate proteins, recognition of, 403–
  405, *see also* Protein(s)
Swordfish bone proteins, 236
Synaptic vessels, Ca^{2+} storage in, 88–89
Synaptosomes
  "fast" and "slow" channels in, 95–100
  functional average for, 101
  K^+-stimulated Ca^{2+} uptake in, 97

T

Terbium, binding of to calmodulin, 127
Terbium ions, titration of octopus calmodulin by, 131
Tetrahymena calmodulin
  activation of particulate guanylate cy-
    clase by, 303–306
  intracellular localization of, 308
  new functions of, 308–319
  properties of, 299–302
  purification of, 299
  similarities and dissimilarities with other
    types, 299–301
Tetrahymena guanylate cyclase, 298
  Ca^{2+} effect on, 305
Tetrahymena pyriformis
  calmodulin in cilia of, 308–313
  calmodulin role in, 297–319
  guanylate cyclase activity of, 304
Tetrahymena thermophilia, 310
Tetraphenylphosphonium, sea urchin
  sperm and, 342
Thermolysin, sequence in, 13
Thin-filament protein interactions, in muscle contraction regulation, 151–155
Thiophosphorylation
  contraction and, 282
  of myosin light chains using ATPσS,
    286–287
Tn, *see* Troponin
TnI, *see* Troponin I
TnT, *see* Troponin T
Torpedo, synaptic vessels isolated from, 88
Translocase isolation, in mitochondrial
  Ca^{2+} transport, 45–46
Transmembrane control, physiological im-
  plication in, 401–403

Transmitter release, "cooperative" models of, 102
Trifluoperazine
  binding of to two domains of calmodulin, 31
  blocking of acrosome reaction in sea urchin sperm, 345
  calmodulin activity and, 292
  as calmodulin inhibitor, 300, 316
  cell shape and, 334
  cellular functions and, 313–319
  ciliary movement and, 318
  in food vacuole formation and excretion, 317
  Tetrahymena cell swimming pattern and, 318
Trigger products, calcium binding rates for, 31
Trigger proteins, cell-organization regions of, 28
Triiodothyronine, in mitochondrial Ca^{2+} efflux, 60–61
Trimethyllysine residue
  in octopus calmodulin, 134
  role of, 133–135
Triton X-100, 206
Tropomyosin
  skeletal and cardiac troponin subunit interactions with, 152
  structure of, 146
Troponin, 145–168
  calcium binding to, 147, 149
  calcium exchange rates in, 167–168
  calcium-induced structural changes in, 155, 163–164
  defined, 146
  NMR studies of, 24
  in thin-filament regulation of skeletal and cardiac muscle contraction, 168
Troponin-actomyocin interaction, model for regulation of, 154
Troponin C, 23–30
  in activation of skeletal enzyme, 131
  calcium binding to, 146–149
  calcium-binding constant in, 27
  calcium-induced structural changes in, 160–163
  defined, 146
  extrinsic probes of Ca^{2+}-induced structural changes in, 160–163

glycogen phosphorylated *b* kinase activation of, 130
  high-resolution x-ray crystallographic analysis of, 156–157
  α-helical structure of, 159
  hinge regions of, 28
  molecular weight of, 155
  myosin changes and, 23–24
  predicted structure of, 156
  proton magnetic resonance in studies of, 3–4, 157–159
  reconstitution of to form fluorescent whole troponin, 163
  sequence in, 13
  sequence homology regions in, 155
  stoichiometric complex with TnI, 153
  structural changes to, in Ca-binding state, 205
Troponin calmodulin, 7, *see also* Calmodulin; Troponin C
Troponin C concentration, in higher vertebrates, 123
Troponin C structural changes
  calcium exchange rates and, 164–167
  extrinsic calcium probes and, 160–163
Troponin C–troponin I, primary interactions between, 153
Troponin C_{DANZ}
  calcium removal from specific sites of, 164
  formation of, 160
  fluorescence of related to protein transient state in, 165
Troponin I
  calcium binding to, 147
  defined, 146
  rapid structural changes in, 167–168
  stoichiometric complex with TnC, 153
Troponin T
  calcium binding to, 146–147
  defined, 146
Troponin T–Troponin C, primary interactions between, 153
Trypsin
  calcium-binding site for, 9
  sequence in, 13
Trypsinogen, 7
  activation of to trypsin, 9, 14
Turkey gizzard myosin light-chain kinases, activation of by octopus and ram testis calmodulins, 135

Tyrosine-containing tryptic peptides, amino acid composition of, 130
Tyrosine fluorescence, calcium binding and, 159

V

Vertebrate CNS presynaptic terminals, Ca^{2+} chemical properties in, 96
Vesicles, in calcium stores, 23
Villin, molecular weight of, 333
Vipera russelli, 229
Viral invasions, calcium and, 21
Vitamin D, *see also* $1,25(OH)_2D_3$ *under letter O*
  intestinal calcium absorption and, 188
  physiological effects of, 188
Vitamin D-dependent calcium binding protein, 175–207

Vitamin K antagonist, prothrombin and, 218
Vitamin K-dependent carboxylation systems, tissues containing, 239
Vitellin, as calcium protein, 7
Voltage-dependent Ca^{2+} influx, in synaptosomes, 101

W

Wasserman protein, sequence in, 13
Wheat germ system, chick intestinal nuclear mRNA and, 200

X

X-ray crystallographic analysis, in troponin C structure determination, 156–157

Molecular Biology

An International Series of Monographs and Textbooks

Editors

HAROLD A. SCHERAGA. Protein Structure. 1961

STUART A. RICE AND MITSURU NAGASAWA. Polyelectrolyte Solutions: A Theoretical Introduction, *with a contribution by Herbert Morawetz.* 1961

SIDNEY UDENFRIEND. Fluorescence Assay in Biology and Medicine. Volume I—1962. Volume II—1969

J. HERBERT TAYLOR (Editor). Molecular Genetics. Part I—1963. Part II—1967. Part III—Chromosome Structure—1979

ARTHUR VEIS. The Macromolecular Chemistry of Gelatin. 1964

M. JOLY. A Physico-chemical Approach to the Denaturation of Proteins. 1965

SYDNEY J. LEACH (Editor). Physical Principles and Techniques of Protein Chemistry. Part A—1969. Part B—1970. Part C—1973

KENDRIC C. SMITH AND PHILIP C. HANAWALT. Molecular Photobiology: Inactivation and Recovery. 1969

RONALD BENTLEY. Molecular Asymmetry in Biology. Volume I—1969. Volume II—1970

JACINTO STEINHARDT AND JACQUELINE A. REYNOLDS. Multiple Equilibria in Protein. 1969

DOUGLAS POLAND AND HAROLD A. SCHERAGA. Theory of Helix-Coil Transitions in Biopolymers. 1970

JOHN R. CANN. Interacting Macromolecules: The Theory and Practice of Their Electrophoresis, Ultracentrifugation, and Chromatography. 1970

WALTER W. WAINIO. The Mammalian Mitochondrial Respiratory Chain. 1970

LAWRENCE I. ROTHFIELD (Editor). Structure and Function of Biological Membranes. 1971

ALAN G. WALTON AND JOHN BLACKWELL. Biopolymers. 1973

WALTER LOVENBERG (Editor). Iron-Sulfur Proteins. Volume I, Biological Properties—1973. Volume II, Molecular Properties—1973. Volume III, Structure and Metabolic Mechanisms—1977

A. J. HOPFINGER. Conformational Properties of Macromolecules. 1973

R. D. B. FRASER AND T. P. MacRae. Conformation in Fibrous Proteins. 1973

OSAMU HAYAISHI (Editor). Molecular Mechanisms of Oxygen Activation. 1974

FUMIO OOSAWA AND SHO ASAKURA. Thermodynamics of the Polymerization of Protein. 1975

LAWRENCE J. BERLINER (Editor). Spin Labeling: Theory and Applications. Volume I, 1976. Volume II, 1978

T. BLUNDELL AND L. JOHNSON. Protein Crystallography. 1976

HERBERT WEISSBACH AND SIDNEY PESTKA (Editors). Molecular Mechanisms of Protein Biosynthesis. 1977

TERRANCE LEIGHTON AND WILLIAM F. LOOMIS, JR. (Editors). The Molecular Genetics of Development: An Introduction to Recent Research on Experimental Systems. 1980

ROBERT B. FREEDMAN AND HILARY C. HAWKINS (Editors). The Enzymology of Post-Translational Modification of Proteins, Volume 1. 1980

WAI YIU CHEUNG (Editor). Calcium and Cell Function, Volume I: Calmodulin. 1980

OLEG JARDETZKY and G. C. K. ROBERTS. NMR in Molecular Biology. 1981

DAVID A. DUBNAU (Editor). The Molecular Biology of the Bacilli, Volume I: *Bacillus subtilis*. 1982

GORDON G. HAMMES. Enzyme Catalysis and Regulation. 1982

CHARIS GHELIS and JEANNINE YON. Protein Folding. 1982

WAI YIU CHEUNG (Editor). Calcium and Cell Function, Volume II. 1982

In preparation

P. R. CAREY. Biochemical Applications of Raman and Resonance Raman Spectroscopies. 1982

GUNTER KAHL and JOSEF S. SCHELL (Editors). Molecular Biology of Plant Tumors. 1982